THE FRONTIERS COLLECTION

The books in this collection are devoted to challenging and open problems at the forefront of modern science and scholarship, including related philosophical debates. In contrast to typical research monographs, however, they strive to present their topics in a manner accessible also to scientifically literate non-specialists wishing to gain insight into the deeper implications and fascinating questions involved. Taken as a whole, the series reflects the need for a fundamental and interdisciplinary approach to modern science and research. Furthermore, it is intended to encourage active academics in all fields to ponder over important and perhaps controversial issues beyond their own speciality. Extending from quantum physics and relativity to entropy, consciousness, language and complex systems—the Frontiers Collection will inspire readers to push back the frontiers of their own knowledge.

More information about this series at http://www.springer.com/series/5342

Philippe Bertrand · Louis Legendre

Earth, Our Living Planet

The Earth System and its Co-evolution
With Organisms

Springer

Philippe Bertrand
Institut National des Sciences de l'Univers
Centre National de la Recherche
Scientifique
Gradignan, France

Louis Legendre
Villefranche Oceanography Laboratory
Sorbonne University
Villefranche-sur-Mer, France

ISSN 1612-3018   ISSN 2197-6619   (electronic)
THE FRONTIERS COLLECTION
ISBN 978-3-030-67775-6   ISBN 978-3-030-67773-2   (eBook)
https://doi.org/10.1007/978-3-030-67773-2

This Springer imprint is published by the registered company Springer Nature Switzerland AG
The registered company address is: Gewerbestrasse 11, 6330 Cham, Switzerland

Preface

We, the two authors of this book, are both oceanographers, but our scientific and geographic backgrounds are quite different. One of us was initially trained as a chemist, and spent most of his career in France. The other was initially trained as a biologist, and spent the first part of his career in Canada before moving to France. In the latter country, one of us was based on the Atlantic Ocean and the other on the Mediterranean Sea. We first met in the multidisciplinary context provided by the French national research organization CNRS (Centre National de la Recherche Scientifique). Talking together, we realized that we shared a common interest in topics that went beyond our own scientific domains.

In 2004, Louis Legendre published a book on the discovery process entitled *Scientific Research and Discovery: Process, Consequences and Practice* (International Ecology Institute, Oldedorf-Luhe) followed by an abridged electronic edition in 2008 (https://www.int-res.com/articles/eebooks/eebook16.pdf). In 2008, Philippe Bertrand published a book on the evolution and functioning of the Earth System entitled *Les Attracteurs de Gaïa* (Publibook, Paris). We were each interested in each other's books, and talked from time to time about some of the general topics following from our works. One theme that emerged from our discussions was the general lack of analyses in the scientific literature of the conditions that had led organisms and ecosystems to take over the Earth System during the course of the history of the planet. We explain in Sect. 1.1.1 what we mean by *takeover* in this context.

Indeed, there is a large body of studies on the emergence of life on Earth and its further diversification into billions of species, both extinct and extant. These studies involve such scientific disciplines as molecular and cell biology, developmental and evolution biology, genetics, taxonomy, and paleontology. There is another large body of research on the functioning of ecosystems and their perturbations by the ongoing climate change, which involve such disciplines as biogeochemistry, ecology, oceanography, limnology, and climatology. However, less attention is devoted to what lies between the two sets of phenomena, that is, the fact that organisms not only appeared and diversified on Earth, but became so abundant that they modified the physical and chemical environment of the planet and largely took over the functioning of the Earth System. This led us to the question: "Which combination of factors led to this apparently unique development in the Solar System?" We thought that trying to answer this question would be an original contribution to Earth System Science.

Philippe Bertrand had already provided some high-level answers to this deep question in his 2008 book, in a form mainly intended for experienced researchers. For the present book, we decided early (in 2011) to address a more general audience. We looked for a possible Publisher for our future work, and found that the books in Springer's *Frontiers Collection* were "devoted to challenging and open problems at the forefront of modern science and scholarship […] written in a manner accessible also to scientifically literate non-specialists". This corresponded precisely to what we had in mind for our book. In turn, the approach underlying the collection largely determined our approach to the book, including the encouragement to "active academics in all fields to ponder over important and perhaps controversial issues beyond their own speciality."

As we wrote our manuscript, we investigated numerous domains of astronomy, geology, geophysics, chemistry, climatology, biology, ecology, systems theory, and other disciplines, in which we revisited the territories we had journeyed in the past and also explored new lands. The resulting information influenced the content and structure of the book, and the resulting changes to our original plans led us to explore new domains of knowledge.

The book thus took a life of its own, but we always reined in our exploration of new areas of knowledge to stay on the track of the conditions that have enabled organisms and ecosystems to take over the Earth System. Accordingly, each of the first eight chapters of the book begins with the identification of a connection between organisms and the Earth System, after which information from several disciplines is combined to explain the occurrence of this global connection. The last three chapters are progressively broader syntheses of the materials presented in the previous chapters, which lead to general mechanisms that govern the functioning of the Earth System (Chapter 9), and its evolution from past to present (Chapter 10) and present to future (Chapter 11).

Experts from the many disciplines cited above may well find that we have not delved into certain topics enough and taken too many shortcuts. However, we have deliberately written our book for non-specialists, namely scientifically literate members of the general public, students, and colleagues from various scientific and non-scientific disciplines who would be fascinated by the interactions between astronomical, geological, environmental, biological phenomena and ecological characteristics of the planet that led to the unique Earth System.

Bordeaux, France   Philippe Bertrand
Nice, France   Louis Legendre

Acknowledgments

We first want to thank our patient and obliging draughtsman, Mohamed Khamla, who drew or modified most of the figures in this book. His contribution was essential, and we thank the Director of the *Institut de la Mer de Villefranche* for allowing him to devote innumerable hours to our book. We are particularly indebted to nine colleagues who each pre-read some chapters of our manuscript, namely Maryam Cousin, Muriel Gargaud, Satoshi Mitarai, André Monaco, Tom Pedersen, Claude Pinel, Richard B. Rivkin, Carolyn Scheurle, and Marie-Thérèse Vénec-Peyré. Their thoughtful suggestions greatly helped us improve our final text. We also thank the colleagues who provided us with information or assistance at various times during the writing of our book, in particular Leif Anderson, Andrew C. Clarke, Francisco Chavez, Katherine Clark, Christina De La Rocha, Peter J. Edmunds, Martine Fioroni, Jean-Pierre Gattuso, David M. Karl, Robert Knox, Nianzhi Jiao, Robbert Misdorp, Purificación López-García, Sophie Rabouille, Florian Rastello, Bruce H. Robison, Christian Sardet, David Schindler, Morgane Thomas-Chollier, and Paul Tréguer. We are also very grateful to the researchers and institutions cited in the figure credits for giving us permission to use their illustrations. Alexis Vizciano has been our kind and helpful Editor from our first contact with Springer to the publication of the book. Last but not least, during the many years between the conception of this book and its publication, we benefited from the constant support and wise advice of Marie-Soline Bertrand and Mami Ueno-Legendre, without whom this book would not have become a reality.

Contents

Chapter 1
The Living Earth: Our Home in the Solar System and the Universe

1.1 The Living Planet Earth

1.1.1 Focus and Organization of This Book

This book investigates the billion-year takeover of planet Earth by its organisms and ecosystems (see Information Box 4.2 for a definition of *ecosystem*). By *takeover*, we mean the progressive changes brought about by organisms and ecosystems to the chemical, geological and/or physical conditions of the Earth's environment, and the feedbacks of the latter into ecosystems (see Information Box 6.2). One example is the oxygenation of the atmosphere by the photosynthetic activity of organisms, which led to the development of the ozone layer, which in turn provided the protection from harmful solar radiation that allowed the occupation of continents by plants, which produced even more oxygen (see Sect. 9.4.3). We use *takeover* as a metaphor to stress the roles of organisms in the Earth System (defined two paragraphs below), in a way similar to Darwin borrowing in 1859 the word *selection* from animal husbandry and plant breeding to characterize the mechanism of biological evolution, although artificial selection in husbandry is guided whereas natural selection is not.

This takeover is unique in the Solar System, and we explain it by the existence of interactions between a small number of key environmental and biological mechanisms. In this book, we take the existence of organisms on Earth since almost 4 billion years for a fact, and we do not consider the origin of life on the planet (1 billion $= 1,000$ millions $= 10^9$; also 1 milliard). We focus on the conditions that allowed the organisms to diversify, and build up the huge biomasses that allowed ecosystems to successfully occupy all the Earth's habitats. Other, very interesting books examine hypotheses concerning the intriguing questions of how and where life appeared on Earth. Here instead, we investigate the conditions that allowed the initially very small number of cells to give rise to the innumerable

© Springer Nature Switzerland AG 2021
P. Bertrand and L. Legendre, *Earth, Our Living Planet*, The Frontiers Collection,
https://doi.org/10.1007/978-3-030-67773-2_1

organisms and huge biomasses that ultimately took over the whole planet. We show that the establishment of organisms and the development of ecosystems on Earth have been conditioned by the conditions that led to the formation of the planet in the Solar System, in our galaxy the *Milky Way*, and more generally in the Universe (Sect. 1.4.1).

The environment we consider is planet Earth within the whole Solar System (see Information Box 3.3 for a definition of *system*). We compare Earth with the billions of astronomical bodies that exist in our Solar System (yes billions, as explained below; see Sect. 1.2.2), and find that our planet has a number of unique characteristics. We look at hidden relationships among the suite of nested systems made of ecosystems, the Earth System, the Solar System, and the Universe. The *Earth System* consists of the atmosphere (air), the hydrosphere (liquid water), the cryosphere (ice), the lithosphere (rocks), and the biosphere (organisms), as well as the physical, chemical and biological processes within and among them. The study of Earth as a complex, adaptive *system* (see Information Box 3.3) is called *Earth System Science* (ESS). This book does not systematically refer to the ESS, but is very much a contribution to it. The book of Shikazono (2012) provides further reading on the ESS.

A first example of the unique characteristics of Earth is the presence of a significant atmosphere with a high concentration of free oxygen (O_2). Some other bodies in the Solar System also have significant atmospheres, but contrary to these bodies where the atmospheric concentration of O_2 is very low or even nil, the percentage of O_2 in the Earth's atmosphere is high: 21% by volume. This reflects the presence on Earth of organisms able to break up water molecules and release the O_2 they contain. It is the high concentration of O_2 in the Earth's atmosphere that allowed ecosystems to occupy the emerged lands. A second example is the generally moderate temperature that prevails at the Earth's surface, which varies within a relatively narrow range during the course of a day, and does not seasonally reach extreme values except in geographically limited regions. This narrow temperature range was one of the conditions that allowed organisms to establish themselves durably on Earth, and ecosystems to prosper. A third example is the existence on Earth of a large amount of liquid water, which is the key fluid of organisms. There is increasing evidence that water is also present on some other planets of the Solar System and also on some moons and dwarf planets and in comets, but contrary to these bodies where water is not liquid at the surface (or, at least, not permanently), more than 70% of the Earth's surface is covered with stable reservoirs of liquid water. A fourth example is the continual modification of the Earth's surface by large-scale motions of the rigid but mobile pieces of the crust (called *plates*) that together make the outermost shell of Earth. The success of ecosystems on Earth is intimately connected with the slow, large-scale motions—called *plate tectonics*—of these slabs of crust, as tectonic activity plays a key role in the recycling of carbon and other chemical elements essential to organisms. This recycling takes place among the ocean, the seafloor, the continents and the atmosphere, at geological time scales of millions of years.

Our analysis of the above and other characteristics of Earth in Chaps. 2–8 uncovers major *hidden connections* between ecosystems, planet Earth, the Solar System and the Universe. These connections have shaped the organisms and ecosystems in the forms they exist on Earth, and involve environmental characteristics of the planet as well as geological and astronomical characteristics of the Earth System and the Solar System. In the following chapters, we explore the principal environmental characteristics of Earth: the presence and retention of the atmosphere; the thermal and overall habitability of the planet for organisms; the prevalence of liquid water; the availability of the chemical building blocks of organisms; and the natural greenhouse effect. The geological and astronomical characteristics we consider include: the mass, rotation, gravity, geological activity and magnetic field of Earth; the motions of Earth around the Sun; the existence of the Earth-Moon system; actions of various bodies of the Solar System on Earth; and the role played by the gravitational pull of stars.

This chapter considers successively five aspects of the Earth System in the context of the Solar System: organisms, which are subject to evolution (Sect. 1.1.3); the Solar System, which is the homeland of Earth in the Universe (Sect. 1.2); Earth with its sister planets and their consorts (Sect. 1.3); brief history of the Universe (Sect. 1.4); and brief history of Earth (Sect. 1.5). The chapter ends with summary of key points concerning the connections between organisms, Earth, the Solar System, and the whole Universe (Sect. 1.6).

Sections 1.2–1.5 provide evidence of the following key interconnections between Earth, its organisms and ecosystems, and the cosmic environment:

Some connections of Earth with asteroids and comets, over distances of hundreds of millions of kilometres and times of billions of years, have largely determined the evolution of organisms and ecosystems.

Different astronomical characteristics of Earth, such as its short day-and-night rotation cycle and moderate axial tilt (angle between the planet's axis of rotation and the plane of its orbit around the Sun), have created conditions suitable for the establishment and the development of organisms and ecosystems on the planet.

Some aspects of the 13.7 billion-year evolution of the Universe explain key characteristics of Earth.

There has been a *co-evolution* of Earth and its organisms over the last 3.8 billion years.

Chapters 2–8 examine seven major hidden connections between ecosystems, planet Earth, the Solar System and the Universe. Chapter 9 explains how the interactions among several components of the Earth System create feedback loops that contribute to stabilize the environment of the planet. Finally, Chap. 10 weaves the threads spun in Chaps. 1–9 into a 4.5 billion-year tapestry called *The Legends of the Eons* (Fig. 10.1), and Chap. 11 looks at conditions of present Earth that contribute to the deterioration of this tapestry, and explore some aspects of the possible future state of the planet.

1.1.2   The Long Geological Times

Planet Earth and the Solar System are very ancient. As a consequence, this book often refers to very large numbers of years, that is, thousands, millions (thousands of thousands), and even billions (thousands of millions) of years. Such long durations are seldom used in day-to-day life, and could sometimes be confusing to readers.

Figure 1.1 summarizes the timescales of billions and millions of years, with reference to some of the key points in the history of the Earth System examined in this book. The billion-year timescale (left) identifies key events in the Earth System (left side) and the four phases of the atmosphere (right side). The million-year timescale (right) identifies the five mass extinctions of organisms documented by fossils (left side) and key events in the Earth System (left side).

Planet Earth is about 4.6 billion years old, and the oldest time generally considered in this book is 4.5 billion years, corresponding to the formation of Earth-Moon system (see Sect. 1.5.1). The references to the long timescales of billions, millions and thousands of years in this book follows some general rules:

When we refer to the very long period of 4.5 billion years, we generally express the time as a fractional number of billions of years, or equivalently hundreds of millions of years. For example, complex organisms began their occupation of emerged lands some time earlier than 0.4 billion or 400 million years before present (see Sect. 1.5.4).

When we refer to periods of millions of years, we generally express the time as a multiple or a fractional number of millions of years. An example of the first is the above reference to 400 million years. An example of the second is the Quaternary Ice Age in which we are currently living and which was marked by a succession of glacial and interglacial episodes that started 2.6 million years ago (see Sects. 1.3.4 and 4.5.2).

When we examine events that occurred in the last millions of years, we generally express the time as hundreds or tens of thousands of years. For example, there have been 30–50 successive cycles of glacial and interglacial episodes since the beginning of the Quaternary Ice Age, and their duration was initially around 40,000 years and then 100,000 years (see Sect. 4.5.4).

The same three units of time (100 million, 10 million, and 100,000) also correspond to the precision of the time values cited in this book. Billions of years have an uncertainty of 100 million (or 0.1 billion) years, for example 4.5 ± 0.1 billion years. Hundreds of millions of years have an uncertainty of 10 million years, for example 400 ± 10 million years. Millions of years have an uncertainty of 100 thousand (or 0.1 million) years, for example 4.6 ± 0.1 million years. These precisions are not stated throughout the book, for simplicity.

Finally, it should be noted that the word *period* has three different meanings in this book, depending on the context. First, a *period* is generally a length or a

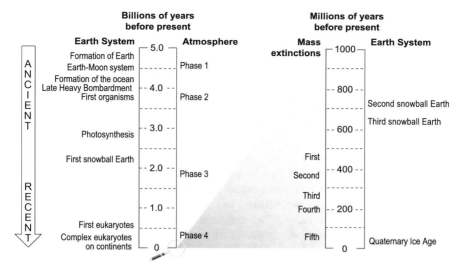

Fig. 1.1 History of the Earth System, with identification of key events described in this and subsequent chapters. Note the two different timescales: billions (*left*) and millions (*right*) of years before present. Credits at the end of the chapter

portion of time, as in the above paragraphs. Second, a *period* can be the interval of time between successive occurrences of the same state or event in cyclical phenomena, such as the movements of Earth and other astronomical bodies on their orbits (see Sects. 1.2.2 and 1.3.2). For example, the *orbital period* of Earth around the Sun is 1 year. Third, the word *period* can also designate a major division of geological time, as in Table 1.3. One example is the *Quaternary period*.

1.1.3 The Characteristics of Organisms

A central characteristic of Earth is the all-pervading occurrence of organisms over the planet. Life forms may well exist in various places in the Universe and even close to us in the Solar System, but Earth is, so far, the only astronomical body where organisms and ecosystems are known (by us) to be present. Furthermore, Earth is the only known astronomical body whose characteristics bear the signature of ecosystems. Organisms are ubiquitous in almost all Earth's environments, and they continually interact with physical, chemical and geological processes of the planet. These interactions between the organisms and the environment take place within and among the oceans, the seafloor, the continents and the atmosphere, and they largely determine the functioning of Earth.

Information Box 1.1 General Features of Organisms on Earth The definition of life is elusive, and many researchers, including biologists, physicists and philosophers have tried in the past to define life with varying success. Further reading on this topic is provided by the books of Schrödinger (2012, first published in 1944), Pross (2012), and Schulze-Makuch and Irwin (2018). Other researchers have proposed to try instead to identify the general features of organisms on Earth, such as in the scheme illustrated in Fig. 1.2.

The Earth's organisms comprise the three essential features shown schematically in Fig. 1.2, and a fourth not in the figure: *first*, the software, namely the genetic information encoded in nucleic acids; *second*, the hardware, consisting of the biochemical compounds (carbohydrates, lipids and proteins) and membranes of the cells, and the components (tissues and organs) and structure (body systems) of the bodies of multicellular organisms (that is, organisms consisting of more than one cell); and *third*, the flux of energy. The latter keeps organisms alive, meaning that organisms maintain their high level of internal organization through a continual intake and dissipation of energy: when the flux of energy stops, they die. An organism is alive only if it exhibits all three components shown in Fig. 1.2: software, hardware, and energy transfer (details on these three components are given in Sects. 6.1 and 6.2). In addition and *fourth*, organisms are capable of replication or reproduction, by either cell division or the production of *gametes* (germ cells in sexual reproduction) that allows them to perpetuate and multiply. Reproduction led to the development of large biomasses of organisms that were subject to biological evolution, which resulted in the massive and diversified ecosystems that progressively occupied the planet. Examples of the above are given in Chap. 6 (see Sect. 6.2.5).

Any list of general characteristics of organisms on Earth, such as those above, raises a number of philosophical questions about the definition of life. For example, viruses outside living cells have the software (genetic instructions) for the construction of new viruses, but they lack the hardware (machinery) to do it and they exhibit no energy flow. Because they lack these two properties (Fig. 1.2), viruses should be considered as inert particles, but they become alive when they infect a cell where they co-opt hardware and energy flows to produce new viruses (reproduction). Such considerations are important philosophically, and could also become crucial someday if humanity is confronted by extraterrestrial life that differs from terrestrial life. However, such concerns are outside the scope here, which focuses on the durable establishment of organisms on Earth and their takeover of the whole planet during about 4 billion years of biological evolution. Readers interested in exploring the general aspects of life can delve into the many books, articles and websites that investigate them.

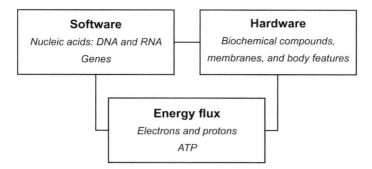

Fig. 1.2 Schematic representation of three key components of organisms on Earth. A fourth component (not illustrated in this figure) is replication or reproduction (see Information Box 1.1). Credits at the end of the chapter

Important biological and philosophical questions concerning the origin of life in the Universe and its evolution on Earth are addressed in many studies and books, among which those of Meinesz and Simberloff (2008) and Whittet (2017) provide further reading on this topic. The present book takes instead the existence of life on Earth for a fact, and focuses on the conditions that allowed organisms (see Information Box 1.1) to establish themselves durably on Earth and develop large biomasses there. During their conquest of Earth over the last billions of years, the ecosystems became one of the main factors that largely governed the evolution of the whole planet, and they will likely continue to determine its fate during billions of years to come. How did the above conditions developed and persisted over the first 4.6 billion years of Earth's existence, and in particular during the last 3.8 billion years in which ecosystems have progressively dominated the planet? This history did not occur elsewhere in the Solar System as far as we know. How is it that it happened on planet Earth?

1.2 The Homeland of Earth in the Universe: The Solar System

1.2.1 The Multifaceted Solar System

The homeland of Earth is the *Solar System*, which consists of the Sun, the eight planets and the billions of other astronomical bodies that move around it. We sometimes tend to imagine that our home planet, Earth, is quite isolated in the Universe, with our astronomical neighbourhood being limited to the Moon and a few closest planets. However, the reality is quite different since the Solar System is populated by billions of objects of many different types. These include the Sun, the planets and their moons (also called *natural satellites*); the word *moon* without

Information Box 1.2 The Astronomical Unit (AU) The distance between the centre of Earth and the centre of the Sun is 149,597,870,700 km. This distance defines what is called the *astronomical unit* (AU). Hence 1 AU = 150 million km.

The distance from the Sun to the outer edge of the Solar System is not certain, and may be around 100,000 AU. In order to develop an intuitive idea of such a distance, let us imagine that the distance between Earth and the Sun is 1 cm. Taking this distance of 1 cm as reference, the outer edge of the Solar System would be located 1 km from the Sun.

capital *m* is a synonym of *natural satellite*, and *Moon* with capital *M* is the name of Earth's moon. The objects found in the Solar System also include dwarf planets, meteoroids, asteroids, comets, centaurs and interplanetary dust. We will become acquainted with all these objects in the remainder of this chapter.

Most or all of the objects in the Solar System likely result from the condensation, more than four billion years ago, of interstellar gas (mostly hydrogen) and dust that previously occupied the region of space that is now the Solar System. In this chapter and following ones, we will see that various types of Solar System bodies played key roles in the past, and continue to do it today, in the development and maintenance of the conditions that allowed organisms to establish themselves durably and prosper on Earth. The interactions of these astronomical bodies and Earth contributed to the progressive build-up of large biomasses of organisms.

All the bodies in Solar System share certain common characteristics since they were all formed from the same *solar nebula* (a nebula is a cloud of gas and dust in outer space). However, other characteristics of these bodies can be very different depending on where and how they formed within the nebula and their history since formation. Similar to what travellers often do when they visit a new city for the first time, we will now take a general tour of the Solar System, before focussing on Earth and its ecosystems.

1.2.2   The Huge, Diverse, and Life-Bearing Solar System

Within the immense Universe, our Solar System homeland is very small. Indeed, the universe contains a very large number of galaxies (perhaps more than 2 trillions), of which our own galaxy, the Milky Way, is only one. The name *Milky Way* describes the galaxy seen from Earth. Indeed, far from city lights, our galaxy is seen in the night sky as a hazy band of light formed from 100 to 400 billion stars. The name comes Greek mythology, where the Milky Way is a trail of milk sprayed by the queen of the gods Hera when she was suckling baby Heracles (Hera and Heracles correspond to Juno and Hercules in Roman mythology).

Information Box 1.3 The Variety of Objects in the Solar System There is a wide variety of objects in the Solar System. The names of some of the most cited Solar System objects in this book are defined here, in alphabetical order. The names of other objects are defined in the text when they are first used.

Asteroid. A non-satellite body that fulfils only criterion (1) of a planet (see *Planet* below). An asteroid is also called *small Solar System body* or *minor planet*.

Centaur. A non-satellite body that has characteristics of both comets and asteroids.

Comet. Small, icy Solar System body that releases gases when passing near the Sun. This release of gases produces a visible atmosphere around the comet (called coma), and also sometimes a tail trailing it.

Dwarf planet. A non-satellite body that fulfils criteria (1) and (2) of a planet, but not criterion (3) (see *Planet* below).

Meteoroid. Same as an asteroid, but smaller.

Moon. An astronomical body that orbits a planet or a dwarf planet. The word *moon* without capital *m* is synonym of *natural satellite*, whereas *Moon* with capital *M* is the name of the moon of Earth.

Planet. Astronomical body that fulfils the following three criteria: (1) it is in orbit around the Sun, (2) it has a sufficient mass to achieve a round shape, and (3) it has cleared the neighbourhood around its orbit of other material (meaning that the planet has become gravitationally dominant around its orbit).

Galaxies are systems of millions to billions of stars, each surrounded with billions of bodies of various sizes and compositions, plus gas and dust, which are held together by gravitational attraction (see Sect. 6.6.1). Our own galaxy contains between 100 billion and 400 billion stars, of which our Sun is only one. And the Solar System contains billions of objects, of which our Earth is only one. These mind-bogglingly large numbers should not, however, distress us. Indeed, Earth is, so far, the only place in the Universe where we know that organisms have taken over a planetary body, and its homeland, the Solar System, is huge and contains a fascinating variety of astronomical objects (see Information Boxes 1.2 and 1.3).

This book focuses on the unique *Living Earth*, an expression explained at the end of this chapter (see Sect. 1.6). Further reading is provided by the book of Vita-Finzi (2016), for a compact history of the Solar System, and that of Cohen and Cox (2019), for an illustrated tour of the planets and other astronomical bodies of the Solar System.

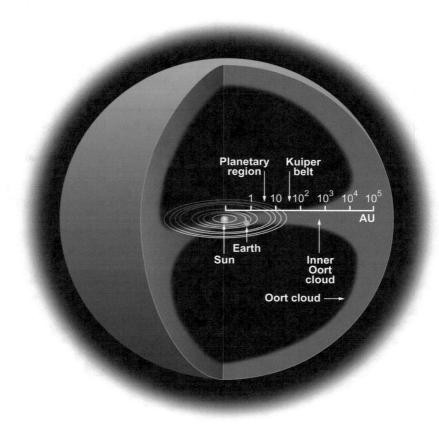

Fig. 1.3 Schematic representation of the Solar System: the Sun, at the centre; the orbits of the eight planets (planetary region); the Kuiper Belt; and the Oort cloud. Each interval in the distance scale from the Sun corresponds to a factor of 10: Earth is 1 AU from the Sun (Information Box 1.2), and the outer edge of the Solar System may be 100,000 AU. Credits at the end of the chapter

The Solar System consists of the *heliosphere*, which is the region of space influenced by the Sun (see Sect. 8.4.2), and the *Oort cloud* (Fig. 1.3). The existence of the Oort cloud beyond the heliosphere in interstellar space is predicted by models, but has not been supported by direct observations so far (Fig. 1.3). The heliosphere comprises the *planetary region*—with the eight planets, their moons, and the asteroid belt—and the Kuiper belt.

The Sun is at the centre of the Solar System, and like the other stars in the Universe, it continually radiates energy coming from the fusion of atoms in its core, primarily hydrogen (see Information Box 3.2). Most of this energy is radiated as light and heat (see Information Box 2.7). The diameter of the Sun is 1.4 million kilometres, or 109 times that of Earth. Its mass is 333,000 times that of Earth, and

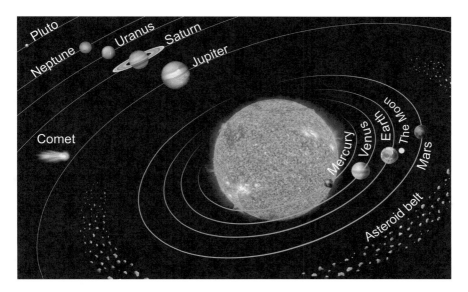

Fig. 1.4 Solar System. Illustrated in this figure: the Sun, the four rocky planets and our Moon, the asteroid belt, the four gas giant planets, dwarf planet Pluto (largest object in the Kuiper belt), and a comet. The scale is very far from reality. Credits at the end of the chapter

represents over 99% of the total mass of the Solar System. The Sun formed from the solar nebula about 4.6 million years ago, and when the first organisms appeared on Earth billions of years ago, the Sun was much cooler than it is now, this phenomenon giving rise to the *faint young Sun paradox* (see Sect. 3.7.1). Since then, the Sun warmed up to its present temperature, and is expected to continue to warm in the future. As a result, the Earth's environment is expected to become too hot for organisms within one to several billion years (see Sect. 11.3.5).

The Sun is orbited by eight planets divided in two groups of four, in order of increasing distance from the Sun: Mercury, Venus, Earth and Mars (*inner planets*, also called *terrestrial* or *telluric planets*), and Jupiter, Saturn, Uranus and Neptune (*outer planets*, also called *giant planets*) (Fig. 1.4). Most of the planets are accompanied by one or several moons.

The four inner and four outer planets are separated by a region of smaller astronomical bodies, the *asteroids*, called the *asteroid belt* (Fig. 1.4). The eight planets and their 205 moons are examined below (see Sects. 1.3.1–1.3.5). The asteroid belt contains more than half a million bodies of various shapes, called *asteroids* (see Information Box 1.3), and is located between about 2.1 and 3.3 AU from the Sun. Some of the large asteroids have their own moons, and almost 300 such moons have presently been identified in the asteroid belt. About half the mass of all the matter in the asteroid belt is made by four large asteroids named Ceres, Vesta, Pallas and Hygiea. Asteroid Ceres possibly contains liquid water (see Information Box 5.4). The objects in the asteroid belt are rocky remnants left over from the early formation of the Solar System about 4.6 billion years ago, and it is thought that Vesta and Ceres narrowly missed becoming planets. If these two asteroids had become large enough to join the planets, there would now be 10 planets in the Solar System.

Outside the orbit of planet Neptune, which is farthest from the Sun, there is a region called the *Kuiper belt*, which is similar to the asteroid belt but 20 times wider. It extends from the orbit of Neptune (30 AU from the Sun) to about 100 AU (15 billion kilometres). It contains some small bodies made of rocks and metals like those in the asteroid belt, but most objects in the Kuiper belt are frozen masses of volatile substances such as methane, ammonia and water. The adjective *volatile* identifies substances that evaporate easily at normal temperature and pressure (see Sect. 2.1.2). The Kuiper belt also contains three known dwarf planets (see Information Box 1.3), called Pluto, Haumea and Makemake, named respectively after the god of the underworld in Greek and Roman mythology, the goddess of fertility and childbirth in Hawaiian mythology, and the creator of humanity and god of fertility in Easter Island mythology. It is interesting to note that Pluto was formerly known as the ninth planet of the Solar System, but was reclassified as a dwarf planet by the International Astronomical Union in 2005 because it did not meet the third criterion of planets (see Information Box 1.3). Even if the Kuiper belt is located very far from Earth and most of its objects are small, more than 1300 bodies have presently been observed individually and identified as *Kuiper belt objects*. The overall number of objects in the Kuiper belt is not known, but it could be hundreds of millions or more.

Comets (see Information Box 1.3) are icy astronomical objects that move between the outer reaches and the inner region of the Solar System. They are remnants from the formation of the Solar System 4.6 billion years ago, and they mostly consist of ice and rocks, often coated with organic material. There are thousands of known comets, but it is thought that there are hundreds of billions of them, originating from both the heliosphere and the Oort cloud (described two paragraphs below). Comets are grouped in two broad types—short period and long period—based on the time taken by a comet to complete one orbit around the Sun, called the *orbital period.*

Short-period comets originate from the Kuiper belt. The qualifier *short-period* comes from the fact that these icy bodies orbit the Sun in less than 200 years. Given their short period, many comets have been recorded repeatedly in historical chronicles for thousands of years, reflecting in part that the passage of comets in the sky was often interpreted by astrologists and people in general as a divine sign. *Long-period comets* take thousands of years to orbit the Sun, and their occurrence led astronomers to propose the existence of the Oort cloud.

The *Oort cloud* is seen as a thick bubble of icy debris surrounding the heliosphere beyond the Kuiper belt. Although the Oort cloud has not been directly observed as yet, it is assumed that it occupies the region of space between 1000 and more than 100,000 AU (15 trillion kilometres) from the Sun, and contains hundreds of billions and even trillions of icy bodies. It has been proposed that long-period comets are icy bodies of the Oort cloud pulled into unique orbits by gravitational perturbations caused by stars passing near the Solar System. Astronomers divide the Oort cloud in a disk-shaped inner cloud and a spherical outer cloud (Fig. 1.3).

The number of astronomical objects of the different types in the heliosphere (not counting dust) is at last several billions, from the large Sun to the smallest rocky and icy debris. Earth is thus in good company. We will see in this book that life-bearing Earth was affected by many of its Solar System neighbours during the course of their long common history.

Despite the fact that asteroids and comets come from very far away from Earth, these Solar System bodies have very special significance for organisms because most of the water on this planet—a condition for the durable establishment of organisms and the development of ecosystems—may have been brought by aster-oids or comets that impacted the surface of Earth a long time ago. While this issue is debated in the scientific community (see Sect. 5.3), it stresses the importance of the distant asteroid and Kuiper belts and Oort cloud as the sources of the water that contributed to the development of ecosystems on Earth.

It has also been hypothesized that comets carried to Earth already made organic compounds, which may have contributed to the emergence of the first organisms. The provision of water and organic compounds by comets could have also hap-pened on other water-rich bodies of the Solar System, where life may thus exists presently or may have existed in the past. If so, comets were one of the main agents that prepared Earth for the widespread existence of organisms on the planet. Hence Earth's ecosystems may have largely benefited from inputs carried by asteroids and comets over huge distances and very long time periods.

1.3   Earth, Its Sister Planets, and Their Consorts

1.3.1   The Neighbourhood of Earth: The Planets

Robotic spacecraft exploration of the Solar System continually brings new infor-mation on its planetary bodies, and what is presented here reflects the current consensus on their composition and functioning, which is likely to change at least partly in the years to come. For example, the books published before 2005 reported nine planets including Pluto, now assigned to the dwarf planet category (see Sect. 1.2.2, above). The remaining eight planets of the Solar System are divided in two distinct groups (Fig. 1.4).

Planet Mercury is closest to the Sun, followed by Venus, Earth and Mars. Earth is the largest of these four relatively small planets. They all have hard, rocky sur-faces and are thus *rocky planets* (Table 1.1).

Farther from the Sun are, in order of increasing distance, Jupiter, Saturn, Uranus and Neptune, the latter being the farthest planet from the centre of the Solar System (Table 1.1). These four planets are mostly made of gas, and because they are very large, they are called *giant planets*. While they probably have rocky cores, the largest two, Jupiter and Saturn, are called *gas giants* because they are

Table 1.1 Key characteristics of the eight planets of the Solar System. Most characteristics of the four rocky planets, which are relatively close to the Sun, are different from those of the two massive gas giants, which are further away, and the two ice giants, which are the most distant. https://solarsystem.nasa.gov/planetinfo/charchart.cfm

Planet[a]	Distance from the Sun (AU)[b]	Diameter (1000 km)	Mass (10^24 kg)	Mean density (kg cm^-3)	Orbital period (Earth years)	Rotation period (Earth days)[c]	Axial tilt (°)	Surface temperature (°C) min/max	Surface atmospheric pressure (atm)
Four rocky planets (also called terrestrial or telluric planets)									
Mercury	0.4	4.8	0.3	5427	0.2	59	0	−173/+427	10^{-14}
Venus	0.7	12.1	4.9	5243	0.6	−243[d]	2.7[e]	+462/+462	92
Earth	1.0	12.7	6.0	5513	1.0	1.0	23.4	−88/+58	1.0
Mars	1.5	6.8	0.6	3934	1.9	1.0	25.2	−153/+20	0.006
Two gas giant planets									
Jupiter	5.2	140	1898	1326	11.9	0.4	3.1		
Saturn	9.5	116.0	568	687	29.4	0.4	26.7		
Two ice giant planets									
Uranus	19.1	50.7	87	1270	84.0	−0.7[d]	82.2[f]		
Neptune	30.0	49.2	102	1638	164.8	0.7	28.3		

[a]Another object of the Solar System, Pluto, was considered as a planet until 2005, but is now classified as a dwarf planet. Pluto's distance from the Sun is 39.5 AU, and its mass is 0.01×10^{24} kg, or 0.002 time that of Earth

[b]Average distance between a planet and the Sun, with corresponds to half of the average length of the longest axis of the elliptic orbit of that planet (called *semi-major axis*)

[c]The planets, except Venus and Uranus, rotate from west to east (as most other Solar System bodies including the Sun)

[d]The negative sign indicates that the planet rotates from east to west

[e]Tilt seen from the north: 177.3°. Because of Venus rotates from east to west, its tilt away from the plane of the ecliptic is $180° − 177.3° = 2.7°$

[f]Tilt seen from the north: 97.8°. Because Uranus rotates from east to west, its tilt away from the plane of the ecliptic is $180° − 97.8° = 82.2°$

mostly composed of gases hydrogen and helium. The other two giant planets, Uranus and Neptune, are called *ice giants* because the materials incorporated into the planets during their formation were in the form of ice. Uranus and Neptune are mostly composed of volatile chemical elements heavier than hydrogen and helium, namely oxygen, carbon, nitrogen and sulphur, probably because the gravity exerted by their smaller masses was less efficient than that of Jupiter and Saturn at capturing light atoms of hydrogen and helium.

The eight planets differ widely in terms of distance from the Sun, size (diameter), mass, density (mass per unit volume), and other characteristics (Table 1.1). These particular and contrasting features contribute to the explanation, in following chapters, as to why Earth, while sharing some of the characteristics of other planets, was the only one on which organisms both established themselves durably and built up large biomasses.

The planets were formed more than 4.6 billion years ago by the accretion of matter from the solar nebula, and by successive collisions and/or fragmentations of the initial bodies. The *accretion* of a planet is the phenomenon by which its mass, under the influence of gravitational attraction, gradually increased by agglomerating surrounding matter. The latter was present in the forms of gas, dust debris and larger-sized objects. The present Earth did not completely form at that time. Indeed, Earth had two parent planets, which both emerged from the solar nebula. The collision of these two planets—called *proto-Earth* (or *Gaia*) and *Theia*—produced the Earth-Moon system about 100 million years after their formation (see Sect. 1.5.1). In the Greek mythology, Gaia was the goddess of Earth, and her daughter Theia was the mother of the Greek goddess of the Moon, called *Selene*. As a consequence, the name of Selene's mother, Theia, was given to the astronomical body involved in the formation of the Moon. The name *Gaia* is also used in the scientific literature to designate Earth viewed as a self-regulating system: see, for example, the book *The Ages of Gaia* of James Lovelock (1995).

Earth and the other seven planets (and most of the other Solar System bodies) revolve around the Sun along elliptic paths called *orbits*. Because the path of a planet around the Sun is not circular, its distance to the Sun changes continually with time, and the amount by which its orbit deviates from a perfect circle is called *eccentricity* (Table 1.1). Hence a circular orbit has an eccentricity of 0 (Fig. 1.5a). Over the course of one Earth year, the distance between Earth and the Sun varies between 147 and 152 million kilometres, with an average value of 150 million kilometres. The eccentricity of the Earth's orbit is calculated as follows: $(152 - 147)/(152 + 147) = 0.0167$. The eccentricity of the Earth's orbit is close to zero, which means that the Earth's orbit around the Sun is almost circular, but not perfectly so.

It is reported in the news from time to time that Earth will nearly encounter (or has nearly encountered) a relatively large object, such as a comet or an asteroid, whose path around the Sun will cross (or has crossed) that of our planet. The eccentricity of the orbit of these *Near Earth Objects* (NEOs) is much larger than that of Earth. For example, the eccentricity of Halley's comet, whose visits to the inner Solar System every 75–76 years have been recorded in archives since a long time (perhaps as early as 467 years before the Common Era), is 0.9671 (Fig. 1.5a). The very large

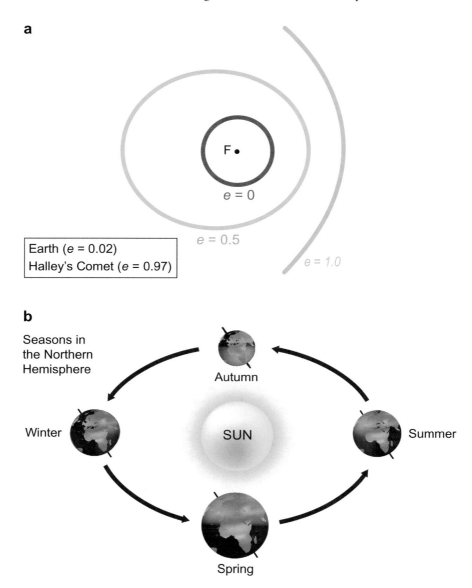

Fig. 1.5 Elliptical orbits. **a** Effect of eccentricity (*e*) on the shapes of three ellipses with the same focus (F) and *e* = 0, 0.5 and 1.0, respectively. The ellipses with *e* = 0 (circle) and 1.0 provide an idea of the real shapes of the orbits (not represented) of Earth and the Halley's comet, respectively. **b** Positions of Earth on its orbit at the time of the summer and winter solstices and spring and autumn equinoxes in the Northern Hemisphere. The elliptical form of the orbit in panel b is strongly exaggerated (compare with panel a). The effects of the positions of Earth on the seasons are detailed in Fig. 3.3. Credits at the end of the chapter

eccentricity of comets is explained by a destabilisation of their initial orbits by gravitational perturbations originating from inside or outside the Solar System.

The approaching NEOs are detected with increasingly powerful telescopes. Because the NEOs have very elongated elliptic orbits, their cycle of appearance within the range of our present means of detection may be very long, and it is thus difficult to calculate precisely the risk that Earth be hit by any one of them. Given that human civilisation could be seriously damaged if Earth were hit by a very large NEO (see Sect. 4.5.3), some space agencies and a number of governments are trying to improve the early detection of NEOs. These agencies and governments are also considering the development of technical means for possibly deflecting the trajectory of a NEO that would represent a major danger for Earth. In the distant past, Earth was often hit by very large NEOs, with consequences described below (see Sect. 1.5.1 and Information Box 1.6).

1.3.2 Rotation Period, Orbital Period, and Orbital Cycles (Eccentricity, Axial Tilt, and Axial Precession)

The day-and-night cycle corresponds to the time it takes for a planet or an asteroid to make one revolution around its axis, and this time is called *rotation period*. The rotation period of Earth is exactly 24 h when the Sun is taken as reference (this defines the duration of the solar day), whereas it is 23 h and 56 min when the duration of day is measured by reference to very distant objects, such as the *fixed stars* (that is, any star other than the Sun). The Earth's rotation period is the shortest among the four rocky planets, and that of Venus is the longest with 243 Earth days (Table 1.1).

The time taken by Earth to make a complete orbit around the Sun defines the duration of one Earth year. In the Solar System, the *orbital periods* of planets around the Sun range between 0.24 Earth year for the planet closest to the Sun, Mercury, and 164.8 Earth years for the planet farthest from the Sun, Neptune (Table 1.1). Three major characteristics in the orbits of the planets—the eccentricity, the axial tilt, and the precession—are subjected to cyclical variations called *orbital cycles*.

Eccentricity. The amount by which the orbit of a planet deviates from a perfect circle is called *eccentricity* (see Sect. 1.3.1). In fact, the current eccentricity value of 0.0167 cited above for Earth is not constant, and the eccentricity of the planet's orbit varies slightly from a quasi-zero value to about 0.058 over hundreds of thousands of years (Fig. 1.5a). The main component of this variability fluctuates with different periods that combine in an *eccentricity cycle* whose period is about 100,000 years. This cycle modifies the seasonal distribution of the solar energy incident on the surface of Earth with a period of about 100,000 years, which influences the long-term climate of the planet (see Information Box 1.4). This explains why the present duration of the glacial-interglacial episodes experienced by Earth since 2.6 million years, known as *Ice Age*, is about 100,000 years (see Sects. 1.3.4 and 4.5.2).

Information Box 1.4 Weather, Climate, and Climate Change The terms weather and climate are often used nowadays in relation with the ongoing global warming. These two words refer to different timescales.

The word w*eather* refers to changes in the conditions of the atmosphere over the short term. The timescales of weather range from minutes to months.

The word *climate* in the narrow sense refers to the average pattern of weather conditions in a given area over a period ranging from one or a few months to thousands or millions of years. The area may be a region, a continent, and even the whole Earth when *climate* is accompanied by the adjective *global*.

Climate in the wider sense is the state of the climate system. The *climate system* is defined in Chap. 3 as the highly complex system consisting of five major components—the atmosphere (air), the hydrosphere (liquid water), the cryosphere (ice), the lithosphere (rocks), and the biosphere (organisms)—and their interactions (see Sect. 3.5.1). The state of the climate system changes under the influence of its own internal dynamics and because of external forcings, which aspects are examined elsewhere in this book.

To avoid any confusion in the present text, we use *climate* for the average long-term pattern of the weather (narrow sense), and *climate system* for the wider sense of the above five major components, their interactions, and their responses to external forcing. The main variables considered by meteorologists and climatologists are temperature, humidity, atmospheric pressure, wind speed and direction, precipitation, and abundance of atmospheric particles.

The expression *climate change* means a change in the state of the climate in general. In the present text, we use *climate change* for changes in the state of the *climate system*. Various chapters examine past episodes of *natural climate change* and the ongoing *anthropogenic climate change*. The adjective *anthropogenic* means "originating in human activity". Anthropogenic climate change and ocean acidification, which are two global effects of human activities, are examined in detail in Chap. 11 (see Sects. 11.1.1 and 11.1.3).

Axial tilt. The axis of rotation of a planet on itself is generally not perpendicular to the plane of its orbit around the Sun (called *plane of the ecliptic*), and the angle between the axis of rotation and the plane of the ecliptic is called *axial tilt* or *obliquity* (Fig. 1.6a). On a planet without axial tilt (such as Mercury), all latitudes receive the same insolation year round. Conversely, on planets having an axial tilt, there are seasonal variations in temperature proportional to the axial tilt (Table 1.1). These variations are very small on Venus, and moderate on Earth and Mars.

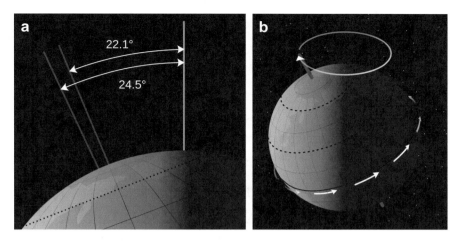

Fig. 1.6 a Axial tilt: the tilt of Earth's axis presently varies between 22.1° and 24.5° over a period of 41,000 years. **b** Axial precession: the orientation of the Earth's axis describes a cone in space over a period of 26,000 years. Credits at the end of the chapter

The current axial tilt of Earth is not constant, and varies between 22.1 and 24.5° over a cycle whose period is 41,000 years. It is thought that the axial tilt of Earth stays close to a value of approximately 23° because of the presence of the Moon. Indeed, the formation of the Earth-Moon system about 4.5 billions years ago (see Sect. 1.5.1) stabilized the Earth's axial tilt, but it is not known if the relatively stable values of the axial tilt were always near 23°.

Precession. The shape of Earth is not perfectly spherical, and the planet is flattened at the poles and bulges at the equator. Because of the presence of the equatorial bulge, Earth behaves as a slowly spinning top, that is, the axis of rotation of the planet changes its orientation by 1° every 72 years and thus describes a cone in space over a period of 26,000 years. This movement is called *axial precession* (Fig. 1.6b). One of the consequences of this movement is a gradual change in the occurrence of the Earth's equinoxes, and during one *precession cycle*, the Earth's equinoxes move progressively earlier in the year all the way back to the starting time at the end of the cycle. This backward movement in the occurrence of equinoxes is called *precession of the equinoxes*.

The *precession period* of Earth *relative to the fixed stars* (defined above), is 26,000 years. However, gravitational forces that other planets exert on Earth make the elliptical orbit of the planet rotate about the Sun, with the consequence that *the precession period experienced on Earth* is not 26,000 years but instead 21,000 years. The later period is, in fact, the combination of two periods, whose values are 19,000 and 23,000 years. It is explained in Sect. 3.3.2 that the precession of the equinoxes does not change the actual dates of the equinoxes in the current Gregorian calendar.

Although the above aspects of celestial mechanics is a bit complicated, the orbital cycles are important within the context of this book because they are integral components of the Milankovitch cycles. The three orbital cycles provide a key explanation for the succession of glacial and interglacial episodes of the current Ice Age (see Sect. 1.3.4).

1.3.3 Effects of Earth's Characteristics on Liquid Water and the Seasons

Two astronomical characteristics of Earth have favoured the long-term presence of water in liquid form on the planet, which was a key condition for the success of organisms. The *short day-and-night rotation cycle* distributes the solar heat quite uniformly and rapidly all around the planet, which generally prevents the occurrence of temperatures that would be too cold or too hot for the existence of liquid water. Conversely, the very long rotation period on Venus (243 Earth days, Table 1.1) may have contributed to the early loss of water in the history of that planet (see Sect. 5.5.4). The *moderate axial tilt* contributes to the lack of extreme seasonal variations in temperature at most latitudes, which also favours the occurrence of temperatures at which water is liquid. These two astronomical characteristics thus strongly contributed to the successful and long-lasting establishment of organisms on Earth and build-up of large biomasses.

In addition, two axial characteristics of Earth—the axial tilt and the axial precession—also affect the seasons.

The *axial tilt* is responsible for the magnitude of the differences in temperature among seasons. Because the tilt of the Earth's axis is the same year-round (that is, regardless of where Earth is in its orbit around the Sun), the Northern Hemisphere is directed towards the Sun on one side of the orbit in May–July, and away from the Sun half an orbit later in November–January, which causes the existing seasonal changes in Earth's temperature (left and right sides, respectively, of Figs. 1.5b and 3.3). Conversely, the Southern Hemisphere is directed away from the Sun in May–July, and towards the Sun in November–January. This explains why the seasons are opposite in the two hemispheres. Although the seasonal changes are large within a year, the variations in the magnitude of the seasonal differences induced by variations in the axial tilt, for example between glacial and interglacial episodes (see Sect. 1.3.4, below), are generally moderate except at the high latitudes of the two hemispheres. The moderate value of the Earth's axial tilt, which varies between 22.1° and 24.5° over a cycle of 41,000 years, favours generally mild seasonal differences in temperature.

The *axial precession* modifies the seasons in which Earth is closest to the Sun and farthest from it, which increases the seasonal contrast in one hemisphere while decreasing it in the other, and this change occurs over a period of 21,000 years (see Sect. 1.3.2). Currently, Earth is closest to the Sun during the

Northern Hemisphere winter (or the Southern Hemisphere summer), which makes the winters in the Northern Hemisphere generally less severe than in the Southern Hemisphere at comparable latitudes.

1.3.4 Milankovitch Cycles

The position of Earth relative to the Sun varies in the long term over different periods (see Sect. 1.3.2, above): the eccentricity of the Earth's orbit varies over periods that combine in a cycle of about 100,000 years; the Earth's axial tilt varies over a cycle with a period of 41,000 years; and the precession cycle experienced on Earth has a period of 21,000 years (Fig. 1.7). The combined effect of these cycles on the solar heat reaching Earth contributes to explain the succession of glacial and interglacial episodes, called *glaciation* or *Ice Age*, which have affected Earth since the beginning of the Quaternary period 2.6 million years ago. This explanation was proposed in 1920 by the Serbian researcher Milutin Milanković (name generally written *Milankovitch* in English), and the three orbital cycles are consequently now known as *Milankovitch cycles* (see Sect. 4.5.2).

Information on the Quaternary Ice Age is documented by various natural records of Earth such as air bubbles trapped in glaciers and the chemical composition and microfossils of marine sediments. Deciphering these natural records is a very difficult task, similar to decoding ciphers or reading ancient languages. Nevertheless, researchers have progressively accumulated data on a number of climate characteristics that include past atmospheric and oceanic temperatures, sea level, gas composition of the atmosphere, and seawater salinity. The information recorded in marine sediments goes back in time to the beginning of the Quaternary Ice Age and beyond, whereas ice-core records from Antarctica do not go farther than one million years.

The relationships between the changes in solar heat corresponding to variations in the three astronomical cycles in Fig. 1.7 and the Quaternary climate variations are generally not direct. This is because the effects of changes in solar heat on the climate are modulated by a large number of interacting factors. These include: the growth and retreat of forests at continental scales; the uptake and release of climate-active gases (CO_2 and others) by oceans and terrestrial peat; and changes in the reflection of solar energy by the snow and ice cover and by clouds. Paleoclimatologists have incorporated the effects of these factors into complex numerical models that simulate the past climate based on the 100,000-, 41,000- and 21,000-year astronomical cycles. Such models show that the Milankovitch cycles are undoubtedly the major cause of the climatic variations that occurred during the Quaternary period, but these cycles do not fully explain all the recorded events, nor their timing or amplitudes. Research continues on the fascinating links that exist between astronomical cycles that take place over millions of kilometres in the Solar System and changes in the climate of planet Earth at much smaller spatial scales.

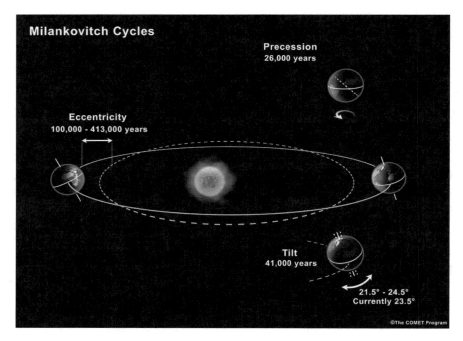

Fig. 1.7 Milankovitch cycles: three combined astronomical cycles cause variations in the amount of solar heat received by Earth. Their dominant periodicities are 100,000 years (eccentricity), 41,000 years (axial tilt) and 21,000 years (precession; this value is different from the 26,000 years stated in the figure for the reason explained in Sect. 1.3.2). The elliptical shapes of the orbits are strongly exaggerated. Credits at the end of the chapter

1.3.5 The Nearest Neighbours of the Planets: Their Moons

Planet Earth is accompanied by a natural satellite, our Moon. In fact, six out of the eight planets of the Solar System planets have moons that orbit around them. Collectively, these six planets have 205 moons, which are unevenly distributed among the rocky and the giant planets. The two rocky planets closest to the Sun, Mercury and Venus, have no moon, Earth has one and Mars two. The remaining 202 moons orbit the four giant planets: Saturn (82), Jupiter (79), Uranus (27) and Neptune (14). Each of the 205 moons has its own name; for example, the Earth's satellite is called *The Moon*, and the two Mars' moons are called *Phobos* and *Deimos*.

Some of the moons are tiny, with a diameter as small as 5 km, and others are very large, the largest being Jupiter's moon Ganymede whose diameter is 41% that of Earth. New moons are discovered from time to time. Some asteroids and dwarf planets also have moons, of which almost 300 are presently known. There are no observations of moons orbiting around moons in the Solar System. Most moons revolve around their planets (or smaller bodies) in the same direction as these rotate on their axes. This is called *prograde motion*, and the reverse is called *retrograde motion*. More generally, the orbits of most objects around the Sun are prograde, that is, in the same direction as the Sun rotates, which is from west to east. Similarly, the orbit of a natural satellite around its *primary* (name given by astronomers to the main physical body of a gravitational system) is also prograde, which means that in the rare case of a primary rotating from east to west, its satellites generally also orbit from east to west.

The mass ratios moon:planet are very different among the three moons of the rocky planets: Moon:Earth = 1:81, Phobos:Mars = 1:60,216,710, and Deimos:Mars = 1:434,798,807. These ratios show that our Moon is very large relative to Earth when compared to the two moons of Mars. The mass ratios moon:planet are also small for the moons of the giant planets because the masses of these planets are very large (Table 1.1). Even in the case of the most massive moon in the Solar System, which is *Ganymede* with a mass 2.0 times the mass of our Moon, the mass ratio to its planet Jupiter is only 1:12,810. In fact, the mass ratio Moon:Earth is the largest moon:planet mass ratio in the Solar System, which largely explains the stabilizing effect of the Moon on the Earth's axial tilt to a small value (see Sect. 1.3.2, above).

Our Moon always shows the same face to Earth, but Earth does not always show the same side to the Moon. Astronomers call this phenomenon *tidal locking*, a name coming from the fact that the same gravitational forces that cause moons to be locked to their planet are also largely responsible for the existence of tides. A tidally locked moon has a prograde orbit, and its *rotation period* is the same as its *orbital period* (see Sect. 1.3.2 for the explanation of these two periods). Interestingly, most of the 205 moons orbiting planets in the Solar System are tidally locked to their planet, meaning that each of these moons always show the same face to its planet.

Life-Bearing Moons? The above paragraphs on moons show that the Solar System is very diverse. Among the almost 500 moons of planets and asteroids presently known, several show some of the life-sustaining characteristics that exist on Earth, such as an atmosphere (Table 2.2) or liquid water (see Information Box 5.4). Hence, it would not be surprising if life forms were discovered on some of these moons in the future, or if it was found that life forms had existed there in the past. However, none of these moons, as far as we know, shows signs of major ecosystem activity at present or in the past, contrary to the situation that has existed on Earth for billions of years. In following chapters, we will compare the conditions that exist on some of the life-suitable moons of the Solar System with those encountered on Earth. These comparisons will help us identify the conditions that allowed organisms to establish beachheads on Earth billions of years ago and subsequently take over the whole planet.

1.3.6   The Eight Planets and Their Moons
Are not Alone in the Solar System

In addition to the Sun, the eight planets and their moons, the Solar System contains rocky bodies called dwarf planets, asteroids and meteoroids, and largely icy objects called comets and centaurs. Like the planets and their moons, these astronomical bodies were formed, directly or indirectly, by accretion of matter from the solar nebula.

Dwarf planets (see Sect. 1.2.2) are planet-like rocky bodies that, contrary to full-fledged planets, have not cleared the neighbourhood around their orbits of other material (see Information Box 1.3; *Planet*). *Asteroids* and *meteoroids* are rocky bodies that orbit around the Sun, whose mass is smaller than that of planets and whose shape is not round. There are millions of asteroids and meteoroids in the Solar System, the meteoroids being smaller than the asteroids. Objects smaller than meteoroids are part of *cosmic dust*, called *interplanetary dust* in the case of the Solar System. *Comets* (see Sect. 1.2.2) are small bodies composed of ice, dust and rocks. When a comet passes close to the Sun, it heats up and releases gases, which create a temporary atmosphere around the comet's nucleus and also sometimes a tail that can be visible from Earth. *Centaurs* have characteristics of both comets and asteroids. Their name comes from the Greek mythology where a centaur was a mixture of horse and human; in the early 2000s, centaurs moved out of the Greek mythology into *Harry Potter* books and movies. Interestingly, some of the centaurs could originate from outside the Solar System, as well as a few other interstellar objects that could have been captured by the Sun 4.6 billion years ago.

Most of the meteoroids that enter Earth's atmosphere vaporize (i.e. burn) before they reach the surface of the planet. The brightest burning meteoroids are called *bolides* (or *fireballs*) and the less bright, *meteors* (or *shooting stars*). A superbolide whose size may have been up to almost 200 m exploded in the air over Siberia, Russia, in 1908. Estimates of its energy range from 3 to 30 megatons of TNT, which corresponds to the energy released by 200–2000 Hiroshima bombs. Some asteroids and meteoroids that enter Earth's atmosphere do not vaporize, or not completely, during their fall and thus hit the surface of the planet.

The rocky remains of the meteoroids that have hit the surface of Earth or other planets are called *meteorites*. The largest meteorite found so far on Earth has dimensions of $2.7 \times 2.7 \times 0.9$ m and a mass of 60 tons. It is called the *Huba meteorite* because it was discovered in Namibia, Africa, on a farm named *Huba West*. The Earth's surface receives annually between 50,000 and 100,000 tons of solid matter from beyond the atmosphere. Most of this matter is small-sized dust, but it has included in the past numerous meteorites and some rocky bodies several kilometres in diameter (see Sect. 1.5.1, below).

All astronomical bodies in the Solar System are continually hit by asteroids, meteoroids and interplanetary dust. On astronomical bodies with no atmosphere or whose atmosphere is thin, most meteoroids reach the surface where they create craters. Examples are the atmosphereless Moon and rocky bodies with a rarefied atmosphere such as planet Mars, whose surfaces are pockmarked by impact craters.

There are many impact craters at the surface of continents on Earth, but not as many as on some other planetary bodies, which is explained by at least three reasons. First, the craters on the Earth's continents are continually erased by erosion (see Sect. 7.3.3), and these processes are not active on bodies without air or running surface water, such as the Moon and Mars. Second, many asteroids and meteoroids vaporize in the Earth's atmosphere and thus do not reach the ground. Third, more than 70% of the Earth's surface is occupied by oceans, with the consequence that two thirds of the meteoroids that reach the Earth's surface fall in water where they do not create impact craters or create submarine craters.

The potential development of organisms on astronomical bodies located far away from the Sun would be strongly influenced by the conditions prevailing in the region of the Solar System where these bodies would be located. For example, the potential occurrence of organisms on a moon of a giant planet may be strongly affected by the distance of this moon from its giant planet. Hence, comparing the environmental conditions that prevail on Solar System bodies located far away from the Sun with those that exist on Earth will require some caution in the following chapters.

1.4   Earth in the Universe: Making a Very Long History Very Short

1.4.1   The Big Bang Cosmological Model

Planet Earth is part of the Solar System, which is part of the *Milky Way* galaxy, which is itself part of a group of about 50 galaxies called the *Local Group*, which is part of the Universe where there may be more than 2 trillion galaxies (see Sect. 1.2.2). The establishment of organisms and the development of ecosystems on Earth have been conditioned by the conditions that led to the formation of the planet in the Solar System, the Milky Way, the Local Group and the Universe. The book of Satz (2013) provides further reading on the limits of the Universe.

The prevailing ideas on the formation of the Universe are based on a largely theoretical model, in which the origin of the Universe is the *Big Bang*. According to this model, the Big Bang occurred 13.7 billion years ago, and was followed by the formation and organization of energy, matter (mass-bearing particles) and space (Table 1.2, Fig. 1.8). Shortly after the Big Bang, the Universe underwent

Table 1.2 Simplified chronology of the main periods/steps in the formation of the Universe. The time units are seconds or years after the Big Bang. Schematic representation is provided in Fig. 1.8

Time after the Big Bang	Astronomical events	Significant events for future organisms and ecosystems on Earth
0 second	Big Bang	Singular point of the Standard Model (arbitrary 0)
10^{-43} second	Planck Wall	Limit of the Universe that can be observed
10^{-43}–10^{-32} second	Inflation phase	Creation of the volume of the Universe, which was then extremely dense and hot
<200 seconds	Primordial nucleosynthesis	Formation of nuclei of H, He and Li
380 thousand years	Free photons	Formation of neutral atoms (nuclei and electrons), which released photons
400 million years	Formation of stars and galaxies	Nucleosynthesis of elements heavier than He in stars
9.14 billion years (beginning of geological times on Earth)	Formation of the Solar System	Favourable conditions for the formation of rocky planets, including Earth
13.7 billion years (present)	Continuing formation of stars and galaxies	Continuing nucleosynthesis in stars

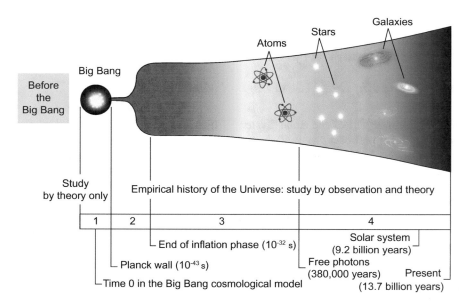

Fig. 1.8 Big Bang cosmological model: schematic history of the Universe. The vertical dimension represents the size of the Universe, which has been increasing since the Big Bang. The main steps are summarized in Table 1.2. Credits at the end of the chapter

both rapid expansion and rapid cooling, from a state where the universe was immensely condensed and immensely hot to the point just before the creation of the first physical particles. This *inflation phase* began 10^{-43} second after the Big Bang and lasted until 10^{-32} second. These timescales are very short (fractions of a second), but the model also considers very long timescales (billions of years), described below.

Following the inflation phase, elementary particles of energy and matter (such as the quarks) became individualized, after which they combined to form more complex entities. These entities were successively the neutrons and the protons, followed by the primordial nuclei of the simplest chemical elements hydrogen, helium and lithium, with 1, 2 and 3 protons, respectively (corresponding to atomic numbers 1–3 in the classification of chemical elements, and resulting from *big bang fusion* in Fig. 1.9). The production of nuclei more complex than those of the lightest isotope of hydrogen is called *primordial nucleosynthesis* (isotopes are explained in Information Box 5.6). The fusion of elementary particles into nuclei of chemical elements was accompanied by the release of other elementary energetic particles, such as photons (light).

Within 380,000 years after the Big Bang, the temperatures were too high for electrons (particles with negative electrical charges) to bind to nuclei (with positively charged protons). Afterwards, electrons and atomic nuclei bound together, forming electrically neutral atoms (see Information Box 1.5). Finally, atoms aggregated and became, much later, organized into stars and galaxies under the effect of gravitation.

This model, which was developed during the second half of the twentieth century, is also called the *Standard Model*. It accounts for a large number of observations, but leaves certain phenomena unexplained. These include the nature of dark matter (a form of matter thought to account for about 85% of the matter in the Universe, and has not yet been observed directly) and dark energy (an unknown form of energy that affects the Universe on the largest scales). As a result, cosmologists and physicists develop theories that go beyond the Standard Model, and are seen as steps toward a *Theory of Everything*.

As stars grew and died, nuclei of heavier chemical elements formed by the merging of lighter nuclei, through various cosmologic processes known to take place within stars (Fig. 1.9). Four of these processes are: the *explosion of massive stars* (stars with masses 7–10 times that of the Sun), the *explosion of white dwarfs* (very dense cores of former stars), the *dying of low mass stars*, and the *merging of neutron stars* (compressed cores of former stars larger than those producing white dwarfs). Gravitation is involved in these processes, since the explosions of massive stars and white dwarfs follow their contraction under the influence of their own gravity, and neutron stars spiral toward each other due to their effect on the curvature of space-time (Fig. 6.10). A different process

The Origin of the Solar System Elements

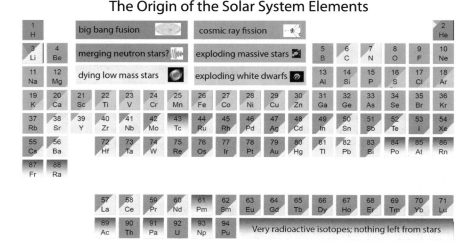

Fig. 1.9 Origin of the nuclei of the chemical elements found in the Solar System, represented in the framework of the periodic table of the elements. The six origins cited in the figure (six different colours) are briefly described in the text. Credits at the end of the chapter

was involved in the formation of some light elements—namely beryllium, lithium and boron (atomic numbers 3–5 in the classification of chemical elements: Fig. 1.9)—called *cosmic ray fission*, which involved hits of heavier atoms by particles coming from the outer space, called cosmic rays (see Information Box 8.5). The very radioactive elements beyond atomic number 94 found in the Universe are the result of local disintegrations of other elements because their radioactive period is so short that the nuclei formed in the stars are long gone (see Information Box 5.6). These *transuranium elements* do not exist naturally on Earth. Indeed, even if the radioactive period of one of the isotopes of the element with atomic number 96 (curium; isotope ^{247}Cm) is relatively long, namely 15.7 million years, this period is very short compared to the 4.6 billion years of the Earth's history and none of the original ^{247}Cm or any other transuranium element is left on the planet.

Information Box 1.5 Electrical Charges of Atoms, Molecules and Ions Atoms generally have equal numbers of negatively and positively charged subatomic particles (called *electrons* and *protons*, respectively), which makes them electrically neutral. The nucleus of an atom also contains particles that are not electrically charged, such as the *neutrons*, which contribute to the mass of the atom but not to its electrical charges, and the *photons*, which have no charge and no mass.

The atoms of some elements can gain electrons and thus become negatively charged, whereas the atoms of other elements can lose electrons and thus become positively charged. An electrically charged atom is called *ion*. Negatively and positively charged ions are called *anions* and *cations*, respectively. The electrical charge of an atom of a given element is represented by a minus or plus exponent to the right of the symbol of the element. For example, H^+ represents an atom of hydrogen with one positive charge, and O^{2-} represents an atom of oxygen with two negative charges.

In nature, the atoms are generally found in molecules, which are made of two or more atoms of one or more chemical elements. For example, a molecule of water (H_2O) is made of two positively charged atoms of hydrogen (H^+) and one negatively charged atoms of oxygen (O^{2-}). The electrical charges of molecules are generally neutral, meaning that their numbers of negative and positive charges are equal, as in the case of H_2O.

Under some circumstances, a molecule may become dissociated in two parts each electrically charged, for example when the molecule is dissolved in a liquid. As in the case of the atomic ions above, each part is called *ion*, more precisely *anion* and *cation* for the parts with the negative and the positive charge, respectively. For example, the molecules of carbonic acid (H_2CO_3) dissociate into bicarbonate (HCO_3^-) and hydrogen (H^+) ions in water (Eq. 6.3).

The anions and the cations in a solution have opposite electrical charges and thus attract each other, which can lead to the formation of new compounds. For example, the sodium cations (Na^+) and chlorine anions (Cl^-) in seawater attract each other and produce sodium chloride ($NaCl$), also called sea salt or table salt, when the water is evaporated. Conversely, when solid $NaCl$ is dissolved in water, its dissociation produces electrically charged Na^+ and Cl^-. Since the organisms on Earth initially developed in seawater, their bodies are largely made of water, and both Na and Cl are among their main chemical constituents (see Sect. 6.1.2).

As the Universe developed and diversified, the number and variety of chemical elements increased until these reached the natural element with the highest atomic number, which is uranium (atomic number: 92). Elements with atomic numbers higher than 92 have been synthesized in the laboratory, but these are not stable and disintegrate rapidly (although small amounts of neptunium and plutonium, with atomic numbers 93 and 94, have been found in uranium ores, where they are continually formed by radioactive disintegration). *Synthetic elements* 92–118 are illustrated in Fig. 6.6. The article of Crockett (2018) provides further reading on the formation of heavy elements.

The Big Bang is a theoretical event that would have occurred 13.7 billion years ago. Phases 2, 3 and 4 in Fig. 1.8 can be studied experimentally (observation), whereas Phase 1 cannot be approached experimentally within the framework of the present theory. Phase 1 is separated from the beginning of the empirical history of the Universe by a theoretical temporal limit called the Planck wall (10^{-43} second after the theoretical 0 represented by the Big Bang), which is derived from quantum theory. The Big Bang cannot be considered as the absolute origin of the Universe. Indeed, the Universe at the moment of the Big Bang was in a theoretical state of infinite density and infinite temperature, which is the point of convergence of models when moving toward the past, but this point has no physical meaning.

The study of ancient astronomical events is based on the observation of very distant objects. This is possible because light travels in space at a constant speed ($300,000$ km s^{-1}), and objects far away from us in distance are thus also far away in time. Hence one could think that by observing the farthest possible astronomical objects using extremely powerful telescopes, it could be possible to see the Universe at it was at the time of the Big Bang. However, theory tells us this is not possible because the Planck wall creates a barrier in time located 10^{-43} second after the Big Bang before which observations are not possible. This means that times closer to the Big Bang than 10^{-43} second (called *Planck era*) will remain hidden forever. Hence the exploration of the Universe at the time of the Big Bang (or before) can only be done through theories, which are still under development and include the superstring and the loop quantum gravity theories. These and other similar theories could be tested in the domain of the observable Universe (younger than the Planck wall), and if they provided satisfactory explanations of observed phenomena, they could then be used to theoretically explore the Universe beyond the Planck wall and even before the Big Bang. The book of Perlov and Vilenkin (2017) provides further reading on cosmology.

1.4.2 Earth Is the Daughter of Stars

In astrochemistry, the expression *primordial elements* refers to the chemical elements synthesised shortly after the Big Bang. These are: hydrogen, helium, lithium and beryllium (atomic numbers 1–4). The formation of rocky planets, including Earth, required the availability of chemical elements heavier than the primordial elements such as carbon, nitrogen, silicon, phosphorus, sulphur, and iron (atomic numbers between 6 and 26). The nuclei of these high-mass elements were formed inside stars, where the enormous temperature and pressure fused together light nuclei of primordial elements into heavier elements (Fig. 1.9). When these stars died, the heavy elements were blown off into space, where they eventually joined nebulae from which new stars grew. This cycle of events has repeated itself continually since the beginning of the Universe, and will likely continue to do so until the end.

The heavy chemical elements that make up the rocky Earth required that countless generations of stars be born, grew and died, during billions of years before the Solar System started to form from the solar nebula. In addition, elements much heavier than iron (atomic numbers higher than 26), especially the radioactive elements, play major roles inside Earth (see Sect. 3.2.3), and these elements were formed in the final, gigantic explosions of massive stars. Hence, Earth is the daughter of stars.

Major aspects of the 13.7 billion-year evolution of the Universe, such as the progressive aggregation of matter from elementary particles into stars and the elaboration of chemical elements from the lightest to the heaviest, explain key characteristics of Earth. These include the rocky nature of the planet, the variety of chemical elements in the Earth's crust, and the presence of heavy, radioactive elements in the deep parts of the planet. We will see in following chapters that all these characteristics were important for the durable establishment of organisms on Earth and the development of ecosystems.

1.5   Earth and Its Organisms and Ecosystems: Making a Long History Short

1.5.1   Formation of the Earth-Moon System

Every day, about 100 tons of materials from space enter the Earth's atmosphere, but only a small fraction actually reaches the ground level (see Sect. 1.3.6, above). This is nothing new as planet Earth was hit by objects from the outer space during its whole history. The most spectacular hit may have been the collision, about 4.5 billion years ago, between the two planetary parents of the present Earth and Moon, namely a planetary body called *proto-Earth* (or *Gaia*) and a smaller, Mars-size astronomical body (called *Theia* by astronomers). It is thought that this giant collision thoroughly mixed the two planetary bodies, and ejected into space a huge amount of matter. This caused the formation of a disc of matter that orbited around the rocky core remaining from the larger of the two bodies. Under the effect of gravitation, part of the fragments in the disc progressively merged with the rocky core, which became planet Earth, and the remainder formed a new body, which became the Moon. The Earth-Moon system was born (see Sect. 1.3.1).

Other such giant collisions between the proto-Earth and large-sized bodies may have occurred before the collision that led to the formation of the Earth-Moon system, but the latter was likely the last giant collision. Indeed, the characteristics of the Earth-Moon system were not upset by further large hits. The mechanism by which the giant hits experienced by the proto-Earth since the beginning of its existence ended about 4.5 billion years ago may be as follows. Until about 4.5 billion years ago, the numerous bodies that had been initially formed in the Solar System often hit each other, and thus progressively merged into the present planets and other, smaller bodies. Given that there were gradually less and less bodies in the Solar System and the largest had removed most material from the path of their orbit, the planetary bodies stopped hitting each other about 4.5 billion years ago.

However, smaller hits continued, as explained above (see Sect. 1.3.6). It is hypothesized that, between 4.1 and 3.9 billion years ago, very large numbers of asteroids and comets have hit the early Earth and its neighbours, that is, the Moon, Mars, Venus and Mercury. However, the existence of the episode of *Late Heavy Bombardment* and the rapid decrease in the number of impacts that would have followed have been called into question. The consequences of the bombardment of Earth by asteroids and comets on organisms and ecosystems are examined in Chap. 5 (see Sect. 5.3.2). Some later hits of the Earth's surface by meteoroids also had major consequences for the ecosystems on Earth (see Information Box 1.6).

It is hypothesised that a major effect of the giant collision between the proto-Earth and Theia was the injection of a large quantity of water into the interior of the embryonic Earth. The addition of this water to the hot, semifluid material in the Earth's interior, called *magma*, might have decreased the magma's viscosity, which together with the water provided later by meteoroids and comets, lead to a chemical differentiation between the crust of continents and that of oceans. The crust of continents, which is dominated by granite rocks (rich in potassium, silica and sodium), began to differentiate from that of oceans, which is dominated by basalt rocks (rich in iron, calcium and magnesium) (see Sect. 6.6.2). Both granite and basalt are *igneous rocks*, which are formed by the solidification of magma deep inside Earth or brought to the surface, respectively (*igneous* comes from the Latin *ignis*, meaning *fire*).

The existence of continental and oceanic crusts with different characteristics combined with the presence of low-viscosity magma in the mantle under the crust favoured the creation and maintenance of active plate tectonics on Earth (see Sect. 7.2.1). It will be explained in following chapters that plate tectonics played a major role in the subsequent evolution of the Earth's environment, and contributed to the global success of ecosystem on the planet (see Sect. 7.4.7).

Information Box 1.6 Consequences of Some Large Collisions: Mass Extinctions After the end of the period of giant collisions, planet Earth and the other large astronomical bodies continued to be hit by objects they met along their trajectories in the Solar System. On Earth, some of these hits not only destroyed every geological structure and organisms in the zones of impact, but also released so much gas and dust in the high atmosphere that they blocked the light and heat from the Sun during years and thus caused long "global winters". These winters could have lasted several years, and some of them may have been responsible for the disappearance of a large fraction of the life forms that then existed on Earth. Such large-scale destruction of life forms is called a *mass extinction.*

A life form (for example a group of individuals defined as a species) becomes really *extinct* when its last individual dies out, which makes descendants impossible. It is estimated that a species becomes extinct, on average, within 10 millions years of its appearance. However, species descending from extinct ones or closely related to them may well exist for very long periods after the extinction of their ancestor or related species. For example, modern birds are the descendants of one group of dinosaurs while all other groups of dinosaurs became extinct 66 million years ago (see Sect. 4.6.2). Some present-day "living fossils" have changed little since they first evolved hundreds of millions of years ago, for example the elephant shark (*Callorhinchus milii*), a fish that inhabits the waters off southern Australia and New Zealand where it feeds on the seafloor at depths of around 200 m, and is the slowest-evolving vertebrate. Hence *species extinction* is a natural phenomenon that occurs continually in the course of biological evolution, and at any given time there are species that become extinct and others appear. A *mass extinction* is different from the usual extinction process in that a very large number of life forms get extinct within a relatively short period (which can span millions of years; see Table 4.4).

One frequently cited example of mass extinction is the disappearance of most dinosaurs 60 million years ago (although the ancestors of modern birds, which are technically dinosaurs, survived). Not only dinosaurs, but many groups of plants and animals also disappeared. This was the *fifth extinction* that had been recorded by fossils in the last 450 million years. However, organisms began to leave easily identified remains in the *fossil record* only about 450 million years ago, although they had been present on Earth for almost 4 billion years. Hence the organisms that existed before 450 million years ago are poorly known, and it is likely that there were many massive losses of organisms during the first 3.5 billion years of the history of organisms on Earth, but these are not documented by fossils. In addition, the oldest fossil-containing marine sediments have been largely destroyed during the course of the Earth's history by erosion (see Sect. 7.3.3) and plate tectonic processes (see Sect. 7.2.3), which contributes to further blur our knowledge of ancient organisms. Hence it is not known to what extent previous massive losses of organisms led to annihilations of entire life forms.

One explanation provided for the fifth extinction is the fall on Earth of an asteroid with a diameter of more than 10 km, which would have created the Chicxulub crater of 180 km in diameter in Mexico, and likely released energy corresponding to one billion times the atomic bombings of the Japanese cities of Hiroshima and Nagasaki in 1945. Some researchers do not agree that the fall of this asteroid was fully or at all responsible for the fifth extinction, and a complementary or alternative explanation is a sequence of massive volcanic eruptions. The two phenomena would have caused a global cooling and reduction of photosynthesis (see Sect. 4.6.2).

1.5.2 From the Beginning of Earth to the Present Time

The geological history of Earth dates back to the formation of the Earth-Moon system about 4.5 billion years ago (Fig. 1.1). From that time on, planet Earth evolved according to both its cosmic environment and internal activity, including the progressively increasing effects of ecosystems. Geologists describe this evolution through major phases summarized in Table 1.3. The broadest time intervals in the Earth's geological chronology are the *eons*, which are divided successively into *eras*, *periods*, *epochs*, and *ages* (the latter division is not included in Table 1.3, for simplicity). The first and last columns of Table 1.3 show that the various geological time intervals have different durations. This is because their boundaries correspond to important environmental and biological events that occurred during the course of the Earth's history, such as the first evidence of organisms on the planet, the onset of atmospheric oxygen, major biological innovations, or the occurrence of mass extinctions, sometimes related to major changes in the positions of ocean basins and continents, phases of exceptional volcanic activity, or impacts of large asteroids.

There are two *eons* in Table 1.3: the oldest (*Precambrian*) lasted more than 4 billion years, and the most recent (*Phanerozoic*) began 542 million years ago. The boundaries of the next divisions of geological times, the *eras*, correspond to very significant events in the Earth's history. Next are the *periods*, whose boundaries reflect geological events that are significant but not as much as those of the eras. The periods of the Phanerozoic eon are divided into *epochs*, which is not possible for the Precambrian eon because its very old rocks do not provide much details on the events that took place so long ago (in Table 1.3, the periods are only detailed into epochs for the Cenozoic era, for simplicity).

Table 1.3 Simplified chronology of geological eons, eras, periods and epochs. The geological times correspond to numbers of years before present. The start date of the proposed *Anthropocene* epoch is debated among specialists

Time	*EONS* and eras	PERIODS and *epochs*	Time
542 Ma to present	*PHANEROZOIC*		
65.0 Ma to present	Cenozoic	QUATERNARY	2.6 Ma to present
		Anthropocene	*Around 0.3 ka to present*
		Holocene	*11.7 to around 0.3 ka*
		Pleistocene	*2588–11.7 ka*
		NEOGENE	23.0–2.6 Ma
		Pliocene	*5.3–2.6 Ma*
		Miocene	*23.0–5.3 Ma*
		PALEOGENE	65.5–23.0 Ma
		Oligocene	*33.9–23.0 Ma*
		Eocene	*55.8–33.9 Ma*
		Paleocene	*65.5–55.8 Ma*
251–65.5 Ma	Mesozoic	CRETACEOUS	145.5–65.5 Ma
		JURASSIC	199.6–145.5 Ma
		TRIASSIC	251.0–199.6 Ma
542–251 Ma	Paleozoic	PERMIAN	299.0–251.0 Ma
		CARBONIFEROUS	359.2–299.0 Ma
		DEVONIAN	416.0–359.2 Ma
		SILURIAN	443.7–416.0 Ma
		ORDOVICIAN	488.3–443.7 Ma
		CAMBRIAN	542.0–488.3 Ma
4.6–0.5 Ga	*PRECAMBRIAN*		
2.5–0.5 Ga	Proterozoic		
4.0–2.5 Ga	Archaean		
4.6–4.0 Ga	Hadean		

Ga: billion years before present; Ma: millions of years before present; ka: thousands of years before present

Geologists, sedimentologists, glaciologists and palaeontologists study planet Earth, the soft sediments and the sedimentary rocks, the ice and the fossils, respectively. Together, these researchers have patiently reconstructed the history of the planet from detailed analyses of rock minerals, sedimentary rocks, ice and fossil remains of organisms (see Information Box 1.7). All these evidences make up the natural record of Earth. Of course, the abundance and the quality of preservation of the Earth's natural record diminish with increasing time into the past.

For example, *sedimentary rocks* are formed by the deposition of materials at the surface of continents and at the bottom of water bodies and their subsequent

---

Information Box 1.7 Stratigraphy The smallest subdivisions of the geological time are the *ages*. The lower and upper boundaries of ages are based on field studies of layers (called *strata*) of sedimentary and layered volcanic rocks, conducted in a branch of geology called *stratigraphy*. Each geological age corresponds to a field-observed stratigraphic *stage*, which is a succession of rock strata that may represent millions of years of deposition with recognizable characteristics. The international reference for each stage is a *stratotype*, which is a stratigraphic section precisely identified in a given region of the world.

For example, one finds in the region of southwestern France called Aquitaine (which is renowned internationally for producing the Bordeaux wines), a stratigraphic section that is a referent stratotype for the Aquitanian geological stage and age. The latter is a subdivision of the Miocene epoch (Neogene period) between 23 and 20.5 million years ago. Because geologists use the same name and temporal boundaries for a given stratigraphic stage, which thus corresponds to the same geological age everywhere on Earth, they can correlate rock stages in different locations.

---

cementation (binding of sediment grains together by crystalline material). As the thickness of sediments and sedimentary rocks increases, local pressure and temperature increase and sedimentary rocks are transformed into *metamorphic rocks*. These are formed differently from the igneous rocks, which were introduced above (see Sect. 1.5.1). Most of the sedimentary rocks formed in the deep ocean are destroyed by tectonic processes that bury them deep in the Earth's mantle within a few hundred million years of their formation, on average, but small pieces of them are incorporated into continents (see Sect. 7.2.3). Also, some of the sedimentary rocks formed in shallow waters are incorporated directly into continents (see Sect. 7.2.3). Once on the continents, the sedimentary rocks exposed to air and water are subjected to erosion, which progressively destroys most of them (see Sect. 7.3.3). As a consequence, old sedimentary rocks are rare and fragmented. Such rocks from the most recent eon, the Phanerozoic era (less than 542 million years ago), are abundant on continents; those dating back to the Upper Proterozoic period (between 1 billion and 542 million years ago) are scarce; and those dating back to older times (Lower Proterozoic period, and Archaean and Hadean eras) are almost absent. The major division of geological times between the Phanerozoic and Precambrian eons (Table 1.3) reflects the relative abundance of sedimentary rocks from these times.

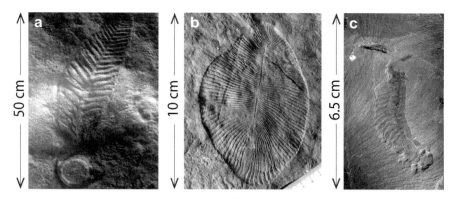

Fig. 1.10 Photos of three fossils from the Ediacaran and Cambrian periods. **a, b** Ediacarian animals *Charniodiscus* and *Dickinsonia*, respectively; **c** Cambrian animal *Opabinia regalis*. Credits at the end of the chapter

1.5.3   The Fossils

The existence of fossils of organisms is closely linked to the existence of sedimentary rocks in which most of them are found. Indeed, most fossils are derived from bodies of organisms that were covered by soft sediments soon after death, followed by one of the different forms of fossilisation. The soft sediments are later transformed into hard sedimentary rocks through processes of *cementation*. The totality of fossils, already found and undiscovered, is called the *fossil record*.

The existence of fossils is also linked to the evolution of organisms on Earth. For example, 540 million years ago or a little earlier, different groups of organisms evolved hard parts (shells, skeletons), which facilitated their fossil preservation. The resulting abundant fossils of marine animals found in sedimentary rocks led to define the Cambrian period, whose beginning was set at 542 million years ago. This is a key date in the geological history of Earth, as it determines the beginning of the Paleozoic era and thus the Phanerozoic eon in which we presently live. The fossils of the Cambrian period present a great biological diversity, and geologists consider that this period was characterized by an exceptional diversification of life forms called the *Cambrian explosion* or *Cambrian radiation*.

However, it is not certain whether the Cambrian diversification was exceptional or not in the evolution of species. Indeed, it is possible that other phases of intense diversification of life forms had happened previously, but the corresponding fossils were not as well preserved as in the Cambrian period, due to the absence of hard parts in the organisms and/or the destruction of the corresponding sedimentary rocks. The hypothesis that the fossil record did not keep track of very ancient, strongly diverse floras and faunas is consistent with the existence of ancient biota with no hard parts in the sedimentary record of the Ediacaran period. This period ended 542 million years ago (not shown in Table 1.3), just before the Cambrian period at the boundary between the Proterozoic and Paleozoic eras, and thus the Precambrian and Phanerozoic Eons (Table 1.3). The Ediacaran biota, which represents the earliest known complex multicellular organisms, was completely different from that of the following Cambrian period. Figure 1.10 compares Ediacarian animals *Charniodiscus* and *Dickinsonia* (from 580 to 542 and 560–555 million years ago, respectively) with Cambrian animal *Opabinia regalis* (approximately 505 million years ago).

1.5.4   From Earth and Organisms to the Earth System

The geological times are first divided into the Precambrian and Phanerozoic eons (Table 1.3). The most ancient of these two eons (*Precambrian*, 4600 to 541 million years ago) ends with the emergence of hard-shelled organisms. The current eon (*Phanerozoic*, 541 million years ago to present) is characterized by the large diversification of plant and animal life observed in the Cambrian fossils. This eon is divided into three eras: Paleozoic, Mesozoic and Cenozoic. The Earth's chronology in Table 1.3 thus includes six eras, that is, three in each of the two eons. The main characteristics of organisms and ecosystems in the six eras are briefly examined in next paragraphs, where it is shown that the first organisms established a strong beachhead on Earth, from which they progressively took over the whole planet. The book of Stanley and Luczaj (2014) provides further reading on the history of the Earth System.

The huge amount of energy released by the giant collision that formed the Earth-Moon system was mostly transformed into heat, causing the Earth's temperature to be initially as high as perhaps 2000 °C. Because of the very high temperature, the surface of the recently accreted Earth was an ocean of hot magma. The degassing of volatile compounds from the magma created the first phase of the atmosphere, which mostly consisted of hydrogen compounds such as water vapour, methane (CH_4) and ammonia (NH_3), similar to the present atmospheres of the gas giant planets. Because of the very high temperature, there was no liquid water on the planet. The magma progressively cooled down and solidified into solid crust about 4.4 billion years ago, and the release of CO_2 by volcanoes contributed to create the second phase of the atmosphere where the concentration of this gas was very high. The density of the atmosphere was so high that the atmospheric pressure at ground level may have been up to 60 times higher than present. The very high concentration of CO_2 caused a very high greenhouse effect (see Information Box 2.5) and the resulting ground temperatures may have reached several hundred degrees Celsius. These conditions were similar to those that currently prevail at the surface of planet Venus, where the very high concentration of atmospheric CO_2 causes both an atmospheric pressure more than 90 times that on Earth and temperatures up to 490 °C (Table 3.2).

During the first era of the Earth's history (*Hadean*, 4.6–4.0 billion years ago), the young planet was initially very hot because its surface had not yet completely solidified. The name *Hadean* comes from the Greek mythological god of the underworld, Hades, and it refers to the hellishly high temperatures that prevailed on Earth at the time. The temperatures just after the formation of the planet were totally unfit for organisms, and they progressively decreased during the Hadean era to values at which the presence of organisms was at least possible and perhaps actually occurred. Indeed, the combination of information recorded in rocks and from genomics indicates that the first organisms may go back to as early in the history of Earth as 4.5 billion years ago, whereas a more conservative date is 3.8 billion years ago. There is no marked limit between the end of the Hadean era and the beginning of the next, Archaean era. For example, the episode of the Late Heavy Bombardment of Earth by asteroids and comets 4.1–3.9 billion years ago (see Sect. 1.5.1, above) covers the two eras. One hypothesis concerning the first organisms is that life forms had appeared on Earth a first time before the Late Heavy Bombardment, disappeared during it, and appeared a second time at the beginning of the Archaean era. Another hypothesis is that some of the life forms that had appeared during the Hadean era survived the Late Heavy Bombardment, and were thus the ancestors of today's organisms.

At some time during the Hadean, perhaps as early as 4.4 billion years ago, the surface of Earth had sufficiently cooled for liquid water to form. Indeed, liquid water could then occur at temperatures higher than 200 °C because the prevailing high atmospheric pressure allowed water to condense at a temperature up to 280 °C (Table 5.2). Part of the atmospheric CO_2 progressively dissolved in the new, warm ocean, where chemical reactions transformed it into huge deposits of solid *calcium carbonate* ($CaCO_3$, also called *limestone*). This caused a major reduction in both the atmospheric pressure and the greenhouse effect, which accelerated the decline in temperatures and the accumulation of liquid water (see Sect. 5.5.4). The reduction of greenhouse effect initiated during the Hadean had thus become self-sustaining.

The second era of the Earth's history (*Archaean*, 4.0–3.8 to 2.5 billion years ago) begins at the hypothesised time of appearance on Earth of the present life forms, and their first diversification in an ocean devoid of dissolved free oxygen, O_2 (see Information Box 1.8), as the Archaean rocks do not show evidence of oxidation. The name *Archaean* comes from a Greek word meaning "archaic". The Archaean atmosphere was also devoid of O_2. The organisms that do not use O_2 for respiration were the sole occupants of Earth for about 2 billion years (until the Proterozoic era, next paragraph), and continue to be very abundant on modern Earth, where they occupy numerous environments without O_2. Because we, the humans respire O_2, we sometimes forget that the food in our guts is digested by organisms that do not use O_2 for respiration, and also that such organisms produce for us a variety of foodstuffs by fermentation that include bread, cheese, kimchi, sauerkraut, soy sauce, and wine (see Information Box 5.2). During that era, some organisms deeply modified the atmosphere by emitting methane (CH_4), a powerful greenhouse gas that contributed to prevent the planet from freezing at a time when

Information Box 1.8 Free Oxygen (or Dioxygen) and Ozone The chemical element oxygen (symbol: O) reacts easily with a large number of chemical elements and compounds present in the environment of planets. This widespread chemical reaction is called *oxidation*. Because O readily combines with other chemical elements, it does not normally exist in nature in the form of the oxygen molecule O_2.

Molecular O_2 is very abundant in the Earth's atmosphere because it is continually produced by the photosynthetic activity of aquatic and continental organisms. Indeed, during O_2-producing photosynthesis, water molecules (H_2O) are split into H and O atoms, and the oxygen is released as O_2 in the environment (see Information Box 3.4). Because the O_2 molecule contains two atoms of oxygen, it is often called *dioxygen*, and because oxygen in the O_2 molecule is not bound to other chemical elements, O_2 is also called *free oxygen*.

In the atmosphere, part of the O_2 is transformed into ozone, O_3. This gas has two different origins and effects depending on where it occurs in the atmosphere. *In the upper atmosphere*, O_3 results from the action of ultraviolet (UV) radiation from the Sun on O_2 molecules, which forms the *ozone layer* (see Information Box 8.2). The latter protects organisms on continents and in the surface layers of water bodies from dangerous solar radiation (see Sect. 2.1.5). *In the lower atmosphere*, O_3 is created by chemical reactions, in the presence of sunlight, between oxides of nitrogen (NO_x) and volatile organic compounds mostly emitted by cars, power plants and various industrial activities. There, O_3 is an air pollutant, which harms sensitive vegetation and people with respiratory problems.

the young Sun was still quite faint (see Sect. 3.7.1). The Archaean ended when, according to certain geochemical indices, O_2 began to be present in the Earth environment, albeit in moderate concentrations, as a consequence of the emergence of O_2-producing photosynthetic organisms much before. This marks the beginning of the Proterozoic era.

The third era of the Earth's history (*Proterozoic*, 2.5 billion to 542 million years ago) begins when O_2 started to accumulate in the environment (see Information Box 1.8). The iron dissolved in ocean waters reacted with the newly available oxygen, causing the formation of insoluble black iron oxides now found in sedimentary rocks. Until about 2.5 billion years ago, most of the O_2 produced by photosynthesis combined with the dissolved iron, and it is only after all the iron had been oxidized that the concentration of O_2 could start to increase, first in the seawater and later in the atmosphere. The latter was the third phase of the

Earth's atmosphere. The Proterozoic was a very long era, which lasted almost two billion years. It experienced many environmental transformations, including the global increase in oxygen called *Great Oxygenation Event* (see Sect. 8.1.2), some quasi-global glaciations called *snowball Earth* (see Sect. 4.5.4), and major phases in the evolution of organisms. One example of the latter is the metabolic innovation of O_2 respiration (see Information Box 3.4). Another example is the development of biological mechanisms of protection against oxygen toxicity, in a world where oxygen was initially a deadly poison to the organisms that had been the sole occupants of Earth for the previous two billion years (and perhaps more) and did not use O_2 for respiration (see Sect. 2.1.5). Oxygen progressively increased to concentrations much higher than present, and the oxidation by O_2-respiring organisms (that is, respiration) of part of the accumulated organic matter contributed to progressively lower the concentration of atmospheric O_2 to the present value.

Other major aspects of the evolution of organisms during the Proterozoic era were the emergence of eukaryotic cells (cells with differentiated nuclei and internal organelles such as mitochondria) between 2.1 and 1.6 billion years ago or even earlier (because some fossils going back to 2.1 billions years may be early traces of colonial eukaryotes) and that of organized multicellular organisms (whose first evidence may be fossilized organic compounds characteristic of the cell membrane of some sponges, found in rocks dating 650–540 millions years ago), which diversified into plants and animals. The Precambrian eon ends with the Proterozoic era.

The fourth era of the Earth's history (*Paleozoic*, 542–251 million years ago) marks the beginning of the *Phanerozoic* eon. The Paleozoic era was characterized by the occupation of continental surfaces by plants and animals, and their subsequent diversification. Simultaneously, the expansion of forests and wetlands had major effects on the global environment, such as an additional increase in atmospheric O_2 leading to the highest values in the history of the planet. The major beachheads of continent occupation occurred in the *Silurian* and *Devonian* periods (from 444 to 359 million years ago), which saw the appearance of the first terrestrial plants and animals, respectively. This was prepared by organisms that existed during the *Cambrian* and *Ordovician* periods (from 542 to 444 million years ago). The Paleozoic era ended with the third (*Permian-Triassic*) mass extinction known from the fossil record (see Information Box 1.6). The *Permian-Triassic* mass extinction took place about 251 million years ago, when large proportions of the marine and terrestrial species of the vertebrate animals that left fossils became extinct. It must be noted that the identification of extinction periods is based on fossils, meaning that the fate of organisms without hard parts during these periods is not known. The causes of this environmental crisis are not satisfactorily elucidated, and they may include a phase of exceptional volcanic activity or the impact of one or several large meteoroids.

The fifth era of the Earth's history (*Mesozoic*, from 251 to 65.5 million years ago) began with the *Triassic* period (251–200 million years ago). The beginning of this period saw a major renewal of the flora and fauna ensuing from the Permian extinction, and it took until the middle of the Triassic period for Earth to recuperate the pre-extinction biodiversity. The Mesozoic era is better known as the *Age of dinosaurs* or *Age of reptiles*, but it is also saw the appearance on the Earth's scene of birds (whose ancestors were dinosaurs), mammals and flowering plants (called *angiosperms* by botanists). Dinosaurs first appeared in the late Triassic, and they dominated the continents during about 135 million years from the early *Jurassic* to the end of the *Cretaceous* periods (from 200 to 65 million years ago). The end of the Mesozoic era, like that of the Paleozoic, corresponded to a mass extinction, which put an end to the dinosaurs (except the ancestors of birds) and several other groups of organisms. However, as in the case of the previous four extinctions, many groups of organisms survived and some of their descendants thrived during the following periods. The possible causes of the *Cretaceous-Paleocene* extinction were examined above (see Information Box 1.6).

The sixth era of the Earth's history (*Cenozoic*, 65.5 million years ago to present) is the last part of the Phanerozoic eon. It begins with the *Paleocene* period, which begins just after the Cretaceous-Palaeocene crisis (65 million years ago), and we are now living in its last period, called *Quaternary*. The Quaternary period is characterized by alternations of glacial and interglacial episodes that have occurred since about 2.6 million years ago (see Sect. 4.5.2), and we are presently living in the latest interglacial episode. This episode is divided in two parts (epochs) called the *Holocene* and the *Anthropocene*. The latter name comes from the Greek word "anthropos" (ἄνθρωπος), which means "human", and reflects the major impacts of human activities on the Earth's environment. The start date of the proposed Anthropocene epoch is debated among specialists, and the date of 300 years ago used in Table 3.1 is only indicative, corresponding to the onset of the *Industrial Age* in the second part of the eighteenth century (see Information Box 1.9).

Co-evolution of Earth and its Organisms and Ecosystems The most remarkable aspects of the history of Earth are the early appearance of life forms on the planet (about 3.8 billion years ago and perhaps earlier) and their continual presence on the planet since then. Organisms and ecosystems not only persisted despite major changes in the physical, chemical and biological conditions of the Earth's environment, but they thrived and diversified enormously as they occupied the whole planet. Doing so, they profoundly modified the characteristics of the Earth System. This success will be explained in following chapters by the fact that organisms took over the planet shortly after their establishment on Earth, modified it and kept it suitable for them. We will show that there was a co-evolution of Earth and its organisms and ecosystems over about 4 billions of years.

Information Box 1.9   Industrial Revolution, Industrial Age, and Information Age On various occasions in this book, we refer to three historical periods of significance for the Earth's climate, namely the Industrial Revolution, the Industrial Age, and the Information Age. We briefly define them here.

The *Industrial Revolution* is the period during which the production of goods in Europe and the United States shifted from being done by hand (of course with tools) to using machines. It lasted from 1760 and 1840. The machines were first powered by running water, then by steam, and later by internal combustion engines and electricity.

The Industrial Revolution marked the onset of the *Industrial Age*, which progressively encompassed an increasing larger number of countries. During the Industrial Age, which began in 1760 and continued until the end of the twentieth century, the economy was dominated by industrial production. The economy of many countries is still largely based on industry.

Industrial production continues in the twenty-first century, but it is commonly believed that human societies moved into the *Information Age* in the late twentieth century. The Information Age is characterized by a shift from traditional industry to an economy based on information technology.

1.6   Key Points: From the Outer Universe to the Living Earth

In Sects. 1.2–1.5, we described many connections between organisms, Earth, the Solar System, and the whole Universe. On the one hand, several processes that occurred in the past and are still occurring in the Universe and in the Solar System favoured both the formation and development of Earth, and the establishment of organisms and development of ecosystems on the planet. These processes include: the supply to Earth of water by asteroids and/or comets and of possibly organic compounds by comets (see Sect. 1.2.2), and the formation in stars of chemical elements used to build Earth and of heavy elements that play major roles inside the planet (see Sect. 1.4.2). On the other hand, organisms and ecosystems were also strongly influenced by characteristics of Earth and processes that occurred on the planet during its whole history. These characteristics and processes include: the short day-and-night rotation cycle and the moderate axial tilt, which favoured the long-term presence of water in liquid form (see Sect. 1.3.3); the rocky nature of Earth, the variety of chemical elements in the crust and the presence of heavy, radioactive elements at depth were important for the durable establishment of organisms (see Sect. 1.4.2); and on the negative side, massive volcanic eruptions may have contributed to cause mass extinctions (see Sect. 1.5.2).

When the organisms and ecosystems became abundant enough on Earth, they increasingly affected the environmental conditions of the planet. Examples of

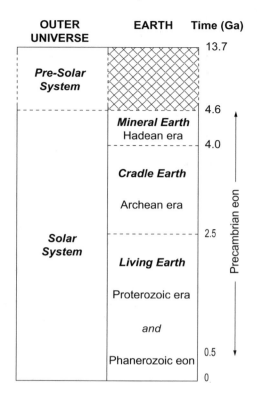

Fig. 1.11 Broad temporal divisions in the history of the Outer Universe (including the Solar System) and Earth pertinent to organisms and ecosystems. The geological eons and eras (Table 1.3) are grouped here in the Mineral, Cradle and Living Earth phases. Ga: billions of years. Credits at the end of the chapter

effects of organisms and ecosystems on the Earth's environment are: the accumulation of O_2 in the atmosphere as a result of photosynthesis; the contribution of O_2 respiration to the progressive stabilization of the concentration of O_2 in the atmosphere; and the effects on the climate of the occupation of continental surfaces by plants and animals (see Sect. 1.5.4).

The co-evolution of the Earth's environment and its organisms and ecosystems over at least 2.5 billion years transformed the planet in such a way that its environmental conditions are now controlled by its ecosystems and its chemical, geological and physical processes. We call this whole Earth system the *Living Earth*.

For the Earth's organisms and ecosystems, the above characteristic and processes belong either to planet Earth, including its ecosystems, or to the *Outer Universe*, meaning the Universe outside Earth including the Solar System. The history of the Outer Universe and Earth started about 13.7 and 4.6 billion years ago, respectively (Tables 1.2 and 1.3, Fig. 1.11). During the first 0.6 billion years of the Earth's history, the environmental conditions of the planet progressively stabilized, but were not suitable for organisms. In Fig. 1.11, we call this phase *Mineral Earth*. During the following 1.5 billion years, organisms developed on Earth, and they progressively became so abundant that they affected the environmental conditions on the planet. One of these conditions was the abundance of methane (CH_4) in the atmosphere, a gas produced in small amounts by geological processes but mostly produced by organisms on Earth. We call this phase *Cradle Earth*. About 2.5 billion years ago, the activity of photosynthetic organisms produced so much O_2 that this gas started to accumulate in the ocean and the atmosphere, and this major change was the sign that the organisms were taking over the Earth's environment (see Sect. 7.4.1). We thus set the beginning of the *Living Earth* phase at 2.5 billion years ago, although ecosystems that did not use O_2 for respiration had influenced key environmental processes, such as the concentration of methane in the atmosphere, much before the advent of O_2-producing photosynthesis (see Sect. 4.7.2).

In other words, an extra-terrestrial astronomer living on an Earth-like planet with O_2-producing organisms who would have looked at planet Earth 2.6 billion years ago could not have discovered that organisms existed on Earth and had inhabited the planet for more than 1 billion years. This is because there was no O_2-related trace of organism's activity in the Earth's atmosphere until about 2.5 billion years ago. However, 100,000 years later, a descendant of this extra-terrestrial astronomer could have detected significant amounts of O_2 in the Earth's atmosphere, and concluded from this chemical imbalance that large biomasses of organisms likely existed on the planet. The presence of O_2 in the atmosphere of a given planet would be a sign of the likely presence of a substantial biomass of oxygen-producing organisms, but the absence of O_2 in the atmosphere of another planet would not necessarily mean that there are no organisms there. Similarly, the presence of other biogenic gases, such as methane in the atmosphere of Saturn's moon Titan, could indicate the presence of organisms (see Sect. 2.1.1).

In Fig. 1.11, the histories of the Outer Universe and planet Earth are divided into broad temporal phases, which are defined by reference to organisms. The phase of the Outer Universe most pertinent to organisms begins with the formation of the Solar System 4.6 billion years ago. Concerning Earth, Fig. 1.11 shows that organisms appeared on the planet 0.6 billion years "only" after its formation. Indeed, this strictly mineral phase represents <15% of the 4.6 billion years of the Earth's history to date, and was therefore quite short. Once organisms and ecosystems got established on Earth, they became major determinants of the environmental conditions within 1.5 billion years, and had remained so during the latest 2.5 billion years (see Sects. 4.7.1 and 4.7.2). These two periods corresponded to more than 30 and 55%, respectively, of the Earth's history. Hence our planet is rightly called *Living Earth*.

Figure Credits

Fig. 1.1 Original. Figure 1.1 is licensed under CC BY-SA 4.0 by Philippe Bertrand, Louis Legendre and Mohamed Khamla.

Fig. 1.2 Original. Based on Fig. 1.1 of Clarke (2014). Figure 1.2 is licensed under CC BY-SA 4.0 by Philippe Bertrand, Louis Legendre and Mohamed Khamla.

Fig. 1.3 This work, Fig. 1.3, is a derivative of https://commons.wikimedia.org/wiki/File:Oort_cloud_lrg.en.png by Nasa.gov, in the public domain. I, Mohamed Khamla, release this work in the public domain.

Fig. 1.4 This work, Fig. 1.4, is a derivative of https://commons.wikimedia.org/wiki/File:Solar_sys.jpg by Harman Smith and Laura Generosa (nee Berwin), NASA's Jet Propulsion Laboratory, in the public domain. I, Mohamed Khamla, release this work in the public domain.

Fig. 1.5a This work, Fig. 1.5a, is a derivative of https://commons.wikimedia.org/wiki/File:Eccentricity.svg by Seahen https://commons.wikimedia.org/wiki/User:Seahen, used under GNU FDL CC BY-SA 3.0 and CC BY 2.5. Figure 1.5a is licensed under GNU FDL and CC BY-SA 3.0 by Mohamed Khamla.

Fig. 1.5b Original. Figure 1.5b is licensed under CC BY-SA 4.0 by Philippe Bertrand, Louis Legendre and Mohamed Khamla.

Fig. 1.6a https://commons.wikimedia.org/wiki/File:Earth_obliquity_range.svg by NASA, Mysid https://en.wikipedia.org/wiki/User:Mysid, in the public domain.

Fig. 1.6b https://commons.wikimedia.org/wiki/File:Earth_precession.svg by NASA, Mysid https://en.wikipedia.org/wiki/User:Mysid, in the public domain.

Fig. 1.7 The source of this material is the COMET® Website at https://www.meted.ucar.edu/climate/climwaterpart1/media_gallery.php of the University Corporation for Atmospheric Research (UCAR), sponsored in part through cooperative agreement(s) with the National Oceanic and Atmospheric Administration (NOAA), U.S. Department of Commerce (DOC). ©1997–2017 University Corporation for Atmospheric Research. All Rights Reserved. Artist: Steven Deyo. With permission from COMET [In order to log into the MetEd website, one must create an account].

Fig. 1.8 This work, Fig. 1.8, is a derivative of https://commons.wikimedia.org/wiki/File:Inflation.PNG by Rogilbert, in the public domain. I, Mohamed Khamla, release this work in the public domain.

Fig. 1.9 This work, Fig. 1.9, is a derivative of http://www.astronomy.ohio-state.edu/~jaj/nucleo/ by Prof. Jennifer Johnson, used under CC BY-SA 4.0. The figure is explained in http://blog.sdss.org/2017/01/09/origin-of-the-elements-in-the-solar-system/ Figure 1.9 is licensed under CC BY-SA 4.0 by Mohamed Khamla.

Fig. 1.10a https://commons.wikimedia.org/wiki/File:Charniodiscus_arboreus.jpg by tina negus from UK https://www.flickr.com/people/84265607@N00, used under CC BY 2.0. Size added to the left of the photo.

Fig. 1.10b https://commons.wikimedia.org/wiki/File:DickinsoniaCostata.jpg by Verisimilus at English Wikipedia, used under GNU FDL, CC BY-SA 3.0 and CC BY 2.5. Size added to the left of the photo.

Fig. 1.10c https://commons.wikimedia.org/wiki/File:Opabinia_smithsonian.JPG by Jstuby at English Wikipedia https://en.wikipedia.org/wiki/User:Jstuby, in the public domain. Size added to the left of the photo.

Fig. 1.11 Original. Figure 1.11 is licensed under CC BY-SA 4.0 by Philippe Bertrand, Louis Legendre and Mohamed Khamla.

References

Clarke A (2014) The thermal limits to life on Earth. Int J Astrobiol 13:141–154. https://doi.org/10.1017/S1473550413000438

Darwin C (1859) On the origin of species by means of natural selection, or the preservation of favoured races in the struggle for life. John Murray, London

Lovelock J (1995) The ages of Gaia: a biography of our living Earth. Oxford University Press, Oxford

Milanković M (1920) Théorie mathématique des phénomènes thermiques produits par la radiation solaire (Mathematical theory of heat phenomena produced by solar radiation). Gauthier-Villars, Paris

Further Reading

Cohen A, Cox B (2019) The planets. William Collins

Crockett C (2018) A crash of stars reveals the origins of heavy elements. Knowable Magazine. https://doi.org/10.1146/knowable-012818-181100, https://www.knowablemagazine.org/article/physical-world/2018/crash-stars-reveals-origins-heavy-elements

Meinesz A, Simberloff D (2008) How life began: evolution's three geneses. University of Chicago Press, Chicago

Perlov D, Vilenkin A (2017) Cosmology for the curious. Springer, Cham

Pross A (2012) What is life? How chemistry becomes biology. Oxford Landmark Science, Oxford University Press, Oxford

Satz H (2013) Ultimate horizons: probing the limits of the universe. Springer, Berlin, Heidelberg

Schrödinger E (2012) What is life? With mind and matter and autobiographical sketches. Reprinted Edition. Cambridge University Press, Cambridge

Schulze-Makuch D, Irwin LN (2018) Life in the Universe. Expectations and constraints, 3rd ed. Springer-Praxis Books, Springer, Cham

Shikazono N (2012) Introduction to Earth and planetary system science: a new view of Earth, planets, and humans. Springer, Tokyo

Stanley SM, Luczaj T (2014) Earth system history, 4th edn. W. H. Freeman, New York

Vita-Finzi C (2016) A history of the Solar System. Springer, Cham

Whittet D (2017) Origins of life. A cosmic perspective. IOP Concise Physics, Morgan & Claypool Publishers, Bristol

Chapter 2
The Atmosphere: Connections with the Earth's Mass and Distance from the Sun

2.1 Organisms and the Earth's Atmosphere

2.1.1 Connections Between the Atmosphere and Ecosystems

The organisms on Earth are in contact with their environment through senses, but no group of organisms has all the possible senses. For example, humans do not have the ability of echolocation, which is used by bats for positioning themselves in the 3-dimensional space, or sensing the Earth's magnetic field, as done by some birds during their large-scale migrations (see Sect. 8.2.2), or detecting movements and vibrations in water with the types of organs fish have in their lateral lines. Invertebrate organisms may have many other senses.

As human beings, we sense our environment through sight, hearing, smell, touch and taste. Because most of the information we acquire through our senses is very repetitive, we generally do not pay much attention to a large part of the sensory data we receive. When we pay attention, we may see clouds drifting in the sky, smell the scent of flowers in the air, hear the tops of poplars singing under the effect of a light breeze, enjoy the wind cooling our skin, or taste on our lips the salt carried by sea spray. These sensations are possible because the air carries moisture and sound as well as fragrant and tasty molecules through its continual movements. However some of our senses do not rely on air, given that light photons do not need the support of any substrate to reach our eyes and are

© Springer Nature Switzerland AG 2021
P. Bertrand and L. Legendre, *Earth, Our Living Planet*, The Frontiers Collection,
https://doi.org/10.1007/978-3-030-67773-2_2

in fact absorbed by atmospheric particles, and molecules that we taste are mostly transported to our sensory papillae by liquids. Nevertheless, air is a major carrier of information to our senses.

The next paragraphs explore the originality of the Earth's atmosphere and some of its connections with ecosystems, and the remainder of this section considers the main atmospheric gases and their interactions with ecosystems. The following sections examine successively: the atmosphere of Earth and the other planets in the Solar System, and that of exoplanets outside the Solar System (Sects. 2.2); astronomical and planetary effects on the Earth's atmosphere (Sects. 2.3); and the connections between the atmosphere and climate (Sects. 2.4). The chapter ends with summary of key points concerning the interactions between the Solar System, Earth, its atmosphere, and its ecosystems (Sect. 2.5).

Because most people spend most of their lives in environments with given atmospheric characteristics, the latter are generally taken for granted. However, people who have experienced changes in atmospheric pressure, for example when climbing a high mountain or arriving by plane in a city located at high altitude, have realized that atmospheric pressure is very important for breathing. Breathing is one of the numerous connections between organisms and the atmosphere, but there are many more. We cite some of these connections in the following paragraphs.

Our most obvious connection with the atmosphere concerns the three key gases involved in our respiration, which are free oxygen, O_2 (see Information Box 1.8), that we inhale and carbon dioxide (CO_2) and water vapour (H_2O) we exhale (see Sect. 2.1.5, below). However, not all organisms take in O_2 (in which case they obtain the oxygen they need differently), and some release, in addition to CO_2, other gases such as methane (CH_4) that find their ways into the atmosphere. Many people know that CO_2, water vapour and CH_4 contribute to the greenhouse effect (see Information Box 2.7), and thus affect the climate (see Information Box 1.4). Many people have also heard of another atmospheric gas, called ozone (O_3), in relation with the *ozone layer*. This gas layer, which is located in the upper atmosphere, is so important for organisms living on continents that an international treaty has banned all countries from releasing in the environment substances that could damage it (see Sects. 2.1.5 and 10.3). It is also generally known that almost all the free oxygen on Earth is produced by photosynthetic organisms (see Information Box 3.4), about half on continents and half in the oceans.

Less known than the above facts is the protection that the atmosphere provides to the Earth's water, which would have been lost to the outer space in absence of this protection, and without liquid water there would be no organisms on the planet (see Sect. 2.1.3, below). Other important facts concerning the atmosphere are as follows: the cumulative uptake of CO_2 by the formation of carbonates (mostly $CaCO_3$) has considerably decreased the very high concentration of CO_2 in the atmosphere of the young Earth, and thus reduced the temperature of the planet to values suitable for organisms (see Sect. 5.5.4); the production of CH_4

Table 2.1 Composition in % of dry air (by volume; see Information Box 2.1) of the main gases in the atmospheres of the four Solar System bodies with a significant atmosphere, that is, Venus, Earth, Mars, and Titan (Table 2.2). The Earth's atmosphere contains a variable proportion of water vapour, whose average is about 0.4% over the entire atmosphere and around 1% at sea level. https://spacemath.gsfc.nasa.gov/astrob/10Page7.pdf

Main gases	Venus	Earth	Mars	Titan
Carbon dioxide (CO_2)	96	<1[a]	95	–
Nitrogen (N_2)	4	78	2.7	97
Oxygen (O_2)	–	21	–	–
Argon (Ar)	–	1	1.6	–
Methane (CH_4)	–	<1[a]	–	2
Other gases	–	<1	0.7	1

[a]Earth. CO_2: 0.04%; CH_4: 0.0002%

by ecosystems in the deep past kept the young Earth warm enough to harbour organisms (see Sect. 2.1.7, below); the uptake of CO_2 by chemosynthesis and photosynthesis contributed to further decrease the concentration of atmospheric CO_2 by storing carbon into fossil organic matter (see Sect. 8.1.1); volcanoes are the main long-term source of CO_2 for the atmosphere, which contributes to maintain the natural greenhouse effect of the planet (see Sect. 7.3.1); and in the last decades, human societies have released as much CO_2 in the atmosphere every year as all volcanoes on Earth normally do during several decades (see Sect. 2.1.6, below).

The last statement shows that some recent aspects of human activities affect the whole planet, here its atmosphere, as much as major geophysical forces, here volcanic eruptions, that have governed Earth since billions of years. Overall, organisms largely control the concentrations of such gases as CO_2, O_2, H_2O (vapour) and CH_4 in the atmosphere, and in return the concentrations of these gases largely determine the fate of organisms. These and other interactions will be described in this section, in the remainder of this chapter, and in following chapters.

Comparing the gas composition of the Earth's atmosphere with that of the three other planetary bodies in the Solar System that have a non-negligible atmospheric pressure shows that the atmosphere of our planet is very unique. In Table 2.1, the atmosphere of Earth is compared with those of Venus and Mars, which are the two sister planets of Earth, and Titan, which is a moon of the giant planet Saturn. Table 2.1 shows that Venus and Mars are quite similar in term of their % atmospheric composition, that is, they are both dominated by carbon dioxide followed by nitrogen. However, they are very different in term of atmospheric pressure, which is 15,000 times higher on Venus than on Mars (Table 2.2). In comparison, the % compositions of the atmospheres of Earth and Titan are very original, being dominated on Earth by nitrogen and oxygen, and on Titan by nitrogen (like on Earth) and methane (different from Earth).

Table 2.2 Key characteristics of the four rocky planets and eight other Solar System bodies with the following three features: mass greater than 1/1000 that of Earth, observable surface (solid or liquid), and detectable atmosphere or exosphere. The four rocky planets and the seven moons are listed in order of increasing distance of the planet from the Sun; Pluto is the farthest body from the Sun in the table (distances are given in Table 1.1). The four moons of Jupiter are listed in order of decreasing mass. The main gases in the atmospheres of Venus, Earth, Mars and Titan are detailed in Table 2.1

Planetary body	Mass ($\times$Earth's mass)	Atmospheric pressure near the surface ($\times$Earth's pressure)	Main gases
Rocky planets			
Mercury	0.055	Exosphere	O_2, Na, H_2, He, H_2O
Venus	0.815	92	CO_2, N_2
Earth	1.000[a]	1	N_2, O_2, H_2O, Ar
Mars	0.107	0.006	CO_2, Ar, N_2
Moons and dwarf planet			
Moon (Earth)	0.012	Exosphere	Ar, He
Ganymede (Jupiter)	0.025	Exosphere	O_2
Callisto (Jupiter)	0.018	Exosphere	CO_2, O_2
Io (Jupiter)	0.015	Exosphere	SO_2
Europa (Jupiter)	0.008	Exosphere	O_2
Titan (Saturn)	0.023	>1.5	N_2, CH_4
Triton (Neptune)	0.004	1×10^{-5}	N_2, CO, CH_4
Pluto (dwarf planet)	0.002	1×10^{-5}	N_2, CO

[a]Earth's mass: 5.972×10^{24} kg

It is known that the oxygen present in relatively large concentration in the Earth's atmosphere is released by the activity of photosynthetic organisms (see Information Box 3.4). It is also thought that the early atmospheres of Venus, Earth and Mars had similarly high concentrations of carbon dioxide, and that a special mechanism reduced the concentration of this gas in the Earth's atmosphere from originally about 95% (as on Venus and Mars) to presently less than 0.1% (see Sect. 5.5.4), whereas Mars progressively lost most of its atmosphere (see Sect. 2.3.4). The latter would explain both the similarity in composition between the atmospheres of Mars and Venus, and the much lower atmospheric pressure on Mars than Venus. Some of the hypotheses proposed to explain the abundance of methane in Titan's atmosphere invoke the possibility of biological production of this gas under the thick atmosphere of Titan, given that microorganisms produce vast amounts of methane on Earth.

Information Box 2.1 Amounts of Gases in the Atmosphere, and of Substances in the Air, Water, Seawater, and Solids This book reports in different chapters the amounts of substances in compartments of the Earth System, environments, and organisms. The various ways to express these amounts are summarized here for the convenience of readers.

In this book, we use the intuitive unit of atmospheric pressure called *atmosphere* (atm) (see Information Box 2.3). The value of 1 atm corresponds to the pressure exerted by the combination of all atmospheric gases at sea level on present-day Earth.

Each of the atmospheric gases contributes to the total atmospheric pressure, and the fraction of the total pressure it exerts is its *partial pressure*. It can be expressed in three different ways:

Mole (or molar) fraction. Number of moles of a gas divided by the total number of moles of all atmospheric gases (moles mole^{-1}), which is often expressed as %. [The *mole* is the SI unit of the amount of substance, and a mole contains 6.022×10^{23} elementary entities (Avogadro number). For example, 1 mol of O_2 contains 6.022×10^{23} molecules of O_2.] The equivalent for atoms is the *atom (or atomic) ratio* or *atom (or atomic) %*, as in Table 6.1.

Volume (or volumetric) fraction. Volume occupied by a gas in the total volume of all the atmospheric gases (volume volume^{-1}), which is often expressed as %. Its value is the same as the mole fraction. For example, the mole fraction or volume fraction of O_2 in dry air is 20.9%.

The volume fraction of a *trace gas* is generally stated in *parts per million by volume* (ppmv). For example, the value of atmospheric CO_2 before the Industrial Revolution was about 270 ppmv, a value more convenient to use than the volume fraction of 0.0270%.

The *concentration of a substance in a fluid* (gas or liquid) is often expressed as its mass (kg, g, or other) or number of moles (mol) per unit volume of the fluid (L, or m^{-3}). For example, 400 ppmv of CO_2 corresponds to 775 mg m^{-3}, that is, 1 m^3 of dry air contains 775 mg of CO_2.

The *concentration of a trace element in water* is often reported in *parts per million by weight* (ppmw), which is similar to ppmv except that the amounts of the trace element and the solution are expressed in weight instead of volume.

For *elements in very small concentrations*, one uses *parts per billion* (ppb, ppbv, or ppbw). Often in the literature, ppmv and ppmw are both abbreviated to ppm, and ppbv and ppbw are both abbreviated to ppb, which can create confusion.

In the case of *seawater*, the total amount of dissolved salts is stated in grams of dissolved salts per kilogram of seawater (g kg^{-1}). *Salinity* does not have units (see Information Box 5.9).

The *concentration of a substance in a solid* (for example, the amount of a given chemical element in a rock, or in organic matter) can be expressed in *mole (or molar) fraction*, or *parts per million (or billion) by weight*, defined above. It can also be reported in *mass %*, that is, the mass of the substance or chemical element divided by the total mass of the sample and multiplied by 100.

2.1.2 Volatile Substances Can Be in Gas or Liquid Forms on Planets

All chemical compounds in the Universe can exist in four main physical states, which are: *solid, liquid, gas,* and *plasma.* The word *plasma* refers to an electrically neutral medium of unbound positive and negative particles (meaning that their overall electrical charge is null). Plasma is encountered in the outer space and the stars, where it is the most abundant form of matter, and closer to us in *auroras* also known as *Northern Lights* and *Southern Lights* (see Sect. 8.4.1), fluorescent lamps and neon tubes, lasers, and some special instruments. There are other states of matter than the four cited here, for example *glass* and *liquid crystals,* some of which only occur in extreme conditions. A compound can experience a *phase change* when there is a change in the environmental conditions of temperature and pressure. The concepts of *state* and *phase* of matter and of *phase change* are explained in Chap. 5 (see Sect. 5.5.1). The book of Fortov (2016) provides further reading on the extreme states of matter on Earth and in the Universe.

A *volatile* substance (often abbreviated to the noun *volatile*) can evaporate easily at normal temperature and pressure, where the qualifier *normal* refers to a temperature and pressure of the same order as those that exist at the surface of Earth. In the cold outer space, dispersed volatile substances are mostly frozen and thus in the solid state (ice). Near the stars and in regions of formation of new stars, which are very dense and very hot, the gases are in the form of plasma. It is only on planets or similarly compact bodies that volatile substances can naturally be in gas or liquid forms.

On planets, various chemical elements or compounds can naturally exist in different phases. On Earth for example, water is a gas in the atmosphere (water vapour), a liquid in the oceans (seawater), and a solid in cold high-altitude and high-latitude environments (ice). In contrast, methane is a gas in the Earth's atmosphere and a solid in cold marine sediments under high hydrostatic pressure, where it forms a compound known as *methane hydrate, methane clathrate,* or *burning ice.* Methane may also occur naturally in liquid phase, not on Earth but on much colder Saturn's moon Titan (-179 °C), at the surface of which large dark areas have been observed and were interpreted as possibly being lakes and oceans of liquid methane. The book of Visconti (2016) provides further reading on the aspects of physics and chemistry of the atmosphere examined in this section and the next ones down to Sect. 2.1.7.

2.1.3 Water in the Atmosphere and Ecosystems

Organisms could not have established themselves on Earth, and ecosystems could not have expanded over the whole planet without the presence of liquid water at the surface of the planet (see Sect. 5.1.1). Furthermore, the abundance of surface liquid water largely explains why organisms on Earth are made of molecules based on carbon and not another chemical element (see Sect. 6.1.2). In addition, water is continually redistributed all over the planet by movements of the atmosphere that carry water vapour from regions of high evaporation to regions of high precipitation, which sustains the wide distribution of ecosystems on the planet. The book of Schulze-Makuch and Irwin (2018) provides further reading on the special role of water for organisms on Earth.

The combination of temperature and pressure determines if water is in solid (ice), liquid or gaseous (vapour) form. Figure 5.3 shows that water at a temperature below 0 °C or a few degrees lower is in the form of ice, and above 0 °C is either liquid or gaseous depending on the atmospheric pressure. The temperature at which water condenses from vapour to liquid or evaporates from liquid to vapour is higher at higher pressure. This shows that if the pressure of the Earth's atmosphere in the past had been significantly lower than it actually was, all the water of the planet would have been in the form of vapour in the atmosphere. If this had been the case, all the Earth's water would have been lost to the outer space a long time ago. This loss would have occurred because of the phenomenon of *atmospheric escape of hydrogen* (see Sect. 5.4.2). Briefly, some of the water molecules (H_2O) that evaporate from the Earth's surface reach the high atmosphere. There, these molecules are dissociated into atoms of hydrogen (H) and oxygen (O) by the ultraviolet radiation of the Sun, which is much stronger in the high atmosphere than below, and some of the free H atoms are lost from the high atmosphere to the outer space. For every two H atoms lost, Earth loses forever one molecule of water (H_2O). Hence, water in the form of atmospheric vapour is in danger of being lost to the outer space, whereas water in liquid or ice forms is protected from this loss. It follows that a key condition for the long-term existence of liquid water on a planet is the right combination of atmospheric temperature and pressure. The article of Catling and Zahnle (2009) provides further reading on the escape of planetary atmospheres.

The Earth's atmospheric temperature and pressure became right for the condensation of water vapour into liquid water about 4.4 billion years ago, and most of the water vapour in the atmosphere then condensed to form the liquid ocean (see Sect. 5.5.4). This transferred most of the water that was in the atmosphere, from where it would have been lost to the outer space, to the ocean, where it was sheltered from this loss. In the following billions of years, the Earth's temperature and atmospheric pressure both decreased, but their combination always allowed the existence of liquid water on the planet even during the cold periods of snowball Earth (see Sect. 4.5.4). Hence the early and continuing combination of atmospheric temperature and pressure ensured the long-term existence of liquid water at the surface of the planet, which allowed the establishment of organisms on Earth and the subsequent expansion of ecosystems to the whole planet including the continents.

Another aspect of atmospheric water vapour is its contribution to the greenhouse effect and thus the regulation of the Earth's climate (see Information Box 2.7). To illustrate the strong effect of water vapour on climate, it was calculated that if all the water on the planet were vaporized into the atmosphere, the surface temperature would become so high that all rocks would melt. The presence of some water vapour in the atmosphere is good for organisms as it prevents temperature from becoming too cold, but too much would be bad as it would make the temperature too high. In addition, if all the water were vaporized, the atmosphere would become so heavy that pressure at the Earth's surface would be multiplied by a factor of 250, which would make the planet uninhabitable.

2.1.4   Nitrogen in the Atmosphere and Ecosystems

The most abundant gas in the Earth's atmosphere is nitrogen, more specifically *dinitrogen* (N_2) that we generally call *nitrogen gas* in this book. It accounts for 78% of the total volume of atmospheric gases. The chemical element N is a key constituent of proteins and nucleic acids in living cells (see Sects. 6.1.3 and 6.2.2). Although N_2 is very abundant in the atmosphere, the availability of N often limits the synthesis of organic matter by organisms. This is because most organisms cannot use the abundant N_2 directly, and only some specific bacteria can do it. Biological N_2 fixation is summarized in Eq. 9.13 (see Sect. 9.3.4). It may have begun between 1.5 and 2.2 billion year ago.

In addition to biological N_2 fixation, there are natural abiotic reactions that transform atmospheric N_2 into biologically available forms. Non-biological N_2 fixation is driven by lightning, which combines atmospheric nitrogen and oxygen into *nitrogen oxides* (NO_x). The chemical reaction of some of the NO_x molecules with water (H_2O) produces *nitrous acid* (HNO_2) and *nitric acid* (HNO_3). These two acids are eventually converted into *nitrate* (NO_3^-), which is used by photosynthetic organisms as a source of nitrogen for the production of organic matter.

In aquatic environments, the N_2-fixing organisms are cyanobacteria, which are generally part of the plankton. It is interesting to note that the ancestors of present-day cyanobacteria are the organisms that "invented" O_2-producing photosynthesis more than 2.5 billion years ago (see Sect. 2.1.5). In continental environments, the N_2-fixing bacteria often live in symbiotic relationship with a host, for example legumes, of which they inhabit the roots. The N-containing organic matter synthesized by cyanobacteria is the source of nitrogen for food webs, directly or after a process called nitrification (See Sect. 9.3.2). After circulating through ecosystems, the N atoms are returned to the atmosphere by decomposition of organic matter (see Information Box 2.2). The latter includes the bacterial processes of denitrification and anammox, which release N_2 (see Sect. 9.3.4). Specialized bacteria fix nitrogen, others nitrify, and still others denitrify, thus carrying out the circulation of nitrogen within ecosystems and between these and the atmosphere. In addition, atmospheric N_2 is used industrially to make artificial N-fertilizers. The total annual production of bioavailable nitrogen currently amounts to 383 teragrams ($Tg = 10^{12}$ g) of nitrogen, of which 53% are the result of natural processes (marine N_2 fixation 37%, terrestrial N_2 fixation 15%, and lightning 1%), and 47% are anthropogenic (fertilizer production 31%, and agricultural N_2 fixation 16%). In addition, the combustion of fossil fuels produces annually 40 Tg of nitrogen oxides (NO_x), which are major atmospheric pollutants. The adjective *anthropogenic*, which is often used in this book, means "originating in human activities".

Nitrogen gas (N_2) does not contribute to trap heat in the atmosphere, and is thus not a greenhouse gas. However, nitrous oxide (N_2O) is among the most powerful greenhouse gases present in low concentration in the atmosphere, together with methane, CH_4 (see Sect. 2.1.7, below). The emissions of N_2O are both natural and from human activities (60 and 40%, respectively). Natural emissions are mainly from bacteria breaking down nitrogenous compounds in soils and oceans, and the N_2O-emitting human activities belonging to agriculture, transportation and industry. The N_2O is naturally removed from the atmosphere by its bacterial use in soils and sediments, and its destruction by ultraviolet radiation or chemical reactions in the atmosphere.

Information Box 2.2 Breakdown of Organic Matter: Vocabulary All the organic matter (see Information Box 3.5) is eventually broken down to simpler, inorganic compounds. The agents of this *breakdown* are either organisms or non-biological processes. Different terms and expressions are found in the scientific literature for different forms of organic matter breakdown. We briefly explain here terms and expressions often encountered.

The three most abundant chemical elements in organic matter are H, O and C (see Sect. 6.1.2), and the corresponding generic organic matter is symbolised [CH_2O]. The chemical formula [CH_2O]$_n$ corresponds to different carbohydrates (sugars) whose n ranges between 3 and 7, for example glucose ($C_6H_{12}O_6$) where $n = 6$.

The word *breakdown* is not specific to organic matter, and is often found in the scientific literature to mean the biological, chemical, or physical decomposition of organic or mineral compounds. In biology, the specific words *decomposition*, *decay* and *rot* refer to the breakdown of organic matter by the action of *microorganisms*, mostly bacteria and fungi.

The breakdown of organic compounds often involves the chemical process of *oxidation*. In this process, parts of the broken organic molecules are combined with oxygen. For example in the breakdown of [CH_2O], the combination of H atoms from [CH_2O] with O atoms from environmental O_2 produces molecules of H_2O (water is an oxide of hydrogen) (see Information Box 3.4; Eq. 3.2).

Two types of oxidation of organic matter are biological O_2 *respiration* and chemical *combustion*. The two processes are represented by the same chemical equation (see Information Box 3.4):

$$CH_2O + O_2 \rightarrow CO_2 + H_2O + \text{energy} \tag{3.2}$$

In O_2 respiration, a process that occur within the cells, the release of energy is slow and under biological control, whereas in combustion, it is sudden and uncontrolled. Another form of respiration, called *denitrification* or NO_3 *respiration* uses nitrate (NO_3^-) as its source of oxygen, and also produces H_2O as in respiration (see Sect. 9.3.4; Eq. 9.14).

In biogeochemistry, the breakdown of organic matter is often called *remineralisation* because it transforms the organic matter into simple, *mineral* forms. For example, O_2 respiration and NO_3 respiration remineralize organic matter to CO_2 and N_2, respectively (Eqs. 3.2 and 9.14).

In ecology, the breakdown of organic matter is often called *recycling* because it cycles the materials contained in the organic matter, and these materials can be used again by organisms. When this process restores the concentrations of chemical elements in short supply in a given environment, it is called *nutrient regeneration*.

2.1.5   Oxygen and Ozone in the Atmosphere and Ecosystems

There are two types of oxygen molecules in the atmosphere (see Information Box 1.8). The most abundant molecules are composed of two atoms of oxygen, and are thus called *dioxygen* (O_2) or *free oxygen* (the latter expression is often used in this book). The other oxygen molecules found in the atmosphere have three atoms of oxygen, and are called *ozone* (O_3). The two types are strongly linked with organisms.

Free oxygen is the second most abundant gas in the Earth's atmosphere (Table 2.1). Almost all the O_2 present in the atmosphere or dissolved in ocean and continental waters comes from the process of O_2-producing photosynthesis, by which plants and phytoplankton use solar energy to split H_2O molecules, use the resulting H atoms for the synthesis of organic matter, and release the remaining O_2 molecules in the environment (see Information Box 3.4). Other types of synthesis of organic matter already existed on Earth before the invention of O_2-producing photosynthesis by aquatic cyanobacteria more than 2.5 billion years ago (see Sect. 2.1.4). These are chemosynthesis, where the source of energy is chemical, and O_2-less photosynthesis, which uses the energy of solar light but does not produce O_2 (see Information Box 5.1). The three types of synthesis of organic matter co-exist on present Earth, where O_2-producing photosynthesis has become dominant.

In addition to the O_2 from photosynthesis, small amounts of atmospheric O_2 also come from chemical reactions in the atmosphere, where the action of the ultraviolet (UV) radiation of the Sun breaks down some oxygen-containing molecules. This process is called *photolysis*, which means "breakdown by light". In the Earth's atmosphere, the photolysis of nitrous oxide (N_2O) and water (H_2O) produces annually 13 and 0.3 teragrammes of O_2 ($Tg = 10^{12}$ g), respectively. The 13.3 Tg produced annually by photolysis are negligible compared to the 290,000 Tg from the annual photosynthesis on land (155,000 Tg) and in the oceans (135,000 Tg).

The development of an O_2-rich environment over billions of years had major effects on organisms in the Proterozoic era, from 2.5 to 0.5 billion years before present (see Sect. 1.5.4), and these contribute to explain the present wide variety of organisms on Earth. On the one hand, many current organisms need oxygen. The O_2-respiring organisms, for example those living on continents, thrive at the present atmospheric O_2 concentration of about 20%, but most O_2-respiring organisms cannot withstand long exposure to O_2 concentrations much higher than 20% because such concentrations are toxic in the long term. The organisms requiring O_2 for respiration are called *aerobic* (see Information Box 5.2). Other organisms, which do not require O_2 for respiration, are called *anaerobic*, and still others, which cannot tolerate O_2 at all, are qualified of *obligate anaerobic*. Large numbers of today's organisms belong to each of these three categories. The toxicity of oxygen comes from the fact the metabolism of oxygen generates dangerous by-products called *reactive oxygen species*. The organisms that use O_2 for respiration have developed specialized enzymes that largely destroy the reactive oxygen species as long as these do not become too abundant. The obligate anaerobic organisms do not have these oxygen-protecting enzymes.

Overall, the various processes that consume O_2 (including O_2 respiration and combustion by forest fires) presently balance the photosynthetic production of O_2 on Earth, and the concentration of atmospheric O_2 is almost constant at the relatively short time scale of decades to centuries (See Sect. 9.4.5). On the other hand, many organisms do not need O_2. Some of them can tolerate O_2, but others are harmed by the presence of this gas. The organisms that do not use O_2 for respiration extract the energy contained in organic compounds by fermentation or various types of respiration that do not require O_2 (see Information Box 5.2).

The photosynthetic production of O_2 is responsible for virtually all the free oxygen on Earth, and this O_2 allows O_2-respiring organisms to thrive on the planet. The downside for many organisms is that O_2 is a poison for them, but these organisms nevertheless continue to prosper on the O_2-rich Earth in environments without O_2, such as the marine sediments deeper than a few centimetres in the seafloor. Organisms that do not use O_2 also often prosper in symbiosis with O_2-respiring organisms, which provide them O_2-free shelters and to which they provide metabolic services, for example the bacteria in the rumens of cattle or in our guts where they break down the ingested food.

Another effect of a high oxygen concentration in the atmosphere is that it favours the spontaneous combustion of organic matter, that is, wildfires or forest fires. The occupation of continents by plants and the following expansion of forests caused a major increase in the concentration of atmospheric O_2 (up to 35% per volume, compared to the present 21%; Fig. 9.17), which led to widespread wildfires, especially that wood was then very abundant. This mostly occurred between about 350 and 250 million years ago, but wildfires also occur nowadays. Under the present O_2 concentration, wildfires require fuel (plant matter) that has low moisture content, and thus mostly occur during dry seasons. In contrast beyond 30% O_2, wildfires were globally distributed, and they occurred even in wet climatic areas because of the frequent lightning strikes. Indeed, the most frequent natural cause of wildfires is lightning. The book of Glikson and Groves (2016) provides further reading on fire and the biosphere.

The genetic material contained in the DNA of organisms that live at the surface of Earth (see Sect. 6.2.2), especially those living on continents, is continually damaged by ultraviolet radiations from the Sun (UV). Although surface dwelling organisms have evolved mechanisms to cope with the harmful effects of UV, they could not survive on continents or in surface waters if most of the UV were not absorbed in the upper atmosphere (20–40 km high) by various chemical substances. The best known of these substances is ozone (O_3), which is formed by the action of UV on atmospheric O_2 (see Information Box 8.2). The progressively increasing concentration of O_3 in the upper atmosphere since about 2.2 billion years ago favoured the development of an O_3 layer there half a billion years ago, and it is only after the establishment of this protective *ozone layer*) that multicellular organisms started to occupy the continents. The continuing existence of organisms on continents and in surface waters depends on the maintenance of the O_3 layer in the upper atmosphere.

The O_2 transformed in O_3 in the upper atmosphere is produced by photosynthesis. Hence the O_3 layer exists because photosynthesis produces atmospheric O_2, and the O_3 layer in return allows the existence on continents and in surface waters of organisms that produce O_2 and consume it. Together with the redistribution of water all over the planet by atmospheric water vapour (see Sect. 2.1.3, above), the formation of the O_3 layer allowed the occupation of continental surfaces by plants and animals and their subsequent diversification during the Phanerozoic eon (from 542 million years ago to present). The book of Fabian and Dameris (2014) provides further reading on the chemical and human aspects of atmospheric O_3.

2.1.6 Carbon Dioxide in the Atmosphere and Ecosystems

The concentration of carbon dioxide in the modern Earth's atmosphere is very small. There is additional CO_2 dissolved in the oceans as CO_2 gas and bicarbonate (HCO_3^-) and carbonate (CO_3^{2-}) ions, whose combined amount is about 50 times that of the atmospheric CO_2. Even in the oceans, the concentration of dissolved CO_2 is very small. Despite these small concentrations, CO_2 has the two major effects on organisms, and in return the organisms are deeply involved in the global cycling of CO_2, as explained in the following paragraphs.

First, CO_2 provides all the atoms of carbon present in organisms. Indeed living cells use, in the processes of photosynthesis or chemosynthesis, the atoms of carbon and oxygen from CO_2 to make up organic matter (see Information Box 5.1). In addition, some organisms use the C and O atoms from CO_2 to make up shells and similar solid structures (for example, coral reefs) by the process of calcification (see Information Box 5.11).

Second, variations in the Earth's temperature are mostly determined by changes in the concentration of the greenhouse gas CO_2 (see Information Box 2.7). Hence, the maintenance of temperatures suitable for organisms over billions of years was largely controlled by the concentration of CO_2 in the Earth's atmosphere. Not enough or too much atmospheric CO_2 would have meant for organisms a global environment too cold or too hot for survival, respectively (see Sect. 7.1.2).

Conversely in cases of increasing atmospheric CO_2, photosynthesis, chemosynthesis and calcification contribute to transfer atmospheric carbon to geological reservoirs. Calcification is involved in this transfer as part of a global process that includes the erosion of rocks (see Sect. 7.3.3). The long-term reservoirs of carbon include carbonate rocks, fossil organic matter and deposits of methane (see Sects. 2.1.7, 3.4.2 and 9.2.1).

It was explained above that the global consumption of O_2 on Earth—which includes O_2 respiration and combustion (forest fires)—were about equal to the global photosynthetic production of O_2. This leads to the conclusion that the uptake of CO_2 by the process of O_2-producing photosynthesis is about equal to the release of CO_2 by O_2 respiration and combustion on current Earth. Overall, photosynthetic and chemosynthetic organisms use the carbon (and oxygen) from CO_2 to synthesize organic matter, and respiration or fermentation returns most of the CO_2 to the environment. These coupled natural processes, which are biological, are called the *fast carbon cycle* (see Information Box 9.1). We wrote "about equal" three sentences above because a small fraction of the organic carbon produced by organisms is buried in soils and marine sediments, and thus escapes the fast cycle. Part of the buried organic matter is eventually transformed into fossil organic matter. Most of the organic matter buried in the past is presently dispersed in sediments and soils, but a small part of it is concentrated in deposits of natural gas, petroleum, oil shale and oil sands, peat and coal, which belong to the slow carbon cycle (next paragraph).

The *slow carbon cycle* is thus named because its processes, which are largely geological, occur at time scales of millions of years (see Information Box 9.1). This cycle begins with the natural dissolution of CO_2 in water (rain or seawater), which makes it acid. The CO_2-acidified water dissolves continental or seafloor silicate rocks (for example $CaSiO_3$), which creates calcium (Ca^{2+}) and HCO_3^- ions carried by rivers to the oceans. The combination of the dissolution process and transport of the resulting substances is called *erosion* (see Information Box 7.1). In the ocean, HCO_3^- and Ca^{2+} ions are combined biologically or chemically in calcareous material ($CaCO_3$ and others), which is formed on or within the seafloor or sinks there. Some of the calcareous material is buried in the sediment, where compaction eventually transforms it into sedimentary rocks. In the modern ocean, most of the calcareous structures are formed biologically through the process of *calcification* (see Information Box 5.11 and Sect. 6.4.2).

Lateral or vertical movements of the Earth's crust slowly move the seafloor with its calcareous rocks toward the continental margins or upwards, and some of the calcareous material is incorporated into the continents (for example, the well-known White Cliffs of Dover, in England; Fig. 2.1), whereas most of it is carried deep in the Earth's mantle where it melts. The molten material reaches the surface through continental and submarine volcanoes in the form of fluid lava and CO_2 gas. The cold lava forms new rocks, and the CO_2 gets into the air or the seawater. The volcanic input of CO_2 over millions of years plays an important role in the long-term regulation of the Earth's natural greenhouse effect and thus the climate (see Sect. 9.2.3).

The present problems of *global warming* and *ocean acidification* (see Sects. 11.1.1 and 11.1.3) result from the fact that human societies, at first unintentionally and now with increasing awareness, have moved massive amounts of carbon from the slow cycle into the fast cycle since the beginning of the Industrial Age in the second part of the eighteenth century. The process was the release of enormous amounts of CO_2 in the atmosphere by the combustion of organic matter, both modern (example: wood) and fossil (example: natural gas), and also by the calcination of $CaCO_3$ to make cement (Fig. 9.5). To get an idea of the magnitude

Fig. 2.1 The White Cliffs of Dover, in England, are made of calcareous remains of cocco-lithophores (planktonic algae; Fig. 7.8a) deposited at the bottom of the sea about 70 million years ago. In some areas, the thickness of the resulting chalk exceeds 500 m. Credits at the end of the chapter

of the human disturbance of the carbon cycle, it is enough to say that human societies emit an amount of CO_2 into the atmosphere *each year* as large as the amount naturally released by all Earth's volcanoes during several decades. The present, ongoing increase in atmospheric CO_2 is so fast that the natural controls of the Earth's climate are overwhelmed. Fortunately, a significant fraction of the CO_2 released by humans continues to be stored in the ocean's deep waters and sediments, although with the collateral damage of ocean acidification.

2.1.7   Methane in the Atmosphere and Ecosystems

Methane is a very powerful greenhouse gas, which is present in the Earth's atmosphere in very low concentration (see Information Box 2.7). The different sources of CH_4, are either natural or anthropogenic. The main natural source is the production of CH_4 gas by archaea (microorganisms) through the process of *methanogenesis* (see Sect. 4.7.1 and Information Box 5.2). The environments in which this process takes place include the guts of termites and ruminants (cows, goats, sheep, deer, antelopes and other animals), and also marine sediments and the permafrost (paragraph after next). Another source of CH_4 is the abiotic production of hydrogen ions (H_2) during the chemical transformation of some rocks, a process called *serpentinization*, followed by the reaction of this H_2 with environmental CO_2. On Earth, this occurs in such environments as O_2-less hydrothermal vents on the seafloor and subduction zones (see Sect. 7.2.3). The presence of CH_4 in the atmosphere of some Solar System bodies (Table 2.2) could thus indicate the presence of CH_4-producing organisms, or it could alternatively be explained by the abiotic, geological formation of CH_4. The main anthropogenic source of CH_4 is intensive agriculture, which may account for more than 40% of all the CH_4 emissions to the atmosphere (see Sect. 11.1.2). Another anthropogenic source is the leakage of CH_4 during the extraction, storage, transportation, and distribution of *natural gas*, which consists mostly of methane.

Methanogenesis has important global roles in the Earth System. On the one hand, this process releases in the form of CH_4 carbon that would otherwise accumulate as products of fermentation in environments without O_2 (for example, marine sediments). On the other hand, the release of CH_4 gas in the environment by termites and ruminants contributes to the greenhouse effect, given that atmospheric CH_4 has a high greenhouse potential. Nowadays, the CH_4 released by large-scale livestock farming contributes to the ongoing global warming of the planet (see Information Box 2.7 and Sect. 10.1.6).

Other microorganisms, called *methanotrophs* (meaning "methane eaters"), use CH_4 as their source of carbon and energy, in environments with or without O_2 that include wetlands, soils, marshes, rice paddies, landfills, and aquatic systems. In *environments with O_2*, methane-oxidizing bacteria combine O_2 and CH_4, to form formaldehyde (CH_2O), which they use to synthesize organic compounds. In *environments without O_2*, other bacteria and some archaea process CH_4 in the absence of oxygen. Examples of the latter environments are *cold seeps*, which are areas of the seafloor where gases such as CH_4, H_2S and various hydrocarbon-rich fluids seep (leak) from the underlying sediments and rocks. These areas often support rich communities of organisms, whose existence is based on chemosynthesis, as is also the case for the communities of the warmer hydrothermal vents (see Sect. 3.1.1 and Information Box 5.1).

In marine sediments without O_2, some archaea transform part of the buried organic materials into CH_4 (see Information Box 5.2), and some of this CH_4 can escape to the seafloor and into the overlying water where most of it is consumed by methanotrophs (see above). When the temperature and pressure conditions are right, this CH_4 does not escape and is instead stored frozen in the sediments as solid "burning ice", whose chemical name is *methane hydrate* or *methane clathrate* (see Sect. 2.1.2, above). When there is a change in temperature and/or pressure conditions, the frozen CH_4 can be destabilized and the resulting CH_4 gas released to the seafloor and the water column. There, most of the CH_4 is consumed by methanotrophs, but if the release occurs in shallow waters or a large amount of CH_4 is released rapidly from a deeper seafloor, part of the CH_4 may escape to the atmosphere and thus contribute directly to the greenhouse effect. Similarly in some cold continental areas, large amounts of organic matter are frozen in the *permafrost*, and when the latter thaws due to increased temperature, the organic matter can be transformed into greenhouse gases by microbial activity in the presence of O_2 (production of CO_2 and N_2O) or its absence (production of CH_4 and N_2O) (see Information Box 5.2). It is feared that as Earth continues to warm, large amounts of the greenhouse gases CO_2, CH_4 and N_2O will be released from marine sediments and terrestrial permafrost, thus accelerating the ongoing global warming. This could create a situation where increased warming would cause increased release of greenhouse gases, thus creating a *positive feedback loop* and accelerating the ongoing global warming (Fig. 9.3f and accompanying text).

High concentrations of atmospheric CH_4 may have contributed to increase the greenhouse effect of Earth on several occasions in the geological past. The most significant of such events for organisms has occurred in the deep past, during the Archaean eon (from 4.0 to 2.5 billion years ago), when the Sun was not as hot as present (see Sect. 3.7.1). Indeed, the output of heat from the faint young Sun was only 70% of the present value, and after the decrease in atmospheric CO_2 that occurred during the Hadean eon (see Sect. 5.5.4), Earth could have become too cold for the existence of liquid water, and thus organisms. However, the continued existence of organisms indicates that Earth did not freeze up, hence the *faint young Sun paradox* for which a number of solutions have been hypothesized. One of them is that CH_4-producing microorganisms (see Information Box 5.2), may have released enough greenhouse CH_4 gas to sustain the Earth's greenhouse effect when the concentration of CO_2 became too low to do so (see Sect. 3.7.1). According to this hypothesis, CH_4 production would have exerted a major control on the atmosphere during the first half of the history of organisms on Earth (Archaean eon) by creating a CH_4-rich atmosphere that would have provided temperatures suitable for organisms.

On planet Earth, almost all chemical elements continually circulate between environmental reservoirs and ecosystems, on time scales ranging from very short to very long. This reflects the co-evolution of the Earth's environment and organisms over almost 4 billion years. The above Sects. 2.1.1–2.1.7 illustrated the results of this co-evolution for the atmosphere and the ecosystems, and the remainder of this chapter will examine some important aspects of this co-evolution in greater details. A chapter written by Legendre (2014) provides further reading on the effects of pelagic marine ecosystems on the cycling of chemical elements.

2.2 The Atmospheres of Earth and the Other Planets in the Solar System, and of Exoplanets

2.2.1 The Earth's Atmosphere

The atmosphere is part of everyday's experience on Earth, but the presence of an atmosphere, Earth-like or not, is not a common feature on the planets of the Solar System nowadays. However, this might have been different shortly after the

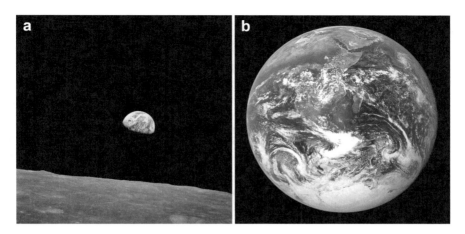

Fig. 2.2 Two historical photographs of Earth viewed from space: **a** *Earthrise* viewed from the Moon (1968), and **b** *Blue Marble* taken from a distance of about 29,000 km (1972). Credits at the end of the chapter

formation of the Solar System more than 4 billion years ago, as we will see in later sections. Space missions conducted since the 1960s and 1970s have allowed humans to see the Earth's environment from outside the Earth System, which was different from seeing it from the surface of the planet as had always been the case until then. In the early days of space exploration, when people saw Earth from space (Fig. 2.2), they became acutely aware of the exceptional nature of the planet. They discovered its partly liquid and partly solid surface, and its thin atmosphere that continually swirls around the globe, both so beautiful and so fragile.

The photo in Fig. 2.2a captured the first *earthrise* ever witnessed by humans as Earth gradually emerged from the shadow of the Moon (the word *earthrise* was created for the occasion, similar to sunrise and moonrise). Called *Earthrise*, this photo was taken from space by the crew of spacecraft Apollo 8 during the first manned flight around the Moon on 24 December 1968. One of the astronauts, Jim Lovell, said: "The vast loneliness [of space] is awe-inspiring and it makes you realize you just what you have back there on Earth". The photo in Fig. 2.2b is one of the few pictures showing an almost fully illuminated Earth disk. It is called *the Blue Marble*, and was taken by the crew of spacecraft Apollo 17 on 7 December 1972. Planet Earth looks blue from space because more than 70% of its surface is covered by liquid water.

Figure 2.3 shows photos taken decades after those in Fig. 2.2, which re-emphasize the apparent fragility and unique beauty of planet Earth. In Fig. 2.3a, small "blue-marble" Earth is seen from space beyond the Moon. The photo, which was taken by the Chang'e 4 lunar probe during a Chinese robotic mission to the Moon in May 2018, contrasts the mineral, dry Moon (seen from its far side) with the blue, ocean-covered Earth. The photo in Fig. 2.3b highlights the beauty and intensely rich colours of Earth (Fig. 2.2b). It was taken by the Russian weather geostationary satellite Elektro-L No.1 in 2020, which scans Earth in visible and infrared wavelengths. The combination of different wavelengths of light emphasizes the planet's features and colours.

Information Box 2.3 The Units of Atmospheric Pressure One finds different units of pressure in the literature. These units include the *pascal*, the *bar* and the *atmosphere*. The International System of Units (SI) uses the *pascal* (Pa), where $1\,Pa = 1\,N\,m^{-2}$ (the newton, N, is the SI unit of force; pressure is therefore a force applied per unit area). The *bar* is mostly used in meteorology and oceanography; this unit is not part of the SI, but is linked to the pascal: $1\,bar = 10^5\,Pa = 100\,kPa$.

The reference atmospheric pressure exerted by all gases together at sea level, on present-day Earth, is 101.3 kPa. This pressure defines the non-SI unit *atmosphere*: $1\,atm = 101.3\,kPa = 1.013\,bar$. Other units are also used, but these are not reported here for simplicity. In this book, we will systematically use the intuitive unit *atmosphere* (atm).

The Earth's atmosphere has a thickness of about 500–1000 km, depending on how the atmosphere is defined. Although atmospheric gases are light, 500–1000 km thick of these gases exert a very high pressure, which is defined as *1 atmosphere* at sea level (see Information Box 2.3). This atmospheric pressure is about the same as the hydrostatic pressure exerted by 10 m of water. Hence, each of us continually bears the equivalent of the weight of a column of 10 m of water. This weight is very heavy, but we do not pay attention to it because our body has evolved to withstand the atmospheric pressure that continually applies on it, and also because the weight is distributed over the entire surface of our body and not only rests on our shoulders.

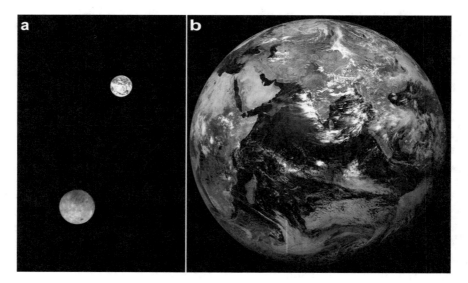

Fig. 2.3 Two photographs taken decades after those in Fig. 2.2: **a** the Moon and Earth seen from space beyond the Moon (2018), and **b** richly coloured Earth (2020). Credits at the end of the chapter

The Earth's atmosphere is stratified into four main layers of decreasing densities. These layers are: the *troposphere*, from 0 to 10 km above ground; the *stratosphere*, from 10 to 50 km above ground; the *mesosphere*, from 50 to 85 km above ground; and the *thermosphere*, from 85 to 500–1000 km above ground.

It is not easy to imagine the very high altitudes of the atmosphere, up to 500 or 1000 km. In order to provide a visual representation, the four layers are depicted in Fig. 2.4, which is not to scale because the thermosphere would occupy most of the height in a correctly scaled diagram. Some important objects cited in other chapters are shown in each layer: in the troposphere, Mount Everest (see Sect. 7.5.2) and a plane; in the stratosphere, the ozone layer (see Sect. 2.1.5); in the mesosphere, meteors (see Sect. 1.3.6); and in the thermosphere, auroras (see Sect. 8.4.1). In the latter layer, there is also a representation of the *International Space Station*, whose altitude ranges between 408 and 410 km and is shown in a different context in Fig. 8.3b. The history of the Earth's atmosphere since 4.5 billion years before present is described in Chap. 8 (see Sect. 8.1.1). The book of Wallace and Hobbs (2006) provides further reading on the physics of the Earth's atmosphere.

2.2.2   The Giant Planets: Atmosphere and Formation

The four planets of the Solar System whose orbits are furthest from the Sun have the largest masses, and are thus called *giant planets* (Table 1.1). They are, in order of increasing distance from the Sun, Jupiter, Saturn, Uranus and Neptune (Fig. 1.4). On these planets, there is no clear boundary between the gaseous atmosphere and the liquid or solid surface. Indeed, because of the large gravitational pull exerted by the huge mass of these planets, their atmospheric pressure increases very quickly with depth into the atmosphere up to the point their most compressed gases have about the same densities as their liquid phases. Hence these planets do not have a well-defined interface between their gas atmosphere and their liquid or solid surface. This is opposite to the situation that exists on the four smaller planets Mercury, Venus, Earth and Mars, whose surface is distinct from their atmosphere (Fig. 1.4).

The physical state of volatile compounds on planets (see Sect. 2.1.2, above) defines the *ice line* in the Solar System, which is also called *frost line* or *snow line*. This line is the distance from the Sun where it is so cold that volatile compounds such as water, ammonia, methane, carbon dioxide and carbon monoxide are condensed into solid ice grains. The giant planets did form in the cold region beyond the ice line of the Solar System, and their accretion likely occurred according to the following general sequence: first, the establishment of a rocky core, in the same way as for the rocky planets (see Sect. 2.2.3, below); second, the accretion of ice grains of volatile compounds; and third when the mass was sufficiently large to create a huge gravitational attraction, the accretion of gaseous hydrogen and helium. The special formation mechanisms of the two classes of giant planets described in Chap. 1 (see Sect. 1.3.1), namely the gas giants and the ice giants, are poorly understood and currently debated among specialists.

Fig. 2.4 The four layers
of the atmosphere, and the
altitude of their upper limits.
Each layer is illustrated by
some characteristic objects.
The thicknesses of the layers
and the sizes of the objects
within the layers are not to
scale. Credits at the end of
the chapter

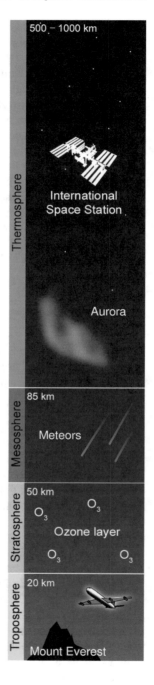

2.2.3   The Rocky Planets: Atmosphere and Formation

Contrary to the four giants, the four smaller planets that revolve closer to the Sun—in order of increasing distance from the Sun, Mercury, Venus, Earth and Mars—are largely made of rocky materials (see Sect. 1.3.1; Fig. 1.4). Because the atmosphere of Venus is opaque, the observation of its surface from Earth had to wait the advent of powerful land-based radars in the 1960s. The first direct contacts with the surface of Venus were the crashing with, and finally landing on, the burning surface of Venus by Russian spacecrafts Venera 3–7 in the late 1960s and early 1970s. In fact, Venera 7 was the first probe ever to survive the landing on another planet (although only 23 min). The first spacecrafts to orbit Venus were Russian Venera 9 and 10 and US Pioneer, all in the 1970s. The Venera program continued into the 1980s, and ended with Venera 16. The soft landings on Venus in the early 1970s were quite remarkable given that some of the probes sent to the Moon or to Mars 50 years later crash instead of landing.

The four rocky planets have different atmospheric pressures (Table 2.2). As already explained above (see Information Box 2.3), the Earth's average atmospheric pressure is 1.0 atm at sea level. In contrast, the atmospheric pressure at the surface of Mars and Mercury is about 0.006 and 10^{-14} atm, respectively, which values are 170 and 10^{14} times smaller, respectively, than the atmospheric pressure on Earth. At the other end of the spectrum of atmospheric pressures is the very dense atmosphere of Venus, which exerts a pressure at ground level of about 92 atm, which is 92 times higher than at the surface of Earth.

The starting point for the accretion of rocky planets was the agglomeration of bodies a few kilometres in size. These *tiny planets* formed in the central plane of the solar nebula, where the dust density was highest, thanks to gravitational instabilities that triggered localized coalescence of dust grains. Some of the tiny planets progressively grew to reach the size of the present dwarf planets (see Information Box 1.3), this stage being called *rocky protoplanet*. The rocky protoplanets and the many tiny planets that probably developed during the formation of the Solar System experienced one of the following three fates: merging with one of the four largest rocky planets, destruction by impact with one of the rocky planets, or expulsion from the Solar System by the gravitational slingshot effect. The latter is the increase in velocity (relative to the Sun) of an object that is caught by the gravitational field of a planet and swings around it, thus acquiring a velocity sufficient to eject it from its orbit around the planet. This way, the flying body acquires some of the kinetic energy of the planet, which increases its velocity considerably (while decreasing the velocity of the planet by an infinitesimal fraction because the mass of the objet is minute relative to that of the planet). This natural effect is regularly used by space agencies as a deliberate manoeuvre to change the velocity of spacecrafts during long flights through the Solar System.

Finally, there is an anomaly in the distributions of masses among Mars, Earth and Venus (Table 1.1). Indeed, although the three planets were formed at the same time and in the same region of the early Solar System, the mass of Mars is about 10 times smaller than that of Earth and Venus (the mass of Venus is 0.8 times that of Earth). This can be explained by changes in the orbits of the giant planets during the formation of the Solar System. It will be shown below that the masses of Mars and Earth must be considered in order to understand the presently rarefied atmosphere of the first planet and the thicker atmosphere of the second (see Sect. 2.3.2). The book of Lissauer and de Pater (2019) provides further reading on the atmospheres of planets including Earth.

2.2.4 Chemical Composition of the Atmosphere of Earth and Venus

Planets Earth and Venus both have a significant atmosphere, although with very different gas composition and atmospheric pressure at ground level. While on both planets light chemical elements (mostly carbon, oxygen and nitrogen) are dominant in the atmosphere, these are in different chemical forms. The atmosphere of Venus is dominated by carbon dioxide (CO_2), a molecular compound that associates carbon and oxygen in a stable form. The adjective *stable* means here that CO_2 does not react easily with other compounds of the Venusian environment, so that the concentration of CO_2 in the atmosphere of Venus has likely remained the same for a very long period, perhaps since the formation of the planet. In contrast, the atmosphere of Earth is dominated by nitrogen and oxygen, the first chemical element being in the form of stable molecules of dinitrogen (N_2) and the second in the form of unstable molecules of dioxygen (O_2). The latter molecules are unstable because oxygen can easily react with a large number of chemical elements and compounds present in the Earth's environment.

It follows from the above explanations that the chemical composition of the atmosphere of Venus is stable in the long term, and that the atmosphere of Earth is potentially unstable. In addition, the huge Venusian atmospheric pressure compared to that of Earth (i.e. 92 times higher; Table 2.2) is largely due to the massive abundance of CO_2 (around 965,000 parts per million, or ppm), whereas this compound exists only in trace amounts in the Earth's atmosphere (around 280 ppm at the beginning of the Industrial Age, and more than 400 ppm nowadays). The difference between the atmospheres of the two planets is striking because it can be assumed that they were similar initially. Indeed, the two planets were formed at about the same time and in the same area of the solar nebula, their masses are similar, and their distances from the Sun are not very different (108 and 150 million kilometres for Venus and Earth, respectively). This is consistent with the fact that, if all the carbon existing on Earth (in rocks, sediments, fossil organic matter, and biomass) were oxidized and the resulting CO_2 released into the atmosphere, the atmospheric pressure at the surface of the planet would be about 60 atm, which is of the same order of magnitude as the atmospheric pressure at the surface of Venus (Table 2.2).

Different Fates of the Atmospheres of Venus and Earth If we accept the hypothesis that the atmospheres of Venus and Earth were initially similar, their present differences indicate that some mechanism removed the carbon initially present as CO_2 in the Earth's atmosphere, and transferred this carbon somewhere else on the planet, whereas this did not happen on Venus. In other words, different environmental conditions on the two planets led Earth to lose most of its atmospheric carbon, whereas Venus retained a highly carbonaceous atmosphere. We examine the possible mechanism below (see Sect. 5.5.4).

2.2.5   Exoplanets

There are planets orbiting other stars than the Sun, called *exoplanets*. Their detection and characterisation have been the subject of very active research since the first exoplanet was discovered in 1995. The number of 1,000 confirmed exoplanets was reached in 2013, and since then the number has continued to increase every year. The vast majority of known exoplanets are very massive bodies, such as the giant planets of the Solar System, given that these large objects are more easily detectable from Earth than smaller bodies. However, these giant planets are not habitable (see Information Box 2.4). Exoplanets of lower mass, similar to the rocky planets in the Solar System, are more difficult to detect than their giant counterparts, but the number of discoveries of such exoplanets increases steadily.

Thanks to the Kepler space telescope, astronomers discovered in 2014 the first exoplanet both close to Earth's size (10% larger) and located in the habitable zone of its star. (The star, which is known as Kepler-186, is located some 500 light years from Earth in the constellation of the Swan.) Does this exoplanet have a significant atmosphere? The latter is likely, but is this atmosphere close to the sweet, cool air of Earth, or the acidic, scorching atmosphere of Venus? The answer to this question will not be known soon, if ever. The progressive discovery of exoplanets and the characterization of their habitability will be one of the great quests of space science in the coming decades.

Information Box 2.4 The Habitability of Planets, and the Goldilocks Principle The *habitability* of a planet or any natural planetary body such as a moon is defined as its potential to offer environments hospitable to organisms. This does not mean that habitable bodies contain(ed) life, or that life could appear there (spontaneously, or coming from elsewhere), but simply that they offer environmental conditions similar to those suitable for organisms on Earth. The planet of reference for habitability is Earth because is the only place where life is presently known (by humans). A habitable planetary body should orbit at the right distance from its star for liquid water to be present at its surface, and have various other characteristics (mass, etc.) discussed in Chaps. 1–8.

The *habitable zone of a star* is the region of space around that star where habitable bodies can be found. The habitable zone is also called *life zone*, *comfort zone, green belt*, or *Goldilocks zone*. The concept of habitable zone of a star is used in studies of both planetary and non-planetary bodies of the Solar System (see Sect. 1.3) and exoplanets (this section). The books of Kasting (2012) and Conrad (2016) provide further reading on planetary habitability.

The word *Goldilocks* used in the previous paragraph refers to the *Goldilocks principle*, which generally means in science "just the right amount". Within the context of astrobiology, the Goldilocks zone is the area around a star where the conditions are just right for organisms.

The Goldilocks principle refers to the children's story *The Three Bears*, in which a little girl named Goldilocks (meaning "with golden hairs") finds a house inhabited by three bears, who just left. In that house, each bear has its own preference of porridge temperature, chair size and bed firmness. After testing the bears' three bowls of porridge, three chairs and three beds, Goldilocks decides which ones are "just right" for her (hence, the Goldilocks principle). For the remainder of the story, read the fairy tale *Goldilocks and the Three Bears* in print or on the internet.

The book of Gargaud et al. (2006) provides further reading on the building of a habitable planet in the Solar System.

2.3 Conditions that Determine the Earth's Atmosphere

2.3.1 The Earth's Atmosphere: Two Special Characteristic

In addition to the eight planets, the Solar System comprises hundreds of moons and several other types of objects that include dwarf planets, most of which are probably still to be discovered (see Sect. 1.3.6). Among all these Solar System objects, Table 2.2 summarizes the mass and atmospheric characteristics of those

> **Information Box 2.5 What is an Exosphere?** Some planetary bodies have an "atmosphere" so rarefied that its molecules do not collide with each other and thus do not behave as a gas. Such an atypical atmosphere is called *exosphere*. Planet Mercury, our Moon and some satellites of Jupiter (Callisto, Europa, Ganymede, and Io) have an exosphere.

that share the following three features: mass greater than 1/1000 that of Earth, an observable surface (solid or liquid), and a detectable atmosphere or exosphere (see Information Box 2.5). These objects are: the four rocky planets; our Moon; the six most massive moons of three giant planets, that is, Jupiter's moons Ganymede, Callisto, Io and Europa, Saturn's moon Titan, and Neptune's moon Triton; and dwarf planet Pluto. There are many other natural satellites and dwarf planets in the Solar System, but they miss at least one of the above three features.

Comparing the atmospheric properties of the planetary and non-planetary bodies in Table 2.2 shows that Earth's atmosphere combines two special characteristics, i.e. it exerts high pressure at the surface of our planet, and it contains a high proportion of O_2. Concerning the first characteristic, the atmospheric pressure at the surface of Earth ranks third, after Venus and Titan, among the rocky planets, moons and dwarf planet Pluto in the Solar System. The atmospheric pressure of Venus is 92 times that of Earth at ground level, and it was explained above (see Sect. 2.2.4) that a similar condition likely prevailed on Earth more than 4 billion years ago. The second characteristic of the current atmosphere of Earth is its exceptionally high O_2 content (21% by volume). Although O_2 has been detected in the atmospheres of three moons of Jupiter (Ganymede, Callisto, and Europa) and in trace concentrations in the atmospheres of other bodies of the Solar System, a high atmospheric pressure together with a significant proportion of O_2 is unique to Earth. This was not always the case during the history of Earth, and the present high O_2 content in the Earth's atmosphere results from the unique biological evolution that took place on the planet (see Sects. 1.5.4 and 8.1.1).

2.3.2 The Atmospheres of the Four Rocky Planets and Titan

In Table 2.2, the four Solar System objects with the largest masses are the four rocky planets, i.e. in order of decreasing masses, Earth, Venus, Mars and Mercury. The masses of Earth and Venus are similarly large, and the atmospheric pressure on these two planets is high. Hence a high planetary mass seems to favour the presence of a significant atmosphere. This idea is supported by the small mass and thin atmosphere of planets Mars and Mercury, but is not consistent with the characteristics of Saturn's natural satellite Titan. Indeed, the atmospheres of Mars and Mercury, whose masses are about 11 and 6% that of Earth, respectively, are rarefied and very rarefied (exosphere), respectively, which links the presence of a thick atmosphere with planetary masses $\geq 10\%$ that of Earth. However, Titan has a heavier atmosphere than Earth but its mass is only 2% that of Earth, showing that the presence of an atmosphere is not necessarily linked with a large planetary mass. Titan is the second largest moon in the Solar System after Ganymede, and this largest moon of Jupiter has a negligible atmosphere.

A first preliminary conclusion, drawn from the atmospheric and mass characteristics of the four rocky planets, is that planetary bodies with small masses do not exert sufficient gravitational pull to generally retain a significant and stable atmosphere. However, this conclusion does not account for the observed high atmospheric pressure of Titan, which indicates that the gravitational attraction exerted by the mass is not the only factor that explains the actual occurrence of an atmosphere. Examining the cases of Mercury, Triton and Pluto will shed additional light on this matter. However, explaining the heavy atmosphere of Titan will require further robotic exploration of this atypical planetary body.

2.3.3   The Atmospheres of Mercury, Triton and Pluto

The atmosphere of planet Mercury is so rarefied that it is, in fact, an exosphere (see Information Box 2.5). This situation is explained by both the small mass of this planet and the fact that, because Mercury is located very close to the Sun, its surface is constantly subjected to a stream of charged particles released from the upper atmosphere of the Sun, called *solar wind* (see Information Box 2.6). The latter, together with the pressure of solar light (photons) and very small meteoroids that frequently strike Mercury, blasts atoms off its surface. These atoms are blown into space by the solar wind, forming a tail of particles trailing Mercury somewhat similar to that of a comet when it passes near the Sun (see Information Box 1.3). Because of Mercury's low gravity and proximity to the Sun (continual solar wind and photon buffeting), this planet never developed or retained a real atmosphere.

Contrary to Mercury, two planetary bodies with very small masses—Neptune's moon Triton, and dwarf planet Pluto—have non-negligible atmospheres (Table 2.2). The reason why these two planetary bodies and also Saturn's moon Titan (see Sect. 2.3.2, above) are able to retain an atmosphere despite their small masses is likely that they are located far away from the Sun. The three bodies are thus counterparts of Mercury, whose closeness to the Sun makes it unable to retain an atmosphere.

A second preliminary conclusion, drawn from the atmospheric and mass characteristics of Pluto, Titan and Triton, is that the distance of planetary bodies from the Sun contributes to their ability to retain an atmosphere. However, the distance from the Sun does not explain by itself the present occurrence of an atmosphere on Titan, Triton and Pluto. In the next section, we combine the mass of a planet and its distance from the Sun to explain the presence or not of an atmosphere.

> **Information Box 2.6 Solar Wind and Solar Magnetic Activity Cycle** The
> Sun continuously emits nearly one million tons (metric) of material per sec-
> ond in the interplanetary medium. This stream of charged particles is called
> *solar wind*. The solar wind is in the form of *plasma*, that is, a state of mat-
> ter consisting mostly of electrons and protons in which there are also ions
> (atoms that have lost one or more electrons).
>
> The solar wind exists in two states, called slow and fast. The *slow
> solar wind* is quite constant, with a speed that ranges between 300 and
> 500 km s^{-1} independently from the solar magnetic activity cycle. The *fast
> solar wind* is related to the solar magnetic activity cycle, and its speed
> ranges between 500 and 800 km s^{-1}. The *solar magnetic activity cycle* is the
> nearly periodic 11-year cycle in the Sun's activity (solar radiation and ejec-
> tion of solar material) and appearance (number and size of sunspots, flares
> and others) (see Sect. 8.3.2).

2.3.4 Planetary Atmospheres: Mass of the Planet and Distance from the Sun

The above examination of the data in Table 2.2 showed that the planetary bodies
without a significant atmosphere (that is, with an exosphere) in the Table either
have a relatively small mass (our Moon, Ganyamede, Callisto, Io, and Europa) or
are located close to the Sun (Mercury). Conversely, a necessary condition to the
presence of an atmosphere appears to be a relatively large mass (Venus, Earth) or
a location not too close to the Sun (Mars, Titan, Triton, and Pluto). The mass of
a planet contributes to the presence of an atmosphere because it determines the
gravity of the planet (see Sect. 6.6.1), and the latter sets the *escape velocity* of gas
molecules from its atmosphere (see Sect. 8.4.1). The presence or not of an atmosphere also depends on the history of each planetary body, as examined in later
chapters. Briefly, the initial atmospheres of Earth, Venus and Mars may have been
similar given that the three planets formed in the same general area of the Solar
System at about the same time. Early it its history, Venus may have lost most of
its water and Earth may have stored most of its atmospheric CO_2 in solid $CaCO_3$,
thus explaining the much lower concentration of CO_2 in the Earth's atmosphere
than in that of Venus (see Sect. 5.5.4). Later, Mars would have lost most of its
atmosphere, explaining the similar concentrations of CO_2 and N_2 in the atmospheres of Mars and Venus and the much lower atmospheric pressure on Mars than
on Venus and Earth (see Sect. 8.4.1).

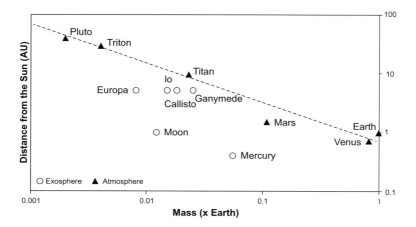

Fig. 2.5 Scatter diagram (mass *versus* distance from the Sun) of the planetary bodies in Table 2.2. Axes: logarithmic scales. Dashed line: linear regression *for the 6 bodies with an atmosphere* (solid triangles): $\log_{10}$ distance $= -0.66 \log_{10}$ mass -0.18 ($r = 0.98$, prob. < 0.001). The distances from the Sun and the masses are from Tables 1.1 and 2.2, respectively. Credits at the end of the chapter

Figure 2.5 illustrates the relationships between the distance from the Sun and the mass for the twelve planetary bodies listed in Table 2.2. The highly significant regression (dashed line) shows that the six planetary bodies with an atmosphere generally follow a similar distance-mass relationship, which is defined by the regression equation in the legend of Fig. 2.5. This equation can be expressed as follows:

$$\text{Distance from the Sun} \times \text{Mass}^{0.66} = 10^{-0.18} = 0.66 \qquad (2.1)$$

In other words, Eq. 2.1 indicates that the presence of an atmosphere on a planetary body is related to a specific combination of its mass and distance from the Sun. Figure 2.5 also shows that, for similar masses, the planetary bodies with an exosphere are generally at a smaller distance from the Sun than those with an atmosphere, and for similar distance from the Sun, they generally have a lower mass than those with an atmosphere.

The information from Fig. 2.5 is consistent with two hypotheses proposed by astronomers concerning the atmosphere of Mercury and Mars, namely that planet Mercury never had an atmosphere because of its close proximity to the Sun, and planet Mars initially had an atmosphere but lost most of it during the course of its history because of its relatively small mass. It is also consistent with the above idea that the presence of an exosphere instead of an atmosphere on a planetary body can be explained by the closeness of the body to the Sun or its small mass. Finally, the information from Fig. 2.5 indicates that two conditions that allow planet Earth to retain its atmosphere are its relatively large mass and its safe distance from the Sun. The astronomical characteristics of mass and distance from the Sun thus contributed to the durable establishment of organisms on Earth and the build-up of the huge biomasses that led them to progressively take over the planet.

2.4 Atmosphere and Climate

2.4.1 Greenhouse Effects on Earth and Venus

The ongoing global warming of Earth is largely caused by an anthropogenic increase of the *greenhouse effect* on the planet (see Information Box 2.7). In order to understand the relation between the greenhouse effect and the temperature of a planet, it is useful to look at the situation on Venus. There, the lower atmosphere has a temperature of +462 °C, and the amount of CO_2 gas in the atmosphere is very high as it makes up 96% by volume and creates an atmospheric pressure of 92 atm (Tables 1.1, 2.1 and 2.2, respectively). The large amount of CO_2 creates a huge greenhouse effect, which generates the observed temperature of +462 °C. Such a temperature is not compatible with the existence of liquid water (see Sect. 5.6.1), and with the development or maintenance of organisms at the surface of the planet. In addition to the dominant greenhouse effect of CO_2, the little water that exists above ground on Venus (the amount of water underground is not known) is present in the form of vapour in the atmosphere (0.002%) where it contributes to the greenhouse effect.

On Earth, several atmospheric gases contribute to the natural greenhouse effect (see Information Box 2.7), which generates the moderate temperatures essential to organisms. Indeed without the natural greenhouse effect, the average surface temperature of the planet would be about −18 °C, which would produce a planet-wide development of ice sheets that would bring the temperature down to about −50 °C. This very low temperature would result from the fact that an average temperature of −18 °C would cause very large areas of the oceans and the continents to become covered by ice, which would increase the *albedo* of the Earth's surface (see Information Box 2.9). The higher albedo would decrease the absorption of solar heat by the ocean and the continental surfaces, which would cause further cooling and ultimately lead to a planet-wide Ice Age. This phenomenon occurred at least once and perhaps several times on Earth, when the planet experienced *snowball Earth* conditions (see Sect. 4.5.4). The interaction between the cooling and the global formation of ice is called *positive feedback loop*, and the resulting global cooling, *runaway cooling* (see Sect. 9.1.3, Fig. 9.1d). The word *runaway* denotes that the change in temperature promoted further change in the same direction.

a

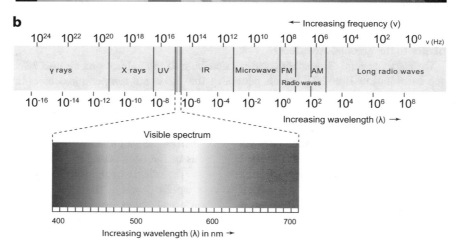

b

Fig. 2.6 The greenhouse effect on Earth: **a** conceptual model illustrating the fate of the solar radiation received by Earth (see Information Box 2.5); **b** spectrum of electromagnetic radiations (see Information Box 2.8). Credits at the end of the chapter

Information Box 2.7 What is the Greenhouse Effect? The solar energy received by any planet including Earth is radiated by the Sun at short wavelengths, predominately in the visible or near-visible (ultraviolet) part of the electromagnetic spectrum (Fig. 2.6b). The Sun emits radiation at these short wavelengths because the radiation emitted by a body peaks at a wavelength inversely related to the temperature of the body, and the temperature of Sun being very high (more than 5 million degrees), its emissions are thus at short wavelengths. Part of the solar energy reaching the planet is reflected directly back to space by the planet's atmosphere and surface. The proportion of the solar light reflected back to space is about 30% for Earth and 75% for Venus, corresponding to fractions of 0.30 and 0.75, respectively, called *albedo* (see Information Box 2.9). The non-reflected energy is absorbed by the atmosphere and by the surface of the planet (yellow arrows in Fig. 2.6a).

Earth radiates back to space, on average, the same amount of energy as it receives (see Sect. 3.2.1), and because the planet is much colder than the Sun, it radiates at much longer wavelengths, primarily in the infrared part of the spectrum (Fig. 2.6b). Given that infrared radiation is heat, the Earth's radiation toward the atmosphere is thermal (two upward red arrows in Fig. 2.6a). Much of this thermal radiation is absorbed by the atmosphere, where molecules of *greenhouse gases* (GHG) and clouds reradiate it in all directions including toward the surface of the planet (orange arrows in Fig. 2.6a). The combined heat radiated by the surface and reradiated by the atmosphere keeps the lower atmosphere warm.

The above phenomenon is called *greenhouse effect*, because it is analogous to the effects of the glass roof in a greenhouse (a building for growing plants). There, the glass roof lets the solar radiation enter and prevents the warmed air from escaping via convection, whereas the greenhouse effect warms the lower atmosphere through the different physical process explained above. An increase in the greenhouse effect can cause an increase in the global mean surface temperature of the planet, called *global warming*.

Different atmospheric gases have different *global warming potential*, which is a measure of the total energy that a gas absorbs over a particular period of time (usually 100 years), compared to carbon dioxide (CO_2). The natural atmospheric gases that have the highest global warming potential (over 100 years) on Earth are, in addition to CO_2, methane (CH_4) and nitrous oxide (N_2O), whose warming potential is 34 and 298 times that of CO_2, respectively. These three gases and other ones belong to the greenhouse gases. The actual importance of various gases in the greenhouse effect on a planet depends on their global warming potential and their abundance in the atmosphere.

Another major GHG on Earth is water vapour, which accounts for about 50% of the greenhouse effect. Overall, CO_2 accounts for 20% and clouds for 25%, the remainder being caused by small particles (aerosols) and minor greenhouse gases in term of their abundance such as CH_4 and N_2O.

Despite its high concentration in the atmosphere, water vapour is not a primary cause of global warming because its concentration depends itself on temperature. Indeed, when the CO_2 concentration increases, the average temperature increases, which increases evaporation and thus the average water vapour concentration, which in return increases the warming. The same effect amplifies the cooling when the CO_2 concentration decreases (Fig. 9.3d and accompanying text). This phenomenon corresponds to a *positive feedback loop* (see Information Box 6.2). The magnitude of the greenhouse effect on Earth is mainly governed by the concentrations of CO_2, CH_4 and N_2O in the atmosphere.

Information Box 2.8 Spectrum of Electromagnetic Radiations The *spectrum of electromagnetic radiations*, in short *electromagnetic spectrum*, covers the whole range of electromagnetic waves (Fig. 2.6b). These include gamma (γ) rays, X rays, ultraviolet radiation (UV), visible light, infrared radiation (heat), microwaves, and radio waves. The colours of the visible light range from violet to blue, cyan, green, yellow, orange, and red.

All radiations behave as waves and are thus characterized by the *distance* between two successive crests or troughs, called *wavelength* (λ), or equivalently by the *number* of crests or troughs occurring *per unit time*, called *frequency* (ν). The unit of λ is the metre (m), and that of ν is the inverse of time (s^{-1}, meaning the *number of events occurring per second*), which is also called the *hertz* (Hz; $1 \, Hz = 1 \, s^{-1}$). Given that electromagnetic radiations travel at the speed of light in vacuum ($c = 300{,}000 \, km \, s^{-1} = 3 \times 10^8 \, m \, s^{-1}$), the frequency of a given radiation is obtained by dividing the speed of light by the wavelength ($\nu = c \, / \, \lambda$). Taking the UV radiation as example (Fig. 2.6b): $\lambda_{UV} = 300 \, nm$ (that is, $3 \times 10^{-7} \, m$), hence $\nu_{UV} = (3 \times 10^8 \, m \, s^{-1}) \, / \, (3 \times 10^{-7} \, m) = 10^{15} \, Hz$.

The energy associated with the electromagnetic spectrum is carried by particles called *photons*. The photon is the minimum amount (called *quantum*) of an electromagnetic radiation, including light. The mass of a photon is zero, and it moves at the speed of light in a vacuum. The *amount of energy* of a photon is directly proportional to its electromagnetic frequency (ν), and is thus inversely proportional to its wavelength (λ). In other words, the higher the photon's frequency, the higher its energy, and the longer the photon's wavelength, the lower its energy.

It follows from the above paragraphs that the natural greenhouse effect can hinder or favour organisms and ecosystems. On the one hand, the extreme greenhouse effect that exists on Venus forbids the existence of organisms at the surface of that planet, although it has been speculated that organisms could perhaps live high in

---

Information Box 2.9 The Albedo The *albedo* is the proportion of the light striking a surface reflected by that surface. On Earth, high albedo is associated with snow, ice, and white sand. Indeed, the surfaces covered by these brightly white substances reflect back most of the solar energy impinging on them. For example, fresh snow can reflect 80–90% of the sunlight that strikes it, and the corresponding albedo is thus 0.8–0.9. Because of their high albedo, the areas covered by ice and snow play important roles in the functioning of the heat machinery of the planet despite the fact that only 2% of the water on Earth is solid snow or ice. Indeed with higher albedo, more solar energy is reflected back to space, which keeps the atmosphere and the Earth's surface cooler. High albedo affects not only the global climate, but also potentially each of us in the presence of strong reflexion of the sunshine by snow, ice or white sand, from which we protect our eyes by wearing sunglasses and our skin by applying high-factor sunscreen.

The upper surface of the clouds also contributes to the global albedo of Earth, and thus to the reflexion of solar radiation back to space, whereas their lower surface contributes to the greenhouse effect, and thus to the trapping of heat in the atmosphere. Given that the ratio between the two effects depends on the type of clouds, the overall role of clouds in the global climate is not well understood.

---

the Venusian atmosphere where temperatures are moderate. On the other hand, the mild natural greenhouse effect of Earth has been a major condition that allowed the durable establishment of organisms and their build-up of large biomasses. However, the ongoing anthropogenic increase in atmospheric greenhouse gases causes an increase in the temperature of the planet that could deeply perturb the organisms and ecosystems most sensitive to temperature changes. This immediately brings to mind the pathetic fate of such iconic species as polar bears and emperor penguins, but developed human societies may be as vulnerable to global warming as the ecosystems of these wild species (see Sects. 11.1.1–11.1.3).

2.4.2 The Atmosphere Redistributes Heat Received from the Sun

In areas with harsh winters, like central and eastern Canada and Russia, there are sometimes very cold episodes in winter when temperatures can fall down to −20 or −30 °C especially during the night. The media and hotels alert the residents and visitors of these countries of the upcoming cold spell, especially if the forecasted low temperatures are to be accompanied by winds. Indeed, the *apparent temperature*, which is the temperature one actually feels, depends not only of the actual temperature but also the wind speed. For example with a wind speed of

Information Box 2.10 Temperature and Heat *Temperature* is the property of matter that reflects the quantity of kinetic and vibrational energy of the atoms and molecules in a substance (gas, liquid, or solid). In other words, it reflects how fast the atoms and molecules in a substance are moving. Faster molecules create higher temperature. Temperature is measured on a very large number of atoms or molecules, and is thus a statistical property. Temperature is independent of the volume of the substance; for example, 100 ml of 50 °C water have the same temperature as 10 ml of 50 °C water.

Heat is the energy transferred from one body to another as the result of a difference in temperature. In other words, heat reflects how many atoms or molecules there are in a body and how much energy each atom has. Heat is dissipated by transferring energy from the warmest to the coldest body. This effect is illustrated by the fact that although it is possible to make two litres of warm water by mixing one litre of hot water with one litre of cold water, it is not possible to do the opposite without an additional input of energy. Indeed in order to obtain one litre of hot water and one litre of cold water from two litres of warm water, one needs to heat half of the warm water on the stove, and to cool the other half in the refrigerator.

The difference between temperature and heat is illustrated by considering two glasses that contain 50 and 100 ml of 50 °C water, respectively. The temperature of the water in the two glasses is the same, that is, 50 °C, but the 100 ml glass contains twice the amount of heat as the 50 ml glass.

12 m s^{-1} (43 km h^{-1}) a temperature of -20 °C will feel like -35 °C. The latter temperature is called the *wind chill index* (Fig. 2.7). At apparent -35 °C, exposed parts of the body (fingers, nose, ears, etc.) can freeze very quickly, with unpleasant consequences. In such conditions, it is recommended to winter visitors of Old Quebec City in Canada or Yekaterinburg in Russia to wear thick gloves and a warm scarf covering the neck and lower part of the face!

Fig. 2.7 Wind chill index as a function of temperature and wind speed. Wind speed of 10 m s^{-1} corresponds to 36 km h^{-1} Credits at the end of the chapter

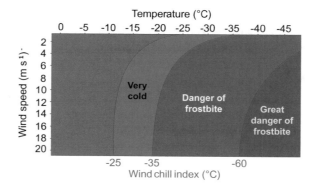

The difference between the apparent and the air temperature reflects the fact that moving cold air absorbs heat from one's body faster than stationary cold air does. The flow of air molecules in contact with the skin increases with increasing wind speed, and each air molecule carries away a small amount of heat due to the large difference between the temperature of the air and that of the skin (see Information Box 2.10). This is the same heat transfer phenomenon that one creates when blowing on a spoon of hot soup or a cup of hot beverage to cool it.

In fact, the apparent temperature should take into account not only the speed of wind but also the relative humidity, but the latter is often not included when computing the chill index because humidity is always low at low temperature. This is not the case in hot weather conditions, and humidity is then included in the calculation of what is called the *heat index*. In humid conditions, the air feels hotter because less perspiration evaporates from the skin.

Whether one feels the cold or hot wind, hears bursts of a whistling storm, sees clouds racing in the sky, shivers in the cold or pants in the heat, all these perceptions reflect the fact that the atmosphere is a fluid in motion that transports heat and exchanges it with the liquids and the solid surfaces it encounters. The atmosphere, together with the ocean, plays a similar role in redistributing over the whole Earth the heat the planet receives from the Sun, and is thus a major component of the Earth System (see Sect. 3.5.4). Indeed, it is at low and mid-latitudes that Earth receives most of the heat that arrives from the Sun, and this heat is largely redistributed to the remainder of the planet by the circulation of the atmosphere and oceans around the globe. Hence the atmosphere is a major climate player not only regionally, through winds and storms, but also globally, through its global circulation.

2.5   Key Points: The Atmosphere and the Living Earth

We conclude this chapter with a brief summary of some of the key points of the above sections. The expression *Living Earth* was introduced in Chap. 1 to characterize the entity resulting from the co-evolution of Earth and its ecosystems over almost 4 billion years (see Sect. 1.6). This co-evolution transformed the planet in such a way that the Earth's ecosystems combined with its chemical, geological and physical processes currently control the environmental conditions on the planet. After the establishment of organisms on Earth, the ecosystems became a major determinant of the environmental conditions. Examples of the numerous and complex interactions between the Solar System, Earth, its atmosphere, and its ecosystems are examined in the following paragraphs in the context of the long geological history of Earth (see Sect. 1.5.4), and more specifically the three phases of the Earth's history illustrated in Fig. 1.11. These are the Mineral, Cradle and Living Earth, which occurred from 4.6 to 4.0, 4.0 to 2.5, and 2.5 to 0 billion years ago, respectively.

Some of the main determinants of the Earth's atmosphere are the relatively large mass of the planet and its distance from the Sun (see Sect. 2.3.4), and these were set at the time of the planet's formation 4.6 billion years ago (*Mineral Earth*). Early in the Earth's history (Hadean eon, 4.6–4.0 billion years ago; *Cradle Earth*), some processes removed a large part of the carbon initially very abundant as CO_2 in the Earth's atmosphere (see Sect. 2.2.4).

Organisms appeared on Earth during the next, very long Archaean eon (4.0–2.5 billion years ago), which was very rich in events and processes that linked together planet Earth, its atmosphere and its ecosystems.

Some of these events and processes were unique to the Archaean, for example the early and continuing interaction between the atmosphere and the ocean on Earth contributed to the overall retention of water on the planet by preventing hydrogen escape, which was a key condition for organisms and ecosystems on Earth (see Sect. 2.1.3), and the possibility that production of CH_4 by microorganisms during that period of faint young Sun may have contributed to create the temperatures that allowed organisms to begin their occupation of Earth (see Sect. 2.1.7).

Other events began in the Archaean but continued in following eons, until now. Examples for *water and CO_2* are: the essential role of the atmosphere and liquid water for the establishment and expansion ecosystems on Earth (see Sect. 2.1.3); CO_2 and H_2O started to be used by organisms to synthesize organic matter, and conversely CO_2 and water were released into the environment by respiration (see Sect. 2.1.6); calcification used C and O atoms from dissolved HCO_3^- ions together with dissolved calcium ions to produce $CaCO_3$ (see Sect. 2.1.6); organisms contributed to store the carbon from excess atmospheric CO_2 in geological reservoirs using photosynthesis, chemosynthesis and calcification (see Sect. 2.1.6); volcanic eruptions were the main source of atmospheric CO_2 (see Sect. 2.1.1); and erosion of continental rocks moved carbon from the continents to the bottom of oceans (see Sect. 2.1.6). Examples for *nitrogen* are: some microorganisms fixed N from environmental N_2 and thus supplied N to food webs, and others decomposed organic matter and returned N to the environment (see Sect. 2.1.4); and ultraviolet radiation and chemical reactions in the atmosphere destroyed N_2O (see Sect. 2.1.4). Examples for the *greenhouse effect* are: atmospheric gases such as H_2O (vapour), CO_2, CH_4 and N_2O, contributed to the greenhouse effect (see Sect. 2.1); the natural greenhouse effect is essential to life on Earth, that is, without it the planet's temperature would be much too low (see Sect. 2.4.1). One major event for the future atmosphere was the "invention" of O_2-producing photosynthesis by cyanobacteria (see Sect. 2.1.4).

The next eon—the long Proterozoic (2.5-0.5 billion years ago)—began with increasing concentrations of O_2 in the atmosphere, resulting from O_2-producing photosynthesis (see Sect. 2.1.5). This was the beginning of the *Living Earth* phase. The innovations of this phase, which all continue to this day, include, for *O_2 and CO_2*: the use of C and O atoms from CO_2 molecules and H atoms from H_2O

molecules by O_2-producing photosynthetic organisms to make up organic matter (see Sects. 2.1.5 and 2.1.6); O_2-producing photosynthesis released into the environment the O_2 resulting from the splitting of H_2O (see Sect. 2.1.5); conversely, O_2 respiration used O_2 and released CO_2 into the environment (see Sect. 2.1.5); organic carbon was stored in the long term in marine sediments and continental soils (see Sect. 2.1.6); natural combustion of organic matter (forest fires) also returned CO_2 to the environment (see Sect. 2.1.6). The innovations also include, for *O_3 and atmospheric water vapour*: the formation of O_3 by the action of solar UV radiation in the upper atmosphere, thus creating the O_3 layer that protects organisms on continents and in surface waters (see Sect. 2.1.5); and the redistribution of water on Earth by the atmospheric water vapour, which sustains the planet-wide distribution of ecosystems (see Sect. 2.1.3). The latter two processes allowed the occupation of continental surfaces by plants and animals, and their subsequent diversification during the next, Phanerozoic eon, in which we are still living (see Sect. 2.1.5).

The complex processes involving Earth, its atmosphere and its ecosystems, which had evolved during the billions of years of the Archaean and Proterozoic eons and continued during the Phanerozoic eon until present, had created an overall equilibrium in the exchanges of materials end energy among the living and non-living compartments of Earth. These allowed organisms to take over the planet. This continued until the Industrial Revolution, which began the second part of the eighteenth century, caused durable changes of a magnitude equivalent to those of the most spectacular geological events over such a short period. For example, human societies release into the atmosphere in a single year as much CO_2 as all the Earth's volcanoes in several decades (see Sect. 2.1.6). The next paragraph addresses the very last epoch of the Earth's geological history, called the Anthropocene, which started about 300 years ago (see Sect. 1.5.4).

During the few hundred years of the Anthropocene epoch, the massive input of CO_2 in the atmosphere by human societies caused global warming, which started to build up with the beginning of the Industrial Age (see Sects. 2.1.6 and 2.4.1). Since then, human societies have released increasing amounts of CO_2 in the atmosphere by burning organic matter and calcining $CaCO_3$, causing the ongoing increase in atmospheric CO_2 (see Sect. 2.1.6). A collateral damage of the natural storage in the ocean of part of the atmospheric CO_2 produced by human societies is ocean acidification (see Sect. 2.1.6). Not only CO_2 but other atmospheric gases are involved in global warming. One of them is methane (CH_4), which either escapes from frozen marine hydrates or is produced by archaea decomposing organic matter in thawed permafrost, which processes may intensify with the increase in temperature (see Sect. 2.1.7). Another of the greenhouse gases is N_2O, which is released to the atmosphere by human activities (see Sect. 2.1.4). This way, human societies are interfering with natural dynamic equilibria Earth had evolved over almost 4 billion years.

Figure Credits

Fig. 2.1 https://commons.wikimedia.org/wiki/File:White_Cliffs_of_Dover_02.JPG by Immanuel Giel https://commons.wikimedia.org/wiki/User:Immanuel_Giel, used under CC BY-SA 3.0.

Fig. 2.2a https://commons.wikimedia.org/wiki/File:NASA-Apollo8-Dec24-Earthrise. jpg by NASA http://www.hq.nasa.gov/office/pao/History/alsj/a410/AS8-14-2383HR. jpg, in the public domain.

Fig. 2.2b https://commons.wikimedia.org/wiki/File:The_Earth_seen_from_Apollo_17. jpg by NASA https://web.archive.org/web/20160112123725/http://grin.hq.nasa.gov/ ABSTRACTS/GPN-2000-001138.html (image link) and https://www.nasa.gov/multi-media/imagegallery/image_feature_329.html, in the public domain.

Fig. 2.3a http://www.cnsa.gov.cn/n6758823/n6758838/c6809578/content.html by the Chinese National Space Administration (CNSA).

Fig. 2.3b http://electro.ntsomz.ru/en/. Rights holder: Roscosmos State Corporation; the data were processed by Russian space systems NTZ OMZ JSC, Electro-L satellite, 29 June 2020, 11:00. With permission from the Roscosmos State Corporation.

Fig. 2.4 This work, Fig. 2.4, is a derivative of https://commons.wikimedia.org/ wiki/File:Atmosphere_layers.svg by NOAA & User:Mysid https://en.wikipedia. org/wiki/User:Mysid, in the public domain. I, Mohamed Khamla, release this work in the public domain.

Fig. 2.5 Original. Fig. 2.5 is licensed under CC BY-SA 4.0 by Philippe Bertrand, Louis Legendre and Mohamed Khamla.

Fig. 2.6a Modified after Figure FAQ 1.3 of Le Treut et al. (2007). With permission from the Intergovernmenal Panel on Climate Change (IPCC).

Fig. 2.6b https://commons.wikimedia.org/wiki/File:EM_spectrum.svg, used under GNU FDL, CC BY-SA 3.0, CC BY-SA 2.5, CC BY-SA 2.0, CC BY-SA 1.0.

Fig. 2.7 This work, Fig. 2.7, is a derivative of https://commons.wikimedia.org/ wiki/File:Windchill_effect_en.svg by RicHard-59, https://commons.wikimedia. org/wiki/User:RicHard-59, used under CC BY-SA 3.0. Fig. 2.7 is licensed under CC BY-SA 3.0 by Mohamed Khamla.

Reference

Le Treut H et al (eds) Climate change 2007: The physical science basis. Contribution of Working Group I to the Fourth assessment report of the Intergovernmental Panel on Climate Change. Cambridge University Press, Cambridge, UK and New York, NY, USA [Can be downloaded free of charge from the IPCC website via https://www.ipcc.ch/]

Further Reading

Catling DC, Zahnle KJ (2009) The planetary air leak. Sci Am 300:36–43. http://faculty.washing-ton.edu/dcatling/Catling2009_SciAm.pdf. May 2009

Conrad PG (2016) Planetary habitability. Cambridge Astrobiology. Cambridge University Press, Cambridge

Fabian P, Dameris M (2014) Ozone in the atmosphere. Basic principles, natural and human impacts, Springer, Berlin Heidelberg

Fortov VE (2016) Extreme states of matter: Springer series in materials sciences, vol 2016. Springer, Cham

Gargaud M et al (eds) (2006) From Suns to life: A chronological approach to the history of life on Earth. Springer, New York, NY

Glikson AY, Groves C (2016) Climate, fire and human evolution. The deep time dimensions of the anthropocene, Modern Approaches in Solid Earth Sciences, vol 10. Springer, Cham

Kasting JF (2012) How to find a habitable planet?. Princeton University Press, Princeton

Legendre, L (2014) Pelagic marine ecosystems and biogeochemical cycles. In: Monaco A, Prouzet P (eds) Ecosystem sustainability and global change. ISTE, London, and John Wiley and Sons, Hoboken, NJ, pp 37–75

Lissauer JJ, de Pater I (2019) Fundamental planetary science: Physics, chemistry and habitability, Updated edn. Cambridge University Press, Cambridge

Schulze-Makuch D, Irwin LN (2018) Life in the Universe, 3rd edn. Springer-Praxis Books, Springer, Cham

Visconti G (2016) Fundamentals of physics and chemistry of the atmosphere, 2nd edn. Springer, Cham

Wallace JM, Hobbs PV (2006) Atmospheric science: An introductory survey. International Geophysics, 2nd edn. Elsevier, Amsterdam

Chapter 3
Thermal Habitability: Connection with the Earth's Motion Around the Sun

3.1 Life and Temperature

3.1.1 Thermal Limits for Growth and Survival

Organisms are very sensitive to temperature, and their successful occupation of Earth has largely been determined by the temperature conditions of the planet. It was explained in Chapter 1 that four key features of organisms on Earth are: the genetic information encoded in nucleic acids (the software); the structural and functional elements of the cells, and the body structure in the case of multicellular organisms (the hardware); the flux of energy; and the replication, or reproduction (see Sect. 1.1.3 and Information Box 1.1). These four features are all affected by temperature. This chapter examines the conditions that keep the Earth's temperature within a range generally suitable for organisms, and reciprocal effects of ecosystems on temperature.

This section and the next are devoted to the thermal limits for growth and survival, and the temperatures on Earth, Venus, Mars and the Moon. The next sections examine in turn: the sources of energy at the surface of Earth (Sect. 3.2); the solar radiation flux that reaches Earth (Sect. 3.3); the fate of the solar radiation absorbed by the Earth System (Sect. 3.4); the effects of the absorbed solar radiation on the climate system (Sect. 3.5); how the atmosphere and the Earth's distance from the Sun contribute to set the range of temperatures that occur on Earth (Sect. 3.6); and the long-term controls on the Earth's temperature (Sect. 3.7). The chapter ends with a summary of key points concerning the interactions between the Solar System, Earth, its temperature, and its ecosystems (Sect. 3.8). The textbook of Wells (2012) provides further reading on several of the topics examined in this chapter.

© Springer Nature Switzerland AG 2021   89
P. Bertrand and L. Legendre, *Earth, Our Living Planet*, The Frontiers Collection,
https://doi.org/10.1007/978-3-030-67773-2_3

Table 3.1 Thermal limits for growth and for survival for the three domains of life: bacteria, archaea, and eukarya. Values from Clarke (2014)

Temperature limits (°C)	Archaea	Bacteria	Eukarya
High-temperature limit for growth	+122	+100	From about +45 to +65
Low-temperature limit for growth	About −16.5	About −20	From 0 to about −20
High-temperature limit for survival	Less than +130	*No data*	From about +70 to about +90
Low-temperature limit for survival	*No data*	About −196	From −2 to about −196

There is a major difference between the thermal limits for the growth and those for the survival of organisms. *Growth* refers to the completion of the life cycle from one cell division to the next, or for complex organisms from conception to reproduction and death, whereas *survival* is the fact for an organism of continuing to live. These two temperatures vary among species. Within the range of *temperatures for growth*, a species can continue to exist generation after generation, whereas the *temperatures for survival* are the lower and upper boundaries beyond which an organism dies and a species may become extinct. The upper temperature limit for the survival of organisms is determined by the thermal stability of their biological molecules, which are destroyed by high temperature. The lower limit for survival is set by the physical properties of water, especially the temperature at which water freezes because ice crystals can damage the cells. However as seen a few paragraphs below, some types of cells and multicellular organisms can survive temperatures lower than freezing by dehydrating and thus avoid damages by ice crystals.

On Earth, some organisms are found in high-temperature environments such as places where almost boiling water emerges from the ground, for example in hot springs and geysers on continents and hydrothermal vents on the seafloor. Other organisms are found in low-temperature continental environments at high latitudes in winter and at high altitudes in mountains. Some organisms are also found in the sea ice, where they inhabit channels filled with brine that can reach −20 °C before without freezing. This brine results from the exclusion of sea salts from the growing sea ice, whose crystal structure is made of pure water.

Observations made in different marine and continental environments indicate that the three domains of life (see Information Box 3.1) do not have the same thermal limits for growth and survival (Table 3.1). In the following two paragraphs, we consider first the highest temperature limits, and next the lowest.

Information Box 3.1 The Three Domains of Life: Archaea, Bacteria and Eukarya; Prokaryotes and Eukaryotes The organisms on Earth belong to three broad groups, called *domains of life*. These are the archaea, the bacteria and the eukarya (or eukaryotes), which coexist successfully on the planet. The *archaea* and the *bacteria* are single-celled (unicellular) microorganisms that do not have a nucleus or membrane-bound organelles in their cells. Such organisms are called *prokaryotes*, which name means "before a nucleus". The names *archaea* and *bacteria* come from Greek words meaning "archaic" and "rod", respectively. The cells of *eukaryotes*, which name means "true nucleus", have membrane-bound organelles, especially a nucleus. These organisms can be unicellular or multicellular. The unicellular eukaryotes are called *protists* or *protozoa*. The multicellular eukaryotes include the *plants* and the *animals*, to which the human beings belong. The eukaryotes also include the *fungi*, some of which are unicellular (for example, yeasts and moulds) and others are multicellular (for example, mushrooms). The amazing variety of prokaryotic and eukaryotic plankton is remarkably displayed and beautifully illustrated in the book of Sardet (2015), recommended as further reading.

Nature does not always completely follow the strict definitions created by biologists. This is the case of multi-layered sheets of prokaryotes called *microbial mats* and of *stromatolites* (see Sect. 6.4.2), which are sometimes considered to be multicellular organisms. However, it is generally recognized that all multicellular organisms belong to the eukaryotes.

The three domains of life are very different from each other, but also show some paired similarities. On the one hand, archaea have in their cell membranes a type of lipids different from that of bacteria and eukaryotes, whereas on the other hand, the enzymes of archaea involved in gene transcription and translation (see Sects. 6.2.4 and 6.2.5) are more similar to the enzymes of eukaryotes than those of bacteria. Hence in some respects, bacteria and eukaryotes are similar to each other and different from archaea, and in other respects, archaea and eukaryotes are similar to each other and different from bacteria.

For archaea and bacteria, the highest temperatures for both growth and survival were observed for strains coming from a marine hydrothermal field and a terrestrial hot spring respectively. For eukarya, the highest temperatures for growth were observed for various groups organisms that included unicellular algae from hot springs, yeasts from terrestrial geothermal sites, a grass growing close to hot springs, and nematodes (a type of worms) in compost heaps. The same compost-heap nematodes exhibited the highest temperatures for survival, which were also observed for polychaetes (a type of marine worms) living in hydrothermal vents on the seafloor.

The lowest temperatures for the growth and survival of archaea were determined by measuring the production of methane in permafrost cores since only archaea produce methane (see Information Box 5.2), and for bacteria by measuring respiration in permafrost cores. For eukarya, the lowest temperatures for growth is assumed to be above -20 °C as no moulds have been observed to grow on food frozen at temperatures lower than -20 °C, and some Antarctic lichens seem to be active at -10 °C. The lowest temperature for survival of eukarya can be extremely low for some of these organisms, which can dehydrate and thus escape cellular damage caused by the freezing of their intracellular water. These organisms include yeasts, lichens and also some terrestrial invertebrates.

The values in Table 3.1 provide overall ranges of temperatures presently known for the different types of organisms. However for a given species, the actual range of extreme temperatures is much narrower than the overall ranges reported in Table 3.1, and is generally limited to few tens of degrees at most. One strategy used by some species to survive extreme temperatures is the production of resting spores or stages, and the species able to do this include bacteria, fungi, eukaryotic algae and the brine shrimp *Artemia salina*. The survival forms include plant seeds, tardigrade tuns (see Sect. 4.1.4), dehydrated nematodes, and *Artemia* dormant eggs (known as *cysts*), which all contain the software (DNA) and the hardware (proteins, membranes) of organisms, but exhibit no energy flux and are thus not alive as such (see Information Box 1.1). However, these survival forms can become alive when the conditions are suitable. Another example is the freezing of gametes and embryos of vertebrates (including humans), which are now commonly preserved in liquid nitrogen (-196 °C) before being thawed and used for assisted reproduction. The book of Clarke (2017) provides further reading on the effects of temperature on organisms and ecosystems.

3.1.2   Temperatures on Earth, its Two Sister Planets Venus and Mars, and the Moon

Given the previous section, organisms could establish themselves durably on Earth and ecosystems to thrive there because of the relatively narrow range of temperature variations at the surface of the planet, except in some specific regions. These variations include changes in temperature at a given location within the course of one day, which do not generally exceed 10–20° C, and during one year, which do not generally reach extremely low or extremely high values except in geographically limited regions (Table 3.2). These exceptions include some tropical or mid-latitude deserts, where temperatures can range from more than +50 °C during the day to less than 0 °C during the night and where there are no or only few organisms, and some polar areas such as the Antarctic ice cap.

The diurnal or annual ranges of temperatures on Earth are suitable for organisms, but not on its sister planets Venus and Mars, or on the Moon (Table 3.2).

Table 3.2 Surface temperature and atmospheric pressure on Venus, Earth, Mars and the Moon

Surface temperature and pressure	Venus	Earth	Mars	The Moon[a]
Average (overall) surface temp. (°C)	+464	+15	−63	−20
Minimum surface temp. (°C)	+446	−89	−133	−173
Maximum surface temp. (°C)	+490	+57	+27	+127
Diurnal range of surface temp. (°C)	~0	10 to 20	Up to > 100	~300
Atmospheric pressure (atm)	93	1	0.006	0

[a] Surface temperature usually refers to a value measured close to the surface of an astronomical body in one of its fluid envelopes, which can be gaseous or liquid. Given that temperature cannot be measured in the Moon' tenuous exosphere (see Information Box 2.5), its surface temperature is that of its solid surface

On Venus, the temperature at ground level is uniformly around +464 °C, which is too hot for biological molecules to exist there, but the high atmosphere of that planet has temperatures less than +100 °C and it has been speculated that it could, in theory, contain organisms. On Mars, the temperatures are generally too low for the growth of organisms, and the diurnal range of temperatures is so wide that temperatures suitable for organisms do not last for very long at a given location during the course of one day. However, it is possible, although not very likely, that some life forms exist in the soil of Mars, and the continuing exploration of the planet by robots may confirm (or not) this exciting possibility in the coming years (see Sect. 5.5.3). On the Moon, the extreme variability of temperature at any given location, from deep-freezing cold to scorching hot, precludes the existence of life forms there except perhaps in a few secluded areas protected from thermal extremes.

Contrary to Venus, Mars and the Moon, organisms thrive on Earth. The next sections of this chapter examine key characteristics of Earth and the Solar System that contribute to keep the temperatures of the atmosphere and the oceans of the planet at levels generally suitable for organisms, and have done so since billions of years.

This chapter examines both the conditions that make the Earth's temperature generally suitable for organisms, which determines the thermal habitability of the planet (see Information Box 2.2), and the effects of ecosystems on temperature. The next chapter (Chapter 4) will deal with the variations in the Earth's overall habitability at timescales ranging from one day to more than one hundred million years.

3.2 The Sources of Energy at the Surface of Earth

3.2.1 The Earth's Radiation Budget

The two main sources of energy at the surface of Earth are the radiation the planet receives from the Sun, and for a very small fraction, heat produced inside Earth.

The temperature at the surface of Earth is mostly determined by the level of the solar radiation reaching the surface of the planet. One example is the succession of worldwide cycles of global cooling and warming that have been occurring since 2.16 million years, called *Quaternary Ice Age* (see Sect. 4.5.2). These cycles largely reflect a cyclical change in the solar radiation flux that reached Earth (Fig. 11.1b, last 800,000 years). Over 100,000 years, the change in this flux has an average magnitude of 15–20% at 60°N in June, and is caused by astronomical phenomena (see Sect. 1.3.4). Any change in the energy Earth receives from the Sun is accompanied by an "almost instantaneous" change in the energy the planet radiates back to space (see Information Box 2.7) that takes place within about 50–100 years (which is instantaneous at the geological timescale).

The Earth's long-term radiation budget has been balanced over billions of years, as shown by the geological and biological evidence that water in liquid form has been present on Earth for more than four billion years. Indeed, if there had been a prolonged imbalance in the radiation budget, Earth would have cooled off or warmed up rapidly, reaching temperature conditions that would have prevented the presence of liquid water. In the case of a prolonged global cooling or warming of Earth, all its liquid water would have become solid ice or gaseous water vapour, respectively, which never happened. Moreover in the case of a prolonged warming, the atmosphere would have become extremely concentrated in water vapour, which would have increased the photochemical dissociation of water molecules (H_2O) into their constituent H and O in the upper atmosphere, and the latter would have in turn increased the rate of loss of the Earth's hydrogen towards space by kinetic escape (see Sect. 5.4.2). Given that hydrogen is a key constituent of water, the escape of hydrogen would have caused Earth to progressively lose its water, which did not happen. This means that the planet never experienced strong warming during billions of years. Hence the Earth's radiation budget has been balanced in the long term, but there were long periods of global cooling or global warming because of changes in the incoming solar radiation or longer-term changes within the Earth System (see Sects. 4.5.2 and 4.5.6, respectively).

The mechanism that maintained the long-term balance in the Earth's radiation budget was as follows: when the planet warmed, it radiated more heat toward space, which re-established the energy balance; and when the planet cooled, it radiated less heat, which also re-established the energy balance. Measurements of the present energy budget of Earth at the short timescale of a few years or a few decades show that the planet absorbs more energy from the Sun that it radiates back to space. The reason for this temporary imbalance is the increasing amount of anthropogenic CO_2 and other greenhouse gases in the atmosphere, which retains more energy in the climate system than is radiated to space (see Information Box 2.7). This causes an imbalance of 0.85 W m^{-2}, which is very small relative to the whole incoming solar flux of 340 W m^{-2} (see Information Box 3.2), but is enough to significantly warm the climate system, and thus cause the ongoing global warming (Fig. 11.1b, since year 1860). The present imbalance between the fluxes of incoming and radiated energy is temporary because

---

Information Box 3.2 Solar Radiation Flux at the Top of the Earth's Atmosphere: Energy and Power Physicists and climatologists measure *energy* in joules. Another unit of energy is the calorie, which is used for food, and 1 cal $=4.2$ joules. In this text, we express energy in joules (symbol: J). The Sun, as the other stars in the Universe, continually radiates energy coming from the fusion of atoms in its core, mostly hydrogen atoms. This phenomenon is called *nuclear fusion*. It has been used in the twentieth century to make thermonuclear weapons (hydrogen bombs), and research is conducted with the objective of domesticating nuclear fusion to produce electricity in the future. All Solar System bodies, including Earth, receive some of the energy radiated by the Sun, and the amount each body receives depends on its distance from our star. The flux of solar energy at the top of the Earth's upper atmosphere, which is called *radiation flux*, is measured in units of joules per square metre and per second ($J\ m^{-2}\ s^{-1}$). It is the amount of energy received by one square metre of the upper atmosphere during one second.

In physics, the amount of energy consumed per unit time to do work is called *power*. Given that the unit of power is the watt (symbol: W) and $1\ W = 1\ J\ s^{-1}$, the radiation flux at the top of the Earth's atmosphere is generally expressed in units of $W\ m^{-2}$ ($1\ W\ m^{-2} = 1\ J\ m^{-2}\ s^{-1}$).

The average value of the radiation flux coming from the Sun is 1,361 $W\ m^{-2}$. The solar radiation flux that *reaches the top of the Earth's atmosphere has a circular area*, but given that the atmosphere around the planet is a sphere around the planet, the solar radiation flux *is received by a spherical surface*. The circular area of the incoming flux is calculated using the geometrical formula for the surface area of a circle, πR^2, where R is the radius of the atmosphere around Earth. The area of the receiving surface is calculated using the geometrical formula for the surface area of a sphere, $4\pi R^2$. It follows that the incoming radiation flux is distributed on an atmospheric area $4\pi R^2/\pi R^2 = 4$ times larger than the area of the solar flux, with the consequence that its average value at the top of the Earth's atmosphere is 1,361 $W\ m^{-2}/4 =$ approximately 340 $W\ m^{-2}$.

---

the increase in temperature will eventually cause an increase in the flux of energy radiated to space. As a consequence, when the atmospheric content in greenhouse gases is stabilized, the flux of energy radiated to space by Earth will balance again the flux received from the Sun, but the global temperature will be higher than at present. However as long as the greenhouse gases continue to increase in the atmosphere, the energy imbalance will persist or increase, and the Earth's temperature will continue to rise. This will affect the ecosystems, and in return many human activities (see Sects. 11.1.1–11.1.3).

The Earth's albedo has a value of 0.30 (see Information Box 2.9 and Table 4.2), meaning that 30% of the 340 W m^{-2} received by Earth, that is, around 105 W m^{-2} are reflected back to space by various processes. The remaining 235 W m^{-2} are absorbed by the atmosphere, and distributed among the various components of the Earth System, whose five main components are liquid water, the atmosphere, the ice, the continental surfaces (soil and rocks), and the ecosystems. The 235 W m^{-2} from the Sun together with less than 1 W m^{-2} coming from inside Earth, called *geothermal energy* (see Sect. 3.2.3), fuel the whole Earth System, including the functioning of organisms and ecosystems. After being used in the Earth System, the 235 W m^{-2} are radiated back to space as infrared radiation (Fig. 3.1).

The Earth System Pays its Energy Debt to the Universe in a Different Currency It follows from the previous paragraphs that the same amount of energy enters and leaves the Earth System, but the quality of the incoming and outgoing energy is different. This is because the Earth System, like all dissipative systems, absorbs usable energy and radiates back the same amount of energy in a less usable form, in this case heat. The significance of the difference between the quantity and the quality (that is, the wavelengths) of the energy absorbed and radiated by Earth is explained in the next section. Readers interested in a more detailed radiation budget than Fig. 3.1 could look at the *global mean energy budget* of Earth in a report from the IPCC or one of the many versions of the budget available on the Internet.

3.2.2   Different Fates of Materials and Energy

Nature uses repeatedly the same materials. In fact, all particles that make up Earth have been used and reused billions of times since the formation of the Solar System, and many of these particles had been forged in faraway stars in the Universe billions of years before the formation of the Sun and Earth (see Sect. 1.4.2). Although materials continually move within the Solar System and are constantly exchanged with the outer Universe, the Solar System and each of its components are essentially finite, and their constituent materials are thus perpetually recycled. For example, most of the atoms that make up the present Earth's organisms have been continually reused on the planet since its formation 4.6 billion years ago. In modern physics, the classical principle of conservation of mass is replaced by conservation of mass-energy as a small fraction of mass is converted into energy during radioactive decay, for example in the Earth's crust and mantle (see Sect. 3.2.3), and nuclear fusion, for example in the Sun (see Information Box 3.2).

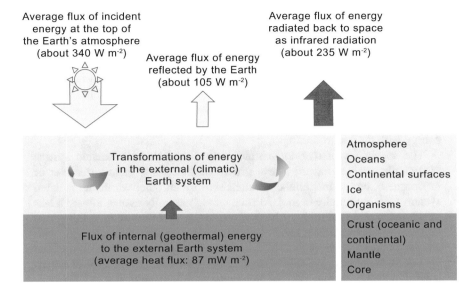

Fig. 3.1 Radiation budget of the Earth System. Most of the energy that fuels the Earth System comes from the Sun, and the geothermal component from inside Earth is very small. Credits at the end of the chapter

Contrary to the materials that make up physical systems, the energy that flows through them, including the organisms, cannot be reused. Indeed, all energy (mechanical, chemical, electrical, etc.) is progressively degraded into heat, from where it cannot be recovered for further use. However even if energy is dissipated, it is not lost as specified by the principle of conservation of energy, which is one of the bases of classical physics. The next paragraph explains what happens to the solar energy received by Earth by reference to the spectrum of electromagnetic radiations (Fig. 2.4b).

It is explained in Chapter 2 (see Information Box 2.7) that on the one hand, the Sun emits its radiation at short wavelengths, which are predominately in the visible and the ultraviolet (UV) parts of the electromagnetic spectrum (200–700 nm), and on the other hand, Earth radiates energy back to space at much longer wavelengths, primarily in the infrared (IR) part of the spectrum (longer than 700 nm, which is heat). It is also explained there that Earth radiates back to space, on average, about the same amount of energy as it absorbs, but at different wavelengths.

It follows from the previous paragraph that, on average, there is no, or little difference between the *quantity* of energy Earth absorbs and radiates to space, but there is a major difference in the *quality* of the two fluxes of energy. Indeed, the energy received from the Sun belongs to the short wavelengths and can therefore be used to do work, whereas the energy radiated back to space is long-wavelength heat, which cannot be used again. As it passes through the Earth System, solar energy is used in the atmosphere, oceans, glaciers, continental soil and rocks, and

Information Box 3.3 Systems and Entropy The branch of physics that deals with energy is called *thermodynamics*. The specialists of thermodynamics distinguish between closed and open systems. A *system* is a group of interdependent parts that form a unified whole. Examples of systems are: airplanes, organisms, and the Solar System. *Isolated system* do not, in theory, exchange any matter or energy with their surroundings, whereas *open systems*, which are those actually found in Nature, exchange matter and/or energy with their surroundings.

The specialists of thermodynamics have found that an isolated system continually evolves toward a less structured state, or a greater disorder or randomness, whose magnitude is called *entropy*. In other words, an isolated system with an initially organized state would naturally evolve toward a less structured state. Contrary to the isolated systems, some open systems can temporarily increase their level of organization by means of exchanges with the surrounding environment. The increase in organization is accompanied by a decrease in entropy. This is the case of organisms, whose level of organization increases, or at least does not decrease, as long as they stay alive.

It was explained previously (see Information Box 1.1) that organisms maintain their high level of internal organization through a continual intake and dissipation of energy. The above considerations on open systems help understand the role of the flux of energy through organisms, namely that the flux of energy keeps them alive, and when the flux stops, they die.

organisms. The use of energy by organisms and its effects on the remainder of the Earth System is central to this book. The various uses of the solar energy on the planet progressively degrade it into long-wavelength heat, which is radiated back to space.

Earth and its components, including organisms, are open systems, which use outside energy to maintain and even increase their organization, or equivalently decrease their entropy (see Information Box 3.3). The outside energy used by organisms on Earth is mostly solar, but it is also partly geothermal. For example some ecosystems, such as those that grow in hydrothermal environments in deep oceans, thrive on geothermal energy. Marine and terrestrial hydrothermal ecosystems do not necessarily use the geothermal energy itself, but they extract chemical energy from compounds carried by the hydrothermal fluid flowing up from deep in the ocean crust, whose buoyancy results from geothermal heating.

The difference in the *quality* of the energy that Earth absorbs and radiates back toward space reflects the success of the Earth System in counteracting the natural tendency of natural systems to continually increase their entropy (see Information Box 3.3). The textbook of Rapp (2014) provides further reading on the Earth's radiation budget and climate change.

3.2.3   Geothermal Energy

It was mentioned above that some of the energy that fuels the Earth System is not of solar origin, but comes instead from the interior of the planet (see Sect. 3.2.1). Most of this energy is in the form of heat, hence its name *geothermal energy*, where *geothermal* means heat from Earth. This energy has two main sources: the primordial heat remaining from the formation of the planet (see Sect. 1.3.1), and the ongoing disintegration of nuclei of radioactive elements in the Earth's crust and underlying mantle. The first source comes from the fact that young Earth was very hot, and this *initial (or primordial) heat* continues to progressively flow from inside the planet as it cools down. This heat resulted from the conversion of the kinetic energy of the impacting bodies that formed the planet (see Sect. 1.5.1) and the friction of the materials that moved inside the young Earth during the planetary differentiation under the influence of gravity (Sect. 6.6.2). The second source of geothermal heat is *radioactivity*, that is, the ongoing disintegration of radioactive isotopes that had accumulated in the crust and the underlying mantle when Earth formed (radioactive isotopes are explained in Information Box 5.6). Four radioactive isotopes are responsible for most of this heat: uranium-238, uranium-235, thorium-232, and potassium-40.

The outer crust of Earth is solid. This is not the case of the underlying mantle, whose upper part can flow in the long term because it is very hot, but the lower part is also solid because of the very high pressure. Deeper into Earth, the centre of the globe is made of a solid inner core, outside which there is a liquid outer core in contact with the mantle (Fig. 6.11). In some areas of the mantle, the material becomes very hot because of the heat that radiates from the core, and the resulting buoyancy makes the material rise, whereas in other areas, cooler (and thus not buoyant) material sinks. The combination of rising and sinking movement is called *convection*, and it can be easily observed in a fluid heated in a pan. Convection probably also occurs in the liquid outer core, where it is hypothesised that convection currents caused by the release of heat from the inner core are responsible for the Earth's magnetic field. By this mechanism, thermal energy of the core, combined with the rotation of Earth and because of the presence of the solid core, is transformed into electromagnetic energy (see Sect. 8.3.1).

The manifestation of the flux of energy from the interior of Earth is multifaceted, and geothermal heat does flow upwards by several mechanisms that include conduction through the crust, convection in the mantle, circulation of water in the crust, and volcanic activity. Because of the geothermal upward *heat flux*, temperature increases by 20 °C or more per kilometre down into the ground on most of Earth. Depending on the thickness of the Earth's crust, the heat flux is dominated by the heating of the bottom of the crust by convection in the mantle (as seems to be the case for the oceanic crust, which is thin) or radioactive activity within the crust (as seems to be the case for the continental crust, which is thick). Mantle convection also contributes to the movements of tectonic plates, which cause earthquakes and thus the release of mechanical energy (see Sect. 7.2.1).

The *hydrothermal circulation of water* within the Earth's crust transports heat upwards, and the hot, buoyant water comes to the surface in hot springs and geysers on continents and through hydrothermal vents in oceans. The hydrothermal water contains not only heat but also very high concentrations of minerals, which fuel terrestrial and marine hydrothermal ecosystems as indicated a few paragraphs above (see Sect. 3.2.2). *Volcanic activity* also transports heat, solid materials, and gases including CO_2. The latter is a major greenhouse gas, and its emission by volcanoes has a key effect on the Earth's temperature (see Information Box 2.7). Geothermal energy thus affects the Earth's crust, ocean, atmosphere and ecosystems in a number of ways. The article of Holden et al. (2012) provides further reading on hydrothermal vent systems.

The total amount of geothermal energy is $87 \, mW \, m^{-2}$ (1 mW is 1/1000 of a watt), which is less than 0.04% of the $235 \, W \, m^{-2}$ of solar energy that reach the surface of Earth. Hence geothermal energy does not play a significant role in the global heat budget. However, the heat and minerals that come from inside the planet are locally important for some ecosystems, which often thrive in the vicinity of hydrothermal sources. In addition, the mechanical and electromagnetic energy generated by heat inside the planet are important for the Earth System, including its living component, as explained in Chapter 7 for tectonic activity and Chapter 8 for the magnetic field. The book of Boden (2017) provides further reading on the geological, environmental and technological aspects of geothermal energy.

In summary, approximately one third of the flux of the incoming solar energy is reflected back to space and two thirds are absorbed by the atmosphere, the oceans, the continental surfaces, the ice, and the photosynthetic organisms, which radiate it back toward space in the form of long-wavelength energy (heat). Earth absorbs and radiates about the same quantity of energy, but the quality (wavelength) of the absorbed and radiated energy is different. This change in quality reflects the fact that the energy absorbed by Earth System is used to maintain the organization of the climate system, which includes the ecosystems (see Sect. 3.5.1).

3.3   Solar Radiation Flux Reaching Earth

3.3.1   Angle of Incidence of the Radiation Flux

The angle between the radiation that strikes a surface (called *incident radiation*) and that surface—here the radiation flux from the Sun and the top of the atmosphere—is called *angle of incidence*. More specifically, it is the angle that incident radiation makes with a line perpendicular to the top of the atmosphere at the point of incidence. In other words, the angle of incidence is 0 when the radiation is perpendicular to the surface, and maximum (close to 90 °) when the radiation grazes the surface.

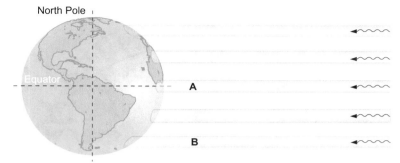

Equal amounts of solar radiation are distributed over a smaller area in **A** than **B**

Fig. 3.2   Effect of latitude on the amount of solar radiation at the Earth's surface at the equinox. Credits at the end of the chapter

The angle of incidence has a major effect on the radiation reaching Earth because as the angle increases the radiation is spread over an increasingly larger area, and the radiation flux (W m^{-2}) correspondingly decreases. When the radiation flux is vertical relative to the top of the atmosphere, its value corresponds to that of the incoming flux from the Sun, which is 1,361 W m^{-2} (see Information Box 3.2). Conversely when the radiation flux grazes the top of the Earth's atmosphere, its value is 0 W m^{-2}. Hence depending on the latitude and on the time of the day and of the year, the actual value of the radiation flux at a given point at the top of the Earth's atmosphere and at a given time varies between 1,361 and 0 W m^{-2}.

The latitude and the time of the year determine the angle with which Earth faces the Sun, and this affects the magnitude of the solar radiation at the surface of Earth at different latitudes. Figure 3.2 represents the situation at the equinox, at which time of the year the planet faces the Sun with no angle (Fig. 3.3). Because of the curvature of the Earth's surface, the same amount of incoming radiation from the Sun is spread over a larger area at a high latitude (region B in Fig. 3.2) than at the equator (region A), and the amount of solar radiation per unit area at the Earth's surface is thus smaller in B than A. This has a deep effect on the spatio-temporal variations in temperature at the Earth's surface, and thus the suitability of various environments for ecosystems.

3.3.2   *Latitudinal and Seasonal Variations*

The axis of rotation of Earth on itself is not perpendicular to the plane of the ecliptic, which is the plane of the Earth's orbit around the Sun, and the corresponding axial tilt of Earth is 23.5° (Fig. 1.6a). Tilted Earth orbits around the Sun over one year (Fig. 3.3a). Because the axial tilt of Earth is constant, summer occurs in a

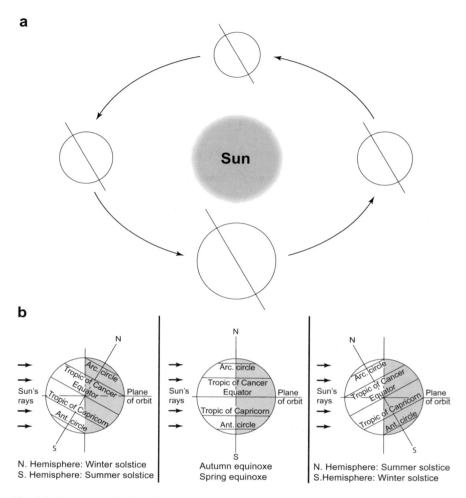

Fig. 3.3 Seasons on Earth: **a** Earth's orbit around the Sun during one year (Fig. 1.5b); **b** corresponding positions of Earth with respect to solar radiation at the solstices (opposite in the two hemispheres) and the equinoxes (same for the two equinoxes in both hemispheres). Credits at the end of the chapter

given hemisphere when that hemisphere is facing the Sun directly, and winter in that given hemisphere takes place when the other hemisphere is facing the Sun directly (corresponding to the right and left sides in Fig. 3.3b). Given its constant axial tilt, the Earth's axis has a maximum inclination relative to the Sun's rays twice a year, these days being called the winter and summer solstices, and is parallel to the Sun's rays twice a year, these days being called the spring and autumn equinoxes, as shown in Fig. 3.3b.

The winter and summer solstices are the first days of the winter and summer seasons, respectively, and the spring and autumn equinoxes are the first days of the spring and autumn seasons, respectively. The dates of the two solstices are 21 June and 21 December and those of the two equinoxes 21 March and 21 September. The current Gregorian calendar, which is used in most countries, is designed to keep the spring equinox in the Northern Hemisphere on 20 or 21 March, forcing the calendar seasons to start the same dates year after year regardless of the *precession of the equinoxes* (see Sect. 1.3.2).

The latitudes of the points at the top of the atmosphere with the smallest and largest angles of incidence change with the seasons (Fig. 3.3b). More precisely, the latitudes of the points with the smallest angles of incidence are: 0° (equator) on the day of the spring equinox, 23.5°N (tropic of Cancer) on that of the Northern Hemisphere summer solstice, again 0° on that of the autumn equinox, and 23.5°S (tropic of Capricorn) on that of the Northern Hemisphere winter solstice. The latitudes of the points with the largest angles of incidence are: 90°N and S (North and South Poles, respectively) on the day of the spring equinox, 90°S (South Pole) on that of the Northern Hemisphere summer solstice, again 90°N and S on that of the autumn equinox, and 90°N (North Pole) on that of the Northern Hemisphere winter solstice.

The angle of incidence of the radiation flux at the top of the atmosphere varies as a function of the latitude and the local (solar) time. Indeed, the flux is perpendicular to the top of the atmosphere (angle of incidence = 0°) only at the latitude where the Earth's surface is facing the Sun directly. Away from this latitude, the radiation flux meets the top of the atmosphere at an increasingly larger angle, which reaches 90° in regions that do not face at all the Sun in winter and thus experience 24 h of darkness (constant night in Fig. 3.4, called *polar night*). The latitude at which the noontime radiation flux is vertical moves back and forth between the two tropics over the course of one year. This latitude corresponds to the equator on the day of the spring equinox, the tropic of Cancer on that of the Northern Hemisphere summer solstice, again the equator on that of the autumn

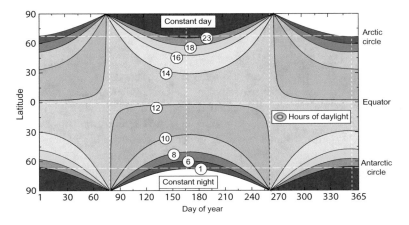

Fig. 3.4   Day length (daylight hours) as a function of latitude and day of the year. Credits at the end of the chapter

equinox, and the tropic of Capricorn on that of the Northern Hemisphere winter solstice. At high latitudes, the incident radiation flux at noontime is strongly tilted year round, even in summer. Beyond the polar circle (67.5°), the Sun no longer appears above the horizon at noontime in winter, hence the polar night.

The present value of the Earth's axial tilt is 23.5° (Figs. 1.6a and 3.3). If the axial tilt were very small, close to 0°, there would be no seasonal variations in temperature. In addition the latitudinal gradient in temperature would be similar year-round to that presently observed on the days of equinoxes, that is, the mid-latitude and the polar regions would be colder than the equatorial and tropical regions year-round. As a consequence, the ice-covered high-latitude areas might be larger than on present Earth, and the strong latitudinal temperature gradient may cause strong circulation in the fluid envelopes of the planet.

For comparison, the axis of rotation of planet Uranus has a tilt of 97° with respect to the plane of its orbit around the Sun, which means that a large part of the planet is subjected to an alternation between long periods of permanent sunshine and long periods of permanent night. As Uranus orbits around the Sun in 84 terrestrial years, the periods of permanent day or night last several decades, that is, 42 years at each of the two poles. Uranus is located at 19 AU from the Sun, and its surface temperatures are consequently extremely low, with a minimum value of -224 °C. Although the incident global solar flux is only about 1/400 that reaching Earth, it weakly warms the face of the planet exposed to the Sun, and this warming is enough to trigger enormous storms that cover areas equivalent to those of terrestrial continents, with winds of the order of 900 km/h. Applied to Earth, an inclination of 97° would result in extreme temperatures in each of the two hemispheres, probably higher than +100 °C in the hemisphere exposed to Sun, and less than -100 °C in the other hemisphere. As a consequence, water would mostly be in the form of water vapour or ice, and the equatorial belt would be affected by extreme atmospheric currents due to the strong thermal gradient between the two hemispheres. Hence if Earth had a large tilt, the resulting environmental conditions would not have favoured the preservation of water on the planet or the long-term establishment of organisms.

The Presence of the Moon Stabilizes the Earth's Climate It is thought that the Earth's axial tilt has been quite constant for millions, and perhaps billions, of years, which maintained temperature conditions on the planet suitable for organisms over a very long period. The constant tilt of Earth is explained by the stabilizing effect of the presence of the Earth's large satellite, the Moon (see Sect. 1.3.2). Indeed, the ratio of the mass of the Moon to that of Earth is 1:81, which is the highest value for all the known natural satellites in the Solar System, and it has been proposed that a much smaller Moon would not have stabilized the Earth's axial tilt. It follows that the presence of the Moon likely favoured the general occurrence on Earth of climatic conditions suitable for organisms since their establishment on the planet about 4 billion years ago.

3.3.3   Diurnal Variations

Since Earth is rotating on itself, the value of the radiation flux received by the planet varies during the course of a day, and the magnitude of this diurnal variation changes with the latitude and the season. This is illustrated in Fig. 3.4, where the four vertical dotted lines represent, from left to right, the Northern Hemisphere spring equinox, summer solstice, autumn equinox and winter solstice; and the white patches in each hemisphere correspond to the polar night, between the autumn and spring equinoxes in the given hemisphere.

Figure 3.4 shows the following characteristics of daytime length. At the equator (0° latitude), the lengths of the day and the night are always equal, and there is 12 h of daylight year round. The lengths of the day and the night are also equal (12 h each) everywhere on the planet on the equinoxes. Between the tropic of Cancer (23.5°N) and the tropic of Capricorn (23.5°S) daytime length ranges between a maximum value of 13.6 h in summer and a minimum value of 10.7 h in winter. In the two polar zones (above 66.5° N and S), daytime length is 24 h during summer, and nighttime length is 24 h during winter. At other latitudes and times of the year, the lengths of daytime and consequently nighttime range between short and long as illustrated in Fig. 3.4.

A few paragraphs above, we used the radiation flux at noontime to compare the angle of incidence among latitudes and seasons (see Sect. 3.3.2). We did this because the angle of incidence of the radiation flux at the top of the atmosphere at a given latitude varies not only during the year but also during the course of one day, that is, between sunrise and sunset. At sunrise, the angle of incidence is very large (in other words, it makes a large angle with the vertical), and the Sun is low in the eastern sky. At noontime, the angle is the smallest of the day, and the Sun is highest in the sky. At sunset, the angle of incidence is very large, and the Sun is low in the western sky.

The Earth's rotation period is the shortest, and thus its rotation is the fastest, among the four rocky planets of the Solar System (Table 1.1). Because of this, the exposure of a given point of the planet to a continually high or low radiation flux is short, except at the height of the summer or in the dead of winter above the polar circle. This is reflected in the number of hours of daylight at different latitudes and seasons (Fig. 3.4). By comparison, a point on planet Venus, whose rotation period is 243 Earth days (Table 1.1), is exposed to a continually high or low solar radiation flux for a very long time.

3.3.4   Combined Effects on Surface Temperatures and Organisms

It follows from the above considerations on the latitudinal, seasonal and diurnal variations of the solar radiation flux that the latter does not reach Earth uniformly in space and time. Indeed, there are diurnal variations in the instantaneous flux and seasonal variations in the diurnal flux, which both increase towards the poles, that is, with the distance away from the equator.

The radiation flux at a given point on Earth varies within each day as a result of the rotation of Earth on itself, and seasonally as a result of the motion of the tilted Earth on its orbit around the Sun. This has consequences that are not the same depending on latitude. On the one hand at low latitudes below the tropics, where the radiation flux is quite vertical year-round, the lengths of the day and the night are about equal, and the intensity of the flux varies during daytime with a maximum value at noontime. Hence the radiation flux at low latitudes is strong, but not continually so during the whole day. On the other hand at high latitudes above the tropics, daytime can be much longer than nighttime in summer and its length can even be 24 h above the polar circle, but the angle of incidence of the flux is small. Hence at high latitudes, even if daytime can be very long during part of the year, the radiation flux is never very strong during any part of the day.

The above conditions have two combined effects on the radiation flux reaching Earth. First, the radiation flux is not continually strong at any latitude, contrary to a body that would not rotate on itself and would thus continually show the same side to the Sun. Second, latitudinal differences in the annual flux are smaller than those that would experience a non-tilted planet. These conditions contribute to create surface temperatures suitable for organisms on most of the planet.

3.4 Fate of the Solar Radiation Absorbed by the Earth System

3.4.1 Reflexion of the Incident Solar Radiation (Earth's Albedo)

Approximately 30% of the radiation flux is reflected directly towards space. The Earth's albedo, which is the proportion of the incident radiation that the planet reflects back to space (see Information Box 2.9), is 0.30. This is an average value that combines the reflection on clouds and aerosol particles in the atmosphere, and also on continental surfaces (including those covered with bright ice and snow, and deserts) and oceans (including the bright sea ice).

The albedo of clouds can range from less than 0.1 to more than 0.9. Generally, thin high-altitude clouds (such as cirrus) tend to transmit solar radiation downwards (which enhances the greenhouse effect) and thus have low albedo, and in contrast, thick low-altitude clouds (such as stratocumulus) mostly reflect solar radiation and thus have high albedo. If the planet were suddenly emptied of its clouds, solar radiation absorbed by Earth would gain on average 20 W m^{-2}. This is because in the present situation, the cooling resulting from the albedo of the thick low-altitude clouds is larger than the warming caused by the greenhouse effect of the thin high-altitude clouds. Removing the two types of clouds would remove more cooling than warming, hence a warmer cloudless Earth.

The albedo of the ocean and the continental surfaces depends not only on their brightness but also on the angle of incidence between the radiation flux and the surface. When the radiation is perpendicular to the surface (angle of incidence of 0°), the reflected fraction of the radiation is minimum and the albedo is small. As the angle of incidence increases, an increasing fraction of the radiation is reflected and the albedo increases. At grazing incidence (close to 90°), the albedo is maximum, which is thus the case at dawn and at dusk. This effect is especially pronounced for bright surfaces.

3.4.2 Absorption of the Solar Radiation

The fraction of the solar flux not reflected back to space (about 70%) is absorbed by the Earth System, mostly as short-wavelength radiation between UV and near IR (see Sect. 3.2.2). This radiation is involved in numerous processes including the absorption of near IR radiation by gas molecules in the atmosphere, the evaporation of liquid water and photosynthesis, and also the direct transmission of radiation to air, water and soil, which absorb it. Each of these processes contributes to degrade the solar radiation received by Earth into heat, and thus shift the absorbed, short-wavelength solar radiation towards long-wavelength IR radiation (see Information Box 2.10), which is radiated back by Earth towards space (see Sect. 3.2.2). The sequence of processes comprises the following four steps (Fig. 2.6a).

First, some atmospheric gases absorb the short-wavelength solar radiation both directly and as reflected by the Earth's surface. These gases also absorb most of the heat radiated upwards by the Earth's surface as long-wavelength (IR) radiation. The gas molecules release the absorbed energy in the form of long-wavelength radiation (heat), in all directions including toward the Earth's surface. The radiation toward the Earth's surface warms up the planet. This phenomenon is known as the *greenhouse effect*, and the gases responsible for it are called *greenhouse gases* (see Information Box 2.7).

Second, some of the solar radiation that reaches the Earth's surface is absorbed by liquid water, which is found mostly in the oceans but also in lakes, rivers, soils and the vegetation. As a result of the absorption of the solar radiation, surface liquid water warms up. In some regions and seasons, such as tropical oceans, lakes and wet soils year-round and mid-latitude surface waters in summer, this may warm the water enough to cause some of it to evaporate. *Evaporation* is the transfer of water from liquid to gas phase (water vapour) in the atmosphere, and it occurs only if the latter is not already saturated with water (see Information Box 5.7). Evaporation removes energy from the liquid water, which thus cools down, and this energy is transferred to the water vapour under the name of *latent energy*. The reverse phenomenon is *precipitation* (rain, snow, hail), which releases

Information Box 3.4 O_2-Producing Photosynthesis, O_2 Respiration, and Combustion The *O_2-producing photosynthesis* (whose scientific name is *oxygenic photosynthesis*) is the process by which plants, algae and cyanobacteria use solar energy carried by photons (light) to split molecules of water (H_2O) and combine the resulting hydrogen atoms with the carbon and oxygen atoms of carbon dioxide gas (CO_2) to synthesize simple sugars:

$$CO_2 + 2H_2O + \text{energy carried by photons} \rightarrow [CH_2O] + O_2 + H_2O \qquad (3.1)$$

where [CH_2O] is the generalized chemical formula for sugar. In fact, the first compound synthesized in the photosynthetic production of organic matter is glucose, $C_6H_{12}O_6$. Equation 3.1 shows that O_2-producing photosynthesis has two by-products (to the right of the arrow): dioxygen, O_2 (see Information Box 1.8), and water. This is why the process is called *oxygenic*, an adjective meaning *oxygen-producing*. There also exists a form of photosynthesis that does not produce O_2, which is known in scientific studies as *anoxygenic photosynthesis*, but is called here *O_2-less photosynthesis*. The O_2-less photosynthesis is not as widely spread among modern organisms as the O_2-producing photosynthesis (see Information Box 5.1).

The two oxygen atoms that form the O_2 molecule (to the right of the arrow in Eq. 3.1) come from the splitting of the two molecules of water ($2H_2O$, to the left of the arrow in Eq. 3.1). This reaction provides virtually all the oxygen molecules (O_2 and O_3) present in the atmosphere. The photosynthetic glucose is further elaborated into energy-rich molecules of glucides (sugars), lipids (fats) and proteins (see Sect. 6.1.3).

When organisms need energy, they break down organic molecules through various processes that occur within the cells (see Information Box 5.2), which include *O_2 respiration* (whose scientific name is *aerobic respiration*, where *aerobic* means *life that uses air*):

$$[CH_2O] + O_2 \rightarrow CO_2 + H_2O + \text{energy} \qquad (3.2)$$

It should be noted that *respiration*, which takes place within the cells, is not the same as *breathing*, which is the intake of air into the lungs of vertebrates and its release from them. The reaction of environmental O_2 with organic molecules (left side of Eq. 3.2) releases energy (right side of Eq. 3.2). This energy is used to synthesize molecules of adenosine triphosphate (ATP), which are distributed to various parts of the cells where they liberate their energy. The latter fuels the chemical reactions involved in the functioning organisms. The oxygen atom in water to the right of the arrow in Eq. 3.2 (H_2O) comes from the oxygen to the left (O_2). Hence O_2 respiration returns to the global pool of water the oxygen atoms originally taken from it by O_2-producing photosynthesis. Equation 3.2 describes not only O_2 respiration but also other forms of oxidation of organic matter that include *combustion* (fire). The key difference between O_2 respiration and combustion is that the first process releases energy slowly, under biological control, whereas the second process releases it suddenly, without control.

The O_2 is very unstable, as it readily combines with many chemical elements (see Information Box 1.8). Nevertheless, its concentration in today's Earth's atmosphere is constant, implying that global respiration is about equal to global photosynthesis. Concerning the increase in atmospheric O_2 that began 2.4 billion years ago and continued during about 2 billion years (see Sect. 8.1.1), Eq. 3.2 indicates that this increase required, among other conditions, the burial of huge amounts of organic matter away from O_2 to prevent the oxidation of this organic matter, and thus allow O_2 to build up in the environment (see Sect. 9.4.5).

to the atmosphere the latent energy absorbed during evaporation. The coupled processes of evaporation and precipitation largely contribute to increase the wavelength of the radiation radiated by Earth toward the IR part of the spectrum (heat).

Third, part of the energy (light) from the solar radiation is absorbed by photosynthesis, a process by which organisms use the energy carried by photons to synthesize organic molecules (see Information Box 3.4). Part of this energy is transformed into heat, which is radiated into the environment. One example of the heating of the surrounding medium by organisms is the warming of surface waters of oceans and lakes by the presence of large amounts of microscopic phytoplankton cells, whose total biomass can be large enough that their absorption of solar energy warms up the surrounding water.

The origin of all the organic matter on Earth is biological (see Information Box 3.5). Most of the organic matter resulting from photosynthesis is progressively modified as it circulates through the food web (see Information Box 5.3), and a small fraction of it becomes fossil organic matter, with a lifetime of millions of years. The fossil organic compounds are mostly dispersed in sediments and rocks, but some of them are accumulated in geological reservoirs in the form of coal, petroleum, natural gas, and other carbon-based materials.

Except for the relatively small amount that will be buried and fossilized, all the organic matter originating from photosynthesis is eventually oxidized by respiration, which is the reverse process of photosynthesis in terms of mass balance (see Information Box 3.4). This means that the atoms of carbon and oxygen (from CO_2) and hydrogen (from H_2O) fixed into organic matter by O_2-producing photosynthesis are eventually returned to the inorganic environment (as CO_2 and H_2O) by O_2 respiration. The key role of respiration is to release the energy absorbed by photosynthesis inside the cells, where it is used for functions such as metabolism, growth and reproduction. Ultimately, the solar energy initially absorbed by photosynthesis is released to the atmosphere and the ocean in the

Information Box 3.5 Organic and Inorganic Compounds, and Particulate and Dissolved Organic Matter There is a distinction between organic compounds and organic matter. In chemistry, almost all chemical compounds that contain carbon are called *organic compounds*. Conversely, chemical compounds that do not contain carbon are qualified of *inorganic*. For historical reasons, some carbon-containing compounds are considered to be inorganic, including the gases carbon dioxide (CO_2) and carbon monoxide (CO) and the dissolved ions carbonate (CO_3^{2-}) and bicarbonate (HCO_3^-). Organisms are not necessarily involved in the formation of organic compounds. For example, some organic compounds were observed on comet 67P/Churyumov–Gerasimenko, and others were found in meteorites collected on Earth, and it has been hypothesized that comets had carried to Earth organic compounds that may have been among the initial building blocks of organisms (see Sect. 1.2.2).

Organic matter includes the organisms and also the organic compounds they produce or are derived from their remains, and their waste products (called *organic detritus*). Contrary to the organic compounds, which may be synthetic, all forms of organic matter involve organisms.

Organic matter can be in particulate or dissolved forms. *Particulate organic matter* (POM) or *organic particles* refer to the organic matter larger than a given threshold. The threshold used by researchers is different for soils and water, that is, in soil studies POM is the organic material larger than 0.053 mm (53 µm) retained on a sieve, and in water studies, the POM is the organic material larger than 0.7 µm (sometimes 0.22 µm) retained on a filter. The *dissolved organic matter* (DOM) in a water sample is the organic material that passes through the filter. The POM fraction contains small organisms and organic detritus, the latter being actively produced by organisms or coming from the breakdown of their remains after death. The DOM fraction contains organic molecules released by organisms, or resulting from the solubilization of POC by enzymes produced by bacteria and archaea called *exoenzymes*. The expressions *particulate organic carbon* (POC) and *dissolved organic carbon* (DOC) are used to express POM and DOM in term of carbon. The article of Hansell et al. (2009) provides further reading on DOM in the ocean.

Bacteria, archaea and fungi do not ingest particles, and they instead take up DOM through their cell membrane. Given that their feeding process involves osmosis, these organisms are called *osmotrophs* and their feeding mode is known as *osmotrophy*.

form of long-wavelength radiation, that is, heat produced by the chemical reactions involved in respiration. Organisms thus contribute directly (because some of the energy of photons is lost to heat during photosynthesis) and indirectly (through respiration) to the flux of heat radiated by Earth to space, although in a very small proportion relative to physical processes.

Fourth, the energy absorbed directly by surface waters and soils is largely released in the form of heat to the atmosphere, from which it is radiated back to space. This occurs, for example, when masses of warm water that have absorbed solar radiation at low latitudes are transported by ocean currents to high latitudes, where they cool by releasing part of their thermal energy content to the atmosphere. It will be explained below that the cooling of surface water in some high-latitude areas in winter plays a key role in the global oceanic circulation (see Sect. 3.5.3). In a similar way, the soils at mid and high latitudes accumulate heat by direct absorption of solar radiation in summer and return it to the atmosphere as heat in winter.

There are latitudinal differences in the annual radiation flux at the top of the Earth's atmosphere, but these are not exceedingly strong (see Sect. 3.3.4). At the Earth's surface, the distribution of the solar radiation flux is more contrasted than at the top of the atmosphere because of the geographical differences in both the albedo and the processes involved in the absorption of the solar radiation described in the above paragraphs. The effects of the unevenness in the distribution of the absorbed solar radiation at the surface of the planet are examined in the next section.

3.5 Effects of the Absorbed Solar Radiation on the Climate System

3.5.1 *Global Redistribution of Heat*

The solar radiation absorbed by Earth fuels the highly complex *climate system*, which consists of five major components—the atmosphere (air), the hydrosphere (liquid water), the cryosphere (ice), the lithosphere (rocks), and the biosphere (organisms)—and their interactions. The climate system changes under the influence of its own internal dynamics and because of external forcings (see Information Box 1.4). The interactions among climate system components include the exchanges of water vapour and heat between the atmosphere and the ocean, which were already examined above in terms of precipitation and evaporation (see Sect. 3.4.2). These exchanges are responsible for such phenomena as precipitations (rain, snow, hail), erosion, and storms.

The absorption of solar radiation is not even at the surface of Earth (see Sect. 3.4.2). A major effect of this unevenness is the existence of large temperature gradients between latitudes in the ocean and the atmosphere, and large seasonal variations in temperature at high latitudes. These latitudinal gradients and seasonal differences in temperature contribute to cause large-scale vertical and horizontal movements in the two fluid envelopes of the Earth System, which are called atmospheric circulation and oceanic circulation. These large-scale movements are also influenced by the positions of the continental masses. They redistribute heat all over the planet, and thus contribute to even up its distribution. Another factor that affects these large-scale movements is the rotation of Earth, as explained in the next section under the name of *Coriolis effect*.

3.5.2 *Atmospheric Circulation*

The *atmospheric circulation* is the large-scale movement of air in the atmosphere. This movement results from the fact that the radiation flux is stronger at the equator than at the poles, which causes hot air at the equator to rise and colder air at the poles to sink. This creates, in the two hemispheres, movements of air from the poles towards the equator in the lower atmosphere to replace the rising tropical air, and converse movements from the equator towards the poles higher in the atmosphere to replace the sinking polar air. The real situation is more complex, and the circulation in each hemisphere is in fact divided in three latitudinal cells from about 0 to 30°, 30 to 60°, and 60 to 90°, called *Hadley, Ferrel* and *polar* cells, respectively (Fig. 3.5a). The first two cells are named after the English lawyer and amateur meteorologist George Hadley (1685–1768) and the American meteorologist William Ferrel (1817–1891), who both contributed to explain the global atmospheric circulation.

The rising of air decreases the atmospheric pressure at the Earth's surface, and conversely the sinking of air increases it. As a consequence, the areas where the air flows upwards or downwards correspond to zones of low or high atmospheric pressure, respectively. These are represented by letters H (subtropical highs) and L (polar and equatorial lows), respectively, in Fig. 3.5 and elsewhere in this chapter. The subtropical latitudes are also called *horse latitudes*, an expression whose origin is debated.

Figure 3.5a shows three circulation cells in each hemisphere separated by zones of high (H) and low (L) atmospheric pressure. Each cell is characterized by dominant surface winds (curved arrows). Overall, the winds flow from zones of high pressure to zones of low pressure, but Fig. 3.5a illustrates the fact that winds do not flow straight from high to low pressures, but are deflected to the right in the Northern Hemisphere and to the left in the Southern Hemisphere. This phenomenon is known as the *Coriolis effect*, and is caused by the Earth's rotation. It influences all *large-scale movements* on the planet, including winds and ocean currents. As a result of the Coriolis effect, winds between latitudes of 0 and 30° dominantly blow from the east (called *trade winds*), between latitudes of 30 and 60° from the west (called *westerlies*, and *roaring forties* in the Southern Hemisphere), and between latitudes of 60 and 90° from the east (called *polar easterlies*). This effect was described for machines with rotating parts by the French mathematician Gaspard-Gustave Coriolis in 1835. It must be noted that winds are characterized by the *direction from which they blow*, for example, westerlies blow from the west. The dominant cause of the atmospheric circulation is the global latitudinal temperature gradient.

Figure 3.5a represents the atmospheric circulation on an idealized planet with a homogeneous surface, whereas the surface of Earth is in reality a mosaic of continents and oceans, where each type of surface responds differently to solar heating. Because continents are generally warmer than oceans in summer and colder in winter, the zones of idealized high and low atmospheric pressures in Fig. 3.5a

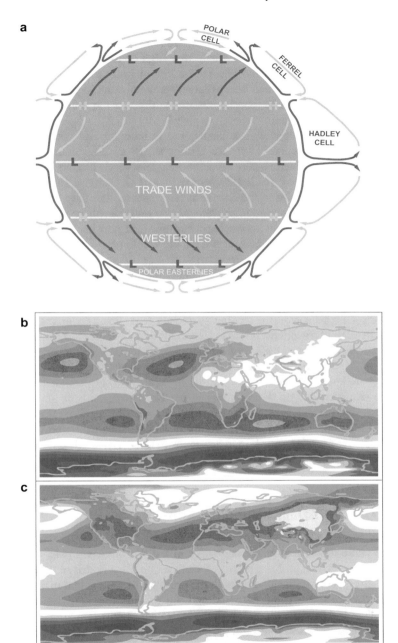

Fig. 3.5 The Earth's atmospheric circulation. **a** Idealized circulation on a planet without continents. High pressure: H (blue); low pressure: L (red). **b** and **c** Observed horizontal distributions of 15-year average atmospheric pressure at sea level (units: hPa) on real Earth (with continents) for **b** June–August and **c** December–February. High pressure: red and yellow; low pressure: blue and purple. Credits at the end of the chapter

actually break down into centres of high and low pressure located over the colder and warmer areas, respectively (Fig. 3.5b,c). Yet, the observed pressure centres are located within wide belts dominated by either high or low atmospheric pressure, which broadly correspond to the idealized zones of highs and lows in Fig. 3.5a. The positions of continents and oceans largely influence the locations of the main centres of high and low atmospheric pressure, and hence the routes of the atmospheric currents between them.

Because of the differences in solar heating in summer and winter, the highs and lows and the atmospheric currents change their locations among seasons, in different ways in each hemisphere (Fig. 3.5b,c). The winds around the *centres of low atmospheric pressure* (called *cyclones*) flow anticlockwise in the Northern Hemisphere and clockwise in the Southern Hemisphere, and they flow in reverse directions around the *centres of high atmospheric pressure* (these winds are called *anticyclones*). The latitudinal atmospheric currents transport heat among latitudes, and thus contribute to redistribute heat over the whole planet, which makes most of the Earth's surface suitable for organisms.

The atmospheric circulation also influences the precipitation regimes, and thus the distributions of continental ecosystems. In the zones of low atmospheric pressure, the rising air becomes cooler, and this cooling generally favours the condensation of air moisture, leading to clouds and precipitations. Conversely, in zones of high atmospheric pressure, the sinking air becomes warmer, and this warming favours the evaporation of air moisture, leading to dry air and blue sky. The latter has direct consequences for organisms on continents given that the warm, dry and sunny conditions of the subtropical highs are responsible for the existence of major deserts that include: in the Northern Hemisphere, the Sahara Desert (Africa), the Arabian Desert (Middle East) and the Sonoran Desert (United States and Mexico); and in Southern Hemisphere, the Atacama Desert (Chile), the Kalahari Desert (Africa) and the Great Victoria Desert (Australia). Desert conditions do not support abundant organisms.

3.5.3 Oceanic Circulation

The large-scale movement of surface water in the World Ocean is called *surface ocean circulation*. The global patterns of surface currents are driven by the winds. They are also deeply influenced by both the presence of continental masses, which act as barriers to the oceanic currents, and the rotation of Earth responsible for the Coriolis effect (see Sect. 3.5.2, just above). Other factors that affect ocean currents are exchanges of heat between the surface ocean and the atmosphere, and the tides.

We saw in the previous section that *winds* are characterized by the direction *from which they blow*. Differently, the *ocean currents* are characterized by the direction *towards which they flow*. For example, one says that dominant winds at latitudes of 30–60° *blow from the west* and the corresponding currents *flow eastwards*.

The speed of surface currents in the open ocean varies from a few centimetres per second to a maximum of about 2.5 m s^{-1} (9 km h^{-1}), although currents can be more rapid in some coastal areas, for example in channels or in relation to large tidal amplitudes. The wind-driven oceanic circulation is as follows (Fig. 3.6):

At the *latitudes of trade winds (0–30°)*, surface waters generally flow westwards north and south of the equator, and are called *North Equatorial Current and South Equatorial Current*, respectively. The two westward Equatorial Currents are separated by the Equatorial Countercurrent, which flows eastwards at about 0°.

At the *latitudes of westerlies (30–60°)*, surface waters generally flow eastwards in the Northern and Southern Hemispheres, and have such names as *North "Ocean" Current and South "Ocean" Current*, respectively (for example, the North Atlantic Current, and the South Pacific Current).

At the *latitudes of polar easterlies (60–90°)*, the situation differs in the two hemispheres because of the different geometry of the landmasses. In the Southern Ocean, the *Antarctic Circumpolar Current and the Antarctic Subpolar Current* flow eastwards and westwards, respectively, around the Antarctic continent. The situation is more complex in the Arctic Ocean because of the presence of landmasses and the restricted connections of that basin with the Atlantic and Pacific Oceans.

The *above latitudinal currents* are linked by currents flowing along the continents, which have specific names such as *Gulf Stream*, *Benguela Current* and *Mozambique Current*.

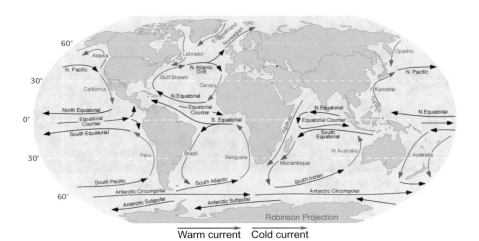

Fig. 3.6 Schematic representation of the main surface ocean currents. Red and blue arrows: warm and cold currents, respectively. Credits at the end of the chapter

   We saw in the previous section that *atmospheric* currents are deeply affected
by the presence of continents, but nevertheless form wide belts all around the
planet (Fig. 3.5). This is not the case of *surface ocean* currents, because these are
physically constrained by the presence of continents (Fig. 3.6). Overall, the surface ocean currents tend to flow from higher to lower surface heights, and because
of the Coriolis effect these currents in the Northern Hemisphere are deflected
toward the right, which causes them to flow clockwise around sea-surface highs
(Fig. 3.7a). Conversely, the surface ocean currents in the Southern Hemisphere are
deflected toward the left, which causes them to flow anticlockwise around sea-surface highs. Hence in both hemispheres as a result of the Coriolis effect, surface
waters flowing westwards are deflected towards the poles and waters flowing eastwards are deflected towards the equator. This causes surface water to converge
at latitudes of about 30°, which generates sea-surface "hills", and to diverge at

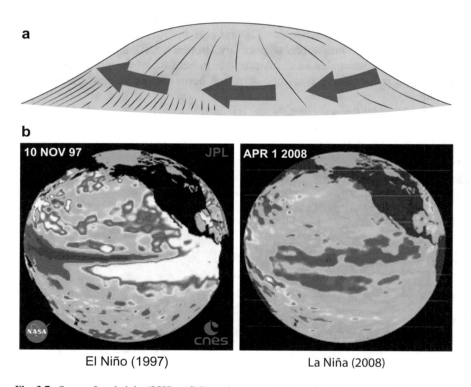

El Niño (1997)   La Niña (2008)

Fig. 3.7   Sea-surface height (SSH). **a** Schematic representation of surface currents (blue arrows)
flowing in a clockwise circulation around a sea-surface topographic high in the Northern Hemisphere (anticyclone). Vertical exaggeration of over 10^6 (1 cm on the vertical corresponds to more
than 10 km on the horizontal. **b** Sea-surface height in the Pacific Ocean during conditions of El
Niño (1997) and La Niña (2008). Green areas: normal SSH; white and red areas: SSH 14 to
32 cm above normal; purple areas: SSH at least 18 cm below normal. Credits at the end of the
chapter

latitudes of about 60°, which creates sea-surface "valleys" (Fig. 3.7b). The height of sea-surface hills and the depth of sea-surface valleys range between a few tens of centimetres to more than one metre.

Figure 3.7b illustrates two well-known conditions of the climate system called *El Niño* and *La Niña*. The El Niño follows the weakening or even reversal of trade winds over the Pacific Ocean, which allows large masses of warm water to move eastwards along the equator and reach the coast of South America (white and red areas in the El Niño map). The displacement of this large amount of warm water increases the evaporation and thus rain along the coasts of North and South America, and modifies the atmospheric circulation around the world. The conditions of La Niña are the opposite of those of El Niño, with trade winds stronger than normal, which pushes to the middle of the equatorial Pacific cold water that usually exists along the coast of South America (blue and purple areas in the La Niña map). This causes less moisture in the air and less rain along the coasts of North and South America, and changes global weather patterns. The Spanish names *El Niño* and *La Niña* mean "the young boy" and "the young girl", respectively. The expression *El Niño* has been used by Peruvian fishers since the 1600s because during El Niño years, water temperature is often maximum along the South American coast near Christmas. The opposite phenomenon was later called *La Niña*. The two phenomena are the warm and cold phases, respectively, of a climatic cycle called *El Niño-Southern Oscillation* (ENSO).

Large anticyclonic gyres can be seen in Fig. 3.6 in five ocean basins, that is, the North and South Atlantic and Pacific and South Indian Oceans (*gyres* are large systems of rotating ocean currents). These gyres correspond to the idealized anticyclonic circulation represented in Fig. 3.7a for the Northern Hemisphere. The vocabulary for oceanic currents is the same as that used for the atmospheric winds (see Sect. 3.5.2), that is, clockwise and anticlockwise oceanic circulation in the Northern Hemisphere is called anticyclonic and cyclonic, respectively (reverse in the Southern Hemisphere).

In the open ocean, surface currents are both driven by winds and largely influenced by the continental masses and the Earth's rotation, which constrain and deflects the flows, respectively. In the coastal ocean, currents may also be strongly influenced by tides and the topography of the seafloor. Overall, the winds reflect the global latitudinal temperature gradient resulting from the interaction between the Sun and Earth (see Sect. 3.5.2), and the tides are caused by the gravitational pull exerted by the Moon and the Sun combined with the Earth's rotation. Hence the surface circulation of the ocean is strongly related to the astronomical characteristics of Earth.

A major effect of the surface ocean circulation is the transport of heat between the warm low latitudes and the cold high latitudes. The transport of heat across latitudes is evidenced by the red and blue arrows in Fig. 3.6, which represent warm and cold surface ocean currents, respectively. The surface ocean circulation contributes, together with the atmospheric currents (see Sect. 3.5.2), to redistribute heat within and among ocean basins, and thus make most of the oceans suitable for organisms.

The surface circulation is linked to ocean depths by large-scale movements of water in the ocean basins called *global ocean circulation*. The *deep ocean circulation* consists of slow horizontal water movements driven by the density of water. Indeed, this circulation results from a phenomenon that takes place in two areas of the World Ocean in the midst of winter, which are located in the northern part of the North Atlantic Ocean and in the Southern Ocean off the coast of Antarctica south of the Atlantic Ocean, respectively (see the two areas of deep water formation in Fig. 3.8). There in winter, the already cold surface water is further cooled by the loss of heat to the seasonally very cold air (the radiation flux at high latitude is very low in winter; Fig. 3.4), which increases its *density*, that is, its mass per unit volume (see Sect. 5.5.1). The very dense water sinks to great depth, and this initiates horizontal movements of water deep in the ocean.

The global ocean circulation is often schematised as a *global conveyor belt*, also called *ocean conveyor belt*. Although the real ocean circulation is actually quite different from the schematic representation in Fig. 3.8, and is much more complex, the global conveyor belt shows the broad connections that exist between the surface (red) and deep (blue) currents. The reddish parts of the conveyor represent the net transport of warm water in the uppermost 1000 m or so, and the

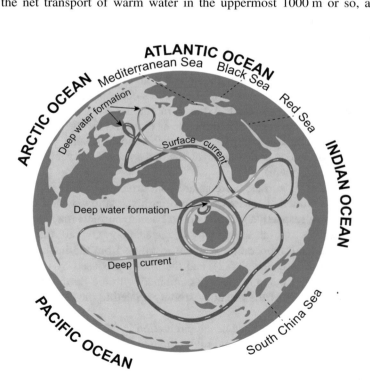

Fig. 3.8 Schematic representation of the global ocean circulation as a global conveyor belt, drawn on a continuous-ocean map. Light and dark colours: oceans and continents, respectively. Ocean currents: see text. Credits at the end of the chapter

blue parts the net transport of cold water below the permanent thermocline. The resulting water movements at the scale of the World Ocean are as follows. On the one hand, surface ocean circulation carries warm tropical surface waters towards the poles, and these waters release heat to the atmosphere along their way. On the other hand, cold waters sinking to depth North and South of the Atlantic Ocean in winter form the ocean deep waters, which circulate through the various ocean basins and progressively return to the surface at lower latitudes. Further reading is provided by the article of Broecker (1991), in which the creator of the great ocean conveyor diagram explains the origin and meaning of his widely known schematic.

The ultimate causes of the global ocean circulation are the latitudinal temperature gradient and the seasonal variations in temperature at high latitudes. More immediately, the ocean conveyor belt is driven by global density gradients created by fluxes of surface heat and fresh water, hence its other name of *thermohaline circulation*, which refers to its temperature (*thermo*) and salinity (*haline*) components.

The deep ocean circulation is very slow, about 1 cm s^{-1} (0.4 km h^{-1}), and its timescale is of the order of 1000 years, meaning that waters presently circulating at depth in the North Pacific Ocean partly reflect the conservative properties (such as temperature and salinity) of waters that left the surface of the Atlantic Ocean more than 1000 years ago (Fig. 3.8). The waters circulating at depth progressively return to the surface along their paths. This occurs in particular in areas where the action of winds makes the deep waters move toward the surface, a phenomenon called *upwelling*, which is characterized by the presence at surface of cold, biologically-rich water. Such water is found on the eastern sides of ocean basins, for example off California-Oregon and Peru in the North and South Pacific, respectively, and off Morocco and southwest Africa in the North and South Atlantic, respectively, these being called *eastern boundary upwelling systems* (sometimes abbreviated EBUS). Coastal upwelling also occurs elsewhere in some seasons, for example off the coast of Somalia in the Indian Ocean during the June-September monsoon. Upwelling also takes place in certain areas of the open ocean, for example along the equator in the Atlantic and Pacific Oceans. The overall effects of the global ocean circulation are to warm up the higher latitudes and cool down the lower latitudes, which reduces the latitudinal range of oceanic and also continental temperatures, thus making most of Earth suitable for organisms. The book of Kämpf and Chapman (2016) provides further reading on the upwelling systems of the world.

3.5.4 Combined Effects of Atmospheric and Oceanic Circulations

The combination of temperature-driven atmospheric and oceanic circulations balances out the uneven distribution of the absorbed solar radiation on Earth by exchanging heat between warmer and colder regions. The mechanisms of heat redistribution over the planet operate as follows.

First, the ocean areas between the tropics (approximately 30°N and 30°S) are the site of strong evaporation because of their high surface temperature and the strong incident solar radiation during the whole year. The resulting water vapour rises in the atmosphere where it causes the formation of clouds, whose whiteness increases the albedo in areas where they are present. The precipitation of the evaporated water occurs both locally, in the intertropical zone, and also in other regions that include continents and oceans above 30°N and 30°S, where water is transported by atmospheric currents. This mechanism of evaporation, atmospheric transport and precipitation transfers both water and heat from low-latitude oceans to higher-latitude oceans and continents.

Second, some of the heat stored in the surface ocean at low latitudes is transported directly by surface ocean currents to higher latitudes, where it contributes to energy exchanges with the surrounding ocean and the overlying atmosphere.

Third on some land surfaces, the combination of evaporation and transpiration from terrestrial plants creates evapotranspiration (see Information Box 5.7), which transfers water vapour and latent heat to the atmosphere. Although evapotranspiration occurs in specific continental areas, the atmospheric currents redistribute the resulting water and heat to the whole planet.

Overall, the redistribution of heat over the planet by the combined atmospheric and oceanic circulation is largely responsible for the presence of rich ecosystems on most of Earth. Further reading on the ocean's role in the climate and in the global water cycle can be found in articles of Schmitt (2018) and Lagerloef et al. (2010), respectively.

3.5.5 Effects of Global Photosynthesis

Photosynthesis does not occur equally everywhere on the planet. In fact, it is highest in regions where water and essential chemical elements are abundant, temperature is suitable, and light is high during at least part of the year. The chemical elements essential for the synthesis of organic matter include hydrogen, which is provided by liquid water (H_2O) for O_2-producing photosynthesis, and carbon and oxygen, which come from the CO_2 gas in the atmosphere or dissolved in water. Other essential chemical elements include nitrogen, phosphorus, iron, and several others depending on the types of photosynthetic organisms (see Sect. 6.1.2). In the natural environment, these chemical elements are contained in inorganic and organic compounds called *nutrients*. In agriculture, the missing nutrients are supplied by *fertilizers*, which may be natural (for example, manure) or synthetic (for example nitrogen fertilizers made from ammonia, NH_3, which is synthesised from atmospheric N_2 and natural gas, CH_4, an energy-intensive process).

Photosynthetic organisms—which include phytoplankton, seaweeds and plants—absorb a small part of the solar radiation when they produce organic matter (see Information Box 3.4). The resulting organic matter is used directly or indirectly as food resource by other organisms in the food web (see Information Box 5.3). Because water is essential for O_2-producing photosynthesis, photosynthetic organisms are found almost everywhere in the lighted parts of aquatic environments, where water is not a limiting factor, whereas the growth of photosynthetic organisms on terrestrial surfaces can be limited by water in some areas (for example, deserts) and/or at some times of the year (for example during dry seasons). Light is another essential requirement for photosynthesis, which often limits the growth of photosynthetic organisms especially in aquatic environments where light is strongly absorbed by water and its intensity thus decreases rapidly with depth.

In the oceans, where the availability of water is unlimited, O_2-producing photosynthesis only occurs near the surface because of light limitation at depth. However, surface waters are continually depleted of nutrients because organic matter tends to sink towards deeper waters. The sinking organic matter includes organic debris and aggregates, faecal material, carcasses, and others. As it sinks, organic matter carries its chemical constituents to deep waters, where they are released by decomposition. As a consequence, surface waters are often adequately illuminated but nutrient-limited, and deeper waters have high concentrations of dissolved nutrients resulting from the decomposition of the sinking organic matter but are light-limited. Hence, photosynthetic organisms will be particularly successful in surface or shallow waters where the influx of dissolved nutrients, from depth or continents, compensates their continual loss to depth.

Offshore, photosynthetic organisms are abundant in areas where water movements favour the frequent or continual replenishment of surface nutrients from deeper waters. These areas include the upwelling zones including the EBUS (see Sect. 3.5.3), where deep water rises to the surface. Photosynthetic organisms are also abundant offshore at times when there is both enough light and nutrients, for example at the end of the winter or in the springtime at mid-latitudes when the water vertically stabilizes after deep winter mixing.

In the coastal ocean, both phytoplankton and seaweeds are often thriving, especially in areas where running water loaded with nutrients from the continents is discharged. However in cases where the supply of nutrients is very high, the production of seaweeds or phytoplankton may be so high that the massive decomposition (respiration) of the resulting organic matter (dead material) may take up all the dissolved oxygen. The resulting low concentration of dissolved oxygen may cause mass mortality or emigration of O_2-respiring organisms. This phenomenon, called *eutrophication*, is observed in areas where excess fertilizers are used on land for agriculture, or much organic wastes are carried in domestic effluents or released by coastal aquaculture.

On the continents, the photosynthesis of terrestrial plants strongly depends on regional and/or local effects of the climate system, which largely determine the availability of liquid water. In some areas or at certain times, the supply of water can be very limited or absent, as in deserts or at very high altitude. In other areas, temperature can be so low that all water is frozen, as in areas permanently covered by ice or at high latitude during winter. Finally, solar light may be too low for net photosynthesis or even null, as in the polar regions in winter. On the continents, vegetation is especially luxuriant in areas where water is abundant, nutrients are plentiful (most of them coming from the on-site recycling of organic matter by decomposers), the sun shines year-round and temperature is constantly suitable for organisms, as observed in well-watered parts of intertropical zones such as the Amazonian basin.

Given that the fraction of solar radiation absorbed by photosynthesis is small, the effect of this biological process on the Earth's radiation budget is not significant at the short timescale. However because of the long-term build-up of biomass on continents and in oceans, photosynthesis has the following three major effects on the radiation budget at long timescales.

First, the photosynthetic release of O_2 that began more than 2.5 billion years ago in aquatic environments progressively modified the Earth's atmosphere, leading to the emergence of O_2-respiring organisms and the occupation of continents by thriving ecosystems. Organisms could occupy the continents because they were protected from dangerous solar UV radiation by the atmospheric ozone layer, where the O_3 is produced from atmospheric O_2 (see Sect. 2.1.5).

Second, the presence of an extensive vegetation cover on continents contributes to decrease the fraction of the solar radiation reflected back to space, because the albedo of vegetation-covered land is much lower than that of deserts or areas covered by snow (Table 3.3). This contributes to set the global albedo of the planet, and thus the amount of solar radiation absorbed at the surface of Earth.

Third, photosynthetic organisms take up CO_2, whose carbon is incorporated into organic matter in the food web (see Information Box 5.3). Most of this organic matter is ultimately decomposed, but a small part of its carbon is progressively sequestered in the long term as fossil organic matter mainly dispersed in sedimentary rocks and also in the form of coal, peat, petroleum, and natural gas. During the course of the Earth's history, the sequestration of carbon (that is, the long-term storage of carbon in natural reservoirs) was compensated by sources of atmospheric CO_2, especially volcanoes (see Sect. 7.3.1). Without the release of

Table 3.3 Albedo of different types of natural surfaces on Earth. Values from Fig. 2.5 of Girard (2009)

Surface type	Average albedo
Forest land	0.10–0.20
Agricultural land	0.10–0.20
Deserts	0.25–0.40
Snow cover (above 60° latitude)	0.60–0.80
Oceans (above 70° latitude, with sea ice)	0.15–0.20
Oceans (below 70° latitude, without sea ice)	0.06–0.10
Clouds (all types)	0.35–0.40
Overall Earth's average	0.30

CO_2 by volcanoes, the concentration of this gas in the atmosphere, and thus the greenhouse effect, would have slowly decreased and the global temperature of the planet would have become freezing (see Information Box 2.7).

3.6   Temperature on Earth: Roles of the Atmosphere and the Earth's Distance from the Sun

3.6.1   Composition of the Earth's Atmosphere and Earth's Distance from the Sun

We explained above that Earth's axial tilt largely contributes to determine temperature conditions suitable for organisms on most of the planet year round (see Sect. 3.3.2). We also reported mechanisms proposed by astronomers to explain how the axial tilt of Earth was maintained over the 4.6 billion years of the planet's history, thus providing temperatures suitable for organisms since they started to occupy Earth. The axial tilt controls the distribution of the radiation flux from the Sun reaching Earth, but there are additional factors contributing to set the actual range of Earth's temperatures on the planet. In the coming paragraphs, we examine two such factors, namely, the composition of the Earth's atmosphere, which is a characteristic of the Earth System, and the Earth's distance from the Sun, which is a characteristic of the Solar System.

Composition of the Earth's atmosphere. The surface temperatures of Earth, its two sister planets, Venus and Mars, and its satellite, the Moon are very different (Table 3.2). A key characteristic that explains this difference is the density and composition of the atmospheres of the four bodies.

On Venus and Mars, the atmospheric pressure is more than 90 times higher and more than 150 times lower, respectively than on Earth, and it is even much lower on the Moon (Tables 2.2 and 3.2). The very high atmospheric pressure on Venus is caused by the very high concentration of gases in its atmosphere, with a high proportion of the greenhouse gas CO_2 (Table 2.1). The combination of a thick atmosphere with a high proportion of CO_2 creates a very strong natural greenhouse effect (see Information Box 2.7) that keeps the surface temperatures on Venus at very high values (around 460 °C at ground level, Table 3.2). Conversely, the very low atmospheric pressure on Mars reflects the very low concentration of gases in its atmosphere. Even if the proportion of CO_2 is high in the Martian atmosphere (Table 2.1), the latter is so thin that the greenhouse effect caused by CO_2 is low. As a consequence, the atmosphere of Mars cannot retain heat when the Sun disappears over the horizon at sunset, or when summer gives way to winter, thus causing wide diurnal and seasonal variations in temperature. The absence of atmosphere on the Moon explains the absence of any greenhouse effect, and temperature variations there even more extreme than on Mars.

In contrast, the Earth's atmosphere has the right density and greenhouse gas composition for creating a moderate natural greenhouse effect. The comparison of the atmosphere of Earth with those of Venus, Mars and the Moon shows that the Earth's greenhouse mechanism is largely responsible for the moderate temperatures that exist at the surface of the planet. These temperature conditions have been suitable for thriving ecosystems on Earth since billions of years and they continue to be so nowadays.

Earth's distance from the Sun. The presence of organisms on Earth is the basis of the definition of the habitable zone of the Sun (see Information Box 2.4). The concept of *habitable zone* was developed for the study of exoplanets, but we apply it here to our star, the Sun, and planets Venus, Earth and Mars and their moons. It is estimated that the *habitable zone of the Sun* extends from 0.5 to 3.0 AU away from the star. On the one hand at distances from the Sun smaller than 0.5 AU, all water is lost by the process of atmospheric escape (see Sect. 5.4.2). Hence bodies as close to the Sun as 0.5 AU, such as planet Mercury, have no water and thus no organisms. On the other hand at distances from the Sun larger than 3.0 AU, adding CO_2 to the atmosphere would fail to create a greenhouse effect strong enough to keep the surface of the planet at a temperature above 0 °C (see Information Box 2.7). Hence planetary bodies father than 3.0 AU from the Sun have no liquid water at their surface, and thus no surface organisms. However, such bodies may have liquid water under their frozen surface (see Information Box 5.4).

Earth is located at 1.0 AU from the Sun (this is the definition of the AU), and is thus within the habitable zone of the star. It follows that Earth's distance from the Sun is one of the factors involved in the existence of temperature conditions at the surface of the planet suitable for organisms and ecosystems.

Planets Venus and Mars, our Moon, and the two satellites of Mars are also located within the habitable zone of the Sun (Table 1.1). This means that these bodies could theoretically harbour life, but their surface temperatures are either too cold (the Moon, Mars and its satellites) or too hot (Venus) for the existence of liquid water, and thus organisms (Table 1.1). However, it is possible that the soil of Mars contains liquid water and thus perhaps harbours organisms (see Sect. 5.5.3). In addition, even if the temperature of +462 °C and the pressure of 92 atm that prevail at the surface of Venus are much too high for organisms, it has been noted that at an altitude of 50 km in the Venusian atmosphere, temperatures range between +30 and +70 °C and the atmospheric pressure is the same as at the surface of Earth. This makes this environment habitable, and it was hypothesized that organisms could exist there. Elsewhere within the habitable zone of the Sun, liquid water may exist on at least one dwarf planet in the asteroid belt (see Information Box 5.4). Even outside the habitable zone of the Sun, liquid water may exist under thick layers of ice on at least five natural satellites of two giant planets (see Information Box 5.4). Hence, there are several environments where organisms could possibly exist in the Solar System. However, the occurrence of suitable temperatures or the presence of liquid water, which are key conditions of habitability, provide no certainty that organisms are actually present. Indeed, no sign has been observed so far that the environment of a body in the Solar System outside Earth was even marginally controlled by ecosystems.

3.6.2   Characteristics that Determine the Solar Radiation at the Surface of Earth

As explained above, the temperature at the surface of Earth is mostly determined by the level of solar radiation that reaches the surface of the planet (see Sect. 3.2.1). This is modulated by four characteristics of the Solar System and the Earth System.

Two of the four characteristics that contributed to keep the temperature of the Earth's atmosphere and oceans within limits suitable for organisms over billions of years are the Earth's distance from the Sun, and the presence of an atmosphere with the right density and greenhouse gas composition to create a moderate greenhouse effect (see Sect. 3.6.1). The two other characteristics are the small tilt of the Earth's axis of rotation, which is maintained by the Earth-Moon system (see Sect. 3.3.2), and the planet's fast rotation on itself (see Sect. 3.3.3).

3.7   Long-Term Controls on Earth's Temperature

3.7.1   The Faint Young Sun Paradox

The fact that organisms built up the huge biomasses that led ecosystems to take over the Earth's environment during the last billions of years means that temperatures during that very long period were not only within the limits of survival of organisms, but were also generally within their limits of growth (Table 3.1). It is explained above that a moderate greenhouse effect had been largely responsible for maintaining a range of temperatures suitable for organisms until the present time (see Sect. 3.6.1). However, astronomers have pointed out a different aspect of the Earth's long-term temperatures, which is related to the fact that the radiation flux of the young Sun was 25–30% lower 4 billion years ago than presently. The lower solar radiation flux implies a lower Earth's temperature, which is known as the *faint young Sun paradox* or the *faint young Sun problem*.

The young Sun was fainter in the deep past than presently because the energy released by a star comes from the nuclear fusion that takes place in its core. As a star matures, its core progressively contracts, which increases the amount of energy it releases. Hence the energy released by the young Sun was lower than at present. Because of the lower solar radiation flux of the faint young Sun 4 billion years ago, the surface temperature of Earth should have then been 21–26 °C lower than the present average 15 °C (assuming the same albedo and the same greenhouse effect as presently). With such a low air temperature, all water at the surface of Earth would have been frozen.

However, geological evidence shows that there was liquid water on Earth 4 billion years ago, which indicates that some mechanisms compensated the effect of the faint young Sun. These mechanisms made the temperature high enough for liquid water to exist at the Earth's surface, which allowed the development of ecosystems. The various mechanisms proposed to resolve the faint young Sun paradox include: a low albedo (see Information Box 2.9), partly caused by fewer clouds than presently because the number of cloud nuclei produced by the fewer organisms was small (see Information Box 6.2); and a strong greenhouse effect, caused by high concentrations of atmospheric CO_2 and possibly other greenhouse gases such as methane (CH_4). More details concerning the CH_4 hypothesis are provided elsewhere in this book (see Sects. 2.1.7, 4.7.2, and 11.4). The article of Broecker (2018) provides further reading on the faint young Sun paradox.

3.7.2 Long-Term Control of the Atmospheric CO_2 Concentration

Carbon is present in the Earth's atmosphere in the form of molecules of carbon dioxide and carbon monoxide (CO_2 and CO, respectively) and various organic gases including CH_4 and dimethyl sulphide, DMS (see Information Box 6.2). Atmospheric gases dissolve in oceanic and continental waters, where the chemical reaction between CO_2 and water produces bicarbonate (HCO_3^-) and carbonate (CO_3^{2-}) ions (see Information Box 6.3), and where biological activity produces both such organic gases as dimethylsulphoniopropionate, DSMP (see Information Box 6.2), and dissolved organic matter (see Information Box 3.5). The carbon present in these various chemical forms is continually exchanged with the carbon in organisms, limestone rocks, soils and sediments. These exchanges take place through the formation of solid carbonates, chemically and by organisms (see Information Box 5.11), photosynthesis, the burial of biological remains in sediments and soils, the erosion of sedimentary rocks on continents (see Sect. 7.3.3), and the oxidation of organic matter (by combustion, respiration, and decomposition) in all environments.

If the losses of carbon due to burial (or sequestration) in sediments (and in soils since the time organisms have occupied the continents less than 500 million years ago) had not been continually compensated by corresponding volcanic inputs of CO_2 into the atmosphere, the natural greenhouse effect would have weakened and Earth would have frozen. Since this did not generally happen, with the exception of the periods of snowball Earth (see Sect. 4.5.4), the losses of atmospheric carbon to the seafloor and soils were generally replaced by equivalent gains of atmospheric CO_2. The buried carbon includes: in the seafloor, calcium carbonate and organic sediments, which are at the origin of continental limestone and fossil organic matter (including petroleum and natural gas); and under the ground, organic matter, peat and coal. The following paragraphs examine mechanisms involved in the control of the concentration of atmospheric CO_2.

The mineral and organic carbon buried in marine sediments and underlying sedimentary rocks moves slowly towards the continents through lateral and vertical movements of the seafloor, called plate tectonics (see Sect. 7.2.1). Part of the carbon-containing materials that reach the continental margins accumulate in continental rocks, and part melt in the mantle beneath volcanoes, a process during which CO_2 is released in the fluid magma. During volcanic eruptions, lava and various gases from the mantle's magma are ejected at the Earth's surface, and the gases released into the atmosphere. These include large amounts of CO_2, which contribute to the natural greenhouse effect that keeps the Earth's temperature suitable for organisms. The continual volcanic input of CO_2 into the atmosphere is balanced by the incorporation of carbon from atmospheric CO_2 and dissolved in natural waters into solid compounds buried in sediments and soils (see Sect. 9.2.2).

One long-term effect of the development of ecosystems, which began at least 3.8 billion years ago, was the progressive accumulation of atmospheric carbon into solid compounds. The solid carbon-containing compounds were both calcareous (including inorganic concretions, microscopic tests, shells, coral reefs, and the resulting calcareous rocks) and organic (including cells and bodies of organisms, and fossil organic matter). During most of the history of Earth, the ecosystems were essentially marine, as organisms started to occupy the continents less than 500 million years ago. Indeed, the occupation of continents had to wait for the establishment of the ozone layer in the upper atmosphere, without which continental organisms could not survive the dangerous UV radiation from the Sun. This only occurred after the building of large concentrations of atmospheric O_2 by photosynthesis before 500 million years ago (see Sect. 2.1.5).

The carbon fixed by the chemical and biological formation of $CaCO_3$ (see Sect. 6.4.2) and by aquatic chemosynthesis and photosynthesis came from dissolved CO_2 in water, and the net effect of the burial of calcareous and organic compounds in sediments and soils was to transfer huge amounts of carbon from atmospheric CO_2 into these reservoirs. The chemical formation of $CaCO_3$ caused the initial decrease in atmospheric CO_2 (Fig. 8.1b), which lowered the natural greenhouse effect and was followed by biological contributions. At the same time, the intensity of the young Sun progressively increased (see Sect. 3.7.1), and this together with the decreasing greenhouse effect and other mechanisms (see Sect. 9.2.3) contributed to keep the Earth's temperature within a range suitable for organisms. During more than 3.5 billion years, all the organisms on Earth were unicellular, and even nowadays microbes dominate most of the biogenic fluxes of chemical elements on the planet.

3.8   Key Points: Temperature and the Living Earth

We conclude this chapter with a brief summary of some of the key points of the above sections, as done in previous chapters (see Sects. 1.6 and 2.5). Here we examine interactions between the Solar System, Earth, its temperature, and its ecosystems. The context of this section is the geological history of Earth (see Sect. 1.5.4), and more specifically the three phases of the Earth's history illustrated in Fig. 1.11. These are the Mineral, Cradle and Living Earth, which occurred from 4.6 to 4.0, 4.0 to 2.5, and 2.5 to 0 billion years ago, respectively.

The temperature conditions at the surface of Earth are mostly determined by the level of solar radiation that reaches the surface of the planet (see Sect. 3.2.1). This is modulated by four characteristics of the Solar System and the Earth System, which are: the Earth's distance from the Sun, the presence of a significant atmosphere, the planet's fast rotation, and the constant axial tilt of the planet (see Sect. 3.6.2). The latter characteristic is explained by the stabilizing effect of the Moon (see Sect. 3.3.2). Since the Mineral Earth phase, there is a balance between the energy Earth receives from the Sun and the energy it radiates back to space, the planet absorbing and radiating energy at different wavelengths (see Sect. 3.2.1). In addition to the energy received from the Sun, geothermal energy also contributes to fuel the Earth System (see Sect. 3.2.3). Overall, the presence of liquid water on Earth for more than four billion years provides evidence of a long-term balance of the radiation budget (see Sect. 3.2.1).

During the Cradle Earth phase, the young Sun was fainter than presently, and various mechanisms have been proposed to explain how temperatures were nevertheless high enough for water to be liquid and organisms to grow. Indeed, the occurrence of suitable temperatures and the presence of liquid water are key conditions of habitability (Sect. 3.6.1). Certain of the proposed mechanisms involve effects of organisms (see Sect. 3.7.1). Later, the increasing intensity of the Sun together with the decreasing concentration of atmospheric CO_2, and thus decreasing greenhouse effect, contributed, together with other mechanisms, to keep the Earth's temperature within a range suitable for organisms (see Sect. 3.7.2).

Some temperature-related characteristics of the present Earth already existed during the Cradle Earth phase. Most importantly, about 30% of the solar energy received by Earth is reflected back to space, and the remaining 70% are distributed among the five components of the Earth System (liquid water, atmosphere, ice, continental surfaces, ecosystems) (see Sects. 3.2.1 and 3.5.1). Temperatures suitable for organisms on most of the planet are created by the combination of the conditions that exist at the top of the atmosphere and the redistribution of heat over the planet by atmospheric currents and the oceanic circulation (see Sects. 3.3.4 and 3.5.4). In addition, the Earth's atmosphere has the right density and greenhouse gas composition for creating a moderate natural greenhouse effect suitable for ecosystems (see Sect. 3.6.1). This is critical because at high temperature, biological molecules are destroyed, and at low temperature, the freezing of internal water damages cells. However, the diversity of organisms is such that some live in high-temperature environments, and others in low-temperature environments (see Sect. 3.1.1). The narrow range of temperature variations at the surface of Earth, except in some regions, allowed the organisms to establish themselves durably on Earth and ecosystems to thrive there (see Sect. 3.1.2).

The O_2-producing photosynthesis, which began during the Cradle Earth phase, absorbs part of the solar energy (light) reaching the Earth's surface. In return photosynthesis contributes to the flux of heat radiated by Earth to space, although in a very small proportion relative to physical processes (see Sect. 3.4.2). However, because of the long-term build-up of biomass on continents and in oceans, photosynthesis has major effects on the radiation budget at long timescales (see Sect. 3.5.5). A small fraction of the organic matter becomes fossil organic matter, whose lifetime can reach millions of years; some of the fossil organic compounds are stored in geological reservoirs, but most exist as dispersed particles in sediments and rocks (see Sect. 3.4.2).

Other temperature-related mechanisms of Earth are characteristic of the Living Earth phase. The occupation of continents began less than 500 million years ago, and the presence of an extensive vegetation cover on continents contributes to decrease the fraction of the solar radiation reflected back to space, called albedo (see Sect. 3.5.5). The ongoing succession of worldwide cycles of global cooling and warming (Ice Age), which began 2.16 million years ago, reflects a 100,000-year cyclic change in the solar radiation reaching Earth caused by a 100,000-year cycle in astronomical phenomena (see Sect. 3.2.1). The release of anthropogenic CO_2 and other greenhouse gases in the atmosphere causes a temporary imbalance between the absorption of solar energy by the planet and its radiation of energy back to space. The resulting global warming will persist or increase as long as the greenhouse gases continue to increase in the atmosphere, thus affecting the ecosystems and in return human activities (see Sect. 3.2.1).

Figure Credits

Fig. 3.1 Original. Figure 3.1 is licensed under CC BY-SA 4.0 by Philippe Bertrand, Louis Legendre and Mohamed Khamla.

Fig. 3.2 Modified after Figure 2-12 of Ackerman and Knox (2017). With permission from Profs. Steven A. Ackerman, University of Wisconsin-Madison, USA, and John A. Knox, University of Georgia, USA.

Fig. 3.3 Original. Figure 3.3 is licensed under CC BY-SA 4.0 by Philippe Bertrand, Louis Legendre and Mohamed Khamla.

Fig. 3.4 This work, Figure 3.4, is a derivative of https://commons.wikimedia. org/w/index.php?curid=9389402 by Jalanpalmer (talk) https://commons.wikimedia.org/wiki/User:Jalanpalmer and (https://commons.wikimedia.org/wiki/User_ talk:Jalanpalmer), in the public domain. I, Mohamed Khamla, release this work in the public domain.

Fig. 3.5a This work, Figure 3.5a is a derivative of https://commons.wikimedia. org/wiki/File%3AAtmosphCirc2.png by DWindrim, used under GNU FDL and CC BY-SA 3.0. Figure 3.5a is licensed under GNU FDL and CC BY-SA 3.0 by Mohamed Khamla.

Fig. 3.5b,c https://commons.wikimedia.org/wiki/File:Mslp-jja-djf.png by William M. Connolley https://en.wikipedia.org/wiki/User:William_M._Connolley, used under GNU FDL and CC BY-SA 3.0.

Fig. 3.6 This work, Figure 3.6 is a derivative of https://en.wikipedia.org/wiki/ File:Corrientes-oceanicas.png by Dr. Michael Pidwirny (see http://www.physical-geography.net), in the public domain, U.S. Government publication. The original position the Kuroshio current was slightly modified. I, Mohamed Khamla, release this work in the public domain.

Fig. 3.7a Original. Figure 3.7a is licensed under CC BY-SA 4.0 by Philippe Bertrand, Louis Legendre and Mohamed Khamla

Fig. 3.7b   https://science.nasa.gov/earth-science/oceanography/physical-ocean/ ocean-surface-topography/the-dynamic-pacific-ocean by NASA's Jet Propulsion Laboratory, in the public domain.

Fig. 3.8 This work, Figure 3.8, is a derivative of https://commons.wikimedia.org/ wiki/File:Conveyor_belt.svg by Avsa https://commons.wikimedia.org/wiki/User:Avsa, used under GNU FDL and CC BY-SA 3.0, and https://commons.wikimedia.org/wiki/ File:Oceans.png by Saperaud, used under GNU FDL and CC BY-SA 3.0. Figure 3.8 is licensed under GNU FDL and CC BY-SA 3.0 by Mohamed Khamla.

References

Ackerman SA, Knox JA (2017) Meteorology: an interactive understanding of the atmosphere, Version 5.0. Top Hat. Available via https://tophat.com/marketplace/ science-&-math/earth-sciences/textbooks/meteorology-an-interactive-understand-ing-of-the-atmosphere-john-knox-steve-ackerman/3067/

Clarke A (2014) The thermal limits to life on Earth. Int J Astrobiol 13:141–154. https://doi. org/10.1017/S1473550413000438

Girard JE (2009) Principles of environmental chemistry, 2nd edn. Jones and Bartlett Publishers, Sudbury, MA, USA

Further Reading

Boden DR (2017) Geologic fundamentals of geothermal energy. Energy and the environment. CRC Press, Boca Raton

Broecker WS (1991) The great ocean conveyor. Oceanogr 4(3):79–89. https://doi.org/10.5670/ oceanog.1991.07

Broecker W (2018) CO_2: Earth's climate driver. Geochem Perspect 7:117–196. https://doi.org/10.7185/geochempersp.7.2

Clarke A (2017) Principles of thermal ecology: temperature, energy and life. Oxford University Press, Oxford

Hansell DA et al (2009) Dissolved organic matter in the ocean: a controversy stimulates new insights. Oceanogr 22(4):202–211. https://doi.org/10.5670/oceanog.2009.109

Holden JF et al (2012) Biogeochemical processes at hydrothermal vents: microbes and minerals, bioenergetics, and carbon fluxes. Oceanogr 25(1):196–208. https://doi.org/10.5670/oceanog.2012.18

Kämpf J, Chapman P (2016) Upwelling systems of the world. A scientific journey to the most productive marine ecosystems, Springer, Cham

Lagerloef G et al (2010) The ocean and the global water cycle. Oceanogr 23(4):82–93. https://doi.org/10.5670/oceanog.20https://doi.org/10.07

Rapp D (2014) Assessing climate change: temperatures, solar radiation and heat balance, 3rd edn. Springer Praxis Books, Springer, Cham

Sardet C (2015) Plankton: wonders of the drifting world. University of Chicago Press, Chicago

Schmitt RW (2018) The ocean's role in climate. Oceanogr 31(2):32–40. https://doi.org/10.5670/oceanog.2018.225

Wells NC (2012) The atmosphere and ocean: a physical introduction, 3rd edn. Wiley-Blackwell, Chichester

Chapter 4
Overall Habitability: Connections with Geological and Astronomical Events and Processes

4.1 Habitability of Earth for Organisms, Populations and Ecosystems

4.1.1 Overall Habitability of Earth

This chapter considers the Earth's *overall habitability*, that is, the proportion of the surface of the planet that is habitable. In Chapter 2, we introduced the concept of *habitability* (see Information Box 2.4), and in Chapter 3, we explained the conditions that contribute to keep the temperatures at the surface of Earth at levels generally suitable for organisms, and we also described effects of ecosystems on the global temperature. The habitability of Earth is largely determined by the suitability of its temperature for organisms, but the latter is not the only characteristic of the planet that makes it habitable. The other components of overall habitability include the presence of liquid water, the access to sources of energy usable by organisms, and the availability of nutritive resources.

The following sections of the chapter describe variations in the extent of the Earth's habitable environments at a wide range of periods: diurnal (Sect. 4.2), annual (Sect. 4.3), within centuries (Sect. 4.4), and from tens of thousands to hundreds of millions of years (Sect. 4.5). The chapter also examines mass extinctions, which are witnesses of major crashes in habitability (Sect. 4.6). Biological innovations and anthropogenic changes to atmospheric gases have also caused or are presently causing major overall habitability changes (Sect. 4.7). The chapter ends with a summary of key points concerning the interactions between the Solar System, Earth, its overall habitability, and organisms (Sect. 4.8). Factors involved in setting the overall habitability of the planet include the Earth's rotation and orbit around the Sun, large volcanic eruptions, the Milankovich orbital cycles, climate changes, changes in the positions of continents, and the impacts of large meteoroids.

© Springer Nature Switzerland AG 2021
P. Bertrand and L. Legendre, *Earth, Our Living Planet*, The Frontiers Collection, https://doi.org/10.1007/978-3-030-67773-2_4

The following paragraphs summarize the global significance of the above three components of overall habitability, namely liquid water, energy, nutritive resources.

Liquid water. The range of ambient temperatures allows most of the water at the surface of Earth to be liquid under the existing atmospheric pressure (see Sect. 5.5.1). This is true now as seawater is liquid between about −2 and +101 °C under the present atmospheric pressure, and it was true 4 billion years ago when water could be liquid between about −10 and +280 °C under an atmospheric pressure 60 times higher than at present (Table 5.2). Liquid water is essential to the growth of organisms, for reasons detailed in Chapter 5 (see Sect. 5.1.1). At the surface of Earth, water is found in the liquid state in oceans, rivers and lakes, which cover more than 70% of the planet (see Sect. 5.2.1). A small fraction of the surface water is frozen solid on continents (ice sheets, ice caps, glaciers, and frozen lakes) and in polar waters (ice pack, which is formed of sea ice). A still smaller fraction of the water is in the gaseous form of water vapour in the atmosphere, where it also exists as ice crystals that form the clouds.

The abundance of liquid water early in the history of Earth (see Sect. 5.5.4) allowed ecosystems to rapidly occupy the oceans, where all organisms were unicellular (see Information Box 3.1) during more than 3 billion years, that is, from 3.8 billion years ago or earlier until the appearance of multicellular organisms 650 to 540 million years ago. The abundance of liquid water on most emerged lands allowed multicellular organisms to start their occupation of the continents during the Silurian and Devonian periods, from 444 to 359 million years ago (see Sect. 1.5.4). This was possible because the presence of the ozone layer in the high atmosphere, which had formed about 100 million years earlier, protected the surface organisms from harmful solar UV radiations (see Sects. 2.1.5 and 8.1.1). The book of Falkowski (2015) provides further reading on the billion-years transformation of the Earth's chemistry by unicellular organisms.

Energy. The autotrophic organisms (see Information Box 4.1) use two different sources of energy to synthesize new organic matter through the processes of chemosynthesis and photosynthesis (see Information Box 5.1). The first type of process uses the energy supplied by chemical reactions, and the second is powered by the energy carried by particles of light, called *photons*. *Chemosynthesis* goes back to the first organisms on Earth, and fuels nowadays a wide variety of ecosystems. It was the only process used by autotrophs to acquire energy until the advent of photosynthesis more than 3 billion years ago. The initial form of photosynthesis did not produce oxygen, and this O_2-*less photosynthesis* was followed more than 2.5 billion years ago by O_2-*producing photosynthesis*. The two photosynthetic processes coexist on present Earth, where O_2-producing photosynthesis is dominant. Their occurrence depends on the amount of light available to photosynthetic organisms, which varies during the course of the day, among seasons, and also over longer timescales.

Both the chemosynthetic and photosynthetic autotrophs synthesize organic matter and release chemicals during the process. Examples of such chemicals are O_2 for O_2-producing photosynthesis, and sulphur (S) for O_2-less photosynthesis or

> **Information Box 4.1 Autotrophy and Heterotrophy, Autotrophs and Heterotrophs** The organisms on Earth can be grouped into two broad categories in terms of the materials and energy they use for their maintenance, growth, and replication (Table 1.1). Some organisms, called *autotrophs* (which mean "self feeding"), are able to synthesize organic matter from carbon-containing and other inorganic compounds, using as their source of energy either solar light or the chemical energy released by reactions of inorganic compounds. The synthesis of organic matter from inorganic compounds is called *autotrophy*.
>
> Other organisms, called *heterotrophs* (which means "feeding on others"), are unable to synthesize organic matter from inorganic compounds, and thus derive their materials and energy from organic matter synthesized by autotrophs and passed on through the food web. This is called *heterotrophy*. Still other organisms, called *mixotrophs*, are able of both autotrophy and heterotrophy. The autotrophs, heterotrophs and mixotrophs are part of *food webs* (see Information Box 5.3). The article of Ortega (2019) provides further reading on mixotrophs.

chemosynthesis (see Information Box 5.1). The solar or chemical energy is potentially contained in both the organic matter, where carbon is in a reduced form with respect to the CO_2 carbon source, and the released chemicals. Examples of the latter chemicals are: oxygen in the O_2 released by O_2-producing photosynthesis (Eq. 5.1), where this chemical is more oxidized than in the H_2O oxygen source; and elemental S released by O_2-less photosynthesis or chemosynthesis (Eqs. 5.2, 5.3 and 5.5), which is more oxidized than in the H_2S source. In these examples, the organisms recombine the reduced forms of carbon (organic matter) with the oxidized forms of oxygen or sulphur present in the environment to release the initial energy (see Information Box 5.2). They use this energy to fuel their maintenance, growth, reproduction (Figs. 1.2 and 5.1), and various activities such as movements … and writing and reading books (in humans, the brain of the average adult in a resting state consumes 20–25% of the body's energy).

Nutritive resources. The nutritive resources of autotrophs are CO_2 and various inorganic compounds that exist in the Earth's environment. The latter include inorganic nutrients, and also water for O_2-producing photosynthesis (see Information Box 5.1). These inorganic compounds are used by autotrophs to synthesize organic matter. The nutritive resource of heterotrophs is the organic matter passed on to them by the food web (see Information Box 5.3). When the first organisms got established on Earth, the key inorganic nutritive resources of autotrophs were readily available in the planet's environment (see Sect. 6.5). The availability of the required resources allowed the build-up of large biomasses over hundreds of millions and billions of years, first in the oceans, and much later at the surface of continents, which led the organisms to progressively take over the Earth System. The book of Middelburg (2019) provides further reading on the effects of temperature, light and nutrients on the production of marine phytoplankton.

Planet Earth is characterized by a wide array of environments that have been hospitable to organisms and ecosystems since a very long time. However, the whole planet is not equally habitable. For example, the wet and sunny regions covered by tropical rain forests are much more habitable than the (almost) lifeless dry desert areas. In addition, the habitability of some regions has changed with time. For example, very lush areas in the northern halves of North America and Eurasia were covered by glaciers several kilometres thick during the last Ice Age until about 10,000 years ago. This event was part of the glacial-interglacial cycles that have repeated themselves since the beginning of the Quaternary period 2.6 million years ago (see Sect. 4.5.2). Hence the overall habitability of Earth, in term of the proportion of the planet able to sustain thriving ecosystems, changed with time. Over the history of Earth, there were large changes in habitability of the various environments, to which ecosystems accommodated through biological evolution. In the present chapter, we consider the overall (spatial) habitability of Earth for periods that range from one day to billions of years (Table 4.1). Further reading on the physics of present-day climate variability is provided by the book of Wallace and Hobbs (2006), on changes in the Earth's climate for a wide range of timescales by the book of Ruddiman (2008), and on the evolution of Earth into a habitable world by the book of Lunine (2013).

4.1.2   Habitats on Earth

The bulk of the global biomass of organisms is found at the surface of the solid Earth (continents and islands, bottom of shallow waters) and in the surface layer of water bodies (oceans and lakes). These habitats are preferred by organisms as they provide a substrate, liquid or solid, where they can grow and multiply, and where there is enough light for the photosynthetic production of organic matter. The latter also depends on the availability of essential inorganic substances such as water (which can be scarce at times on land) and nutrients (which sometimes limit photosynthesis in surface waters and also on land).

Non-photosynthetic microbes and animals are present and sometimes very abundant in association with photosynthetic organisms. They can also thrive in other environments such as the ocean's water column and seafloor, where they mostly live on organic matter coming from surface waters; deep in the sediment, where bacteria that do not use O_2 for respiration produce organic matter using chemosynthesis; and the guts of animals, where microorganisms often use fermentation to digest the organic matter eaten by their host, for the benefit of both the symbionts and their hosts (*symbiosis* is the interaction between two different organisms that live in close physical association, for the advantage of both).

Table 4.1 Dominant (or average) periods of variability in the Earth's environment, from 1 day to billions of years, with related changes in habitability, and known or possible causes. Numbers in brackets: section of this chapter where the phenomenon is described

Dominant period of variability (years)	Changes in Earth's habitability	Causes: known or possible (sections of this chapter)
0.003 (1 day)	Day-night cycle: equal habitability of all longitudes	Earth's rotation (4.2.1)
1	Annual changes in day length and temperature; annual changes in environmental conditions: cyclones and seasonal sea ice	Annual changes in the latitudinal distribution of the solar radiation (4.3.1 and 4.3.2)
Less than 100	Nine volcanic eruptions with large climatic effects since year 1450: some significant changes in habitability at regional or global scales	Large volcanic eruptions (4.4)
100,000 (40,000 before 1 million years ago)	From 30 to 50 interglacial-glacial cycles in the last 2.6 million years: large changes in latitudinal habitability	Cyclical changes in the Earth's orbital parameters (4.5.2)
100,000 to 100 millions	Periods of global cooling and mass extinctions	Very large volcanic eruptions and impacts of large asteroids (4.5.3)
100 millions	Five mass extinctions in the last 542 million years: very large changes in habitability at the global scale	Climate change, large volcanic eruptions, impacts of large meteoroids (4.6)
More than 100 millions	Three snowball Earth episodes between about 2,200 and 640 million years ago: very large changes in habitability at the global scale	Low atmospheric CO_2 caused by of supercontinent break-up, volcanic activity and erosion of silicate rocks, or oxidation of CH_4 by global oxygenation (4.5.4)
150 millions	Several long periods of high and low temperature in the last 542 million years: large-scale effects on overall habitability	Changes in concentration of atmospheric CO_2, due to changes in the rate of seafloor spreading or continental erosion (4.5.5)
290 millions	Greenhouse and icehouse Earth since 4.5 billion years: very large changes in habitability at the global scale	Changes in the concentration of atmospheric CO_2, perhaps resulting from changes in tectonic activity (4.5.6)
Billions	One-time changes in overall habitability	Biological innovations (4.7.1 and 4.7.2)
Billions	One-time change in overall habitability	Anthropogenic changes to atmospheric gases (4.7.3)

Large biomasses of organisms developed in the numerous Earth's habitats. This indicates that environmental conditions in these habitats not only allowed organisms to survive, but favoured the development of robust populations and thriving ecosystems (see Information Box 4.2).

Information Box 4.2 Population, Community, and Ecosystem Ecologists distinguish between the following three terms: population, community, and ecosystem. A *population* refers to the organisms of a given species that live in a particular geographic area and have the capability of interbreeding. For example, the population of haddock in the North Sea includes all individual haddock fishes in that sea. An ensemble of populations of different species is called a *biological community*.

In the ecological literature, the word *ecosystem* refers to the combination of a community with its environment. For example, the Himalayan ecosystem includes the community of all the species found in the Himalaya Mountains together with the components of the Himalayan environment. Similarly, the wetland ecosystem includes the communities of all the species that inhabit wetlands together with the characteristic components of wetlands worldwide. In this book we look at hidden relationships among a series of nested systems, which are the ecosystems, the Earth System, the Solar System, and the Universe (see Sect. 1.1.1).

4.1.3   Habitability of Earth for Organisms and Their Populations

The thermal limits for the growth and survival of organisms were described in Chapter 3 (see Sect. 3.1.1). The following paragraphs consider the effects of these limits on the habitability of Earth for organisms and their populations.

Living cells can individually survive over a very wide range of temperatures (Table 3.1), from about $-196\ °C$ (possible survival of certain bacteria or eukaryotes) to almost $+130\ °C$ (possible survival of some archaea). These same organisms not only survive but also grow from $-20\ °C$ to $+122\ °C$. These abilities are related, for high temperatures, to the thermal stability of the biological molecules making up cells and, for low temperatures, to the ability of organisms to overcome the freezing of their cells' internal water and thus avoid major damages to their cell structures.

The above extremely cold conditions ($-196\ °C$) were created during laboratory experiments, and do not exist in the Earth's environment. In addition, growth at the lowest thermal limit of $-20\ °C$ is only possible under very high-pressure conditions where the water is supercooled (see Sect. 4.5.1). An individual unicellular organism may be able to temporarily maintain its structure and activity at such low temperatures, but the development of a population of this organism is not possible when the external environment is frozen. Indeed a frozen environment impedes several major biological functions such as: the uptake of nutrients, which are mostly dissolved in water; the excretion of internal waste to the surrounding medium, which is not any more in the fluid state; and exchanges among individuals, except very limited.

The above high temperature limit (+122 °C or higher) can only be reached outside liquid water, which evaporates at +100 °C, or in liquid water under very high pressure, which evaporates at higher temperature. In the first case, a few unicellular organisms can temporarily survive without liquid water, but no significant population can develop in such condition. The habitats corresponding to the second case are located in the deep ocean near hydrothermal vents (up to 5000 m depth), where the hydrostatic pressure is so high (up to 500 atm) that the very warm water emerging from the ocean floor (up to more than +400 °C) can be liquid (Fig. 5.3). In this environment, chemosynthetic bacteria oxidize the hydrogen sulphide (H_2S) contained in the water streaming from the vents to sulphur, and use the resulting energy to synthesize organic matter, which is sometimes abundant enough to sustain thriving ecosystems of large organisms. The latter include snails, clams, mussels, crabs and shrimps, but these systems are very small oases in the midst of the generally low-production ocean floor, whose depths greater than 1500 m represent more than 60% of the Earth's surface. On land, the bacteria and archaea found in the geothermal waters of hot springs and geysers have optimal growth temperatures generally below +100 °C. The article of Holden et al. (2012) provides further reading on hydrothermal vent systems.

It was proposed that some of the extremely warm or extremely cold environments had played important roles in biological evolution. For example, some researchers think that the first organisms on Earth might have appeared in high-temperature environments such as hydrothermal vents. This has the implication that other hot environments in the Solar Systems could harbour life forms. Other researchers found in the laboratory that dilute solutions of ammonium cyanide (NH_4CN) frozen at −20 and −78 °C during 25 years yielded substantial amounts of N-containing bases and simple amino acids, which are the building blocks of nucleic acids (Table 6.2). This suggests the possibility of an evolutionary development of organisms in very cold environments, which is especially interesting because such an environment exists on Europa, a moon of Jupiter, where frozen water possibly overlies an ocean of liquid water (see Information Box 5.4).

An example of very warm climate is provided by the hot desert conditions around the town of Reggane, which located in the Sahara desert in southern Algeria (Table 4.2). Temperatures there can be extremely high in summer, that is, +47 °C on average with some record values higher than +55 °C. The hot Sahara desert is one of the least habitable places on Earth, and organisms are only present in a few rare oases, although their virtual absence elsewhere in the desert is more due to the lack of water than to the high temperature. Indeed, many bacteria are able to live at temperatures above 60 °C, and some archaea up to 115 or 120 °C (Table 3.1).

Conversely, the Arctic polar site of Alert Point, in Canada (Table 4.2), is among the coldest places on the planet where organisms do survive. Temperature varies between −30 and −37 °C in winter, under conditions of permanent night, with record lows of about −50 °C. In the High Arctic during summer, the continual sunlight allow temperatures to become high enough to thaw some of the snow and ice, creating small amounts of liquid water. The few terrestrial organisms adapted

Table 4.2 Examples of diurnal and seasonal variations in surface temperature at climatically-contrasted sites. Temperatures are in °C

Site	Reggane	Manaus	Alert Point	Sydney	Xi'An	La Paz	Nice	Mediterranean Sea off Nice	
Country or depth	Algeria	Brazil	Canada	Australia	China	Bolivia	France	43° 41' N 7° 19' E (2015)	
								Surface	Depth
Latitude	26° 43' N	3° 08' S	82° 29' N	33° 52' S	34° 15' N	16° 30' S	43° 42' N		
Longitude	0° 10' E	60° 1' O	62° 20' O	151° 12' E	108° 57' E	68° 08' O	7° 15' E		
Altitude (m)	207	92	63	58	405	3650	5	−1m	−100 to −500m
Climate	Warm desert	Wet equatorial	Polar Arctic	Coast. wet subtropical	Cont. wet subtropical	High mountain	Coastal mid-latitude	Western Mediterranean Sea	
Mean temperatures									
Highest value during the warm season	47	33	6	27	32	15	27	28.5	13.5
Lowest value during the warm season	33	23	1	20	22	5	18	28	13.5
Highest value during the cold season	23	31	−30	17	−5	12	12	13.5	13.5
Lowest value during the cold season	9	23	−37	8	−5	−1	4	13.5	13.5
Mean diurnal temperature amplitudes[a]									
During the warm season	14	10	5	7	10	10	9	0.5	0

(continued)

Table 4.2 (continued)

Site	Reggane	Manaus	Alert Point	Sydney	Xi'An	La Paz	Nice	Mediterranean Sea off Nice
During the cold season	14	8	7	9	0	13	8	0
Record temperatures								
Record high	>55	37	20	46	43	30	38	
Record low	5	23	−50	0	−17	−11	−4	
Mean seasonal temperature amplitudes[b]								
Of highest temperatures	24	2	36	10	37	3	15	15
Of lowest temperatures	24	0	38	12	27	6	14	14.5

[a]Difference between the day and night temperatures

[b]Différence between the highest [lowest] value during the warm [cold] season and the highest [lowest] value during the cold [high] season

to this type of environment develop very rapidly during the warm season and use overwintering forms to survive through the cold season. One example of these organisms is the Antarctic midge *Belgica antarctica*, whose adults live only one week and larvae survive the deep winter conditions in dehydrated form. Another example is provided by mosquitoes in continental Greenland, whose pupae survive winter in diapause in water under the ice.

The habitability of environments is an important property for populations. Indeed, the survival of individual organisms may depend on their ability to withstand extreme environmental conditions. Because individual organisms either divide or allocate part of their resources to the production of gametes for sexual reproduction at some point in their growth process, the growth of individual organisms may result in the development of populations. As a consequence, the environmental limits for the growth of individuals determine the habitability of the environment for the populations.

Through biological evolution, the selected species progressively occupied the Earth's habitats, and in turn the presence of large populations of organisms modified the outer envelopes of the planet. The co-evolution of species and the Earth's habitats during billions of years led organisms to occupy almost all habitats at the surface of the planet, even those with quite extreme conditions. This evolution was not a continual development from the earliest times to present since, on several occasions, large parts of the existing species were destroyed during mass extinctions (see Information Box 1.6). After such events, new species evolved, and the resulting species and ecosystems were generally very different from those that had been destroyed.

4.1.4 Survival Strategies of Populations

The ability of a population to survive and develop depends on the individual abilities of some of its members, seeds or embryos, to survive during adverse periods until the return of suitable environmental conditions. These conditions are those that define the habitability for the population. Hence habitability is linked not only to the biological ability of individuals to survive extreme conditions (which can be seen as individual achievements), but also to the survival strategies developed through evolution by which the genes (which are located in individuals, seeds or embryos) are passed on from one suitable period to the next.

Through biological evolution, a number of strategies were progressively selected that allowed populations to survive periods of unsuitable conditions of temperature or resource (food, water). These strategies include four main categories: special metabolic and/or behavioural adaptations, overwintering and dormancy, short life cycle, and resting forms.

Metabolic and/or behavioural adaptations. Special metabolic and/or behavioural adaptations are found in both polar regions and hot deserts. In *polar regions,* several warm-blooded animals remain active during winter, when temperatures are often several tens of degrees below freezing and food resources are scarce. Examples are polar bears, reindeers, Arctic foxes and Antarctic penguins, whose bodies have layers of thermo-protective fat. Their food resources depend on the presence of liquid water: either liquid fresh water on the continent during the summer, which allows the growth of vegetation and prey organisms; or liquid seawater under or close to the ice pack, which is the habitat of prey, for example seals for polar bears and fish for Antarctic penguins. *In hot deserts*, where temperature can reach high values, a major problem for many species is the retention of water. There, plants and animals use various mechanisms to prevent the loss of water, such as small leaves or spines that minimize evaporation, the production of very concentrated urine, or activity restricted to the cool night period.

Overwintering, hibernation, or dormancy. Other species use strategies of overwintering or dormancy. *Overwintering* means surviving the cold, resource-poor season by migration or hibernation. Indeed, one way to survive that season is to *migrate* to areas with milder climate, as do many bird populations. Another way is *hibernation*, a condition in which the body slows down its metabolic functions to minimize energy consumption and live on reserves accumulated during the previous season. Well-known examples bears and marmots in cold areas. This type of strategy is also found in plants that decelerate or stop growth in winter, a state called *dormancy*. The latter is also encountered in deserts, where some plants produce seeds that lie dormant in the soil and some animals remain in a state of dormancy for long periods until the occasional rainfall.

Short life cycle. Another strategy in very cold, very hot, or very dry environments is a short life cycle, whereby all active phases of the life cycle are completed within the short periods of favourable conditions, and the population is in a dormant stage during the long periods of harsh conditions. For example, some annual desert plants germinate, bloom and die within the course of a few weeks after a rainfall, and insects (such as mosquitoes) in the Arctic go from egg to aquatic larva and pupa and to flying adult in a few weeks as soon as the snow melts and the ground thaws. Those who have visited Arctic or sub-Arctic regions in summer remember the dense clouds of blood-hungry female mosquitoes and black flies that constantly surrounded them.

Resting stages or resting forms. Finally, individual cells of bacteria, archaea and protists, and some multicellular organisms or their embryos are able to survive extended periods of freezing or drought, well beyond a year, through resting stages or resting forms. These stages or forms are characterized by a total lack of metabolic activity, which is possible because resting forms contain "antifreeze" molecules and are also largely dehydrated. Dehydration blocks metabolic exchanges with the surrounding environment, and prevents the cells from being damaged by the creation of internal ice crystals. The dehydration mechanism is found in lichens, some trees, and various animals that belong to tardigrades, nematodes and arthropods.

Tardigrades (Fig. 4.1) are eight-legged microscopic invertebrates that can naturally survive unfavourable periods for centuries and even millennia in a desiccated state, in which the tardigrade is known as a *tun* (this name referring to the fact that

Fig. 4.1 Scanning electron microscope picture of tardigrade *Milnesium tardigradum* in active state. The total length of the body does not exceed 0.7 mm. Credits at the end of the chapter

the animal then has a rounded barrel-like shape). It was found that in this resting stage, the internal clock of the tardigrade stops, so that the animal does not age any more. Because of their shapes, tardigrades are sometimes called *water bears* or *moss piglets*, and their 1500 species form the separate phylum Tardigrada. In a different way, the reproductive cells and embryos of animals and humans can now be preserved frozen in liquid nitrogen (−196 °C), which requires the addition of products that prevent freezing and dehydration, and thus optimize the duration and survival of the preserved material.

The above paragraphs showed that the co-evolution of species and the Earth's environments led to the establishment of ecosystems in most environments of the planet, even those with extreme conditions provided that liquid water was present, at least temporarily. The habitability of environments determines the presence or absence of populations of various species on a wide range of timescales. Four astronomical and planetary conditions contribute to maintain at the surface of Earth temperatures within a range suitable for organisms: the Earth's distance from the Sun, the presence of a significant atmosphere on the planet, the tilt of the Earth's axis of rotation maintained by the Earth-Moon system, and the planet's fast rotation (see Sect. 3.6.2). It follows that these astronomical and geological conditions determine the habitability of the planet.

The above sections examined various aspects of the overall habitability, and how organisms and their populations progressively occupied most environments at the surface of Earth. The remainder of the chapter examines variations that occurred in the extent of the Earth's habitable environments for periods ranging from one day to billions of years (Table 4.1). The textbook of Goosse (2015) provides further reading on past variations in the Earth's climate for periods that range from a few years to billions of years.

4.2   Diurnal Changes in Environmental Conditions

4.2.1   Effects of Earth's Rotation

In this section and the next two, we examine the effects that changes with a periodicity of one or a few days have on the overall habitability of the planet (Table 4.1). The Earth's rotation is the fastest among the four rocky planets in the Solar System (Table 1.1), with the consequence that the exposure of a given point of the planet to a continually high or low radiation flux is short, except at the height of summer or in the dead of winter above the polar circle (see Sect. 3.3.3). Because of this, most of the surface of Earth is located at latitudes where the periods of continuous light (day) or darkness (night) are not overly long at any time of the year, except above the polar circles where the days and nights can last 24 h depending on the season (Fig. 3.4).

The 24 h Earth's rotation thus makes all longitudes equally habitable. To really appreciate the effect of rotation on habitability, one can try to imagine what would be the situation at the surface of Earth if the planet always showed the same face to the Sun, in the same way as the Moon always shows the same face to Earth. This phenomenon, called *tidal locking*, comes from the fact that the Moon is located close to Earth and its mass is much smaller than that of Earth (see Sect. 1.3.5). Hence a planet located close to its star would be tidally locked to it, and thus have one face continually exposed to light and heat, and the other face in continual darkness. It has been speculated that if such a planet had an atmosphere, the latter could transport heat around the planet and thus reduce the temperature differences between the two faces, whereas different hypotheses about the environment of the planet led to the conclusion that there would be major temperature contrasts between the two faces. In any case, the habitability of a tidally locked Earth would certainly be very different from that of the rapidly rotating Earth.

4.2.2   Diurnal Variations

The Earth's rotation creates an alternation between days and nights over most of the planet. As a consequence, the solar flux that reaches a given location of the lower atmosphere and the ground in 24 h varies with the length of the day and with the season (see Sect. 3.3.3). In polar areas, the Sun remains continually above or under the horizon during several months, and the long period with continual sunlight during the summer is called *polar day*, whereas the corresponding period without sunlight during the winter is known as *polar night*. At latitudes below the polar circles, the duration of daylight is shorter than 24 h year round. Despite the short period of daylight, diurnal variations in temperature can be large. For example the mean diurnal temperature amplitude can be up to 14 °C at the Reggane site in the Sahara desert, and is close to 10 °C at several other sites in Table 4.2.

While the length of the day is presently 24 h, it was much shorter in the deep past because the rotation of Earth was faster. The solar day was about 6 h when Earth formed (4.5 billion years ago), 12.3 h at the end of the Archaean (2.5 billion years ago), 17.7 h during the Middle Proterozoic (1.2 billion years ago), and 20.5 h in the Cambrian (500 million years ago). It follows that changes in solar flux during the daily cycle were much faster on ancient Earth, especially in the low and mid-latitudes where there are presently no 24 h days or nights at any time of the year. The short duration of the days and nights at these latitudes caused the heat gain during the day and heat loss during the night to be small, which made the diurnal thermal range narrower than presently especially in dry deserts. This effect had less impact in humid areas where the thermal amplitude was limited by the humidity of the air.

The diurnal changes in temperature affect both the atmosphere and the ocean. The atmosphere generally shows higher diurnal temperature variability than the ocean, as observed in the Mediterranean coastal city of Nice (France) where the mean diurnal temperature amplitudes are more than 10 times wider on the land that in the adjacent sea (Table 4.2). This because it takes 4 times more solar radiation to heat a given mass of seawater by 1 °C than the same mass of air. Conversely, the cooling of air is much faster than that of seawater. Because of this, the ocean is a major reservoir of heat, and as a consequence, diurnal variations of temperature in the air over the ocean are smaller than over continental surfaces.

Given the low diurnal temperature variability of oceans and large lakes, organisms living in these environments are not subjected to large diurnal temperature differences. These organisms may experience some small diurnal variability in temperature in the upper few centimetres of the water bodies, especially where there is strong density stratification. In the deep layers of the ocean and in the sediment, there are no temperature differences between day and night.

Contrary to their aquatic counterparts, the organisms on land surfaces may experience strong daily variability in temperature. This is also the case of organisms in very small volumes of water, such as puddles and interstitial soil water. However, this temperature variability does not create a problem of habitability in itself in any of these environments, even in those where liquid water disappears temporarily. The latter can occur in both hot and cold environments, as explained in the next two paragraphs.

In *hot deserts* where water disappears during the day, liquid water is generally present in small amounts during the night in the form of condensation droplets, which is enough to sustain the few species that live in these low-habitability environments. There, the organisms survive the high temperature and water scarcity during the day by minimizing their losses of water during the hottest hours, either by specific metabolic adaptations such as very concentrated urine (see Sect. 4.1.4) or by sheltering from the high solar radiation and heat during the day and being active during the night. The recurrent availability of some liquid water at some time during the 24 h cycle is sufficient to ensure a minimum level of habitability. In fact, the scarcity of organisms in low-habitability environments is mostly due to the scarcity of food resources, which is itself often linked to the low abundance of liquid water over very long periods. The latter is a climatic problem, in which the diurnal cycle plays no role.

In *cold environments* where liquid water temporarily freezes, the diurnal variability of temperature does not influence much the habitability. During polar winters, the frozen water does not thaw during the day, and the organisms living in these regions have the ability to cope with long periods of intense cold through either complete hibernation (plants and some animals whose internal metabolic activity is almost stopped), semi-hibernation (warm-blooded animals whose internal metabolic activity is slowed down), continuation of predatory activity (warm-blooded predators such as polar bears, foxes, wolves, and penguins), or feeding on hibernating plants (warm-blooded herbivores such as reindeers and elks).

Although temperature variability within one day can be quite high (Table 4.2), it does not seem to affect much, or at all, the habitability of most oceanic or continental environments. Concerning the Earth's overall habitability, the main environmental characteristic within one day is the 24 h Earth's rotation, which makes all longitudes equally habitable as explained above (see Sect. 4.2.1).

4.2.3 Variability Among Days

At the timescale of a few days, the local temperature may vary because of changes in weather (see Information Box 1.4). Such changes sometimes cause record high or record low temperatures, as those cited in Table 4.2. Some exceptional, very large deviations from average values can be greater than the diurnal or seasonal amplitudes, as shown at the Sydney site. Indeed in Sydney, there is a difference of 19 °C between the high-temperature record (+46 °C) and the highest average warm-season value (+27 °C), a difference much larger than the mean diurnal and seasonal amplitudes of the order of 10 °C. Large variations among days can partly be caused by the presence or absence of the cloud cover during several days, which considerably modify the solar flux that reaches the lower layers of the atmosphere and the ground.

Temperature variations among days are mostly related to movements of air in the atmosphere, which include horizontal winds, vertical movements of air, average trajectories of depressions and cyclones, and changes over timescales of a few days. Figure 4.2 illustrates the overall transport of heat from equatorial to higher

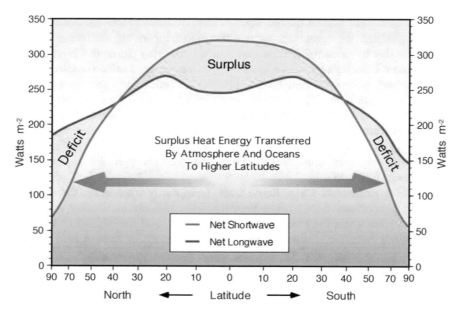

Fig. 4.2 Annual values of the net shortwave (incoming) and net longwave (outgoing) solar radiation between the North Pole (left) and the South Pole (right). Credits at the end of the chapter

latitudes. Between 35° North and 35° South, the incoming absorbed solar radiation exceeds the outgoing emitted terrestrial radiation, and there is thus a surplus of energy. The situation is reversed from 35 to 90° North and South, where there is a deficit of energy. The resulting gradient of heat between the low and high latitudes causes heat transfers from the equator towards the poles.

The combined atmospheric and oceanic circulations (see Sect. 3.5.4) are responsible for the *meridional transport* of energy. The adjective *meridional* comes from the fact that the lines of equal longitudes on the planet are termed *meridians*, so that the transport of substances or properties along these lines is called *meridional*. If there were no meridional transfers of energy, the polar latitudes would be 25 °C cooler than observed, and the equatorial latitudes 14 °C warmer.

At low latitudes, the solar radiation warms up the lower atmosphere, and part of the resulting heat is accumulated in surface waters of the oceans. Heat is transported directly to higher latitudes by marine currents, and is also transferred from the ocean to the atmosphere along the way of marine currents and subsequently transported polewards by atmospheric currents. The transfer of heat from the ocean to the atmosphere is mainly done by evaporation of water, which stores thermal energy in the air masses in the form of moisture and clouds (see Sect. 3.4.2). The atmospheric circulation transports part of these moist air masses to colder regions, where precipitation (rain, snow, hail) releases the heat stored in the moisture and clouds. The net effect of these processes is to increase the overall habitability of Earth among latitudes by evening out the temperature and the availability of water among the low and the high latitudes. This evening out of habitability among latitudes complements the effect of the Earth's rotation, by which the habitability is evened out among longitudes (see Sect. 4.2.1).

4.3   Annual Changes in Environmental Conditions

4.3.1   Solar Radiation and Heat Transport

In this section and the next, we examine the effects of annual changes on the overall habitability of the planet (Table 4.1). Solar radiation reaching Earth is strongest at low latitudes year-round and at mid-latitudes in summer. Because of the inclination of the Earth's axis of rotation, the angle of incidence of the solar radiation varies during the annual orbit of the planet around the Sun, which modifies the day length over a period of one year (Table 4.1). Because of this astronomical phenomenon, the amount of solar radiation, and thus heat, received during 24 h at each point at the surface of the globe changes seasonally (see Sect. 3.3.3). The astronomical mechanism of this seasonal change is illustrated in Fig. 3.3, and its effect on day length in Fig. 3.4.

Part of the heat received at low latitudes is transported towards high latitudes by the combination of atmospheric circulation and ocean currents (see Sect. 3.5.4). From the equator (0°) to 20° North and South, the ocean and the atmosphere carry similar amounts of heat polewards, and beyond 20° latitude the atmosphere becomes the main carrier (Fig. 4.3). At mid-latitudes, a large part of the heat carried by the atmosphere comes from the ocean, this heat being transferred to the lower atmosphere by evaporation of water from the upper ocean (see Sect. 3.4.2). Because of the transfers of heat from the ocean to the atmosphere, the geographic distributions of surface ocean temperatures largely contribute to determine the intensity and trajectory of the atmospheric circulation and the atmospheric phenomena associated with them (cyclones, storms, depressions, rainfalls).

The diagram in Fig. 4.3 combines all longitudes around the planet into global latitudinal values of polewards heat transport. Globally, the heat transported jointly by the oceanic (blue line) and atmospheric (red line) circulations from lower to higher latitudes reduces the seasonal temperature variability. The atmospheric transport of heat can be further divided into transport due to the movement of dry air (dashed red line) and transport associated with the evaporation and condensation of water vapour (dotted red line; latent heat; see Sect. 3.4.2). A net transport of energy due to the movement of water vapour occurs when evaporation and condensation take place at different latitudes.

The actual distribution of polewards heat transport between 90 °S and 90 °N at any longitude may be quite different from the global values in Fig. 4.3, because the combined oceanic-atmospheric heat transport does not influence in the same way the oceans and the continents and also different parts of the continents.

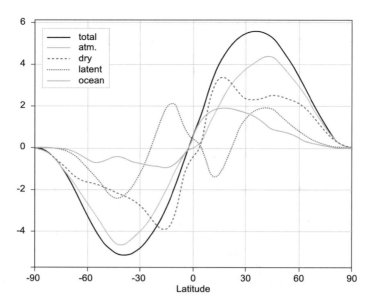

Fig. 4.3 Heat transport northwards (positive values) and southwards (negative values) between 90 °S (−90) and 90 °N (90): total transport is divided into oceanic and atmospheric transport, and the latter is divided into transport by dry air and latent heat (evaporation and condensation). Units PW: petawatts, 10^{15} watts. Credits at the end of the chapter

Indeed, seasonal temperature variations are generally much higher within the continental masses than in the marine coastal areas. An example of this difference is found in Table 4.2, where comparing continental Xi'An and coastal Sydney—which are located at similar latitudes north and south of the equator (34° 15' N and 33° 52' S, respectively) and have a similar wet subtropical climate—shows that the mean seasonal amplitudes of the highest and lowest temperatures are much greater in continental Xi'An (37 and 27 °C, respectively) than in coastal Sydney (10 and 12 °C, respectively).

Seasonal variability is mainly due to changes in the inclination of the Earth's axis of rotation relative to the plane of its orbit around the Sun, which cause the Northern and Southern Hemispheres to be alternatively more oriented towards the Sun during the annual cycle (Fig. 3.3). At inter-tropical latitudes (between 23.4 °S and 23.4 °N), the seasonal contrast is generally small, as seen at Manaus and La Paz, which are close to the equator and not far from the tropic of Capricorn, respectively (mean seasonal temperature amplitudes of 0–2 and 3–6 °C, respectively; Table 4.2). At mid-latitudes, the largest flux of solar energy (longest day of the year, summer solstice) is in June in the Northern Hemisphere and in December in the Southern Hemisphere. In polar areas, the solar flux is almost null in winter as the Sun does not come over the horizon for several months (polar night) and is maximum in summer when the sun does not go below the horizon for several months (polar day), as reflected in the very large mean seasonal amplitudes of both the highest and lowest temperatures at Alert Point (36 and 38 °C, respectively, in Table 4.2).

The oceanic and atmospheric currents transport heat and also water. This is not only the case for oceanic currents, which obviously transport water, but also for atmospheric currents, which receive water in the form of moisture in regions of strong evaporation and evapotranspiration and transport it elsewhere. The horizontal transport of water by oceanic and atmospheric currents also moves nutrients dissolved in seawater (oceanic currents) and in suspended aerosols (atmospheric currents). In addition, the atmospheric activity of storms and tropical cyclones (next section) and the large-scale movements of the ocean conveyor belt (see Sect. 3.5.3) contribute to mix the ocean vertically at different timescales and thus bring deeper, nutrient-rich waters to the surface of the ocean. These combined processes increase the global availability of water and nutrient resources, and thus contribute to even out the geographic distributions of these key resources for ecosystems, which sustains the overall habitability of Earth.

4.3.2   Examples of Drastic Annual Changes in Environmental Conditions

Despite the global redistribution of heat and nutrients by atmospheric and oceanic transport, there are some major perturbations of environmental conditions that recur every year in some areas. Two examples are tropical cyclones and the seasonal sea ice (Table 4.1).

Tropical cyclones, which are also called tropical storms or hurricanes (in Western countries) and typhoons (in Eastern countries) cause drastic changes in environmental conditions. These storm systems result from heat transfer from the ocean to the atmosphere, in which the temperature of the surface ocean plays a major role. The starting point of a cyclone is a warm ocean's surface layer (>26 °C over at least 50 m), which causes intense evaporation and warming of the air above the ocean. Because the warm air is lighter than the overlying colder atmosphere, it rises rapidly, like a hot-air balloon. The water vapour contained in the rising warm air cools at high altitude, which produces intense rainfall. The Earth's rotation transforms this vertical air movement into the system of rotating winds called *cyclone* (see Sect. 3.5.2). Because the effect of the Earth's rotation is small near the equator, cyclones only form at latitudes higher than 5–10°. Cyclones keep on as long as there is a large temperature difference between the air and the ocean's surface, which depends on the thickness of the warm surface ocean layer. The strength of cyclones decreases rapidly when they moves over continental surfaces, away from their oceanic source of heat. Because cyclones depend on ocean's heat, they mostly occur on the western sides of oceans where surface waters are especially warm from late spring to early autumn. The phenomenon is annual, but the strength of cyclones and the duration of the cyclone season vary among years.

Cyclones can be extremely violent, with winds sometimes up to more than 250 km^{-1}. The resulting waves can severely damage coastal marine ecosystems, especially coral reefs, as well as human infrastructures and communities. Offshore, the stirring of surface waters has been reported to locally and temporarily increase phytoplankton production by pumping up nutrient-rich deeper waters. Overall, marine ecosystems in regions regularly affected by cyclones do not seem to be very different from those in cyclone-free regions, although they may experience temporary setbacks (for example nearshore coral reefs, which may take decades to recover from very severe storms) or boosts (for example offshore phytoplankton, which may experience temporary blooms).

Seasonal sea ice is another example of a recurrent drastic change in environmental conditions. It concerns the high latitudes, where parts of the ocean become covered by sea ice during winter. Other parts of high-latitude oceans are covered by sea ice year round, and the combination of seasonal (or *first-year*) and *multi-year* sea ice forms the *ice pack*. Sea ice comes from the freezing of seawater in the ocean, which is different from the masses of floating ice detached from continental glaciers or ice sheets called *icebergs*. The growth of the sea ice cover in winter and its melt in summer cause drastic changes to the marine polar environment annually. In the Arctic, the historical and present maximum and minimum extents of the ice pack are of the order of 16–14 (winter) and 8–4 (summer) million km^2, respectively, and the corresponding values for the Antarctic are 19–18 and 3–2 million km^2, respectively.

In recent years, the rapid decrease of the extent of the ice pack has been one of the numerous effects of the ongoing climate change (see Information Box 1.4). Polar ecosystems were adapted to the seasonal change in the ice pack, and the populations of many organisms took advantage of the succession of ice-covered and open waters. For example, polar bears in the Arctic hunted their seal prey at the boundary between the seasonal sea ice and open waters during summer and early autumn. Unfortunately, polar bears may not survive the ongoing disappearance of the seasonal sea-ice cover.

Concerning the global climate, the growth of the sea-ice cover in winter seasonally reduces to a considerable extent the surface area of open waters in polar seas, which diminishes the exchange of gases between the atmosphere and the ocean. Hence the seasonal change in the ice cover deeply affects the air-sea exchanges of CO_2 in polar regions. However, the global effect on the greenhouse gases and thus the climate (see Information Boxes 2.4 and 4.3) is null or small for two reasons: first, the total surface covered by sea-ice in the combined two hemispheres at any time of the year does not exceed 30 million km^2, which represents less than 10% of the surface of the global ocean (around 360 million km^2); and second, the seasonal variations in the extent of the sea ice cover in one hemisphere tend to compensate those in the other hemisphere. Today, one major environmental concern is the progressive disappearance of the permanent ice pack as a consequence of global warming, which would make the polar seas nearly ice-free in summer.

Climate Change Affects Overall Habitability It appears from the above that the natural annual changes in environmental conditions did not significantly affect the overall Earth's habitability regionally or globally. This is likely because the populations and ecosystems in regions affected by such regular changes evolved with them, and were thus adapted to either withstand these changes or even take advantage of them. However, present ecosystems may not be able to cope with rapid decreases in overall habitability caused by climate change, resulting for example from changes in annual phenomena such as the ongoing disappearance of the seasonal sea-ice cover or the increasing strength and/or frequency of tropical cyclones.

4.4 Major Volcanic Eruptions Since Year 1450

In this section, we examine the effects that changes with a periodicity of several decades have on the overall habitability of the planet (Table 4.1). Volcanic eruptions have a number of different effects. These include losses of human lives in the vicinity of volcanoes, damages to the economy over a wider area, and climate perturbations at continental or global scales. Hence, different types of eruptions may be qualified of "major" depending on the criterion considered. In the context of the present chapter, which deals with the overall habitability of the planet, we consider as major the past eruptions that caused a global cooling of at least 0.2 °C. According to this criterion, there were nine major volcanic eruptions that significantly affected the global climate since year 1450, and these eruptions thus had an average periodicity of less than 100 years (Table 7.2).

The cooling effect of such eruptions mostly resulted from the emission of sulphur dioxide gas, SO_2 (see Sect. 7.4.6). When this gas reaches an altitude between 10 and 50 km in the atmosphere, it causes the formation of sulphur-rich aerosol particles that reflect part of the incoming solar radiation back into space and thus cause a global cooling of the planet. During the last 400 years, such major eruptions occurred between one and four times per century. Drops in global temperature of 0.2 °C or more affect the Earth's habitability. This effect is not permanent, but it can often be significant. For example, the explosive eruption of Mount Pinatubo (Philippines) in 1991 was followed by a climate cooling of about 0.5 °C that lasted about 6 months.

Volcanic eruptions that occur with a periodicity of 100 years or more generally have moderate effects on the overall habitability of the planet, but some affect Earth's habitability at the regional and even global scale. These are different from the very large volcanic eruptions, which periodicities of 10,000 to 100,000 years, which can cause severe disturbances to the Earth's overall habitability (see Sect. 4.5.3).

4.5 Tens of Thousand- to Hundreds of Million-Year Changes in Environmental Conditions

4.5.1 Long-Term Changes in Habitability

Some major environmental changes have affected the overall Earth's habitability in the long term. In the following sections, we examine the effects that various changes, with average periodicities ranging from tens of thousands to hundreds of millions of years, had on the overall habitability of the planet (Table 4.1).

First, we look at the interglacial-glacial cycles affecting the planet since 2.6 million years, and are known as *Quaternary* (or *Pleistocene*) *glaciation* or *Ice Age* (Sect. 4.5.2). We currently live in the most recent interglacial episode of this ice age. Second, we examine the effects on the overall habitability of very large volcanic eruptions and the impacts of very large asteroids, whose average periodicities range between 10,000 and 300,000 years or longer (Sect. 4.5.3). Third, we move far in the deep past, where we consider three long periods during which the Earth's surface became entirely or near-entirely frozen, between 2,200 and 640 million years ago. These are known as episodes of *snowball Earth* (Sect. 4.5.4). Fourth, we focus on the *Phanerozoic climate variations*, which refer to multiple long episodes of high and low temperatures that took place during the last 542 million years (Sect. 4.5.5). Fifth, we investigate the succession of 16 very long periods that marked the Earth's history since the cooling of the planet 4.5 billion years ago, during which Earth was alternatively without and with glaciers or ice sheets. These are known as *greenhouse* and *icehouse Earth*, respectively (Sect. 4.5.6). We consider these different scales of variability one after the other because the present Quaternary Ice Age and the three ancient snowball Earth episodes are among the nine periods of icehouse Earth that occurred since the formation of the planet.

It must be noted that, as one goes back into the past, especially the deep past hundreds of millions and even billions of years ago, there is an increase in the smallest time interval for which a difference between two events can be identified. This smallest time interval is called *temporal resolution*. For the very recent past, it is sometimes possible to identify differences between events that occurred days, hours, minutes or even seconds apart. This level of resolution is not possible for older events. For example, the resolution of ancient historical facts is generally of the order of decades and even centuries, even if a few of the oldest existing texts, which were written more than 4,000 years ago, report important events on a yearly basis. Similarly, measurements made on Antarctic and Greenland ice cores provide paleotemperatures with an almost annual resolution for the last 10 thousand years or longer, but the length of the interval resolved increases rapidly with increasing time. For example, the temporal resolution of temperatures between 542 and 65 million years ago (Paleozoic and Mesozoic eras; Table 1.3) is seldom better than 500,000 years. For the Proterozoic or Archaean eras (up to 2.5 and 4.0 billion years ago, respectively; Table 1.3), the temporal resolution is not better than 10 million years. The increase in the length of temporal resolution with increasing time results from the fact that the physical and chemical traces left by ancient events in rocks, sediments and ice sheets, and other natural records become scarcer and fainter with increasing distance in the past.

Researchers gain knowledge about the past conditions of Earth through physical, chemical and biological characteristics of the past called *proxies* (see Sect. 9.2.5). Another type of information is provided by testimonies on the natural environment in the forms of paintings and engravings left on stone by people who lived on Earth tens of thousand years ago and called *rock art*. Artwork done on cave walls or large blocks of stone is called *parietal art*, from the Latin word *paries* meaning *wall of a cavity*. Paintings and engravings more than 20,000 years old have been discovered in Australia, East and Southeast Asia, India, the Franco-Cantabrian region (southern France and northern Spain), and Southern Africa. This artwork often includes representations of wild animals and thus provide, in addition to their cultural significance, information on the environment in which the people lived.

For example, the animals represented on the walls of caves in southern France and northern Spain include such steppe-tundra species as reindeers, bison, woolly mammoths and woolly rhinoceros (the last two species are now extinct), whereas representative large animals found at present in the Franco-Cantabrian region are brown bears, wolves and wild boars. The latter animals are representative species of temperate climate, whereas the fauna represented on the walls of caves shows that the climate of southern France and northern Spain more than 20,000 years ago was similar to that of present Siberia. This is because large parts of Eurasia and North America were then covered by ice sheets. During the succession of 30 to 50 glacial and interglacial episodes that occurred since the beginning of the Quaternary Ice Age 2.6 million years ago, the fauna of the Franco-Cantabrian region likely alternated between the two extremes represented by the animals known by people who lived there in a glacial climate more than 20,000 years ago and the present residents who live in an interglacial climate. The changes in habitability experienced by people during the last 20,000 years in one region help us imagine longer-term changes in overall habitability.

We explained above that the temporal resolution corresponding to very recent events can range between seconds and days, whereas it is at best one year, 500,000 years, and 10 million years for events that occurred thousands, hundreds of millions, and billions of years ago, respectively. The dating techniques used by researchers continually improve, and so does the resolution with which past events are known. Even with improved techniques, the interval that can be resolved between events will continue to be wider for events deeper in time, and this affects both our knowledge of ancient environmental variations, as indicated above, and the possibility of explaining these variations.

For example, researchers have proposed quite precise explanations for the succession of glacial and interglacial episodes during the last 2.6 million years, but found it more difficult to identify the causes of some of the major environmental changes that occurred millions or billions of years ago (see Sects. 4.5.2 and

4.5.4–4.5.6, respectively). The causes they invoked for the latter events were often one or two possibilities among a short list of major catastrophes, such as the impact of a very large asteroid, a large-scale volcanic eruption, or a major change in atmospheric gases. This is not because researchers lacked imagination, but simply because the traces left by potential causes are scarcer and fainter or more difficult to interpret with increasing distance in the past.

4.5.2   Quaternary Ice Age (Since 2.6 Million Years Ago)

Planet Earth has experienced 30 to 50 successive cycles of glacial and interglacial episodes since the beginning of the Quaternary period 2.6 million years ago. This succession of glacial-interglacial cycles is known as the *Quaternary glaciation* or *Quaternary Ice Age* (see Information Box 4.3). During an interglacial, the average temperature was about 10 °C higher than during the preceding glacial maximum, the latter being the time period when which ice sheets were at their greatest extension in the Northern Hemisphere (Fig. 4.4) and temperature was coolest (Fig. 4.5a). Overall, about 80% of the time in the last 2.6 million years was spent in glacial, colder conditions, and only 20% in interglacial, warmer conditions (Fig. 11.1b for the last 800,000 years). We presently live in interglacial conditions.

The duration of each glacial-interglacial cycle has been around 100,000 years since about 1 million years ago, and had previously been around 40,000 years. There are also shorter periodicities within the glacial-interglacial cycles, especially a dominant period around 20,000 years. These three periodicities are related to

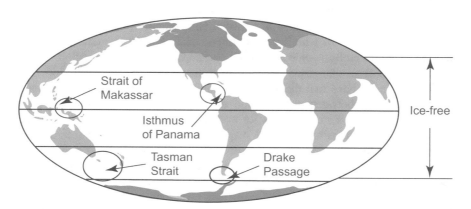

Fig. 4.4 Map showing the high-latitude seaways that opened more than 30 million years ago, creating the Drake Passage and Tasman Strait, and the equatorial seaways that closed 3–4 million years ago, forming the Isthmus of Panama and diverting ocean circulation to the Strait of Makassar (see Information Box 4.3). Darker blue areas: estimated ice cover in the Northern Hemisphere and Antarctica during the Last Glacial Maximum. Credits at the end of the chapter

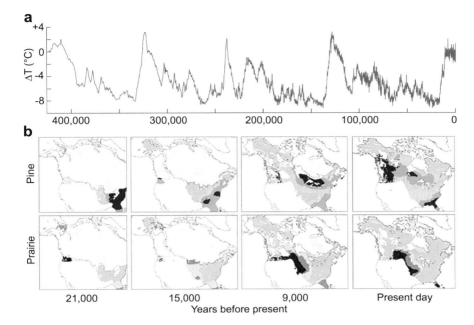

Fig. 4.5 Quaternary Ice Age. **a** Reconstructed surface temperatures for the last 420,000 years, from Antarctic ice-core data (ΔT: temperature difference between past and present). **b** Shift in the locations of pine trees (upper panels) and non-grass prairie vegetation (lower panels) in North America from 21,000 years ago to present. Coherent pale-blue mass: ice sheet. Increasing colour intensities: increasing concentration of pollen in sediments, which is proportional to the abundance of the species in a given area. Credits at the end of the chapter

three types of variations in the Earth's orbital parameters: 100,000-year variations in the eccentricity of the Earth's orbit; 41,000-year variations in the Earth's axial tilt, which is also called obliquity; and variations in axial precession, whose periodicity is 19,000 or 23,000 years (see Sect. 1.3.2). These cycles in orbital parameters are known as *Milankovitch cycles* (see Sect. 1.3.4). The present succession of glacial-interglacial cycles, which began 2.6 million years ago, is a very recent phenomenon in the 4.6 billion years of geological history of Earth, and it is explained below there were other ice ages in the distant past (see Sects. 4.5.4–4.5.6).

Each of the 30 to 50 glacial episodes that occurred during the Quaternary likely began with the incomplete melting in summer of the snow that had accumulated the previous winter. The progressive accumulation of snow year after year led to the following large-scale changes in the Earth's environment: formation of continental ice sheets in the northern parts of North America and Europe; progression of the ice sheets that already existed in Antarctica; development of seasonal and multi-year sea ice in polar seas; and growth of glaciers in high mountains all over the world. The cooling phenomenon was amplified by the increasing albedo of the planet that resulted from the increasing extent of the snow and ice cover all over

Information Box 4.3 Possible Causes of the Present Ice Age The beginning of the present Ice Age, 2.6 million years ago, defines the geological period called *Quaternary* or the geological epoch called *Pleistocene* (Table 1.3). The difference between the two is that the Pleistocene epoch ended with the last glacial episode about 11,500 years ago, which marked the beginning of the Holocene epoch, whereas the Quaternary period includes the Pleistocene and Holocene epochs and continues today.

Before the Quaternary Ice Age began 2.6 million years ago, there was no succession of glacial and interglacial episodes. It is explained in the neighbouring text that the succession of glacial-interglacial episodes since the beginning of the Quaternary is related to variations in the Earth's orbital parameters. These orbital variations existed since the formation of Earth 4.6 billion years ago, but the Quaternary Ice Age began "only" 2.6 million years ago. What were the changes in the Earth's conditions that caused the initiation of this Ice Age?

It was proposed that the beginning of the Quaternary Ice Age was related to a succession of changes that occurred in the positions of continents during the last 35 million years, as a result of plate tectonics (see Sect. 7.2.1). These changes are the opening of two seaways between large continental masses and the closing of two seaways, which together caused large changes in the oceanic circulation. The latter would have made planet Earth ready for the succession of glacial and interglacial episodes of the Quaternary period.

The first two major events that would have led to the Quaternary Ice Age millions of years later occurred at high latitudes in the Southern Hemisphere about 30 million years ago. These were the opening of the Tasman Strait and the Drake Passage, which cut the Antarctic off from the other continents (Fig. 4.4). This led to the formation of the powerful Antarctic Polar Current (Fig. 3.6), which isolated the Antarctic continent from the heat of the global ocean and thus favoured the development of the massive Antarctic ice sheets that still exist today. The two other major events took place at low latitudes 3 to 4 million years ago. These were the closings of the Panama and Indonesian seaways. The first created the Isthmus of Panama, which now links North and South America and prevents exchanges of water between the Atlantic and Pacific Oceans, and the second diverted the circulation of water between the Pacific and Indian Oceans northwards to the Strait of Makassar, which is located between the Indonesian islands of Borneo and Sulawesi (formerly known as Celebes) (Fig. 4.4).

The resulting changes in ocean circulation and evaporation in the Atlantic and Pacific Oceans would have made the Earth's climate responsive to changes in solar radiation, such as those caused by changes in the Earth's orbital parameters (see Sect. 1.3.4). These would have made the planet ready to experience the first of the glacial-interglacial episodes that followed one another since 2.6 million years. During the Quaternary Ice Age, the planet remained continually ice-free between the maximum southernmost extent of the ice sheets in the Northern Hemisphere and the ice-covered Antarctic continent (Fig. 4.4).

The above changes took place in the context of a global cooling of Earth that has been progressing since about 50 million years, and it was proposed that the Quaternary Ice Age of the last 2.6 million years was simply part of this long-term cooling trend. Hence, some specialists think that the ice age of the last 2.6 million years was mostly caused by the observed long-term decrease in the concentration of atmospheric CO_2 (Fig. 9.8b), and thus in the greenhouse effect. The long-term cooling trend of the last 50 million years corresponds to the latest icehouse (see Sect. 4.5.6).

the world (see Information Box 2.9), which contributed to further reduce the surface temperatures. This created a positive feedback loop, which caused a runaway cooling of the planet (see Sect. 9.1.3; Fig. 9.3b and accompanying text).

Over thousands of years of cooling, the ice sheets in Eurasia and North America progressed considerably to the south (left panels of Fig. 4.5b), and their thickness reached several kilometres. Once this happened, Earth entered a new glacial episode. In the Southern Hemisphere, there was no northward progress of ice sheets because the Southern Ocean isolated the southern continental masses from the ice-covered Antarctic continent (see Information Box 4.3). The cooling that occurred after the initiation of a glacial episode was not continual, and was marked by accelerations and decelerations at the scale of a few thousand years, and even brief reversals at the scale of a few tens of years.

During each glacial episode, the continental ice sheets expanded and their growth moved enormous amounts of water from the oceans to the continents, with the consequence that sea level dropped considerably. Conversely during each following interglacial episode, the ice sheets retracted and the sea level rose. The difference between the lowest and highest sea level during a glacial-interglacial cycle was sometimes more than 130 m. This value was not the same everywhere on Earth because the continents themselves moved down under the immense weight of the advancing ice sheets, and moved up upon their retreat. For example, geologists found that past marine shorelines in Scotland lochs are up to 40 m above the current sea level, which indicates that the continent rose by more than 40 m due to the retreat of glaciers after the *Last Glacial Maximum* (the ice sheets were at their greatest extension about 21,000 years ago).

Modern humans belong to the species *Homo sapiens,* whose human ancestors belonged to different species of the genus *Homo.* Genus *Homo* appeared on Earth about 2 million years ago, meaning that the earliest humans and their descendants, including our species *Homo sapiens,* have lived through tens of glacial and interglacial episodes. One characteristic of humans is the controlled use of fire, which may go back to one million years when people started to *use fire* from wild sources, such as fire created by lightning. Later, about 400,000 years ago, humans developed the ability to *make fire*. One of the oldest evidences of the domestication of fire in Europe is a fireplace discovered by archaeologists in a prehistoric human settlement called *Terra Amata*, which is located on a fossil beach near the French city of Nice on the Mediterranean Sea. The age of this fireplace is 380,000 years, which corresponds to an ancient interglacial episode. The archaeological site and its fossil beach are located 26 m above the present level of the Mediterranean Sea, which indicates that the present sea level is 26 m lower than 380,000 years ago when *Terra Amata* was located on a beach. This is one example of the effect of the change in sea level on people during the Quaternary Ice Age.

During the Quaternary period, ice sheets advanced and retreated on continents in the Northern Hemisphere and on the Antarctic continent, glaciers did the same in mountains all over the world and sea ice in polar seas. These large-scale environmental changes caused major shifts in habitability. Indeed, it is estimated that about one-third of the surface of the planet was covered by ice during the glacial episodes, and the zones of permafrost extended hundreds of kilometres southwards from the edge of the ice sheet in Eurasia and North America. Hence the 30 to 50 glacial-interglacial cycles that occurred during the last 2.6 million years repeatedly changed the habitability of more than one-third of the planet. Populations of animals and plants moved out of the areas affected by the advance of ice sheets and permafrost on land and that of sea-ice in the ocean, and back in again upon their retreats. An example of the latter is shown in Fig. 4.5 for pine trees and non-grass prairie vegetation in North America between the Last Glacial Maximum (21,000 years ago) and present. During these 21 millennia, the ice sheet that covered a large part of North America (thickness up to 3 km) progressively retreated northwards. These major changes in vegetation reflect major changes in habitability during the last 21,000 years. The article of Broecker (2018) provides further reading on the glacial-interglacial cycles of the last 800,000 years.

Based on some astronomical models, the next glacial episode could begin in about 50,000 years, and it is debated whether the ongoing global warming would delay the beginning of the next glacial episode or even prevent it (see Sects. 11.3.2–11.3.5). We further examine below the Quaternary Ice Age in the context of climate variations during the Phanerozoic eon (see Sect. 4.5.5) and the whole climate history of Earth (see Sect. 4.5.6).

4.5.3 Very Large Volcanic Eruptions and Impacts of Very Large Asteroids

A report on extreme geohazards published by the *European Science Foundation* in 2015 ranks explosive volcanic eruptions and the impacts of large asteroids as the two highest natural hazards on Earth. Other natural hazards considered in

the report include tsunamis, floods, droughts, and earthquakes (this report is recommended as further reading). We briefly describe here the two types of events
together because their effects on habitability are somewhat similar, although their
periodicities are quite different.

Examples of *very large volcanic eruptions* are provided in Table 7.1. This table
shows that one of the two largest eruptions in the Quaternary period, which covers
the last 2.6 million years, occurred at the site of the present Lake Toba (Sumatra
Island, Indonesia) about 70,000 years ago (see Sect. 7.4.6). It deeply affected the
Earth's climate during several decades. It was hypothesized that the Toba eruption
had caused a global winter that lasted six to ten years, and a global cooling that
could have lasted up to 1,000 years, but the latter theory is not accepted by all
researchers. Massive volcanic eruptions have also been invoked to explain mass
extinctions, which have a periodicity of about 100 million years (see Sect. 4.6.2).

Closer to the present time, the Little Ice Age is a relatively cool period that
lasted about 600 years between about 1250 and 1850 (Fig. 4.6). It followed the
Medieval Warm Period in the North Atlantic region, which had occurred between
about 950 and 1250. The cooling of the Little Ice Age, which was possibly global,
may have begun with the expansion of the ice pack in the North Atlantic around
1250. This expansion was possibly triggered or enhanced by the massive eruption
of the now-extinct Mount Samalas on the island of Lombock (Indonesia) in 1257,
followed by three smaller eruptions of the same volcano during the following decades. A second pulse of cooling may have been triggered by the eruption of the

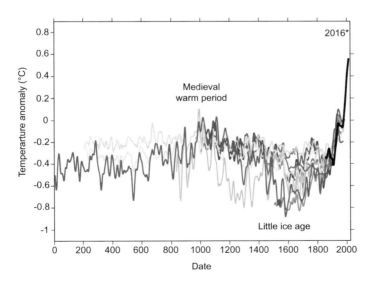

Fig. 4.6 Temperature anomalies over the last two millennia (temperature differences with the
1950–1980 average). Each curve is a different reconstruction of past temperatures based on a
different set of historical and/or paleoclimatic data. Past is to the left, and present time is to the
right. Credits at the end of the chapter

submarine volcano Kuwae (Vanuatu; western South Pacific Ocean) in 1452–1453. This stresses the likely effect on the climate of repeated events of volcanic activity accompanied by the release in the atmosphere of climate-cooling ash and/or SO_2 gas (see Sect. 7.6).

Still closer to the present time was the eruption of Mount Tambora (Indonesia) in 1815, which was followed by a global cooling that lasted several years, and the year after the eruption was called "year without a summer". During that very cool year, crops failed and livestock died in the Northern Hemisphere, and Europe experienced the worst famine of the nineteenth century.

Could a large volcanic eruption cause nowadays a famine as widespread as the one that occurred in Europe in 1816? One way to answer this question is to look at the present yearly supply, utilisation, and stocks of cereals (wheat, rice, and coarse grains such as maize and barley), which are staple foods of both people and livestock worldwide. Currently, the yearly utilisation of cereals is about equal to the supply, which are both of the order of 2,500 million tons (metric) per year, and the stocks are about 800 million tons, that is, about 30% of the yearly supply and utilisation. In the unlucky case of a "year without a summer" during which the world supply of cereals would be reduced by more than 30%, the deficit in cereals would be larger than the existing stocks and a famine would ensue.

The report of the *European Science Foundation* (2015) considered that very large eruptions with periodicities of 10,000 to 100,000 years can trigger global catastrophes, which are defined as events in which more than a quarter of the world's human population would die and which would place civilisation at serious risk. The report also indicated that the largest eruptions, which occur with a periodicity of one million years or longer, can kill more than a quarter of all organisms on Earth and cause mass extinctions (see Information Box 1.6 and Sect. 4.6.1).

The second highest natural hazard on Earth is the impact of large asteroids. Because the *Near Earth Objects* (NEOs) have very elongated elliptic orbits, it is difficult to calculate precisely the risk that Earth be hit by any one of them (see Sect. 1.3.1). Given that human civilisation could be seriously damaged if Earth were hit by a very large NEO, some space agencies and a number of governments are trying to improve the early detection of NEOs, and to develop technical means for possibly deflecting the trajectory of a NEO that would represent a major threat for Earth.

The *impacts of asteroids* have different effects on overall habitability depending on their sizes, and also if the asteroid impacts a solid continent or fall in the liquid ocean. Indeed, given that oceans occupy 71% of the planet, only 29% of the asteroids that reach the surface of the planet will impact a continent. Most of the asteroids and meteoroids that enter the Earth's atmosphere vaporize partly or completely during their fall, but some hit the surface of the planet (see Sect. 1.3.6). Asteroids with a diameter from 100 m to 1 km generally cross the atmosphere and produce craters that vary in size from 1.5 to 6 km. However, we focus here on the impacts of even larger asteroids and their effects on overall habitability.

Very large asteroids (diameter larger than several kilometres) have impacted Earth in the past. One example is the Chicxulub crater in Mexico, which resulted from the fall of an asteroid with a diameter of more than 10 km 66 million years ago, and may have been responsible for the fifth mass extinction (see Sect. 4.6.2). More than 2,000 km away, in what is now North Dakota in the USA, tiny beads of glass resulting from the asteroid impact were found clogging the gills of fish fossilized 66 million years ago. The same fossilized graveyard contained, in addition to fish stacked one atop another, burned tree trunks, conifer branches, mammal remains, bones of mosasaur (giant marine reptile), insects, the partial carcass of a *Triceratops* dinosaur, marine dinoflagellates (phytoplankton), and ammonites (snail-like marine cephalopods), whose burial presumably resulted from the fall of the asteroid.

It is expected that if a very large asteroid were to strike land or shallow water, this would cause massive ejection of dust into the atmosphere that would reduce, together with widespread wildfires, the incoming solar radiation for at least one year, leading to global-scale crop failures. The large-scale destruction of vegetation could lead to a mass extinction of plants and animals. In the case of an ocean impact, the release of large quantities of water vapour to the atmosphere would strongly enhance the greenhouse effect, causing a global warming, and huge tsunamis would occur globally, with heights of several hundreds of metres. As described for very large volcanic eruptions in an earlier section, the impact of a very large asteroid could conceivably kill billions of people, as consequence of widespread water pollution and diseases (Sect. 4.4). The impacts of very large asteroids on Earth have an average periodicity of 300,000 years or longer.

The above paragraphs showed that the effects of the two highest natural hazards on overall habitability, namely very large volcanic eruptions and the impacts of very large asteroids, are somewhat similar. Their periodicities generally range between those of the glacial interglacial cycles (40,000 years more than 1 million years ago, and now 100,000 years) and the episodes of snowball Earth, which had a periodicity of more than 100 million years (see Sects. 4.5.2 and 4.5.4, respectively).

4.5.4   Episodes of Snowball Earth (Between 2,200 and 640 Million Years Ago)

We now move far in the deep past, and consider long episodes during which the Earth's surface became entirely or near-entirely frozen. Three such episodes likely occurred around 2,200, 710 and 640 million years ago (Fig. 1.1) . This astonishing climatic phenomenon is called *snowball Earth* or *ice ball Earth*. The information that led researchers to hypothesize that the Earth's surface froze entirely or nearly entirely in the past was the discovery of sediments of glacial origin in areas considered to have been tropical in the past.

The last Snowball Earth episode occurred around 640 million years ago, and may have lasted more than 10 million years. During that period, the open ocean was perhaps restricted to a small band in the low latitudes between huge ice packs,

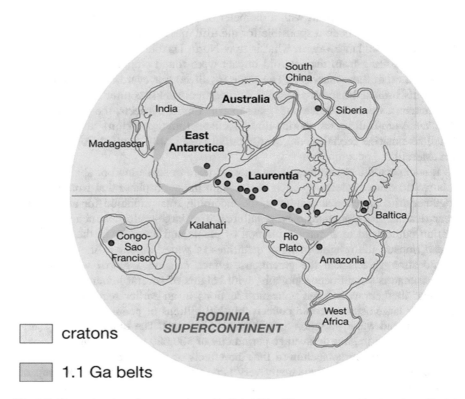

cratons

1.1 Ga belts

Fig. 4.7 Reconstruction of supercontinent Rodinia 750 million years ago. Cratons: large blocks of the crust that formed the nuclei of modern continents (the latter are identified on the map). Green zones: formation of mountain ranges 1.1 billion years ago (1.1 Ga). Red dots: granite intrusions. Credits at the end of the chapter

but the actual extent of the open waters is debated among specialists and some think that it may have been quite large. The latter configuration is called *slushball Earth*. The last Snowball Earth episode began in the geographical setting of supercontinent Rodinia, which existed between 1,300–900 and 750–633 million years ago. The supercontinent was centred on the equator, and extended between 60 °N and 60 °S (Fig. 4.7). It started to break up about 800 million years ago, possibly under the action of volcanic hotspots, which are regions of the crust where a volcano is fed by magma coming from a hypothetical abnormally hot area in the Earth's mantle (see Sect. 7.4.4).

The break-up of Rodinia was accompanied by the creation of very extensive zones of volcanic activity, which produced huge amounts of continental lava that solidified into silicate rocks ($CaSiO_3$). The chemical alteration of these rocks over millions of years led to a major reduction of atmospheric CO_2, through a mechanism described in Chapter 7 (see Sect. 7.3.3). The grouping of landmasses in the tropics might have created intense evaporation and tropical rainfall, the latter accelerating the chemical alteration of rocks. In addition, the albedo of land is higher than that of the ocean (Table 3.3), and the preponderance of continents in the tropics might have favoured strong reflection of solar radiation back to space. The decreasing atmospheric CO_2 and accompanying greenhouse effect caused a global cooling of the Earth's climate, which initiated the growth of ice sheets and glaciers on land and the ice pack in the ocean. As in the case of the Quaternary Ice Age, the resulting increase in the albedo of the planet amplified the cooling, and this positive feedback led to the runaway snowball Earth (see Sect. 9.1.3; Fig. 9.3b,d and accompanying text). Later on, the expansion of the ice sheets on continents had the effect of reducing the surfaces of $CaSiO_3$ rocks exposed to erosion, this negative feedback having the consequence that the progress of the ice sheets inhibited the mechanism causing the cooling (see Information Box 6.2).

The cooling mechanisms that led to the different episodes of snowball Earth are debated in the research community, but whichever mechanisms initiated the processes, these had the same global result: once Earth started to freeze, the presence of extended glaciers and sea ice increased the overall albedo of the planet, which further cooled the climate leading to an extension of glaciers, ice sheets and sea ice, and so forth until Earth turned into a snowball (or ice ball), entirely or with some water free of ice in equatorial regions either during summer or permanently. It is estimated that the longest episode of snowball Earth lasted "only" two million years, meaning that Earth always managed to break out of the snowball Earth condition relatively rapidly (geologically speaking).

An additional mechanism invoked for the initiation of the first snowball Earth around 2.2 billion years ago is a decrease in the atmospheric concentration of methane (CH_4), which is a very powerful greenhouse gas (see Information Box 2.7). This decrease would have been caused by the massive oxidation of CH_4 by the oxygen (O_2) released into the atmosphere by photosynthetic organisms at the time of the *Great Oxygenation Event*, which occurred between 2.4 and 2.0 billion years ago (see Sect. 8.1.2).

The mechanisms that allowed Earth to escape from the snowball condition are debated. Some of the proposed mechanisms involve the injection of CO_2 in the atmosphere by volcanic eruptions during the period when the Earth's surface was covered by ice. The accumulation of CO_2 in the atmosphere generated a greenhouse effect large enough to trigger the melting of the ice and thus allowed Earth to break out of the global glaciation (see Sect. 7.4.6).

Some specialists think that there were already multicellular, photosynthetic eukaryotes on land at the times of the last two snowball Earth episodes, although it is generally thought that there were no multicellular eukaryotes on continents before half a billion years ago. If this had been the case, it could be assumed that most of these organisms did not survive on the ice-covered continents, but some unicellular organisms such as snow algae could have been present since they only need a film of water and some nutrients to grow. In the oceans, the presence of an extended thick ice cover reduced the area of the ocean where solar light reached surface waters, which likely reduced significantly the overall photosynthetic production of organic matter and thus the global activity of food webs in the whole water column. This amount of photosynthetic production depended on the actual area of the ocean covered by permanent thick ice. In deep waters, ecosystems based on chemosynthetic production probably continued to exist and even thrive.

Overall, the habitability of the planet was reduced in major ways during the episodes of snowball Earth. We further examine the three episodes of snowball Earth below in the context of the whole history of the Earth's climate (see Sect. 4.5.6). Further reading is provided by the article of Poppick (2019) in which one of the pioneers of snowball Earth research tells about his life work, and by the article of Broecker (2018) which provides more information on snowball Earth.

4.5.5   *Phanerozoic Climate Variations (Last 542 Million Years)*

During the last 542 million years, called the Phanerozoic eon (see Sect. 1.5.4; Table 1.3), planet Earth has experienced several long periods of global high and low temperature, with an average periodicity of about 150 million years (Fig. 4.8). Temperatures derived from $\partial^{18}O$ measurements, as in Fig. 4.8, are not very precise, but there were clearly hot maxima in the 480, 370, 290–220, and 55 million years before present (early Ordovician, late Devonian, Permian and Triassic, and Eocene respectively). Overall, there were three long periods of warm climate, the first ending 450 million years ago, and the two others ending 420–360 and

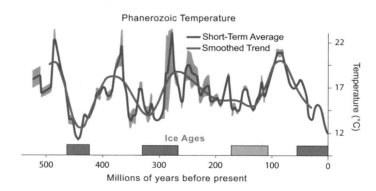

Fig. 4.8 Changes in the Earth's climate during the last 542 million years (Phanerozoic eon). Present time is to the right of the figure (0 year before present), and the past is to the left. Blue bars: periods when temperature was cold enough to cause an ice age (pale blue bar centred on 150 million years before present: occurrence of an ice age is uncertain). The temperatures were derived from $\partial^{18}O$ measurements on fossils (see Information Box 8.7). Grey surface: 95% statistical uncertainty in the moving average (smoothed trend). Credits at the end of the chapter

Fig. 4.9 Map of the World during the Late Cretaceous, 65 million years ago. The present Mediterranean Sea is the last remnant on modern Earth of the Cretaceous Tethys Ocean (see Sect. 7.2.4). Credits at the end of the chapter

260–234 million years ago. There were also cold troughs from 450 to 420, 360 to 260, and since 34 million years ago. The last cooling led to the alternations of cold glacial and warmer interglacial episodes that Earth experiences since the beginning of the Quaternary period 2.6 million years ago (see Sect. 4.5.2, above).

The Cretaceous period extended from 146 to 65 million years ago (Fig. 4.9), and its second half, from around 100 million years ago, was characterized by a warm climate and narrower temperature differences between the tropical and polar regions than today. Dinosaurs, which reached their maximum diversity during the Cretaceous, roamed the planet. Populations of fossil plants and animals show that northern regions equivalent to present-day Alaska and Greenland were experiencing a warm climate. The surface temperatures of tropical oceans could have been more than 35 °C, much higher than the present-day 28 °C. Geochemists believe that the average temperature of Earth about 80 million years ago could have been about 5 °C higher than today's 15 °C, that is, 20 °C (Fig. 4.8). It is thought that this overall warm climate was due for the most part an intense regional or global volcanic activity above the subductions zones, likely resulting from an enhanced rate of seafloor spreading, which released into the atmosphere large amounts of the greenhouse gas CO_2.

The Cretaceous period ended with the *fifth mass extinction* during which many groups died out, including all non-avian dinosaurs (the avian dinosaurs were the ancestors of modern birds). The warm part of the Cretaceous period was followed by the *Late Paleocene—Early Eocene Torrid Age*, between 65 and 55 million years ago, the hottest part of which being the *Paleocene-Eocene Thermal Maximum (PETM)* about 55 million years ago (the exact age and duration of this event are uncertain). The average global temperature reached around 30 °C, which is 15 °C warmer than present. Because of the strong evaporation, the climate was probably very wet, and several groups of mammals, including horses and primates (which included the ancestors of humans), appeared and spread on continents tens of thousands of years after the beginning of the PETM. In contrast some groups of deep-sea organisms became extinct while other groups flourished, which may be due to lower oxygenation of the deep waters related to changes in the ocean circulation caused by the higher temperature (see Information Box 9.2).

The climatic variations of the Phanerozoic eon show that, over 500 million years, Earth shifted at least three times between long phases of warm environments and long phases of cold conditions with ice ages. During the warm phases, the overall habitability of Earth was likely maximum. In contrast, during the cold phases and especially the ice ages, there were major shifts in habitability caused by the advance and retreat of ice sheets on continents, of glaciers on mountains, and of sea ice in the polar seas, as occurred during the Quaternary Ice Age (see Sect. 4.5.2). As in the Quaternary Ice Age, each cycle of glacial and interglacial conditions was likely accompanied by movements of populations of plants and animals out of the areas affected by the advance of ice sheets and permafrost on land and of the sea ice in the ocean, and back in again upon their retreats (see Sect. 4.5.2).

4.5.6 Greenhouse and Icehouse Earth (Since 4.5 Billion Years)

The successive long periods of warm and cold climate during the last 542 million years (Fig. 4.8) represent fluctuations between warm and cold conditions, which the two dominant states of the Earth's climate. These two contrasted, dominant states are called *greenhouse Earth* and *icehouse Earth*, respectively.

The words *greenhouse* and *icehouse* refer to two different types of buildings. A *greenhouse* is a structure designed to protect growing plants from cold weather, and an *icehouse* is a structure designed for storing natural ice throughout the year. Icehouses were common before the invention of refrigerators, to store ice cut from frozen lakes or rivers in wintertime for its use in iceboxes during summertime. The reason why climate scientists use the word *greenhouse* is explained in Chapter 2 (see Information Box 2.7). In climate studies, *greenhouse* is sometimes replaced by the word *hothouse*, which comes from agriculture where a hothouse is a heated greenhouse.

A *greenhouse Earth* is a long period during which the average temperature is high enough to prevent the occurrence of ice sheets or glaciers on continents (see Sect. 4.5.1). Conversely, an *icehouse Earth* is a long period during which the average temperature is low enough to cause the occurrence of one or more ice ages, which do not necessarily occupy the whole period. For example, the icehouse in which we presently live began 34 million years ago, but the present ice age (Quaternary period) began only 2.6 million years ago. During the course of the Earth's history, greenhouse Earth conditions often lasted more than one hundred million years, whereas icehouse Earth conditions generally did not exceed a few tens of millions of years. The average duration of the 16 greenhouses and icehouses in Table 4.3 is 290 million years. However, the planet was in icehouse conditions less than 15% of the time since 4.5 billion years, showing that greenhouse conditions were the most usual climate of Earth during its history until now.

During the course of the history of the planet, the Earth System has switched between the two *stable states* of greenhouse and icehouse (Table 4.3). The expression *stable state* means, in this context, that after the planet entered a given state, it stayed there as long as major changes in conditions did not force it out of that state. The transition from one stable state to another is called a *shift*. Hence, when the climate was in one of the two states, it remained there during generally millions of years before shifting to the other state, and the shift from one state to the other sometimes required a long time. One example is the 40 million-year shift that took place between 305 and 265 million years before present from the icehouse that existed from 360 to 260 million years before present to the greenhouse that extended from 260 to 34 million years before present (see Sect. 11.3.4).

The number of greenhouses and icehouses and their dates vary among studies, because the indices of climate conditions in the deep past are both few and faint. The long succession of greenhouses and icehouses began shortly after the initially very high atmospheric temperature, pressure and CO_2 concentration began to decrease to lower values more than 4 billion years ago (see Sect. 5.5.4).

An example of period during which the climate was generally warm is the last greenhouse Earth, which extended from 260 to 34 million years ago and included the Cretaceous period and the following torrid age described above (see Sect. 4.5.5). An example of icehouse Earth is the period that began 34 million years ago and in which we presently live, which is experiencing an ice age since 2.6 million years (see Sect. 4.5.2 above). Other examples of icehouse climate are the three episodes of snowball Earth described above (see Sect. 4.5.4).

Shifts from greenhouse to icehouse conditions and back are very spectacular events both by the magnitude of the reorganization of the global climate and their long duration of many million years. The fact that icehouses occupied less than 15% of the history of Earth could indicate that once in icehouse condition, the planet tends to come back relatively rapidly to the more usual greenhouse Earth. Figure 7.1 illustrates a different succession of greenhouse and icehouse Earths from that of Table 4.3, but also indicates that icehouse conditions occupied a shorter fraction of the Earth's history than greenhouse conditions.

The cause most often invoked by researchers to explain the major shifts between greenhouse and icehouse climate is a change in the concentration of atmospheric CO_2, that is, a decrease in CO_2 concentration in the case of a global cooling, and an increase in the case of a global warming. Decreases in atmospheric CO_2 could be caused by higher uptake of this gas resulting from stronger erosion of silicate rocks, which for example accompanied intense build-up of mountains by tectonic activity. Increases in atmospheric CO_2 could be caused by higher release of this gas by volcanoes, resulting from stronger tectonic

Table 4.3 Alternation between greenhouse and icehouse conditions since 4.5 billion years ago, with corresponding global climate and key biological events. An *ice age* is characterized by a succession of glacial and interglacial episodes, and a *snowball Earth* by the planet being (nearly) entirely covered by ice for a long period. Mass extinctions: see Table 4.4. The number of greenhouses and icehouses and their dates vary among studies

Million years ago	Greenhouse or icehouse	Global climate	Key biological events
4,500[a]–2,900	Greenhouse	Warm despite the faint young Sun	Diversification of archaea and bacteria; chemosynthesis and photosynthesis
2,900–2,780	Icehouse	Ice age	Earliest stromatolite reefs in coastal waters; first microbes on land
2,780–2,400	Greenhouse	Warm	Major increase in atmospheric O_2 (Great Oxygenation Event)
2,400–2,100	Icehouse	Snowball Earth	Possible mass extinction of organisms that did not use O_2 for respiration
2,100–715	Greenhouse	Warm	First eukaryotes; first protozoa; earliest land eukaryotes (photosynthetic and multicellular)
715–680	Icehouse	Snowball Earth	–
680–650	Greenhouse	Warm	–
650–635	Icehouse	Snowball Earth	–
635–580	Greenhouse	Warm	–
580	Icehouse	Ice age	–
580–450	Greenhouse	Warm	First large aquatic multicellular organisms; Cambrian explosion; first land plants
450–420	Icehouse	Several glacial periods	First mass extinction in the fossil record
420–360	Greenhouse	Warm	First land animals; second mass extinction
360–260	Icehouse	Two glacial periods	Large land-dwelling animals, flight
260–34	Greenhouse	Torrid age 65–55 million years ago	Three mass extinctions; flowering plants; rise and fall of dinosaurs
34–present	Icehouse	Ice age (Quaternary glaciation)	Diversification of mammals and birds; extinction of many large land mammals

[a]Formation of the Earth-Moon system

activity (see Sect. 7.3.1). The oxidation of atmospheric methane (CH_4) was also invoked for the initiation of the first snowball Earth, caused by the accumulation in the atmosphere of the oxygen (O_2) released by photosynthetic organisms (see Sect. 4.5.4, above).

Is It Better to Live During a Greenhouse or an Icehouse Earth? The alternate greenhouse and icehouse Earths represent two different states in the habitability of the planet, whose extreme examples are the torrid age of around 55 million years ago and any one of the snowball Earths, respectively. During the torrid age, most of Earth was likely habitable, whereas during the episodes of snowball Earth, the habitable solid surfaces and liquid volumes were probably minimum. This is not to say that habitability of an icehouse is continually poor. Indeed, we now live in an icehouse Earth, and millions of species thrive—including *Homo sapiens*—and interact in a multitude of ecosystems—including our globally connected society. The present success of species and ecosystems testifies that conditions provided by the current interglacial climate are suitable for them, and thus create a high level of overall habitability. However, it is possible that Earth's habitability was even higher during the last greenhouse when dinosaurs roamed the planet.

4.6   Mass Extinctions

4.6.1   Characteristics of Mass Extinctions

In the previous sections, we examined variations in environmental conditions with periodicities ranging from one day to more than one hundred thousand million years, and we tried to assess in each case the known or possible effects of these variations the overall habitability of Earth. In this section and the next, we use the opposite approach, that is, we begin with known major crashes in habitability that occurred in the past, and look at the changes in environmental conditions invoked by researchers to explain them (Table 4.1). These crashes in Earth's habitability correspond to the five mass extinctions that occurred during the Phanerozoic eon of the last 542 million years, with an average periodicity of about 100 million years.

A mass extinctions is a large-scale destruction of the life forms that existed on Earth at the time (see Information Box 1.6). Such events are only documented for the last 450 million years because older organisms did not leave remains in the fossil record (see Sect. 1.5.3). Organisms have existed on Earth for about 4 billion years, and there is no reason to think that no extinctions took place during the first 3.5 billion years of presence of organisms on the planet. The main reason why these possible extinctions were not documented in the fossil record is the general paucity of fossils older than 450 million years. In addition, the oldest fossil-containing sedimentary rocks have been largely destroyed during the course of the Earth's history by erosion (see Sect. 7.3.3) and plate tectonic processes (see Sect. 7.2.3), which contributes to further blur our knowledge of ancient ecosystems. Another complicating factor is the fact that the vast majority of organisms on Earth are microbes, whereas most of the fossils used to identify episodes of mass extinction belong to relatively large organisms, although there are calcareous and siliceous microfossils of unicellular organisms, and chemical traces of small organisms can be found in rocks. Furthermore, microbes may be less prone to mass extinction than complex organisms. Hence the mass extinctions documented in the fossil record concern mainly, but not exclusively, species of multicellular organisms, and are limited to the last 450 million years.

Table 4.4 reports the five major mass extinctions documented in the fossil record. These extinctions go back to about 450 million years before present. The dates of these extinctions indicate that the major mass extinctions occurred with an average periodicity of about 100 million years during the Phanerozoic eon. Applying this rate to the almost 4 billion years of presence of organisms on Earth would give a value of about 40 major mass extinctions. In fact, this value may underestimate the number of mass extinctions that actually occurred because the environmental catastrophes that may have caused mass losses of species, examined in the next section, were likely more frequent in the deep past than during the last hundreds of millions of years. In addition, there were often a number of smaller extinctions between the major episodes of extinction. All the above numbers are very approximate, especially that the very notion of *species* is debated among specialists.

Outside the episodes of catastrophic extinctions, a species normally dies out within 10 million years of its appearance, or more rapidly. This normal rate of species replacement combined with the potentially large number of extinctions explains why more than 99% of all species that ever existed on Earth are thought to be presently extinct. In other words, the 10 to 20 millions of species found on the planet at present may be the descendants of 5 billion species that would have existed during the course of the Earth's history. It must be noted that some researchers think that the numbers of species, present and past, are orders of magnitude larger than indicated in the previous sentence. This is because it is very difficult to assess the biological diversity of microbial organisms, with the result that biological diversity could be much higher than generally recognized.

Estimating the duration of an extinction event is not an easy task because it is done by studying fossils found in rock strata (see Information Box 1.7). In order to assess the duration of an extinction event, researchers must identify the rock strata that correspond to the beginning and the end of the event, and use very sophisticated methods to determine the approximate ages of the rocks in these strata. Here are two examples of very different extinction durations from Table 4.4. The mass extinction that occurred 252 million years ago may have taken place over 60,000 years, which is a very long period for humans but is very short at the geological timescale. In contrast, the previous mass extinction, between 375 and 360 million years ago, was marked by a succession of extinctions that covered a total of 5 million years, which is quite long.

After each mass extinction of the Phanerozoic eon, it took 5 to 10 million and even 30 million years for the biodiversity to recover, that is, to come back to the pre-extinction level. However, recovery of diversity does not mean the recovery of the previous life forms. Indeed in all instances, the flora and fauna that evolved after an extinction event were very different from those that existed before the extinction. One of the reasons why Nature did not repeat itself may be the

Table 4.4 The five mass extinctions documented in the fossil record, which goes back 542 million years (Phanerozoic eon). There were no multicellular organisms on continents before about 440 million years ago. Ma: million years before present

Number	Time (Ma) Periods	Percent extinctions	Examples of extinct groups	Some of the proposed causes
1	450–440 Ordovician-Silurian	27% of families 57% of genera 60–70% of species	All major taxonomic groups (which were then marine only)	Ocean cooling, because of climate change and/or increased volcanism
2	375–360 Late Devonian	19% of families 50% of genera >70% of species	Marine groups only, including almost all reef-building organisms	Changes in sea level and lower ocean O_2, triggered by global cooling or increased oceanic volcanism
3	252 Permian-Triassic	57% of families 83% of genera >90–96% of species	All trilobites (marine arthropods), and insects (terrestrial)	Global cooling following asteroid impact, and/or massive volcanic eruptions
4	201 Triassic-Jurassic	23% of families 48% of genera, 60–75% of species	Almost all the large non-dinosaurian fauna	Long-term warming from increased volcanic activity, or release of CH_4 and runaway warming
5	66 Cretaceous-Paleogene	17% of families 50% of genera 75% of species	All ammonites (marine); almost all tetrapods >25 kg, all non-avian dinosaurs	Impact of massive asteroid; very large volcanic eruptions

difference between the environmental and biotic conditions that prevailed after the extinction and those that had existed before, as the new conditions selected organisms different from those that had evolved before the extinction.

For example during the period called Cretaceous, just before the last extinction 66 million years ago, there were a wide variety of reptile groups on land, in the oceans and in fresh waters. Today, only a few groups of reptiles are present on Earth, and the most diversified and numerous descendants of Cretaceous reptiles are the birds. The widely diversified dinosaurs, which dominated the land fauna before the mass extinction, are all gone as are also most of the marine reptiles, which are now represented by only a few species of marine turtles, crocodiles and lizards and a variety of marine snakes. The life forms that disappeared during the last mass extinction did not come back afterwards. This is true not only for the last mass extinction, but also for the previous extinctions. Hence most likely, the 5 billions of life forms that disappeared during the last 4 billion years will not reappear during the next billions of years, and billions of new species that we cannot even imagine will evolve in the future. The book of Glikson and Groves (2016) provides further reading on mass extinctions of species.

4.6.2   The Five Mass Extinctions of the Phanerozoic Eon (Last 542 Million Years)

We now briefly describe the five mass extinctions that occurred during the Phanerozoic Eon, and examine the main catastrophic changes in environmental conditions, and thus in the overall habitability, invoked by researchers to explain them (Table 4.4). It must be remembered that available fossils (see Sect. 1.5.3) mainly provide information on groups of eukaryotes with hard parts, and almost nothing is known about the fate of most prokaryotes or soft-body eukaryotes from the past. Some unicellular organisms with hard structures have left abundant fossils, including various types of planktonic algae (coccolithophores, carbonate, Fig. 7.8; and diatoms and silicoflagellates, silica; Fig. 7.9) and planktonic and benthic protozoa (foraminifera, carbonate). One spectacular example of these fossils is provided by the White Cliffs of Dover, in England, which are entirely made of calcareous remains of coccolithophores called *chalk* (Fig. 2.1).

The *first mass extinction* of the Phanerozoic eon occurred between 450 and 440 million years ago, and included two major extinction events. This extinction affected almost all major groups of multicellular organisms, which were essentially marine beings because most of the complex multicellular organisms then lived in the ocean. Before the extinction, the first fishes had appeared in the sea, which had only been populated with invertebrates until then. It is estimated that 60 to 70% of the marine species disappeared during the extinction, which was followed by the diversification of jawed and bony fish.

The first mass extinction occurred in icehouse conditions (Table 4.3), in which global cooling events could have caused mass mortality of organisms, which were used to warmer conditions. The cooling could have been caused by a decrease in atmospheric CO_2 resulting from higher uptake of this gas by stronger erosion of silicate rocks, as a consequence of both higher production of these rocks by volcanoes (see Sect. 7.2.3) and more intense build-up of mountains by tectonic activity (see Sect. 7.3.1).

The *second mass extinction* of the Phanerozoic eon occurred between 375 and 360 million years ago, and included at least two and perhaps more extinction events. Before the extinction, the warm waters of the ocean were occupied by massive reefs, fish thrived in open waters, and the lands were covered with huge forests. The extinction only affected the aquatic fauna, where the reef-building organisms and large fish and tetrapods were wiped out. Also, many groups of aquatic invertebrates became extinct. It took more than one hundred million years before modern corals began building large reefs again. The after-extinction fauna included large organisms, among which the largest land invertebrates and flying insects ever, very diverse land tetrapods (amphibians and reptiles), and abundant fish.

The causes of the second extinction event are not easy to pinpoint, especially that it occurred over a long period of 15 million years. Some of the possible explanations are related to the massive growth of plants on continents, which evolved into large forests during the period. The increased photosynthetic activity could have caused a large decrease in atmospheric CO_2 and thus a strong cooling, detrimental to many species. Another consequence of the massive growth of plants could have been increased erosion of continental rocks through the breakdown of the bedrock and the production of acid compounds (see Sect. 7.3.3), which would have released nutrients. The high concentrations of nutrients could have been detrimental to aquatic organisms in two ways: first, by causing the development of large biomasses, whose decay would have depleted the concentrations of dissolved oxygen, and second by allowing the growth of competitors of reef-forming organisms. In this case, organisms and not changes in the environment could have been the cause of the extinction, which is unique among the five extinctions.

The *third mass extinction* of the Phanerozoic eon occurred 252 million years ago, and was especially devastating. It killed more than 95% of all marine species of multicellular organisms, including the very successful trilobites (arthropods), and 70% of land species, including many species of insects and large herbivores. The earliest-known calcareous-walled foraminifera (protozoa), called *Fusulinids*, also got extinct. Recovery from the extinction was slow, and may have taken as long as 30 million years. In the ocean, the proportion of free-living organisms increased relative to those attached to the seafloor. On land, the ancestors of dinosaurs and crocodilians progressively displaced other large organisms, whereas the ancestors of mammals were small insectivores.

The causes proposed to explain this extinction include the possible impact of a large asteroid, which could have caused massive volcanic eruptions. Such eruptions, which could have been related or not to an asteroid impact, would have had the same catastrophic effect as explained below for the fifth mass extinction, that is, a global cooling. There is also evidence of a long-term development of low-O_2 conditions in the ocean, which may have been caused by an increased erosion of continental rocks due to the warm climate (see Sect. 7.3.3) and the accompanying release of nutrients, as explained above for the second mass extinction.

The *fourth mass extinction* of the Phanerozoic eon occurred 201 million years ago. It saw the disappearance of almost all the large non-dinosaurian fauna, leaving the dinosaurs with little competition to dominate the continental environments during the following 100 million years and more. The situation was different in aquatic environments, where non-dinosaurian reptiles continued to thrive, including turtles, crocodiles and snakes.

Some researchers are explaining this extinction by a possible long-term increase in the greenhouse effect caused by a rising concentration of atmospheric CO_2, itself resulting from increased volcanic activity. Instead of this long-term climate change, other researchers think that a general warming of the atmosphere and the ocean caused the thawing of the permafrost and the destabilization of methane hydrate frozen in the seafloor, which rapidly released in the atmosphere large amounts of methane (CH_4), a powerful greenhouse gas (see Information Box 2.5) and thus accelerated the global warming (see Sect. 2.1.7). This runaway warming would have been a global *positive feedback* (see Sect. 9.1.3; Fig. 9.3f and accompanying text).

The *fifth mass extinction* of the Phanerozoic eon occurred 66 million years ago. It was marked by the disappearance of about three-quarters of all species of multicellular organisms, including all tetrapods weighing more than 25 kg and all dinosaurs except the ancestors of modern birds. In the ocean, the very diversified and abundant ammonite molluscs (a group of shelled cephalopods a bit similar to the modern *Nautilus*) became extinct, together with many groups of fish and other invertebrates. Many species of unicellular organisms present in the fossil record also got extinct, especially calcareous plankton organisms (coccolithophores, foraminifera, and others). The groups that disappeared were replaced by others that had been minor players until then, the best-known example being the replacement of dinosaurs by mammals, to which our own species belongs.

A popular explanation for the fifth extinction is the fall on Earth of a large asteroid in Mexico, which released an amount of energy corresponding to one billion times the atomic bombs that fell on the Japanese cities of Hiroshima and Nagasaki in 1945 (see Information Box 1.6). This would have created a dust cloud that blocked the sunlight for up to a year, creating freezing temperatures and inhibiting photosynthesis, which would have affected all food webs. A number of researchers do not agree that the fall of that asteroid was fully or at all responsible for the fifth extinction, and they invoke in addition or instead the cooling effect of the massive emission of SO_2 gas in the Deccan Traps volcanic province that occurred between 68 and 60 million ago (see Sect. 7.4.6).

Many researchers are concerned that a *sixth mass extinction* may be taking place now, whereas others think it is not occurring now but is nevertheless looming (see Sect. 10.1.9). Indeed, the rate of species extinctions is more rapid at present than it has ever been in the geological record, and seems to be increasing. This sixth extinction would be due to the direct obliteration of numerous species by human activities (hunting and fishing, use of pesticides, etc.) and also the destruction of their environment (for agriculture, urbanization, etc.), both being human-driven reductions in the Earth's overall habitability. Readers preoccupied by this possible sixth extinction could easily find books and articles, both printed and electronic, on this topic.

4.7 Biological Innovations and Anthropogenic Changes to Atmospheric Gases

4.7.1 Biological Innovation: O_2-Producing Photosynthesis

In the previous sections, most of the causes evoked for changes in overall habitability of the planet were of astronomical or Earth environmental nature. Two major exceptions were the possible involvement of biological factors in the initiation of the first snowball Earth around 2.2 billion years ago and as possible causes of the second mass extinction (see Sects. 4.5.4 and 4.6.2, respectively). This section and the next one describe biological effects on the overall habitability of Earth. Here we consider three effects of atmospheric O_2.

The first effect concerns the initiation of the first snowball Earth (see Sect. 4.5.4), According to one hypothesis, the cooling of the planet that triggered the runaway cooling could have been caused, totally or partly, by a massive reduction in the concentration of atmospheric methane (CH_4), which is a very powerful greenhouse gas and whose decrease would have drastically lowered the Earth's greenhouse effect. At that time, which corresponds to the *Great Oxygenation Event* that occurred between 2.4 and 2.0 billion years ago, the accumulation oxygen (O_2) in the atmosphere would have massively oxidized the abundant atmospheric CH_4 (see Sect. 8.1.2). The build-up of atmospheric O_2 was itself a consequence of the invention of O_2-producing photosynthesis by aquatic cyanobacteria more than 2.5 billion years before present, and the subsequent accumulation of this gas in the atmosphere over several hundred millions of years (see Sect. 2.1.5). The massive oxidation of atmospheric CH_4 could have been the cause of the global cooling that triggered the first snowball Earth.

The above chain of reactions connects the biological innovation of O_2-producing photosynthesis more than 2.5 billion years before present with the build-up of atmospheric O_2 over hundreds of million years, the resulting massive oxidation of atmospheric CH_4, and the first snowball Earth around 2.2 billion years before present. This would have led to a reduction in the overall habitability of Earth during the hundreds of thousands of years of the snowball Earth (Table 4.3).

The second effect of atmospheric O_2 concentration on overall habitability was the *Great Oxygenation Event*, which occurred between 2.4 and 2.0 billion years ago (see Sect. 8.1.2). This global crisis affected the aquatic ecosystems, as there were no organisms on emerged lands during the first billion years of the Earth's history. Before the oxygenation crisis, there was no or very little free O_2 in the Earth's environment, and consequently no organisms used O_2 for respiration (see Information Box 5.2). In fact, many of the existing organisms did not tolerate O_2, which was a poison from them. Hence when environmental O_2 increased, the latter organisms saw their habitat reduced to the environments with low or null O_2 concentration, such as the sediments. Simultaneously, the presence of O_2 in aquatic environments led to the evolution of organisms able to use O_2 for respiration, which occupied the newly oxygenated environments in oceans and lakes, and much later on continents. This chain of reactions connects the biological innovation of O_2-producing photosynthesis with a major shift in the overall habitability of Earth, caused by the transformation of the planet from an environment devoid of free O_2 and where no organisms used O_2 for respiration, to a mosaic of environments with and without free O_2, shared by organisms that used and did not use O_2 for respiration, respectively.

The first two effects of atmospheric O_2 on overall habitability occurred about one billion years after the advent of O_2-producing photosynthesis. The third effect of atmospheric O_2 on overall habitability took place more than two billion years after the occurrence of this biological innovation. This effect was the creation of the ozone (O_3) layer in the upper atmosphere, about 500 million years ago. The ozone layer resulted from the increasing concentration of O_2 in the atmosphere, and the transformation of O_2 into O_3 by the ultraviolet radiation from the Sun (UV) 20 to 40 km above the ground (see Sect. 2.1.5). The O_3 layer in the upper atmosphere absorbed most of the solar UV, and this allowed complex multicellular organisms, for which high doses of solar UV would have been lethal, to begin occupying the continents mainly during the Silurian and Devonian periods 444 to 359 million years ago (see Sect. 1.5.4). Until then, complex multicellular organisms could only live in aquatic environments where they were sheltered from dangerous solar UV, so that the creation of the O_3 layer increased the overall habitability of Earth for them by about 40% (which is the ratio of emerged lands, 29%, to oceans, 71%). This chain of reactions connects the biological innovation of O_2-producing photosynthesis with the creation of the O_3 layer in the upper atmosphere, which significantly increased the overall habitability of the planet for complex multicellular organisms.

The above three changes in overall habitability took place hundreds of millions of years after the appearance of O_2-producing photosynthesis, stressing the fact that some responses of the Earth System to environmental changes occur over very slowly (Table 4.1). Each of these three changes in habitability occurred only once in billions of years, and their effects on the Earth System were tremendous.

4.7.2   Biological Innovations: Calcification, Silicification, Methanogenesis, O_2 Respiration, and Multicellularity

The previous section described how the oxygenation of the planet, which was a consequence of the biological innovation of O_2-producing photosynthesis, affected the overall habitability of Earth deeply and in the long term. The present section briefly describes the effects on the planet's overall habitability of five other major biological innovations, which are: calcification, silicification, methanogenesis, O_2 respiration, and multicellularity.

Calcification is the biomineralization process by which organisms form mineral carbonates, mostly calcium and magnesium carbonate ($CaCO_3$ and $MgCO_3$, respectively), by combining calcium (Ca^{2+}) or magnesium (Mg^{2+}) and bicarbonate (HCO_3^-) ions dissolved in water (see Information Box 4.4). This process accounts for the formation of most of the carbonates in the modern ocean, and may go back at least one billion years (see Sect. 6.4.2).

Information Box 4.4 The Different Meanings of Calcification and Silicification The process by which living organisms produce solid minerals is called *biomineralization*. It includes calcification and silicification, two words that have different meanings in different disciplines.

The word *calcification* refers to carbonate minerals. It has different meanings in chemistry, geology, biology, and medicine.

In *chemistry*, the process of calcification is the formation of carbonate minerals, such as calcium carbonate ($CaCO_3$) and magnesium carbonate ($MgCO_3$), from dissolved ions.

In *geology*, a calcification is a rock formation made of carbonate minerals.

In *biology*, calcification is the biomineralization process by which organisms produce mineral carbonate structures, mainly $CaCO_3$, but also $MgCO_3$ and other carbonates, from dissolved ions (see Information Boxes 5.11 and 6.3).

In *medicine*, the process of calcification is the accumulation of calcium in a body tissue—either usually (for example, the bones and the teeth), or pathologically (for example, kidney stones), or after an injury (for example, bone calcification after a fracture)—and *a calcification* is the tissue resulting from this process.

Figuratively, calcification is the intellectual process in which ideas, systems or methods become fixed or difficult to change.

In the present text, *calcification* refers to the *biomineralization* process by which organisms produce mineral carbonate structures, mainly $CaCO_3$, but also $MgCO_3$, from dissolved ions.

The word *silicification* refers to silica minerals. It has different meanings in geology, paleontology, and biology.

In *geology*, silicification is the process in which a solution rich in silicon ions, usually in the form of silicic acid, $Si(OH)_4$, causes the $CaCO_3$ contained in a rock to be replaced by silica (SiO_2).

In *paleontology*, silicification is the process by which the organic matter of a fossilized organism has been replaced by SiO_2.

In *biology*, silicification is the biomineralization process of formation of mineral SiO_2 structures by organisms, from dissolved $Si(OH)_4$ ions (see Sect. 6.4.1).

In the present text, *silicification* refers to the *biomineralization* process of formation of mineral SiO_2 structures by organisms, from dissolved $Si(OH)_4$ ions.

In aquatic environments, *calcification* is performed by organisms that create various $CaCO_3$ and $MgCO_3$ structures such as coccoliths, shells and coral reefs (Fig. 7.8). There are also structures made of layers of minerals (mostly $CaCO_3$) interleaved with microbial biofilms known as *stromatolites* (see Information Box 3.1 and Sect. 6.4.2). *Silicification* is performed by a variety of organisms, such as those illustrated in Fig. 7.9. The two processes have major effects in the Earth System (see Sect. 4.7.2).

Calcification is part of the negative feedback loop that dampens the long-term variations in Earth's temperature (see Sect. 9.2.3). Briefly, the erosion of $CaSiO_3$ rocks on continents and on the exposed seafloor by CO_2-rich water produces ions of Ca^{2+}, HCO_3^- and $Si(OH)_4$ (silicic acid), which combine in the ocean to produce solid $CaCO_3$ and SiO_2 minerals (Sect. 7.3.3). Parts of these two minerals are buried in sediments and eventually form rocks. For each molecule of $CaCO_3$ buried as a result of this process, one molecule of CO_2 is withdrawn from the environment by the alteration of $CaSiO_3$ rocks, and the ensuing sequestration of carbon (that is, its long-term storage in the seafloor) regulates the Earth's greenhouse effect at the timescale of 400,000 years. In this way, calcification is involved in the maintenance of the Earth's overall habitability at the timescale of hundreds of thousands of years.

The calcifying organisms include a wide variety of eukaryotes, both unicellular and multicellular (Fig. 7.8). Coccolithophores (algae) and foraminifera (protozoa) are unicellular calcifying organisms, whereas coralline algae, echinoderms, molluscs, crustaceans and corals are multicellular. All echinoderms are marine animals, and this group includes such organisms as sea stars, sea urchins, sand dollars, crinoids, sea cucumbers and brittle stars. Most echinoderms are benthic, but some sea-lilies (crinoids) can swim at great velocity for brief periods. Molluscs are mostly aquatic, but they are also present in continental environments. They include the gastropods (examples: snails and slugs), bivalves (example: clams), cephalopods (examples: squids, cuttlefish and octopus), and other groups such as the chitons. Most live on the seafloor or on the ground, but some are planktonic (the pteropods; example: sea butterflies). Crustaceans include such animals as crabs, lobsters, crayfish, shrimps, prawns, krill, woodlice and barnacle. Most crustaceans are aquatic, but woodlice are also found on continents. Most crustaceans live on the seafloor or close to it or on the ground, but some are very abundant in the water column (examples: krill and copepods). Last but not least are the reef-building corals, which are colonies of sac-like animals, called *polyps*, Each polyp creates a calcareous exoskeleton, and these exoskeletons form together a structure whose shape is typical of the species. Soft corals do not have a solid exoskeleton and thus do not form reefs, but some have small calcareous secretions in their tissues. Most coral reefs are found in shallow, warm waters, but some reefs also exist in deep waters.

Silicification is the biomineralization process by which organisms form mineral silica (SiO_2) from ions of silicic acid, $Si(OH)_4$, dissolved in water (see Information Boxes 4.4 and 7.3). As explained in the last paragraph but one, solid SiO_2 is buried together with solid $CaCO_3$ in the seafloor. There, the two minerals eventually become rocks, which are slowly subducted and molten in the mantle, from where their components are emitted by continental and submarine volcanoes in the form of lava and CO_2 gas, a process that takes tens to hundreds of millions of years (see Sect. 7.2.3). Upon cooling and solidification of the lava, the chemicals elements it contains combine to form new $CaSiO_3$ rocks.

Siliceous remains of eukaryotes found in sedimentary rocks indicate that siliceous sponges and radiolarians (protozoa) have actively formed silica structures in aquatic environments since around 500 million years before present. These remains also indicate that the formation of biogenic silica had been dominated by diatoms, both phytoplanktonic and benthic, since about 150 million years. The fact that there are no biogenic Si fossils older than about 500 million years does not necessarily mean that no organisms produced silica structures earlier, and some specialists have proposed that biologically mediated production of silica by prokaryotes and unicellular eukaryotes (protozoa) may have occurred much earlier. In summary, the combination of the chemical elements contained in seafloor SiO_2 and $CaCO_3$ rocks forms new CO_2 gas and $CaSiO_3$ rocks through the agency of subduction and volcanic emissions. The CO_2 emitted by volcanoes increases the greenhouse effect, and the erosion of $CaSiO_3$ decreases it by sequestering carbon in $CaCO_3$ (previous paragraph). Hence, silicification and biomineralization are involved together in the maintenance of the Earth's overall habitability at the timescale of tens to hundreds of millions of years.

Methanogenesis refers to the microbial production of methane (CH_4), which is a powerful greenhouse gas (see Information Boxes 2.7 and 5.2). Because of its warming effect, CH_4 may have exerted a major control on the Earth's greenhouse effect during the first half of the history of organisms on the planet (Archean eon, from 4.0 to 2.5 billion years ago) by providing suitable temperatures for them. Indeed during that period, the heat output of the faint young Sun was only 70% of its present value and the temperature of Earth should have consequently been quite low, but organisms nevertheless thrived on the planet (see Sect. 3.7.1). In order to resolve this *faint young Sun paradox*, it was proposed that the production of CH_4 released enough CH_4 into the atmosphere to sustain the Earth's greenhouse effect (see Sect. 2.1.7). According to this hypothesis, methanogenesis largely contributed to maintain the overall habitability of the planet since the appearance of organisms on Earth at least 3.8 billion years ago. This came to an end about 2.2 billion years ago when the increasing concentration of atmospheric O_2 massively oxidized the atmospheric CH_4 and thus caused the initial cooling of the first snowball Earth (Sect. 4.5.4).

On present Earth, CH_4 is only produced by archaea (see Information Box 5.2). However, it was proposed that before the oxygenation of the planet, CH_4 could have also been produced by microbial ecosystems based on O_2-less photosynthesis (see Information Box 5.1). Nowadays, methane may again have a significant effect on the overall habitability of the planet, within the context of global warming. This is because two major natural sources of CH_4 on today's Earth are the permanently frozen soils, called *permafrost*, and marine sediments (see Sect. 2.1.7).

Permafrost is presently thawing at an increasing rate, and microbial activity turns part of its newly available organic matter into CH_4. Similarly, deposits of frozen *methane hydrate* in the seafloor and under the permafrost could be destabilized by the rising temperature. The resulting increase in atmospheric CH_4 will accelerate the ongoing climate warming, thus contributing to lower the overall habitability of Earth (see Sect. 11.1.3).

The process of O_2 *respiration* appeared in the ocean about 2 billion years before present. This process was evolutionarily very successful because it produced more energy per unit of sugar consumed than respiration without O_2 (see Information Box 5.2). The oxygenation of the planet created new habitats, of which the O_2-respiring organisms, which included both autotrophic and heterotrophic life forms, took advantage first in aquatic environments, and later on continents after the establishment of the O_3 layer (see Sect. 4.7.1, above). The spreading of the O_2-respiring organisms in aquatic and terrestrial environments deeply modified the existing habitats, as shown by the "greening" of continents after their occupation by plants.

The O_2 respiration is the main process that uses environmental O_2 to oxidize organic matter, which includes the decomposition of organic matter by microorganisms. On continents, wildfire also consume large amounts of atmospheric O_2 (see Information Box 3.4). A key effect of O_2 respiration on the environment is to prevent the accumulation of O_2, and even to decrease its concentration in some instances. An present-day example of the latter is the *eutrophication* of natural waters, where the decomposition of unusually high algal or plant biomass causes O_2-depletion sometimes accompanied by fish kills or mass mortality of other aquatic organisms (see Sect. 10.3).

More generally, O_2 respiration prevents the increase of O_2 concentration in the Earth's habitats. However and despite the O_2 respiration, O_2 slowly accumulated in the atmosphere over hundreds of millions of years starting more than 2 billion years ago until about 300 million years ago, after which it decreased to the current average of about 20% (see Sect. 4.7.1; Fig. 8.1b). A key factor that allowed the build-up of O_2 to occur was the burial of huge amounts of organic matter, first in the seafloor and later under the ground, which prevented the buried organic matter from being O_2-respired or burned (see Sect. 9.4.5). The accumulation of O_2 in the environment durably modified the habitats of the planet.

Multicellularity is the characteristic of multicellular organisms. It developed in the ocean between a billion and 500 million years before present, following the biological innovation of eukaryotes that had taken place between 2.1 and 1.6 billion years ago or perhaps earlier. All   multicellular organisms respire O_2. The occupation of continents by multicellular organisms, which began between 500 and 400 million years ago, progressively transformed the emerged lands from bare rocks to luxurious prairies and forests, which created new habitats where a wide variety of animal species evolved. Today and excluding the microbes, about 80% of the known 1.5 million species are terrestrial, 15% are marine, and 5% are found in fresh waters.

During the Carboniferous period (from 359 to 299 million years before present), the burial of large amounts of vegetation, which became the coal beds typical of that period, allowed the accumulation of a large amount of O_2 in the atmosphere, up to 35% by volume (see Sect. 9.4.1). These highest concentrations of atmospheric O_2 in the Earth's history created a habitat where terrestrial invertebrates evolved to giant sizes (see Sect. 10.1.9). Today, the massive burning of this coal by (multicellular) humans is one of the major causes of climate change, which is modifying the habitats of the planet for tens of thousands of years and perhaps more (see Sects. 11.3.2–11.3.5).

4.7.3   Anthropogenic Changes to the Earth's Atmosphere

Human activities are presently causing exceptionally rapid changes in the gas composition of the Earth's atmosphere (Fig. 9.5). For example, the ongoing increase in the concentration of atmospheric CO_2 is unprecedented in the last 800,000 years (Fig. 4.10a). By reference to the much longer period of millions of years, it was shown that humans are presently releasing CO_2 at about ten times faster than the most rapid natural event of anytime in the last 66 million years or longer. Human activities are releasing in the atmosphere many greenhouse gases in addition to carbon dioxide (see Information Box 2.7), and as a consequence, the global temperature of Earth has risen by more than 1 °C since the early 1900s. This trend is accelerating, and a very rapid increase in global temperature is feared for the coming decades. The Earth's temperature had sometimes been higher in the past than currently, but the rate at which temperature presently increases is unprecedented (Fig. 4.10b).

It is predicted that the presently occurring anthropogenic changes will have long-term effects on the overall habitability of the planet. These effects concern the glacial-interglacial cycles that the planet has been experiencing for 2.6 million years. We presently live in interglacial environmental conditions, and models indicate that the current episode of anthropogenic global warming will significantly delay the beginning of the next glacial-interglacial cycle, and thus the reduction in overall habitability that takes place when the ice sheets progress (see Sect. 4.5.2). Four hypotheses are described in Chapter 11 and summarized in the next paragraphs.

The *first hypothesis* is based on the amount of CO_2 that has already been released in the atmosphere by human activities, and predicts that the resulting

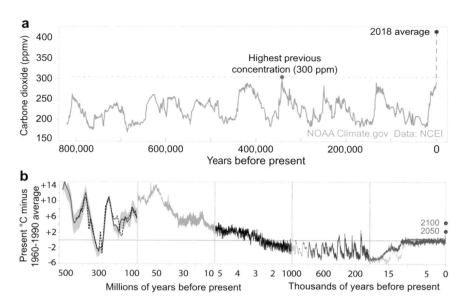

Fig. 4.10 Long-term changes in two global characteristics of Earth: **a** atmospheric CO_2 concentration since 800,000 years, and **b** reconstructed global temperature since 500 million years (temperature difference with the 1960–1990 average). Note the five different timescales (different colours). The values until present (2015) are based on data, and those for 2050 and 2100 come from model predictions. The nature and sources of the data are given on the website. Credits at the end of the chapter

global warming will contribute to delay the initiation of next glacial-interglacial cycle to at least 50,000 years (see Sect. 11.3.2). The *second hypothesis* assumes that the amount of greenhouse gases emitted in the Earth's atmosphere would be so large that it would prevent the occurrence of several glacial-interglacial cycles, in which case the return to a CO_2 concentration below which a glacial episode could occur would not be reached before 500,000 years or more (see Sect. 11.3.3). The *third hypothesis* explores the (unlikely) situation where all the coal deposits would be combusted to CO_2, which could lead to a shift in the condition of the planet from the present icehouse conditions to a new greenhouse Earth (see Sect. 11.3.4). This would be similar to last shift from icehouse to greenhouse, which occurred about 300 million years ago over a period of 40 million years (Table 4.3). The *fourth hypothesis* examines the possibility that a shift from icehouse to greenhouse Earths could occur if the Earth System crossed a planetary threshold, which could be as low as 2 °C above the preindustrial temperature (see Sect. 11.3.5). These chains of reactions connect the present anthropogenic increase in atmospheric greenhouse gases with changes in the glacial-interglacial cycles or in the icehouse-greenhouse conditions that would modify the overall habitability of the planet in 50,000 years, 500,000 years, or more.

The effects of the present human activities on the overall habitability of Earth in thousands of years may appear very far in the future. However, a situation that could have led to a major change in the overall habitability of the planet already occurred in the twentieth century, and humanity then found itself on the brink of a major disaster so striking that the international community reacted decisively to prevent its occurrence. This disaster was the impending destruction of the Earth's protective O_3 layer by the release of human-made *chloroflurocarbons* (CFCs) in the atmosphere (see Sect. 10.3). These synthetic molecules were largely used from the mid-1900s as refrigerants in domestic refrigerators and industrial refrigeration systems. The O_3 layer protects land-dwelling organisms from DNA-destructive solar UV radiation, and the disappearance of this natural protection would have caused a catastrophic lowering of continental habitability that would have threatened complex land organisms, including humans (see Sect. 4.7.1).

Briefly, the crisis began when atmospheric researchers found in the 1970s and 1980s that the breakdown of CFC molecules by solar UV in the upper atmosphere led to the destruction of the O_3 layer. This was evidenced by abnormally low O_3 concentrations over polar regions, which were then called *ozone holes*. The real possibility of the disappearance of the O_3 layer led to the signing of the international *Montreal Protocol on Substances that Deplete the Ozone Layer*, which entered into force in 1989 and phased out the production of O_3-depleting substances all over the world. The destruction of the O_3 layer by human-made synthetic compounds would have caused a loss of overall habitability that would have cancelled out the corresponding gain that had followed from the natural creation of that layer half a billion year earlier. The latter was a consequence of the biological innovation of O_2-producing photosynthesis more than two billion years before. This chain of reactions connected the production of synthetic CFCs by humans in the twentieth century with the creation of the O_3 layer half a billion years earlier. Humanity avoided the depletion or even destruction of the O_3 layer, which would have drastically reduced the Earth's overall habitability, because countries then agreed on an international treaty that phased out the production of O_3-depleting substances. Additional details are provided in Chapter 10 (see Sect. 10.3).

The above paragraphs described two major anthropogenic changes to the overall habitability of the planet. The first could occur as a consequence of a delay in the initiation of the next glacial episode in tens of thousands of years or more, and the second could have resulted from the depletion or destruction of the O_3 layer, but did not take place thanks to the action of human societies in the twentieth century. Given the relatively recent appearance of humans on Earth, the anthropogenic changes to the natural environment are one-time events in the billion-year history of Earth. This single occurrence of anthropogenic changes to the Earth System is similar to that of the changes driven by biological innovation described in the previous section, and their effects on future overall habitability of the planet are potentially as significant as those caused in the past by these natural changes.

4.8 Key Points: Overall Habitability and the Living Earth

We conclude this chapter with a brief summary of some of the key points of the above sections, as done in previous chapters (see Sects. 1.6, 2.5, and 3.8). Here, we examine interactions between the Solar System, Earth, its overall habitability, and organisms. The context of this section is the geological history of Earth (see Sect. 1.5.4), and more specifically the three phases of the Earth's history illustrated in Fig. 1.10, namely the Mineral, Cradle and Living Earth. These three phases occurred from 4.6 to 4.0, 4.0 to 2.5, and 2.5 to 0 billion years ago, respectively.

The components of the Earth's overall habitability include temperature, the presence of liquid water, access to a source of energy usable by organisms, and availability of nutritive resources (see Sect. 4.1.1). Early in the Mineral Earth stage, before continents started to form, the range of ambient temperatures already allowed most of the water at the surface of Earth to be liquid under the existing atmospheric pressure (see Sect. 4.1.1), and the effect of the Earth's rotation made all longitudes equally habitable (see Sect. 4.2.1).

Later during the Cradle and Living Earth phases, the meridional transport of energy by the combined atmospheric and oceanic circulations evened out temperature and water availability among the low and high latitudes, which increased the overall habitability among latitudes (see Sect. 4.2.3), and the oceanic and atmospheric currents increased the global availability of water and nutrient resources, which improved the overall habitability of the planet (see Sect. 4.3.1). There are nevertheless some major perturbations of environmental conditions that recur every year in some areas, of which two examples are tropical cyclones and the seasonal sea ice (see Sect. 4.3.2).

Large volcanic eruptions with a periodicity of 100 years or more generally have moderate effects on the overall habitability of Earth (see Sect. 4.4). At periodicities of 40,000 to 100,000 years, the 30 to 50 glacial-interglacial cycles that occurred during the last 2.6 million years repeatedly changed the habitability of more than one-third of the planet (see Sect. 4.5.2). At periodicities that generally ranged between those of the glacial interglacial cycles (40,000 to 100,000 years) and the episodes of snowball Earth (more than 100 million years), very large volcanic eruptions and the impacts of very large asteroids strongly reduced the overall habitability of Earth (see Sect. 4.5.3). At the periodicity of more than 100 million years, the occurrences of snowball Earth reduced the overall habitability of the planet in major ways (see Sect. 4.5.4). At the average periodicity of 290 million years, most of Earth was likely habitable in greenhouse climate, whereas the habitable surfaces and volumes were reduced at times under icehouse conditions (see Sect. 4.5.6). Nevertheless, the present success of species and ecosystems testifies that the conditions provided by the current interglacial climate create a high level of overall habitability, but it is possible that the habitability of Earth was even higher during the last greenhouse when dinosaurs roamed the planet (see Sect. 4.5.6).

The characteristics and processes of the Solar System and Earth cited above affected the overall habitability of the planet, and some characteristics and processes of the overall habitability affected the ecosystems in the Cradle and Living Earth Phases. Indeed, the abundance of liquid water early in the history of Earth allowed ecosystems to rapidly occupy the oceans, where all organisms were unicellular for more than 3 billion years, and much later (less than 500 millions years ago), the abundance of liquid water on most emerged lands allowed organisms to start occupying the continents after the formation of the ozone layer (see Sect. 4.1.1). When the first organisms got established on Earth almost four billion years ago, the key inorganic nutritive resources to be used by autotrophs to synthesize organic matter were readily available in the environment. This allowed the build-up of large biomasses, first in the oceans and later at the surface of continents, leading organisms to progressively take over the Earth System (see Sect. 4.1.1).

Despite generally suitable conditions, the Earth's life forms were subjected to large-scale destructions, known as mass extinctions, which are only documented for the last 450 million years because older organisms did not generally leave remains in the fossil record (see Sect. 4.6.1). In order to explain the mass extinctions, researchers invoked various catastrophic changes in environmental conditions, that is, in overall habitability (see Sect. 4.6.2). The first mass extinction occurred in icehouse conditions, and the mass mortality of organisms could then have been caused by the global cooling resulting from a decrease in atmospheric CO_2, following higher uptake of this gas by stronger erosion of silicate rocks. The causes proposed to explain the third and fifth mass extinctions include the impact of a large asteroid, which could have produced strong cooling, directly or through increased volcanic eruptions. Differently, the fourth mass extinction may have been caused by a long-term increase in the greenhouse effect, resulting from either a rising concentration of atmospheric CO_2 or methane (CH_4), which are both powerful greenhouse gases.

The organisms and ecosystems were not only deeply influenced by changes in overall habitability, but they actively modified it, especially during the Living Earth phase. For example, one of the mechanisms invoked for the initiation of the first snowball Earth around 2.2 billion years ago is a decrease in the atmospheric concentration of methane, resulting from the massive oxidation of CH_4 by the oxygen released into the atmosphere by photosynthetic organisms (see Sect. 4.5.4). Another example is the second mass extinction, which may have been caused by the massive growth of plants on continents that would have created a large decrease in atmospheric CO_2, responsible for a strong cooling (see Sect. 4.6.2).

Several biological innovations had deep effects on the overall habitability, especially the biological innovation of O_2-producing photosynthesis more than 2.5 billion years before present (see Sect. 4.7.1). It caused an increase in atmospheric O_2 over hundreds of millions of years, which may have resulted in the massive oxidation of atmospheric CH_4 and thus the first snowball Earth around 2.2 billion years before present. The latter led to a reduction in the overall habitability of Earth during hundreds of thousands of years. This biological innovation also caused a major shift in the overall habitability by transforming the planet from an environment devoid of free O_2 to a mosaic of environments with and without free O_2, shared by organisms that used and did not use O_2 for respiration, respectively. In addition, O_2-producing photosynthesis led to the creation of the O_3 layer in the upper atmosphere, which made the continents suitable for complex multicellular organisms and thus significantly increased the overall habitability of the planet. Other major biological innovations also affected the planet's overall habitability: calcification, silicification, methanogenesis, O_2 respiration, and multicellularity (see Sect. 4.7.2).

Some human activities modify the overall habitability of the planet. One example was the production of synthetic CFCs by humans in the twentieth century, which threatened to deplete and even destroy the ozone layer in the upper atmosphere. This would have drastically reduced the Earth's overall habitability by making the surface of the planet unsuitable for complex multicellular organisms. Fortunately, an international treaty phased out the production of ozone-depleting substances (see Sect. 4.7.3). However, human societies are not always that wise, and a sixth mass extinction may be looming or is even already taking place. This sixth extinction would be due to the direct obliteration of numerous species by human activities or the destruction of their environment, that is, a human-driven reduction in the Earth's overall habitability (see Sect. 4.6.2).

Figure Credits

Fig. 4.1 Extracted from Figure 1 of Schokraie et al. (2012) https://journals.plos.org/plosone/article/figure?id=10.1371/journal.pone.0045682.g001, used under CC BY 4.0.

Fig. 4.2 Figure 7j-1 of Pidwirny (2006) http://www.physicalgeography.net/fundamentals/7j.html. Figure produced by Dr. Michael Pidwirny, University of British Columbia, Canada, and reproduced with his permission © Michael Pidwirny.

Fig. 4.3 This work, Fig. 4.3, is a derivative of a figure in Lecture 19.1 "Advanced topic: Heat transport decomposition. 7. Calculating the partitioning of poleward energy transport into different components" of Rose (2020) https://brian-rose.github.io/ClimateLaboratoryBook/courseware/advanced-heat-transport.html by Dr. Brian Rose, State University of New York at Albany, USA, used under CC BY 4.0. Figure 4.3 is licensed under CC BY 4.0 by Mohamed Khamla.

Fig. 4.4 Modified after Figure 2A of Smith and Pickering (2003). With permission from Prof. Kevin T. Pickering, University College London, England.

Fig. 4.5a This work, Fig. 4.5a, is a derivative of https://commons.wikimedia.org/wiki/File:Vostok_Petit_data.svg by NOAA (Vostok-ice-core-petit.png) https://commons.wikimedia.org/wiki/File:Vostok-ice-core-petit.png), derivative work used under CC BY-SA 3.0 and GNU FDL. Figure 4.5a is licensed under CC BY-SA 3.0 and GNU FDL by Mohamed Khamla.

Fig. 4.5b This work, Fig. 4.5b, is a derivative of https://earthobservatory.nasa.gov/Features/BorealMigration/boreal_migration3.php by NASA's Earth Observatory, in the public Domain. I, Mohamed Khamla, release this work in the public domain.

Fig. 4.6 This work, Fig. 4.6, is a derivative of https://commons.wikimedia.org/wiki/File:2000_Year_Temperature_Comparison.png by Robert A. Rohde https://en.wikipedia.org/wiki/User:Dragons_flight, used under GNU FDL and CC BY-SA 3.0. Figure 4.6 is licensed under GNU FDL and CC BY-SA 3.0 by Mohamed Khamla.

Fig. 4.7 https://commons.wikimedia.org/wiki/File:Rodinia_reconstruction.jpg by John Goodge, United States Antarctic Program, in the Public Domain. Change in the positions of the two colour legends.

Fig. 4.8 This work, Fig. 4.8, is a derivative of https://commons.wikimedia.org/wiki/File:Phanerozoic_Climate_Change.png by Robert A. Rohde https://en.wikipedia.org/wiki/User:Dragons_flight, used under GNU FDL and CC BY-SA 3.0. Figure 4.8 is licensed under GNU FDL and CC BY-SA 3.0 by Mohamed Khamla.

Fig. 4.9 This work, Fig. 4.9, is a derivative of https://commons.wikimedia.org/wiki/File:LateCretaceousMap.jpg by Philip D. Mannion, Figure 3 in Mannion et al. (2014), used under CC BY-SA 3.0. Figure 4.9 is licensed under CC BY-SA 3.0 by Mohamed Khamla.

Fig. 4.10a https://www.climate.gov/sites/default/files/paleo_CO2_2018_1500.gif, from https://www.climate.gov/news-features/understanding-climate/climate-change-atmospheric-carbon-dioxide by NOAA, in the public domain. Some indications inside the figure removed, and titles of two axes rewritten.

Fig. 4.10b This work, Fig. 4.10b, is a derivative of https://commons.wikimedia. org/wiki/File:All_palaeotemps.svg by Glen Fergus https://commons.wikimedia. org/wiki/User:Glen_Fergus, used under CC BY-SA 3.0. Figure 4.10b is licensed under CC BY-SA 3.0 by Mohamed Khamla.

References

Mannion PD et al (2014) The latitudinal biodiversity gradient through deep time. Trends Ecol Evol 29:42–50. https://doi.org/10.1016/j.tree.2013.09.012

Pidwirny M (2006) Fundamentals of physical geography, 2nd edn. Viewed 20 May 2020. http://www.physicalgeography.net/fundamentals/7j.html

Rose B (2020) The climate laboratory. A hands-on approach to climate physics and climate modeling. Available via https://brian-rose.github.io/ClimateLaboratoryBook/home

Smith AG, Pickering KT (2003) Oceanic gateways as a critical factor to initiate icehouse Earth. J Geol Soc, London 160:337–340

Schokraie E et al (2012) Comparative proteome analysis of *Milnesium tardigradum* in early embryonic state versus adults in active and anhydrobiotic state. PLoS ONE 7(9):e45682. https://doi.org/10.1371/journal.pone.0045682

Further Reading

Broecker W (2018) CO_2: Earth's climate driver. Geochem Perspect 7:117–196. https://doi.org/10.7185/geochempersp.7.2

European Science Foundation (2015) Extreme geohazards: reducing the disaster risk and increasing resilience. European Science Foundation, Strasbourg. http://archives.esf.org/fileadmin/Public_documents/Publications/Natural_Hazards.pdf

Falkowski PG (2015) Life's engines: how microbes made Earth habitable. Princeton University Press, Princeton

Glikson AY, Groves C (2016) Climate, fire and human evolution. In: The deep time dimensions of the anthropocene, Modern Approaches in Solid Earth Sciences, vol 10. Springer, Cham

Goosse H (2015) Climate system dynamics and modelling. Cambridge University Press, Cambridge

Holden JF et al (2012) Biogeochemical processes at hydrothermal vents: microbes and minerals, bioenergetics, and carbon fluxes. Oceanography 25(1):196–208. https://doi.org/10.5670/oceanog.2012.18

Lunine JI (2013) Earth: evolution of a Habitable World, 2nd edn. Cambridge University Press, Cambridge

Middelburg JJ (2019) Marine carbon biogeochemistry. A primer for Earth System scientists. Springer Briefs in Earth Sciences. Springer, Cham

Ortega RP (2019) Mixing it up in the web of life. Knowable Magazine. https://doi.org/10.1146/knowable-013119-1

Poppick L (2019) The story of snowball Earth. Knowable Magazine. https://doi.org/10.1146/knowable-031919-1

Ruddiman WF (2008) Earth's climate: past and future, 2nd edn. W. H. Freeman and Company, New York

Wallace JM, Hobbs PV (2006) Atmospheric science: an introductory survey, 2nd edn. International Geophysics. Elsevier Academic Press, Amsterdam, Boston

Chapter 5
Liquid Water: Connections with the Outer Reaches of the Solar System

5.1 Water, Organisms and Ecosystems

5.1.1 Key Processes in the Cells of Organisms, and Water

Organisms are made of cells, whose materials are enclosed within membranes (see Information Box 1.1). These membranes isolate the cells and their internal structures, called *organelles*, from their immediate environment. However, this isolation from the environment is not total but only partial. Indeed, in order to stay alive, grow and replicate, cells continually exchange materials and energy with their environment through their membranes.

It is explained in the following paragraphs that water is of key importance for organisms and ecosystems, and the next section investigates connections between key cellular processes. The remainder of the chapter considers successively: the types of water at the Earth's surface (Sects. 5.2); the sources of the Earth's water in the Solar System (Sects. 5.3); water on Earth, Venus and Mars, different abundances and histories (Sects. 5.4 and 5.5, respectively); and the characteristics of the planet that keep most of the Earth's water in liquid form (Sects. 5.6). The chapter concludes with a summary of key points concerning the interactions between the between the Solar System, Earth, water, and organisms (Sect. 5.7).

In part A of Table 5.1, the organisms belonging to the three domains of life (see Information Box 3.1) are assigned to the broad categories of *autotrophy* or *heterotrophy*, based on the processes they use for their synthesis or acquisition of organic compounds (see Information Box 4.1). Some of the archaea and bacteria and all phytoplankton and plants synthesize their own organic matter, and are thus autotrophs (see information Box 5.1). The other archaea and bacteria and the protozoa, fungi and all animals use organic matter synthesized by other organisms, and are thus heterotrophs. Part B of Table 5.1 identifies the processes—respiration or

© Springer Nature Switzerland AG 2021
P. Bertrand and L. Legendre, *Earth, Our Living Planet*, The Frontiers Collection,
https://doi.org/10.1007/978-3-030-67773-2_5

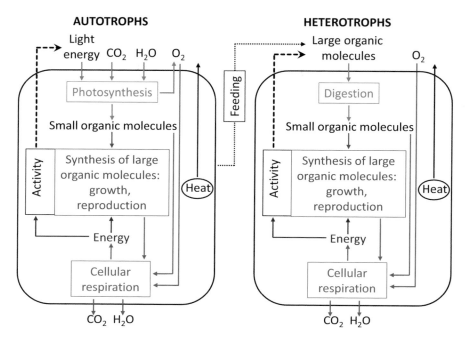

Fig. 5.1 Connections among key cell processes and activities of O_2-producing photosynthetic autotrophs and O_2-respiring heterotrophs. The heat generated by these processes is dissipated in the environment. Solid arrows: circulation of materials and energy; bold dashed arrows: activities for acquiring materials and energy; light dotted arrow: feeding of heterotrophs on organic matter synthesized by autotrophs. Colours: different processes and associated flows. Credits at the end of the chapter

Table 5.1 Biochemical pathways used by organisms (archaea, bacteria, or eukarya) to (A) synthesize or acquire the organic compounds they need, or (B) break down molecules into smaller ones to acquire energy. Details on these pathways are given in the text (see Information Boxes 5.1 and 5.2)

Biochemical pathways	Archaea	Bacteria	Eukarya
A. Synthesis or acquisition of organic compounds			
Autotrophy			
O_2-producing photosynthesis	None	Cyanobacteria	Algae and plants
O_2-less photosynthesis	None	Some	None
Chemosynthesis	Some	Some	None
Heterotrophy	Some	Some	Protozoa, fungi and animals
B. Breakdown of organic molecules to obtain energy			
O_2 respiration	Some	Some	All
Respiration without O_2	Some	Some	None
Fermentation	Some	Some	Animals: lactic fermentation

Information Box 5.1 Photosynthesis and Chemosynthesis The autotrophic organisms use O_2-producing photosynthesis and other cellular processes to synthesize organic matter.

Phytoplankton and plants both use *O_2-producing photosynthesis* (scientific name: *oxygenic photosynthesis*) to build organic matter. It is the most common form of photosynthesis on Earth. The formula of O_2-producing photosynthesis is:

$$CO_2 + 2H_2O + \text{photons} \rightarrow [CH_2O] + O_2 + H_2O \text{ (see Information Box 3.4)} \quad (5.1)$$

where $[CH_2O]$ stands for sugar, which is the first compound formed during this synthesis of organic matter. The free oxygen (O_2) released to the environment comes from the splitting of water molecules (H_2O), which also provides the H found in $[CH_2O]$, while the C and O atoms come from CO_2. The energy carried by photons of light is absorbed by photosynthetic pigments located in chloroplasts, which are the cell organelles where photosynthesis takes place. Only the energy of photons whose wavelength corresponds to the absorption spectrum of photosynthetic pigments is used. The O_2-producing photosynthesis was invented, during the course of biological evolution, by cyanobacteria. The only organisms producing O_2 on Earth belong to cyanobacteria and eukaryotes (algae and plants). No other bacteria and no archaea perform *O_2-producing* photosynthesis.

Some groups of bacteria and archaea use instead *O_2-less photosynthesis* (scientific name: *anoxygenic photosynthesis*), where there is production of organic matter but no production of O_2. In this process, the hydrogen used to build organic compounds is supplied by other compounds than water. The following formula illustrates the case where the hydrogen comes from hydrogen sulphide H_2S (the gas that smells rotten eggs):

$$CO_2 + 2H_2S + \text{photons} \rightarrow [CH_2O] + 2S + H_2O \quad (5.2)$$

The O_2-less photosynthetic bacteria and archaea capture photons using other pigments than chlorophyll. Because the archaea use light energy for other metabolic functions than fixing carbon, their activity is sometimes called *phototrophic* instead of *photosynthetic*.

Other autotrophic organisms, which include archaea and bacteria, produce organic matter using *chemosynthesis*. In this process, the energy is supplied by a reaction between the strong oxidative molecule O_2 and such substances as H_2 or H_2S. Here are three examples of chemosynthetic pathways, among many others:

$$CO_2 + 4H_2S + O_2 \rightarrow [CH_2O] + 4S + 3H_2O \quad (5.3)$$

$$CO_2 + 6H_2 + 2O_2 \rightarrow [CH_2O] + 5H_2O \quad (5.4)$$

$$CO_2 + 2H_2S \rightarrow [CH_2O] + 2S + H_2O \qquad (5.5)$$

Comparing Eqs. 5.5, 5.2 and 5.1 shows that chemosynthesis is the ancestor of O_2-less photosynthesis, and the latter is the ancestor of O_2-producing photosynthesis. The two types of photosynthesis use the energy carried by photons, and thus build organic matter more efficiently than chemosynthesis. A major advantage of O_2-producing photosynthesis is the abundance of water, which is much more available than H_2S in most environments.

fermentation—by which the organisms (autotrophic or heterotrophic) break down organic molecules to acquire the energy they need to stay alive, grow and replicate (see Information Box 5.2).

The various processes in Table 5.1 use inorganic and organic substances found in different environments of Earth. Organisms find in the atmosphere such gases as O_2 and CO_2, and also inorganic and organic substances transported by aerosols. They can also find the same gases dissolved in natural water bodies or in water that circulates in the soil, together with dissolved inorganic nutrients that contain essential chemical elements (for example, nitrogen and phosphorus) and with organic substances. Organic matter also exists in aquatic and terrestrial environments in the form of other organisms and detritus (see Information Box 3.5). Liquid water plays a key role in supplying organisms and ecosystems with the substances they require.

Organisms do not assimilate or incorporate into their cellular material all the inorganic and organic substances they take up, and they reject in the environment the substances they do not use. The latter include: *gases*, for example O_2 and CO_2, which are by-products of O_2-producing photosynthesis and O_2 respiration, respectively; *liquids*, for example H_2O, which is rejected by O_2 respiration, evapotranspiration and transpiration; *dissolved compounds*, inorganic and organic, and ions rejected by cells when they are in excess, such as Ca^{2+}, Na^+, Cl^-; and *solids*, for example unassimilated organic matter in the form of faeces. These unused substances are released in the surrounding air, water or soil. Simultaneously, the energy degraded by organisms is released as heat into the environment (see Sect. 3.2.2).

Information Box 5.2 Respiration and Fermentation All organisms break down organic molecules to obtain energy. Autotrophs break down molecules they have themselves synthesized, and heterotrophs break down organic molecules derived from those of autotrophs. In order to do so, the organisms use O_2 respiration or other cellular pathways.

Some groups of archaea and bacteria and all eukaryotes use O_2 *respiration* (whose scientific name is *aerobic respiration*):

$$[CH_2O] + O_2 \rightarrow CO_2 + H_2O + \text{energy (see Information Box 3.4)} \qquad (5.6)$$

The O_2 respiration releases in H_2O molecules (right side of Eq. 5.6) the H atoms contained in organic matter ($[CH_2O]$) combined with O atoms provided by the O_2 taken from the environment.

Other groups of archaea and bacteria cannot tolerate O_2 and thus use other pathways than O_2 respiration. These pathways, which already existed before the build-up of significant quantities of O_2 in the environment, are *fermentation* and *respiration without use of O_2* (scientific name: *anaerobic respiration*). Some archaea and bacteria can use O_2 respiration when there is O_2, and an alternative pathway when O_2 is low or absent.

Different chemicals (called *electron acceptors*) can act as oxidizing agents in respiration. Examples are: O_2 (Eq. 5.6), nitrate (NO_3^-; denitrification, Eq. 9.14), CO_2 (Eq. 5.8), iron (Fe^{3+}, iron reduction), and sulphur (S^0, sulphur reduction; SO_4^{2-} sulphate reduction Eq. 5.11).

An example of fermentation without O_2 is the production of ethanol, $C_2H_5(OH)$, from sugar (glucose), $C_6H_{12}O_6$:

$$C_6H_{12}O_6 \rightarrow 2C_2H_5(OH) + 2CO_2 + \text{energy} \qquad (5.7)$$

Yeasts, which are heterotrophs, use this pathway to produce energy using organic compounds elaborated by autotrophs. One such compound is sugar, whose fermentation in absence of O_2 releases the by-products alcohol (ethanol) and CO_2 in the environment of yeasts (which can be a beer or wine barrel). Some yeast can also perform fermentation in the presence of O_2.

An important reaction, called *methanogenesis*, is the production of methane (CH_4) gas:

$$CO_2 + 4H_2 \rightarrow CH_4 + 2H_2O \qquad (5.8)$$

It is only done by archaea, and can use other substrates than CO_2 and H_2, such as formate and methyl compounds. It plays important roles in the Earth System (see Sect. 2.1.7).

One type of fermentation converts sugar into lactic acid and cellular energy:

$$C_6H_{12}O_6 \rightarrow 2C_3H_6O_3 + energy \qquad (5.9)$$

People use this fermentation to produce such food specialties as yoghurt, sauerkraut and kimchi. The same pathway is also found in the muscles of animals, in which case it is not called fermentation but *respiration*. Many people who practice sports know the (supposed) connection between exercise and the production of lactic acid (muscle ache). The further oxidation of lactic acid to CO_2 and H_2O in animals requires oxygen, and thus normally takes place after the physical effort.

The O_2 respiration produces more energy per unit sugar than the pathways without O_2. The higher efficiency of O_2 respiration explains its evolutionary success after O_2 began to build up on Earth about 2 billion years ago.

The above exchanges of materials and energy, which occur between unicellular and multicellular organisms and their aquatic or terrestrial environments, ultimately take place through the cell membrane. It will be explained below (see Sect. 5.5.2) that liquid water is the most powerful natural solvent, meaning that it can dissolve almost any natural or man-made compound that exist on Earth, given time. Because of this unique property, most of the substances used by the cells or released by them are transported by liquid water.

In addition, water is a key molecule in many of the processes that take place within the (tiny) cells of organisms and have global effects on Earth. A spectacular example is the O_2-producing photosynthesis, which splits molecules of water (H_2O) and uses the resulting H atoms to build-up a large part of the organic matter that sustains the aquatic and terrestrial food webs (see Information Box 5.3). This process simultaneously releases the O_2 that makes up almost all the free oxygen in the Earth's atmosphere and dissolved in aquatic environments (see Information Boxes 3.4 and 5.1). In addition, the O_2 produced by photosynthesis is the source of ozone (O_3) in the high-atmosphere, which protects terrestrial organisms from harmful UV radiation (see Sect. 2.1.5).

Information Box 5.3 Food Chains and Food Webs The well-known concepts of food chain and food web are fundamental to ecology. The food chain can be seen as a simplified form of the food web. They both combine autotrophic, heterotrophic and mixotrophic organisms (see Information Box 4.1).

A *food chain* describes the feeding links between different functional groups of organisms. These functional groups include: the autotrophs, which synthesize organic matter from inorganic substances (see Information Box 5.1), and are often called *primary producers*; the heterotrophs that feed on autotrophs, called *herbivores*; other heterotrophs that prey on herbivores, called *carnivores*; and so on. A very simple example would be the two-level food chain going from plant products, such as rice or tomatoes, to humans, who are then herbivores. A more complex, three-level example would go from corn to cattle (herbivores) and to humans, who are then carnivores. Hence the diet of many humans is both herbivorous and carnivorous, humans thus being *omnivores*. This raises the following conceptual difficulty: how many levels are there in the food chains of omnivorous humans? two or three? This question stresses the fact that the food-chain concept is useful for describing simple situations, but often cannot account for the real complexity of Nature.

The production of organic matter by primary producers is called *primary production*. Similarly, the production of organic matter by herbivores and carnivores is called *secondary production* and *tertiary production*, respectively. We will not use this specialized ecological vocabulary in this book.

The complex interactions among species or groups of species are described by the *food web*, which interconnects several food chains. Some organisms are both herbivores and carnivores (called *omnivores*), other feed on detritus from both autotrophs and heterotrophs (called *detritivores*), other decompose organic matter from various sources (called *decomposers*), other live as *parasites* of either autotrophs or heterotrophs (parasites take advantage of their *hosts*), in some cases an autotroph lives inside an heterotroph and both benefit from the presence of the other (called *symbionts* and hosts, respectively, for example the microscopic algae that live inside coral hosts and are essential symbionts to the corals), and so forth. The network of such complex interactions among species or groups of species is the food web.

Biologists who study the possibility of life beyond planet Earth have considered the hypothetical cases of organisms whose matter could be based on other chemical elements than carbon. These other elements are silicon, boron, nitrogen, phosphorus, sulphur or germanium, and the corresponding organic compounds could theoretically develop in environments dominated by other solvents than water. These hypothetical compounds and the role that water played on Earth in setting carbon as the key chemical element for organisms are examined in Chapter 6 (see Sect. 6.1.2). In any case, liquid water is a key condition for organisms on Earth and, as far as presently known, would also be essential for life forms on other astronomical bodies. This explains why the present search for extraterrestrial life is largely a search for liquid water. The review article of Ball (2017) provides further reading on the many varied and subtle roles of water in cells.

5.1.2 Connections Between Key Cell Processes

Several of the cell processes explained in previous chapters and/or examined in the present chapter are summarized in Fig. 5.1, which illustrates in a parallel manner some key cell processes of the O_2-producing photosynthetic autotrophs and the O_2-respiring heterotrophs. Figure 5.1 shows that the organisms of the first type acquire their nutritive resources by O_2-photosynthesis and those of the second type by feeding followed by digestion, these two sets of processes thus playing corresponding roles in the two types of organisms. The nutritive resources of the heterotrophs come from the autotrophs. Aside from this major difference between autotrophs and heterotrophs, the cell processes of the two types of organisms are basically the same.

For simplicity, Fig. 5.1 considers only three chemical elements, namely C, H and O. It summarizes the following processes:

The autotrophic organisms use the energy of light to synthesize small organic molecules such as sugar, $[CH_2O]_x$, by combining C and O from CO_2 and H from H_2O. The O_2-producing photosynthesis releases in the environment O_2 molecules, whose oxygen atoms come from H_2O (see Information Box 5.1). Differently from autotrophs, the heterotrophic organisms produce small organic molecules by digestion of the large organic molecules obtained by feeding on autotrophs or other heterotrophs (see Sect. 6.1.3). For single cells using dissolved substances, the processes corresponding to feeding and digestion are the uptake and assimilation of dissolved organic matter (see Information Box 3.5).

The small organic molecules provide the cells of autotrophs and heterotrophs with the materials for building large organic molecules. These are utilized for cell growth and division (see Information Box 1.1).

The cells of autotrophs and heterotrophs use the process of O_2 respiration to obtain energy from organic molecules by chemical reaction with O_2 (see

Information Box 3.4), and they use part of this energy to build large organic molecules from smaller ones (see Sect. 6.1.3).

Another part of the energy obtained by respiration is used to fuel various activities, such as movements. Some of these activities can improve the acquisition of materials and energy, for example, some continental plants reposition their leaves to optimize their exposition to light, some phytoplankton migrate vertically in the water column to access light near the surface and nutrients at depth, and some predators actively chase preys.

All the above processes transform part of the energy to heat, which is dissipated in the surrounding environment.

In Fig. 5.1, the autotrophic synthesis of organic matter is illustrated by the *O_2-producing photosynthesis,* but a large part of the autotrophic organisms use instead one of the following two other pathways to synthesize organic matter (see Information Box 5.1). First, some bacteria and archaea use *O_2-less photosynthesis,* where the energy is provided by light but the source of H is another molecule than H_2O. For example, the source of H can be hydrogen sulphide (H_2S), in which case the organisms release sulphur (S) instead of O_2 as by-product. Second, many microorganisms synthesize organic molecules using *chemosynthesis,* which is fuelled by the chemical energy released by reactions of such compounds as H_2S (inorganic) or methane (CH_4, organic) with other molecules (see Sect. 2.1.7). There are various forms of chemosynthesis, and this process is encountered in environments that include marine and lake sediments, marine hydrothermal systems (hot hydrothermal vents and cold seeps), continental soils and caves.

In Fig. 5.1, the release of energy from organic molecules is illustrated by *cellular O_2 respiration,* but a large part of organisms use different pathways (see Information Box 5.2). Indeed, many bacteria and archaea can release energy from organic compounds without using O_2, by the processes of *fermentation* or *respiration without use of O_2.* These processes produce a variety of compounds that contain C, H and O atoms coming from the organic compounds used. The general audience book of Kratz (2020) provides further reading on the functioning of cells.

5.2 Types of Water at the Earth's Surface

5.2.1 States of Water on Earth

Seen from space, Earth is overwhelmingly a planet of liquid water (Fig. 2.2). Indeed, the oceans cover more than 70% of the surface of Earth, and the surface of continents is speckled with rivers and lakes. Under the surface of continents and thus unseen from people, ground waters run in giant underground aquifers, at depths usually less than 750 m but that can reach thousands of metres. The word *aquifer* designates both the geological formations that contain still groundwater and those in which groundwater flows. The oceans make up 94% of the total volume of water on Earth, the underground reservoirs 4%, and icecaps and glaciers the remaining 2%. Lakes, rivers and atmospheric vapour account for very small fractions of the total, although they have important roles in the functioning of the Earth System.

In addition, there may be huge amounts of water stored in rocks very deep underground, in the transition zone between the crust and the mantle located about 400 to 650 km below the surface. Surface water is carried to these depths by the vertical movements of tectonic plates called *subduction* (see Sect. 7.2.3), and the atoms of hydrogen and oxygen that make up water molecules are chemically bound to the structure of certain minerals in rocks. The magnitude of this potential very deep reservoir is still largely unknown, and consequently the values cited here for the different states and reservoirs of water do not take into account the water that may be stored very deep below the surface. The present chapter only considers surface water and, to a lesser extent, atmospheric and ground water.

More than 98% of the water at the surface of Earth and under the ground is *liquid*. The average thickness of liquid water in the oceans is 3,700 m, which is impressive. Even more impressive is the maximum thickness of water in the deepest spot of the oceans, which is the Mariana Trench in the Western Pacific, whose depth is more than 11,000 m.

Water at the surface of continents is largely liquid, but 10% of the continents are covered by ice and snow that form the polar *ice sheets* and *ice caps* (names used for the masses of ice that extend over more than 50,000 km^2 and less than 50,000 km^2, respectively) and the *glacier* on high mountains. Ice also covers 5% of the oceans, as seasonal (or first-year) and old (or multi-year) *sea ice*. The ice pack is formed of sea ice not attached to the shore and is thus free to move. Ice sheets, ice caps and glaciers result from the accumulation of snow on continents, whereas sea ice is frozen from seawater. *Icebergs*, which are found in the oceans, are pieces of continental ice sheets that broke apart and are drifting at sea. Ice sheets, ice caps, glaciers, icebergs and sea ice are all made of fresh water. The *solid water* can reach a thickness of almost 5 km in some Antarctic ice sheets, and a more modest 5 m in the ice packs of the Arctic and Antarctic (Southern) Oceans.

Some water is also present in the atmosphere in *gaseous form*, called water vapour, which only represents 0.001% of the water on our planet but plays an important role in the Earth's heat machinery (see Sect. 3.5.1). Clouds are mostly made of ice crystals and water droplets in various concentrations. Thick dense clouds made of ice-crystal are brilliant white, and their albedo can reach the high values of 0.7–0.9 (see Information Box 2.9). Clouds thus prevent part of the solar radiation from reaching the Earth's surface, which is good in the overall context of global warming, but it not always welcomed by people who live in regions often cloudy or with long periods of cloud cover.

Finally, all organisms, from single cells to whales, are mostly made of water. Indeed, water constitutes more than 60% and up to 95% of the body masses of organisms (see Sect. 6.1.2). Hence, living matter is mostly *living water*. Indeed, organisms take in liquid and sometimes gas water from their environment and

---

Information Box 5.4 Liquid Water in the Solar System Robotic space missions have provided indications that large reservoirs of liquid water may exist under the surface of three of the 79 moons of Jupiter, namely Ganymede, Europa and Callisto, and two of the 82 moons of Saturn, namely Enceladus and Titan (see Sect. 8.3.2). In addition, there may be liquid water under the surface of dwarf planet Ceres, which is located in the asteroid belt between Mars and Jupiter. Closer to Earth, liquid water has been detected in the soil of planet Mars (see Sect. 5.5.3). Further exploration of the numerous bodies of the Solar System could perhaps find liquid water on several of them. However, the occurrence of liquid water does not necessarily mean that organisms are present. The book of Hand (2020) provides further reading on the possibility of liquid water elsewhere than on Earth in the Solar System.

---

combine part of it with other compounds in their cells to build organic matter. The atoms of hydrogen and oxygen from water included in organic compounds are not any more liquid or gas water, but they are ultimately returned to the environment as liquid water or water vapour.

As far as we presently know, planet Earth is the only place in the Solar System where water occurs in the three states of liquid, solid and gas. On Earth, the presence of large amounts of liquid water was one of the conditions that allowed organisms to establish themselves durably and build up the huge biomasses that led them to progressively take over the whole planet (see Sect. 1.2.2). The same may be possible for some of the other bodies of the Solar System with large reservoirs of liquid water (see Information Box 5.4). Exploration of the Solar System in the coming decades should show if some large astronomical bodies other than Earth contain sizeable amounts of liquid water, which is a crucial aspect of the near-Earth Universe.

5.2.2 Fresh and Salt Water

Fresh water represents less than 2.5% of all water on Earth, and almost 99% of the fresh water is either frozen in ice or flows in underground aquifers. It follows that liquid fresh water at the surface of continents (lakes, rivers, streams) accounts for less than 1% of all fresh water on the planet, and less than 0.025% of all water on Earth. Despite its relatively small amount, surface fresh water has a major role in the geological dynamics of Earth. Indeed, running fresh water slowly erodes (i.e. dissolves) continental rocks, including the highest mountains, as water from

Information Box 5.5 Large Size of the Oceans Because we, the humans, live on continents, it is difficult for us to imagine the huge amount of water contained in the oceans. In order to understand the very large size of the ocean, let us imagine what would happen if we decided to dig out all the continents and deposit the resulting materials in the oceans, with the idea of filling the ocean basins with solid materials from continents. Calculations show that after shovelling all the emerged lands into the oceans, the latter would still be there. If digging and shovelling continued until the solid surface of Earth became of uniform depth, there would be no continents left, and the resulting global ocean would be more than 2400 m deep. In other words, if the continents and oceans were spread evenly over the whole surface of Earth, the resulting ocean would cover the whole planet with an average thickness of more than 2400 m. The corresponding volume of the oceans is 1.35 billion km^3.

rain and melting snow continually moves downhill on continents (see Information Box 7.1). The steady, long-term erosion of continental rocks by fresh water is key to the processes that determine the long-term chemical characteristics of the ocean and the atmosphere, and thus contributes to natural changes in the climate at different timescales (see Sect. 7.3.3). In addition, access to fresh water is an essential daily requirement for terrestrial organisms and human societies. Hence it is beyond comprehension that human societies do not always treat very carefully the scarce and vital (renewable) resource of liquid fresh water. This aspect is considered in Chapter 10 within the context of water insecurity (see Sects. 11.1.1 to 11.1.3).

The most abundant form of water on our planet is *salt water*. Most of the salt water is in the oceans, but some is also found in the deep parts of underground aquifers. The volume of oceans is extremely large, that is, 45 times that of the continents above sea level (1,350 and 30 million km^3, respectively), and their average depth is 3,700 m (see Information Box 5.5).

5.2.3 Isotopic Forms of Water

We address below the question: "Where does the Earth's water come from?" (see Sect. 5.3). In order to provide an answer this question, researchers take advantage of the fact that water (H_2O) in the Universe is naturally made of different isotopes of its two chemical constituents, hydrogen (H) and oxygen (O). The notion of isotopes refers to the structure of atoms (see Information Box 5.6).

Information Box 5.6 Radioactive and Stable Isotopes The word *isotope* refers to the numbers of protons and neutrons (subatomic particles with a positive electrical charge and with no charge, respectively) present in the nuclei of atoms. Each chemical element is characterized by a fixed number of protons, for example the numbers of protons of the hydrogen (H), helium (He) and oxygen (O) atoms are 1, 2 and 8, respectively. However, the numbers of neutrons of these atoms can differ, and the different forms of a given chemical element with different numbers of neutrons are called *isotopes*.

The combined numbers of protons and neutrons of an atom set its *atomic mass*. For example, most atoms of oxygen (O) have 8 protons and 8 neutrons, and their atomic mass is thus $8 + 8 = 16$ (the mass of electrons is negligible retative to that of protons and neutrons). The number of protons (p) and the atomic mass (m) of a chemical element (C) are denoted as follows: $_p^m C$. There is normally no need to write the index p to the left of the symbol of a chemical element because each chemical element has a set number of protons (for example, writing $_1 H$ would not provide more information than simply writing H).

All the isotopes of a chemical element have the same chemical properties, but some isotopes are radioactive and others are stable. The adjective *radioactive* denotes a chemical element that spontaneously transforms itself into another one by changing its number of protons, a phenomenon called *radioactive decay*. Most isotopes do not decay, and are thus called *stable*. As example of radioactive isotope, carbon-14 decays into stable nitrogen-14: $_6^{14}C \rightarrow _7^{14}N$. The time required for half the atoms of a radioactive isotope to become a different element is called *radioactive period* (or *half life*). For example, it takes about 5,700 years for half the atoms of a given amount of $_6^{14}C$ to decay into atoms of $_7^{14}N$.

The two constituents of water, H and O, each have three isotopes in nature. The following paragraphs explain that there are 18 possible combinations of these isotopes into molecules of H_2O, to which correspond 18 isotopic forms of water.

Hydrogen has three naturally occurring isotopes: the "usual" hydrogen or *protium*, with one proton and no neutron ($1 + 0 = 1$, symbol ^1H or H); *deuterium*, with one proton and one neutron ($1 + 1 = 2$, symbol ^2H or D); and *tritium*, with one proton and two neutrons ($1 + 2 = 3$, symbol ^3H or T). Protium accounts for $> 99.98\%$ of all H atoms in the natural environment.

Hydrogen isotopes ^1H and ^2H are stable and ^3H is radioactive. Four other isotopes of hydrogen, all radioactive, have been produced in the laboratory but do not exist in nature (^4H, ^5H, ^6H, and ^7H). Radioactive ^3H has 1 proton and 2 neutrons ($_1^3H$). One of its neutrons may spontaneously decay into a proton and an electron, so that the resulting atom has 2 protons and 1 neutron ($_2^3He$; the chemical element with 2 protons is helium). Hence the result of the radioactive decay of ^3H produces a different chemical element, He, and the emission of the electron by the ^3H atom during decay is one form of radioactivity. The radioactive period of ^3H is 12.3 years.

Oxygen has three naturally occurring isotopes, which are all stable: ^{16}O (8 protons + 8 neutrons), ^{17}O (8 protons + 9 neutrons) and ^{18}O (8 protons + 10 neutrons). The isotope ^{16}O accounts for 99.8% of all oxygen atoms in the natural environment.

It follows from the above that there are six isotopic forms of H_2O for the 6 pairs of H (these are HH, HD, HT, DD, DT, and TT) and three isotopic forms of H_2O for ^{16}O, ^{17}O and ^{18}O. Hence, a total of 6 x 3 = 18 isotopic forms of H_2O in the natural environment. Half of the isotopic forms of water contain T and are thus radioactive, and half do not and are thus stable.

Given that water is made of two atoms of hydrogen and one atom of oxygen, there are 18 natural isotopic forms of H_2O resulting from the various combinations of the different natural isotopes of H and O (see Information Box 5.6). To simplify the matter, we consider here only two isotopic forms of water, which are based on their composition in term of the two stable isotopes of hydrogen (protium, H, and deuterium, D), that is, without considering the third isotope of hydrogen (tritium) or the three isotopes of oxygen. The two isotopic forms of H_2O considered here are *light water*, HHO, and *heavy water*, DDO. Researchers use the relative abundances of H and D in the Earth's water to identify its sources either on our planet or elsewhere in the Solar System. The origin of the present Earth's water goes back billions of years, and the sources of this water are examined below (see Sect. 5.3).

5.3   Where Does the Earth's Water Come from?

The two chemical constituents of water, namely hydrogen and oxygen, are very abundant in the universe. Accordingly, astronomical observations and space exploration in the last decades (mostly by robotic probes) have shown that there is plenty of water throughout the Solar System, mostly as ice or vapour. Some researchers have proposed that this "space water" was at the origin of Earth's water.

Planets, asteroids and comets in the Solar System were formed by accretion of matter from the solar nebula (see Sect. 1.3.6). Because there was much water in the solar nebula, all astronomical bodies formed from this nebula initially contained various amounts of water. However, most of the Solar System bodies lost part of their initial water during their billion-year exposure to solar heat in space.

The book of Hanslmeier (2011) provides further reading on water not only on Earth and in the Solar System but also in the whole Universe.

The origin of the water that presently exists on Earth has been debated among specialists for several decades. The relative importance of the three main potential sources of the Earth's water described here is still uncertain.

The first mechanism proposed to explain the abundance of water on Earth goes back to the formation of the proto-Earth, before the last giant collision that formed the Earth-Moon system 4.5 billion years ago (see Sect. 1.3.1). The solar nebula contained much water, and this abundant cosmic water may have been one of the sources of the present Earth's water, and perhaps the main source according to some researchers. The delivery mechanism from the solar nebula would have been the absorption of water onto dust grains present in the nebula, and the accretion of the water-soaked dust into the proto-Earth. Some geological evidence supporting the idea that the solar nebula was the source of the present Earth's water has been found in the oldest rocks that exist on Earth.

However, it is often thought that the initial Earth's water provided by the solar nebula was lost early in the history of the planet when this water was released into the early atmosphere, because of the high temperature and strong ultraviolet radiation that prevailed at the time. This led to the idea that most of the present Earth's water was brought from space after the atmosphere of our planet had cooled down, during the episode called *Late Heavy Bombardment*, which is thought to have occurred approximately 4.1 to 3.9 billion years ago (see Sect. 1.5.1). During this episode, Earth as well as the Moon and the other rocky planets of the Solar System would have been hit by very large numbers of asteroids and comets, which brought with them huge amounts of water. This supply of water from the Solar System would have been the source of the water in the present Earth's oceans, glaciers, aquifers, lakes and rivers. This supply could have started before the late heavy bombardment, as early as 4.4 billion years ago.

Assuming that Earth indeed experienced the above supply of water from the Solar System, the question becomes: Which interplanetary bodies were mostly responsible for supplying water to Earth: asteroids or comets? Specialists have debated this question during the last decades, and the key to this debate has been the isotopic composition of the oceans' water. As explained above (see Sect. 5.2.3), water can have different hydrogen isotopic composition. In the Earth's oceans, the deuterium-to-hydrogen ratio (D/H) is 1.56×10^{-4}, meaning that one out every 6,400 atoms of hydrogen in seawater molecules is deuterium. What about the D/H ratio of the water in asteroids and comets?

Because the nuclei of comets contain large proportions of water, these Solar System bodies were initially thought to be the main source of the Earth's water. However, astronomical observations and space exploration progressively showed that the D/H ratio of comets' water was farther away from that of Earth's water than the D/H ratio of asteroids' water. This shifted the opinion of researchers towards asteroids as being the main source of the Earth's water. This, until it was reported in 2011 that the D/H ratio of water in a comet named 103P/Hartley 2 was close to that of the Earth's ocean. This reversed the opinion in favour of comets … until observations from the robotic space mission Rosetta to comet 67P/Churyumov-Gerasimenko showed, in 2015, that the D/H of water on that comet was very different from the value in the Earth's oceans. Other observations supported or denied the possible role of comets as sources of the Earth's water, and the debate will likely continue as more data become available. This field of astronomical and space research moves very fast!

In summary, the abundant water that presently exists on Earth comes from the original solar nebula, directly or indirectly. Three possible pathways have been proposed for the origin of Earth's water. The first is the accretion, more than 4.5 billion years ago, of solar-nebula dust grains with their adsorbed water, leading to the formation of the proto-Earth with most of its present water. The two other possible sources are the bombardment of young Earth by water-containing asteroids and/or water-rich comets, which would have mostly taken place between 4.1 and 3.9 billion years ago. The three sources perhaps contributed in various proportions to the present Earth's water. It is expected that continuing study of Earth and the Solar System in decades to come will provide additional information on each of the three potential sources of the Earth's water. The chapter of Pinti and Arndt (2015) provides further reading on the origin of the Earth's oceans, and the book of Ball (2001) provides further reading on water's origins, history and fascinating pervasiveness on Earth, and examines its existence elsewhere in the Solar System.

5.4   Fate of Water on Earth, Venus and Mars

5.4.1   Earth, the Only Planet with Abundant Surface Water

During the Late Heavy Bombardment episode between about 4.1 and 3.8 billion years ago, not only planet Earth but also the other rocky planets of the Solar System would have been hit by very large numbers of asteroids and comets (see Sect. 1.5.1). However, although Earth and its two sister planets, Venus and Mars, were formed from the same water-containing solar nebula and at about the same time more than 4.5 billion years ago, and likely received the same bombardment of water-rich asteroids and comets 4.1 to 3.8 billion years ago, Earth is the only planet among the three that presently shows abundant surface water. This emphasises the originality of Earth in the Solar System, but opens the question of explaining the present difference in the abundance of surface water between Earth and its two sister planets.

The cosmic and physical-chemical processes that led to the great abundance of liquid water on Earth, mostly as seawater in the oceans, are examined in the remainder of this chapter. The question considered here is: How to explain that Venus and Mars do not show the same high abundance of surface water as Earth, although the three planets were formed under similar conditions in the Solar System and evolved under to the same laws of physics and chemistry? Assuming that the three planets had similar amounts of water in their early history, the higher abundance of surface water on Earth than on Venus and Mars implies that the latter two planets lost most of their water whereas Earth did not lose its water or not as much as them. What are the mechanisms by which planets lose their water? And how was the Earth's water successfully sheltered?

5.4.2   Loss of Water from a Planet to the Outer Space

The main mechanism by which planets lose water to space is as follows. High in the atmosphere, the action of solar radiation breaks apart some of water molecules, which releases hydrogen atoms (H^+) and molecules (H_2) from the H_2O molecules. Because of the high altitude in the atmosphere and their very small masses, the H^+ atoms and H_2 molecules are only weakly subjected to the gravitational pull of the planet, so that they are almost free to go, and given their individual kinetic energy, a fraction of them are lost to the outer space. Because the gravitational pull is proportional to the masses of the bodies involved in the interaction, H^+ atoms and H_2 molecules would have been only weakly retained by Venus and even less by Mars, whose masses are about 80% and 10% that of Earth, respectively (Tables 1.1 and 2.2).

One important mechanism of the *atmospheric escape of hydrogen* is the collision of individual molecules with one another, which makes them gain or lose kinetic energy, and the molecules that gain most energy leave the atmosphere (see Sects. 2.1.3 and 8.4.1). Given that H_2 molecules are very light, H_2 can escape the atmosphere very fast (see Information Box 8.4). For example, it has been estimated that Earth presently loses about 3 killogrammes of hydrogen every second, which corresponds to about 95,000 tons of hydrogen and thus 855,000 tons of water per year. Simultaneously, Earth gains annually about 20,000 to 40,000 tons of interplanetary dust, which contain water molecules and thus partly (but not fully) compensate the annual loss of hydrogen. The loss of water by hydrogen escape was probably very high during the early history of Earth, Venus and Mars because most of the water on the three, hot planets was then in the form of atmospheric water vapour.

It will be explained below that a planet-wide ocean appeared very early on Earth (see Sect. 5.5.4). This moved water from the atmosphere to the ocean, that is, from the gas phase where its hydrogen could escape to the outer space to the liquid phase in the ocean, where it was safe from escape. This may be the key process explaining the successful retention of water by planet Earth over more than four billion years, until now. The lack of such a liquid ocean on very hot Venus and the possible loss of an initial ocean by Mars, the latter for reasons not fully understood, would explain their present water paucity relative to Earth.

5.5   Water on Earth, Venus and Mars: Different Histories

5.5.1   *Physical Properties of Water*

All chemical compounds in the Universe can exist in four main *physical states*, which are the four forms that matter takes on: *solid, liquid, gas, and plasma*. The word *plasma*. refers to an electrically neutral medium of unbound positive and negative particles, whose overall electrical charge is null (see Sect. 2.1.2). Water does not exist as plasma within the range of the Earth's surface temperatures and pressures, and we thus focus here on the solid, liquid and gas states of water. The state of matter, and thus water, depends on the ambient temperature and pressure.

Physicists and chemists distinguish between the states and the phases of matter. A *phase of matter* refers to matter that has uniform physical and chemical properties, and the *states of matter* are the forms in which different phases exist. For example, water in an ice-covered lake is present in *two phases*: ice and liquid water, which each have uniform physical and chemical properties. Indeed, the frozen water is in the *solid state*, and the underlying water is in the *liquid state*. When the ice melts, the frozen water undergoes a *phase change* (also called *phase transition*) from ice to liquid. In this book, we refer to the phases of matter and to phase changes (see Information Box 5.7)..

Fig. 5.2 Processes of phase change (or phase transition) between solid, liquid and gaseous water. Credits at the end of the chapter

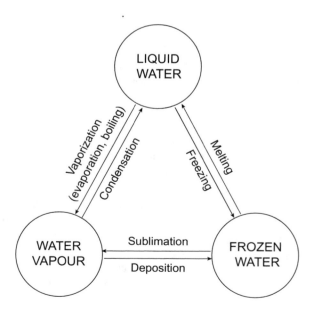

Information Box 5.7 Water Phase Changes Water can change phase between solid, liquid and gas in six different ways. There are three pairs of reverse processes, namely freezing/melting, condensation/vaporization, and sublimation/deposition (Fig. 5.2):

freezing of liquid water produces ice or snow (a change from liquid to solid phase); for example, the freezing of the surface of a lake in winter;

melting of ice or snow produces liquid water (a change from solid to liquid phase); for example, the melting of the Arctic Ocean's first-year ice in summer;

condensation of water vapour produces liquid water (a change from gas to liquid phase); for example, the formation of *fog* when humid air is cooled near the ground, or *dew* when moist air comes in contact with relatively cold surfaces, such as ground objects in early morning; the fall of raindrops condensed in the atmosphere is called *precipitation*;

evaporation and *boiling* (also called *ebullition*) of liquid water (the general, technical term for the two processes is *vaporization*) produces water vapour or *steam* in the case of boiling (a change from liquid to gas phase); for example, the loss of water by the ocean's surface in warm regions; the sum of evaporation from continental and ocean surfaces plus transpiration from terrestrial plants is called *evapotranspiration*;

sublimation of ice or snow produces water vapour directly without going through liquid water (a change from solid to gas phase); for example, the loss of snow or ice in high mountains under the combined action of strong wind, intense sunlight and very low air pressure;

deposition of water vapour produces ice or snow directly without going through liquid water (a change from gas to solid phase); for example, frost on a cold morning in winter.

Although the present section focuses on water, any substance in nature can freeze, melt, condense, evaporate, boil, sublimate or be deposited. A phase change occurs when critical conditions of temperature and pressure are reached, and these values can be very low or very high according to substances. The freezing and melting temperatures are often called *freezing point* and *melting point*, respectively; the temperatures of condensation and evaporation are often called *dew point* or *frost point* and *boiling point* respectively; and the temperatures of sublimation and deposition are often called *sublimation point* and *deposition point*, respectively. One may think that, for a given substance and pressure, the two temperatures in a phase change pair are the same, but these two temperatures are in fact slightly different. However, these differences are relatively small, and they therefore do not need to be taken into consideration for the planetary and cosmic phenomena discussed in this book.

Table 5.2 Temperatures of change of phase for pure water, seawater and brine corresponding to different atmospheric pressures. Ga: billions years before present

Water salinity and condition, and ambient pressure (atm)	Freezing and melting points (°C)	Vaporization and condensation points (°C)
Pure water (salinity 0)		
Triple point: 0.006	+0.01	+0.01
Mars, present, average: 0.006	0	0
Mars, present, highest: 0.012	0	+10
Earth, present: 1.0	0	+100.0
Earth before 4.0 Ga: 60	−7	+280
Venus, present: 92	−7	+280
Seawater (salinity 35)		
Earth, present: 1.0	−1.9	+100.6
Earth before 4.0 Ga: 60	−9.5	+280
Highly concentrated sodium chloride brine (salinity higher than 233)		
Earth, present: 1.0	Maximum −21.1	Maximum +108.7

Under special conditions, there is no phase change when temperature reaches the boiling or the freezing point, in which case there is superheating or supercooling of the liquid, respectively. *Superheating* (also called *boiling retardation* or *boiling delay*) occurs when a liquid is heated up above its boiling point without vaporization. This phenomenon occurs in cases where a perfectly homogeneous liquid (even with impurities) is heated in a container free of sites where gas bubbles can form, such as scratches or other imperfections. It can take place when a liquid (water, coffee) is heated in a container with a smooth inner surface in a microwave oven, and part of the superheated liquid can violently turns to steam when stirred or otherwise disturbed (watch out!) Conversely, *supercooling* (also called *undercooling*) occurs when a liquid is cooled below its freezing point without solidification or crystallization. This phenomenon occurs when the liquid that is cooled does not contain crystalline impurities that can act as seeds to trigger crystallization. One example is provided by the supercooled droplets of water just below their freezing point that accumulate in high-altitude clouds:.

Superheating and supercooling of a given liquid should not be confused with differences in the boiling and freezing points of liquids with different chemical compositions. Examples of the latter are pure water (no dissolved salts), seawater (on average, 35 g of dissolved salts per kilogramme of seawater) and *brine* (more than 50 g of dissolved salts per kilogramme of brine), whose changes of phase take place at different temperatures (Table 5.2)..

Under sophisticated laboratory conditions, pressure can be increased or decreased to very high or very low values, respectively. The results are illustrated in the phase diagram of pure water (Fig. 5.3). At the very high pressure of 217.7 atm and high temperature of 374 °C (called the *critical point* of water; red cross in the upper right part of the diagram), the liquid and gas (vapour) phases are indistinguishable from each other, and the water is then said to be *supercritical*. At

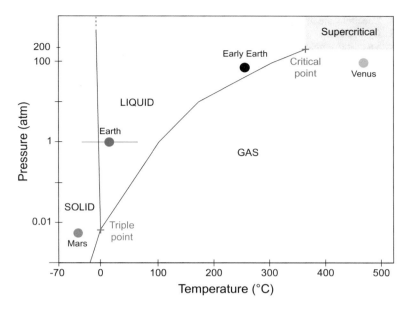

Fig. 5.3 Phase diagram of pure water at pressures lower than 250 atm and temperatures lower than 500 °C. Dots: present pressure and temperature conditions at the surface of Mars (red), Earth (blue) and Venus (orange), and conditions on Earth about 4.4 billion years ago (black). The triple point (blue cross) and critical point (red cross) are explained in the text. Credits at the end of the chapter

the very low pressure of 0.006 atm, the boiling and freezing temperatures of pure water are the same, namely 0.01 °C. The unique combination of 0.006 atm pressure and 0.01 °C temperature is called the *triple point of water* (blue cross in the lower left part of the diagram) because water at this pressure and temperature can coexist in three phases, namely ice, liquid and vapour. Below 0.006 atm, water can only exist as ice or vapour, and because water cannot be liquid, the only phase-change processes are sublimation and deposition.

In the phase diagram of pure water, the three lines that separate the solid, liquid and gas phases meet at the triple point of water (Fig. 5.3). Water is in supercritical state for combinations of pressures and temperatures beyond its critical point (red cross). The positions of planets Earth, Venus and Mars relative to the triple point of water (blue cross) make it easy to understand that water on Earth is mostly liquid or solid (ice), on Venus is in the form of vapour, and on Mars is present as underground ice.

The phase of a substance (gas, liquid, or solid) depends on the interactions between its molecules, which are affected not only by the ambient temperature and pressure, but also by the size of the gaz molecules. Let us take as examples the five halogen elements, which are, from the smallest to the largest: fluorine (F), chlorine (Cl), bromine (Br), iodine (I) and astatine (At) (column before last in the periodic table of the elements; Fig. 1.9 or 6.6). The two elements with the smallest molecules (F_2 and Cl_2) are gases at room temperature, the element with the next largest

Fig. 5.4 The molecule of water (H_2O) is strongly polar, with a negative pole ($\partial-$) located near the nucleus of the oxygen atom (red), and two positive poles ($\partial+$) near the nuclei of the two hydrogen atoms (grey). Dotted lines between the negative and positive poles: hydrogen bonds that link molecules together in the liquid phase. Credits at the end of the chapter

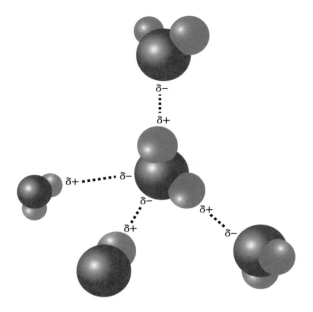

molecule (Br_2) is a liquid, and the two elements with the very largest molecules (I_2 and At_2) are solids.

Given the small size of the water molecule, water should be in the gas phase (water vapour) at the average temperatures that prevail on most of the surface of Earth. However, 98% of the Earth's water is liquid and only 0.001% is in the form of atmospheric vapour. How is this possible? The explanation lies in the fact that the H_2O molecule has a negative electrical pole located near the nucleus of the oxygen atom, and positive poles located near the two nuclei of the hydrogen atoms. Due to this polarization, *hydrogen bonds* form between water molecules, these bonds being caused by the attraction between the positively charged hydrogen atoms and the negatively charged oxygen atoms of different water molecules (Fig. 5.4). The binding of water molecules to each other causes water to be in the liquid phase under the average temperature and pressure conditions that presently exist at the Earth's surface, as it also did under the conditions that existed more than 4 billion years ago.

Values in Table 5.2 show that, as atmospheric pressure increases, the freezing and melting points of both pure and seawater decrease, and their vaporization and condensation points increase. This means that the range of temperatures within which water is liquid widens as pressure increases. At sea level on Earth (1 atm), this range is presently about 100 °C (that is, from 0 to +100 °C for pure water, and from −1.9 to +100.6 °C for seawater), whereas at the atmospheric pressure of 60 atm that existed on Earth more than 4 billion years ago (see Sect. 5.5.4, below), the corresponding range for water to be liquid was almost 300 °C (that is, from −7 to +280 °C for pure water, and −9.5 to +280 °C for seawater). Pressure cookers provide an everyday example of the effect of pressure on the boiling point of water (see Information Box 5.8).

Information Box 5.8 Pressure Cookers Pressure exerts a major effect on the different phase-change processes, as shown in Table 5.2 for Earth, Venus and Mars. One of these effects can be easily experienced when using a pressure cooker for preparing food. How does a pressure cooker works? Heating up the water or the broth contained in a pressure cooker creates water vapour, whose confinement inside the cooking vessel raises the pressure to twice the normal atmospheric pressure. As a consequence, water in the pressure cooker boils at a temperature of 120 °C instead of the usual 100 °C. Because the temperature inside the vessel is 120 °C, the food cooks faster than in the usual boiling water or broth whose temperature cannot rise beyond 100 °C.

Conversely, because atmospheric pressure decreases with increasing altitude (Fig. 5.5), mountain climbers know that the boiling point of water drops with higher altitude. The drop is 1 °C for each 300 m of altitude, so that the boiling point of water at an altitude of 3,000 m is 90 °C. This relatively low cooking temperature may leave some boiled foods undercooked, for example pasta. To compensate this effect, mountaineers often carry special lightweight pressure cookers in order to cook their food properly.

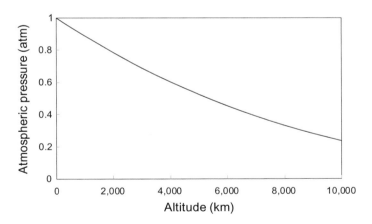

Fig. 5.5 Decrease of atmospheric pressure with increasing altitude on Earth. Credits at the end of the chapter

Values in Table 5.2 also show that, as the salinity increases, the freezing point of seawater and brine decreases and their vaporization and condensation points increase at normal ground-level atmospheric pressure. Water with salinity higher than that of seawater is qualified of *hypersaline*, and highly hypersaline water is called *brine*. Some natural brine on Earth contains more than five times the salt content of seawater. Brine is found in different natural environments,

---

Information Box 5.9 Salinity, and Concentrations of Substances Dissolved in Seawater Seawater has an average *salinity* of 35 (this quantity has no physical units), meaning that a kilogramme of seawater contains 35 grams of dissolved salts. The most abundant salt in seawater is sodium chloride (NaCl), a compound with one atom of sodium (Na) and one of chlorine (Cl), known as *sea salt* or *table salt*.

Because one kilogram of seawater contains 35 grams of dissolved salts, the *density* of seawater is higher than that of pure water. Indeed, the average density of seawater at 0 °C is 1.035 kilogrammes per litre (1.035 kg L^{-1}) or 1,035 kilogrammes per cubic metre (1,035 kg m^{-3}). It follows that concentrations of substances dissolved in seawater (gases, ions) can be expressed by reference to either the volume or the mass of seawater. For example, the concentration of dissolved oxygen is stated in μmol m^{-3} in Fig. 9.13 and in μmol kg^{-1} in Figs. 9.14 and 9.18. In order to transform concentrations stated in μmol m^{-3} into μmol kg^{-1}, the first values are divided by the density of seawater (kg m^{-3}).

---

which include the following four categories. *Saline lakes*, such as the Aral Sea (Kazakhstan and Uzbekistan), lake Assal (Djibouti), the Dead Sea (Israel and Jordan), the Great Salt Lake (USA), and the hypersaline lakes of the McMurdo Dry Valleys (Antarctica). *Deep-sea brine pools*, whose salt content is much higher than that of the overlying seawater and whose source of salt is the dissolution of underlying salt deposits by vertical movements of seawater in the seafloor; such pools are found at the bottom of different water bodies, for example the northern Gulf of Mexico, the Eastern Mediterranean Sea, and the southern Red Sea. *Hypersaline coastal lagoons* are bodies of water partially isolated from an adjacent sea by a sedimentary barrier; in regions where evaporation exceeds precipitation, the coastal lagoons are often hypersaline. *Polar oceans* at the time of formation of sea ice, where brine is formed because the salts contained in the liquid seawater are excluded from the ice matrix as it forms (see Sect. 3.1.1).

Finally, the *density* of liquid water (that is, its *mass per unit volume*) depends on its temperature and *salinity* (see Information Box 5.9). The density of pure water is close to 1 kilogramme per litre (1 kg L^{-1}) or 1,000 kilogrammes per cubic metre (1,000 kg m^{-3}) at 0 °C at sea-level pressure. The density of pure water increases with increasing temperature from 0 to 4 °C, where it reaches its maximum value, which means that the volume occupied by a given mass of pure water is smallest at 4 °C. As temperature increases above 4 °C, the volume of pure water also increases and, consequently, its density decreases. In other words, the density of pure water increases as it cools down until it reaches 4 °C, below which it decreases. This is not the case of seawater, whose density increases continually with decreasing temperature until it freezes, and also increases with increasing salinity. Density plays an important role in the *global ocean circulation*, and thus in the climate (see Sects. 3.5.3 and 3.5.4).

The fact that liquid water cannot exist at very low pressure stresses the importance of the atmospheric pressure for organisms. Indeed, these require liquid water, and thus an atmospheric pressure high enough to allow water to be liquid. Hence adequate combinations of temperature and atmospheric pressure existed on Earth during billions of years for the establishment of organisms and their build-up of large biomasses. The book of Lécuyer (2013) provides further reading on the physico-chemical and biological properties of water on Earth.

5.5.2   Chemical Properties of Water

In addition to the above physical properties, water has important chemical properties for organisms, and some of these properties are quite unique among chemical compounds. Because pure liquid water is odourless, colourless and tasteless, it is not a very glamorous compound for most people. However, the chemical properties of water are such that this compound plays key roles on Earth for both the climate and the ecosystems, and is vital for organisms.

The most spectacular chemical property of water is its power to dissolve almost all other compounds, given time. This property is due to the small size of the water molecule and its two electrically charged poles (Fig. 5.4). Liquid water is the most powerful natural solvent, and because of this property, it always contains some dissolved substances. As a consequence, pure liquid water does not actually exist in the natural environment, except in the form of liquid droplets (in clouds) or newly formed raindrops. Even in the laboratory, there is no material to make a bottle that could contain pure liquid water without being slowly dissolved by it.

Because water has the ability to dissolve a wide range of substances, seawater contains almost all the chemical elements in various amounts, and the volume of this solution is enormous (see Information Box 5.5). On the continents, water can dissolve all rocks, and given time the highest mountains are eroded to small hills, the small hills to plains and the plains to deep canyons, and the eroded materials are ultimately transported into the ocean. In the natural environment, liquid water ranges between the saltiest lake and marine waters, with high concentrations of dissolved compounds, and the relatively pure water created by evaporation (atmospheric vapour, clouds, and rain droplets), and includes continental fresh waters that dissolve rocks and transport their components to the ocean. The high diversity of aquatic habitats favoured the diversification of organisms and ecosystems since their establishment on Earth.

Another chemical property of aqueous solutions is their concentration in hydrogen ions (H^+), which is called *power of hydrogen* or *potential of hydrogen* (abbreviated as *pH*). The pH of seawater is introduced in Chapter 6 (see Information Box 6.4) and examined in Chapter 9 (see Sect. 9.2.6). Its ongoing changes as a consequence of human activities are known as *ocean acidification* (see Sects. 11.1.1 and 11.1.3).

5.5.3 Phases of Water on Earth, Venus and Mars

Pure water at the surface of Earth freezes at 0 °C and vaporizes at 100 °C under the present atmospheric pressure of 1 atm. However, these values are not truly representative of liquid water on Earth because the most abundant form of water at the surface of Earth is seawater, whose average salinity is 35 (see Information Box 5.9). Because most water at the surface of Earth is salty, it freezes at about −2 °C and vaporizes at about +101 °C (Table 5.2). What about liquid water on the two sister planets of Earth, namely Venus and Mars? The temperature and atmospheric pressure of the three planets determine their position in the phase diagram of water (Fig. 5.3).

The atmosphere of a planet is a major determinant of the presence or not of liquid water at its surface, and the general atmospheric characteristics of the rocky planets are described in Chapter 2 (see Sect. 2.2.3). The atmospheric pressure on planet Venus reaches the very high value of 92 atm, which is more than 90 times the atmospheric pressure at the surface of Earth. Under the high Venusian atmospheric pressure, water could be liquid on that planet between −7 and +280 °C, but given that the mean atmospheric temperature at Venus' surface is +464 °C, the small amount of water on that planet is in the form of water vapour. In the Venusian atmosphere, water vapour accounts for only 0.002% by volume of the total gases, which largely consist of CO_2 (96.5% by volume).

Contrary to Venus, Mars has a very rarefied atmosphere, and the average atmospheric pressure at the surface of the planet has the very low value of 0.006 atm (Tables 2.2 and 5.3). This value is 170 and 15,000 times smaller than the atmospheric pressure at the surface of Earth and Venus, respectively. The atmosphere of March is comprised of 96% CO_2, which is almost the same as the proportion of this gas in the atmosphere of Venus. However, the amount of atmospheric CO_2 on Mars is very small because of the very thin atmosphere (very low atmospheric pressure) on that planet. The atmospheric pressure at the surface of a planet decreases with increasing altitude (Fig. 5.5), and the highest mountain in the Solar System, Olympus Mons, is located on Mars (see Information Box 5.10). Atmospheric pressure at the top of that 22 km mountain is extremely low (around 0.0003 atm), in sharp contrast with the highest Martian atmospheric pressure of 0.012 atm found in the lowest part of the largest and deepest impact crater on that *planet.*

Atmospheric pressure on most of Mars is near or below the pressure of 0.006 atm of the triple point of water (Table 5.2 and Fig. 5.3), where the critical temperature for the occurrence of liquid water is +0.01 °C. Given that the average monthly temperatures on Mars range from −23 to +4 °C during the day and −88 to −68 °C during the night, these temperatures are mostly below the critical temperature of +0.01 °C. As a consequence, one would think that water does not generally exist in the liquid phase at the surface of Mars, or is liquid in only the warmest areas during daytime. Indeed, large amounts of frozen water were discovered on Mars, in both the permanent polar ice caps and buried underground. Hence, it was generally assumed that water on Mars was usually not liquid … until robots sent to Mars may have discovered some liquid water in the Martian soil.

Given the low atmospheric pressure and generally low temperature that prevail at the surface of Mars, how could liquid water exist in the Martian soil? Values

> **Information Box 5.10 The Highest Mountain in the Solar System** The highest mountain in the Solar System is *Olympus Mons* on planet Mars. The height of the incredibly tall Olympus Mons is 22 km, which is two and a half times as high as the two tallest mountains on Earth, *Mauna Kea* (in Hawaii, whose summit reaches 10 km above the Pacific Ocean seafloor) and *Mount Everest* (Nepal, which reaches 8.8 km above the sea level). Martian *Olympus Mons* should not be confused with *Mount Olympus*, which is the highest mountain in Greece and rises to 2.9 km above sea level.

Table 5.3 Planetary characteristics of Venus, Earth and Mars related to the temperature at the surface of the three planets (1 atm $= 1.013 \times 10^5$ Pa $= 1.013 \times 10^5$ kg m^{-1} s^{-2})

Planetary characteristics	Venus	Earth	Mars
Atmospheric temperature, observed, surface average, T_{atmos} (°C and K)	+464 +537	+15 +288	−63 +210
Distance from the Sun (million km)	108	150	227
Solar radiation reaching the planet, average, S_{ave} (W m^{-2})	662	342	145
Albedo, (dimensionless)	0.75	0.30	0.25
Effective temperature, calculated, T_P (°C)	−46	−19	−63
Observed minus effective temperatures (°C)	+510	+34	0
Atmospheric pressure, surface, P_{atmos} (atm)	92	1	0.006
Atmospheric composition near surface (major gases, % volume)	CO_2 96%, N_2 4%	N_2 78%, O_2 21%	CO_2 95%, N_2 3%, Ar 2%
Density of atmosphere, surface, ρ_{atmos} (kg m^{-3})	65	1.2	0.02
Mass of atmosphere, M_{atmos} (10^{16} kg)	48,000	515	2.5
Surface area of the planet, A_{planet} (10^6 km^2)[a]	460	510	144
Mass /surface area, M/A_{planet} (10^4 kg km^{-2})	104.3	1.01	0.017
Gravity at surface, g_{planet} (m s^{-2})	8.9	9.8	3.7

[a]$A = 4\pi(D/2)^2$; planet diameters (D) are listed in Table 1.1

in the bottom rows of Table 5.2 provide a clue about the answer to that question: in order to be liquid at the surface of Mars, water there should be very salty. Much saltier than seawater on Earth; so salty that it should be *brine*. The range of Martian temperatures cited above indicates that temperatures on Mars are often high enough to allow brines to be in liquid phase in the soil during the day, but not during the night. Liquid water is a key topic of Mars scientific exploration because researchers wonder if there is, at the present time, enough permanently liquid water on that planet for the existence of unicellular organisms.

5.5.4   Different Early Histories of Earth and Venus

After examining the phases of water on Earth, Venus and Mars, let us consider the conditions on Earth a few million years after the giant collision that created the Earth-Moon system about 4.5 billion years ago (see Sect. 1.5.1). The geological era that goes from the formation of the proto-Earth about 4.6 billion years ago and ends about 600 million years later is called *Hadean* (see Sect. 1.5.4). The name *Hadean* comes from the Greek mythological god of the underworld, Hades, and refers to the hellish conditions that prevailed on Earth during its first geological era. Indeed, the young planet was initially very hot because its surface had not yet completely solidified, was the site of continual volcanic activity, and was subjected to frequent impacts of asteroids and comets (see Sect. 1.5.4). The next paragraphs explain that environmental conditions on Earth evolved rapidly until the end of the Hadean about 4.0 billion years ago.

Most researchers presently think that, during the first part of the Hadean, Earth experienced atmospheric conditions somewhat similar to those observed on Venus today. As on Venus nowadays, the Earth's atmosphere would have then been mostly comprised of CO_2, and exerted a surface pressure of about 60 atm, compared with 92 atm on Venus today (Table 5.2). It is thought that for tens of millions of years after the creation of the Earth-Moon system, the atmospheric temperature at the surface of Earth was above 2,000 °C, after which it progressively cooled down to about 200 °C or slightly more. However, the atmosphere of the young Earth could not cool below 200 °C because the high concentration of atmospheric CO_2 caused a very strong greenhouse effect (see Information Box 2.7). Despite the high temperature of 200 °C or slightly more, a liquid-water ocean may have formed 4.4 billion years ago or perhaps earlier, because water could be liquid at temperatures between −7 and +280 °C under the prevailing atmospheric pressure of about 60 atm (Table 5.2, and black dot in Fig. 5.3).

The 200°C ocean was too hot for the emergence of organisms, but not for a key chemical reaction that determined the later fate of organisms on the planet. This chemical reaction was the formation of solid calcium carbonate ($CaCO_3$) in the liquid ocean. In the warm Hadean ocean, this reaction was favoured by the high concentration of atmospheric CO_2, which dissolved in seawater where it formed large amounts of HCO_3^- (bicarbonate) and CO_3^{2-} (carbonate) ions (see Information Box 6.3). In seawater, $CaCO_3$ is formed by the reaction of HCO_3^- ions with dissolved calcium ions (see Information Box 5.11).

Information   Box 5.11   Formation   and   Dissolution   of   CaCO₃:
Calcification In the natural environment, calcium carbonate ($CaCO_3$, a solid compound) is formed by the combination of two dissolved ions that naturally occur in seawater and fresh water, that is, Ca^{2+} and HCO_3^- (see Information Box 6.3 for the latter). Chemists use the word *precipitation* to describe the formation of a solid substance from a solution, but we will not use the word 'precipitation' with this specialized chemical meaning in the present book. The formation of solid $CaCO_3$ results from the combination of calcium and bicarbonate ions:

$$Ca^{2+}_{(aqueous)} + 2HCO_3^-{}_{(aqueous)} \rightleftharpoons CaCO_{3(solid)} + H_2O + CO_{2\,(aqueous)} \quad (5.10)$$

This equation contains an equilibrium double arrow that indicates a balance between the Ca^{2+} and HCO_3^- ions on the left side of Eq. 5.10, and the molecules of $CaCO_3$, H_2O and CO_2 on the right side. The five substances coexist in the medium, and according to the environmental conditions, the process will favour either the formation of solid $CaCO_3$ (equilibrium shifted to the right) or its dissolution (equilibrium shifted to the left).

Reading Eq. 5.10 from left-to-right shows that the formation of one molecule of solid $CaCO_3$ (right side) requires two molecules of dissolved HCO_3^- (left side). As a result of the formation of $CaCO_3$, half of the carbon and oxygen atoms in the dissolved HCO_3^- ions (right side) become part of the solid $CaCO_3$ and half are released in the surrounding water as CO_2 gas (left side). The formation of solid $CaCO_3$ affects the climate because it releases CO_2 into the surrounding water, and most of this CO_2 will degas to the atmosphere. The latter will increase the concentration of CO_2 in the atmosphere, and thus increase the Earth's greenhouse effect (see Information Box 2.7).

When Eq. 5.10 is read from right-to-left, it describes the *dissolution* of solid $CaCO_3$. The equation shows that the dissolution of one molecule of solid $CaCO_3$ takes one molecule of dissolved CO_2 from the surrounding water (right side) and produces two HCO_3^- ions (left side).

On early Earth, the formation of solid $CaCO_3$ occurred as a strictly chemical reaction, but on modern Earth, most of the $CaCO_3$ is formed by organisms through the biological process of *calcification* (see Sect. 6.4.2). However, some $CaCO_3$ is still produced chemically, in two different ways. First, in warm, shallow marine and fresh waters supersaturated in Ca^{2+} and HCO_3^- ions, there is formation of mall spherical $CaCO_3$ grains called *ooliths*. Second, in certain marine sediments, there is production of large amounts of HCO_3^- either by the oxidation of organic matter (CH_2O) coupled with bacterial sulphate reduction:

$$2[CH_2O] + SO_4^{2-} \rightarrow 2HCO_3^- + H^+ + HS^- \quad (5.11)$$

or by the oxidation of methane (CH_4) in absence of O_2:

$$CH_4 + SO_4^{2-} \rightarrow HCO_3^- + HS^- + H_2O \qquad (5.12)$$

The high concentration of HCO_3^- in subsurface sediments favours the chemical formation of *authigenic* $CaCO_3$ (meaning "$CaCO_3$ formed in its present position"), which may account for as much as 10% of the global carbonate accumulation on the seafloor. Calcification would account for the other 90%.

The formation of solid $CaCO_3$ occurs in oceans, lakes and rivers, but because the bulk of it globally takes place in the ocean, we only consider the ocean's $CaCO_3$ formation in this book. Changes in the formation of $CaCO_3$ during the Earth's history are described in Sect. 6.4.2.

Given that continents did not exist at the time, or just started to appear, $CaCO_3$ was formed by the reaction of dissolved CO_2 with the silicate rocks ($CaSiO_3$) of the young ocean crust, and thus essentially occurred within the seafloor. The process of $CaCO_3$ production from $CaSiO_3$ is detailed in Chapter 7 (see Sect. 7.3.3; Information Box 7.4). The solid $CaCO_3$ accumulated in the seafloor, where it sequestered (stored) huge amounts of carbon of atmospheric origin in carbonate rocks. This went on for tens of millions of years, and caused a major decrease in the concentration of atmospheric CO_2. This decrease lowered the global greenhouse effect, and consequently the Earth's surface temperature.

The temperature of the ocean at the end of the Hadean era or at the beginning of the next, Archaean era was perhaps around +70 °C or lower, and the ocean's temperature was then suitable for the development of organisms (see Sect. 3.1.1). Some specialists think that the maximum temperature acceptable for organisms is around 110 °C, a value consistent with some of the highest temperatures for growth cited in Table 3.1. Such maximum temperature means that organisms could have appeared quite early after the formation of the ocean, but these high-temperature organisms would not be the direct ancestors of the current high-temperature organisms.

If the concentration of CO_2 in the Earth's atmosphere had remained as high as when the warm-water ocean was formed in the Hadean era, the temperature at the Earth's surface could still be today close to +200 °C, which temperature would have been too high for organisms. Instead, the early formation on Earth of a planet-scale liquid ocean initiated the giant chemical reaction explained above, which lasted tens of millions of years. During that long period, most of the oceanic and atmospheric CO_2 was progressively transformed chemically into $CaCO_3$ and stored in this form within the seafloor, from where part of this $CaCO_3$ was moved to the Earth's interior by plate tectonics (see Sect. 7.2.3). There, $CaCO_3$ was incorporated into the magma, which liberated CO_2 that returned to the ocean and the atmosphere through volcanoes (see Sect. 7.3.1). In this long cycle, there was continual formation and recycling of $CaCO_3$, but the creation of a huge reservoir of solid sedimentary carbonates drastically reduced the concentration of atmospheric CO_2.

The evolution of the atmosphere of Venus is still poorly known, for several reasons. These include the thickness of Venus' atmosphere and the presence of sulphuric acid clouds, which prevent the visual observation of the surface of the planet from Earth, and the harsh conditions of temperature and atmospheric pressure, which delayed the robotic exploration of the planet. It was hypothesized, given that Earth and Venus were formed from the same solar nebula and have a relatively similar mass, that the two planets had similarly cooled and developed an initial ocean. However, the sequestration (storage) of most of the atmospheric CO_2 into solid $CaCO_3$ that occurred on Earth would not have taken place on Venus, and as a consequence of the high concentration of CO_2 in its atmosphere and the associated very strong greenhouse effect, Venus would have eventually warmed up and lost most of its water.

However, the hypothesis of an early ocean on Venus is not accepted by all specialists. The reason is that Venus is located close to the Sun and would thus have cooled very slowly, and during the long cooling period, the planet would have lost most of its water by atmospheric escape of hydrogen (see Sect. 5.4.2). The long rotation period of the planet (243 Earth days) may have also contributed to the early loss of water (see Sect. 1.3.3). Hence, Venus would have lost almost all its water before becoming cool enough to allow the presence of liquid water. Future robotic exploration of Venus will provide data to decide between these two hypotheses, or perhaps propose a different one. The book of Taylor (2014) provides further reading on the knowledge acquired about Venus through the scientific exploration of the planet.

> **Different Fates for the Atmospheres of Earth and Venus** The above paragraphs show that the atmospheres of Earth and Venus may have been initially similar, but Earth rapidly lost most of its atmospheric CO_2 and Venus may have rapidly lost most of its water. The early formation of a planet-wide ocean, in which atmospheric CO_2 was rapidly trapped into solid $CaCO_3$, largely determined the fate of Earth, and the lack of a similar mechanism on Venus would have determined the completely different fate of this sister planet of Earth.

5.6 Most of the Earth's Water Is Liquid

5.6.1 Conditions that Favour Liquid Water on Earth

It was explained above that the temperature and atmospheric pressure at the surface of a planet and the salinity of water determine the presence or absence of surface liquid water (see Sect. 5.5.1). Indeed, water can only occur in the liquid phase between its freezing and vaporization points, and these two critical points are themselves set by the atmospheric pressure that prevails at the surface of the planet and the concentration of salts dissolved in water (Table 5.2). These three factors could explain the occurrence of liquid water on Earth and not on its two sister planets Venus and Mars. On Venus, there is almost no water, and the combination of high atmospheric temperature and pressure keeps this water in the form of atmospheric vapour. On Mars, the generally low temperature and atmospheric pressure keep most of the water in the form of underground ice, but the possible presence of highly concentrated brines in the surface soil may allow the occurrence of liquid water there. On Earth, current atmospheric pressure makes seawater liquid between -1.9 and $+100.6$ °C, and surface temperatures range between -89 °C (at the Vostok research station, Antarctica, in winter) and $+57$ (in Death Valley, USA, in summer) with a mean value of $+15$ °C. The relatively moderate temperatures at the Earth's surface (with a few regional exceptions) allow most of the water to be liquid.

It follows from the previous paragraph that the key question concerning the wide occurrence of liquid water on Earth is: What are the planetary characteristics that generally maintained the atmospheric pressure and temperature of Earth within ranges that allowed most of its surface water to be liquid during billions of years? Concerning the atmospheric pressure, it is explained in the next section that this characteristic of a planet is determined by the mass of its atmosphere and the acceleration due to gravity on the planet, which characteristics favoured the presence of liquid water on Earth. Concerning the temperature, Chapter 3 identified and discussed the combined characteristics of Earth that contributed to keep the temperature of its atmosphere and oceans at values generally suitable for organisms over billions of years (see Sect. 3.6.2). These characteristics are re-examined here within the context of the large incidence of liquid water on Earth (see Sect. 5.6.3).

5.6.2 Primary Planetary Characteristics for the Occurrence of Liquid Water

Planets Venus, Earth and Mars are distant from the Sun by 108, 150 and 227 millions of kilometres, respectively, and the chief source of heat at their surface is currently the energy radiated by the Sun (Table 5.3). Because the amount of energy received from the Sun decreases as the square of the distance from the source, planet Venus, which is located two times closer to the Sun than Mars, receives four times $(2^2 = 4)$ the solar energy reaching Mars. It could thus be thought that the increasing distance of the three planets from the Sun is the main factor responsible for the temperatures observed at their surfaces (Table 5.3). However, if the temperature of a planet was only determined by its distance from the Sun and its albedo, then Venus, Earth and Mars should have solid-surface temperatures of -47, -19 and -63 °C, respectively. The latter values correspond to the *effective temperature* of these planets, which is the temperature they would have without an atmosphere but the same albedo as presently, although their albedo could be different without the presence of the atmosphere (see Information Box 5.12). Hence without the insulation provided by their atmosphere and with their present albedo, Venus and Earth (Mars has almost no atmosphere) would experience average temperatures of -47 and -19 °C instead of the observed $+464$ and $+15$ °C, respectively (Table 5.3). It follows that the temperature of a planet is not only determined by its distance from the Sun and its albedo, but also very much by the characteristics of its atmosphere.

Information Box 5.12 Effective Temperature The *effective temperature* (also called the *equivalent black body temperature*) of a planet is the theoretical temperature that the planet would have if it had no atmospheric greenhouse effect but the same albedo as it actually does. Albedo is defined in Chapter 2 as the fraction of the incident radiation received from the Sun that the planet reflects back to space (see Information Box 2.9). The albedo of Venus (0.75) is very high compared to those of Earth and Mars (0.30 and 0.25, respectively; Table 5.3), because Venus is surrounded by a permanent layer of highly reflective clouds containing droplets and crystals of sulphuric acid. This high albedo makes Venus the third brightest object in Earth's sky after the Sun and the Moon. The high albedo of Venus also explains the very low effective temperature of this planet ($-47\,°C$) despite its proximity to the warm Sun.

Brilliant Venus appears in the sky sometimes just after sunset and is then called *Evening Star*, and at other times just before sunrise and is then known as *Morning Star*. It interesting to note that Roman people called planet Venus *Lucifer*, a name that literally means "light bearer" (from the Latin *lux*, light, and *ferre*, to carry), whereas this name now refers to the Devil for Christians. Roman Catholics call Mary the mother of Christ *Morning Star*. The two names of Venus, Lucifer and Morning Star, thus have both astronomical and religious meanings.

For readers interested in equations, the effective temperature, T_P (K, kelvin), is calculated as follows:

$$T_P = [S_{ave}(1-\alpha)/\sigma]^{1/4} \qquad (5.13)$$

where S_{ave} is the average solar radiation reaching a planet (W m^{-2}) and α is its albedo (Table 5.3), and σ is the Stefan-Boltzmann constant (5.67×10^{-8} W m^{-2} K^{-4}).

On Mars, the effective and observed temperatures are the same because the planet has no significant atmosphere. In contrast, the effective and observed temperatures of Venus and Earth differ by 510 and 34 °C, respectively (Table 5.3). Interestingly, the difference between the effective and observed temperatures on Mars, Earth and Venus is directly proportional to their atmospheric pressure (Fig. 5.6). This relationship indicates that atmospheric pressure strongly influences the temperature that exists at the surface of the three planets. Atmospheric pressure reflects the combined effects of properties of the atmosphere and the solid planets, that is, thickness (or volume) and density for the atmosphere, and size and mass for the solid planet (see Information Box 5.13). Hence atmospheric pressure has two major, distinct effects on the presence or not of liquid water on Venus and Earth: on the one hand, it sets the range of temperatures within which seawater can be liquid (between the freezing and vaporization points, provided in Table 5.2), and on the other hand, it largely determines whether the actual average temperature at the surface of the planet falls or not within that range (Fig. 5.6).

Information Box 5.13 Atmospheric Pressure at the Surface of a Planet The atmospheric pressure (P) at the surface of a solid planet is determined by the force exerted by the atmosphere (F) and the surface area of the solid planet (A):

$$P = F/A \text{ (see Information Box 2.3)} \qquad (5.15)$$

Surface area A reflects the size of the planet. Force F is the product of the mass of the atmosphere (M_{atmos}) and the *acceleration due to gravity* (or *gravitational acceleration*) on the planet (g_{planet}). The relationship is as follows (based on Eq. 5.15):

$$P = \left(M_{atmos} \times g_{planet}\right) / A_{planet} = M_{atmos} \times \left(g_{planet}/ A_{planet}\right) \qquad (5.16)$$

The value of g_{planet} is proportional to the mass of the solid planet. The mass of the atmosphere (M_{atmos}) reflects its volume (V_{atmos}) and the average density of its gases (ρ_{atmos}):

$$M_{atmos} = V_{atmos} \times \rho_{atmos} \qquad (5.17)$$

Hence:

$$P = (V_{atmos} \times \rho_{atmos}) \times \left(g_{planet}/ A_{planet}\right) \qquad (5.18)$$

Equations 5.16 and 5.18 show the atmospheric pressure (P) at the surface of a solid planet depends on characteristics of the atmosphere, which are its mass (M_{atmos}) or its volume (V_{atmos}) and density (ρ_{atmos}), and the solid planet, which are its surface area (A_{planet}) and acceleration due to gravity (g_{planet}).

The atmospheric pressures (P_{atmos}) at the surface of Venus and Earth can be computed with Eq. 5.16 using the values of M/A_{planet} and surface g_{planet} in Table 5.3. The resulting surface P_{planet} values for Venus and Earth are 9,290 and 99 kPa, corresponding to 92 and 1 atm, respectively. These two calculated values are the same as those observed at the surface of the two planets (P_{atmos} in Table 5.3: 9 and 1 kg m^{-3} for Venus and Earth, respectively), confirming that P_{atmos} at the surface of a planet is determined by the above characteristics of the atmosphere and the solid planet.

Fig. 5.6 Difference between observed and effective average temperatures plotted as a function of the atmospheric pressure at the surface of Mars, Earth and Venus, Earth (from left to right). The values are in Table 5.3. Credits at the end of the chapter

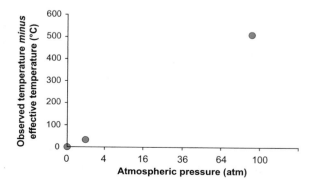

An important property of the atmosphere of a planet is its density (ρ_{atmos}), which contributes to determine the atmospheric pressure (Eq. 5.18). The following paragraphs explain that density is largely determined by the nature of the gases that make up the atmosphere of a planet, and that some of these gases are also responsible for the planet's greenhouse effect.

Density of atmospheric gases. The atmosphere of Venus contains 96% of CO_2 and 4% of N_2 (molecular masses, MM: 44 and 28 g mol^{-1}, respectively) whereas that of Earth is made of lighter gases, which are 78% of N_2, 21% of O_2 and 1% of Ar (MM: 28, 32, and 40 g mol^{-1}, respectively). Combining these values provides the mean MM for the atmospheres of the two planets: 43 and 29 g mol^{-1} for Venus and Earth, respectively. The density of a gas (ρ) depends on its MM and its temperature (T) and pressure (P):

$$\rho_{calc} = (P \times MM) / (T \times R) \qquad (5.14)$$

where the units of P, MM and T are atm, g mol^{-1} and K (kelvin), respectively, and R is the universal gas constant (0.000,082,1 m^3 atm K^{-1} mol^{-1}).

The densities (ρ_{atmos}) for the atmospheres of Venus and Earth can be computed with Eq. 5.14 using the surface values of T_{atmos} and P_{atmos} in Table 5.3 and the above mean MM for the two planets. The resulting surface atmospheric ρ_{calc} values for Venus and Earth are 66 and 1.2 kg m^{-3}, respectively. These two values are the same as the observed densities at the surface of the two planets (ρ_{atmos} in Table 5.3: 65 and 1.2 kg m^{-3} for Venus and Earth, respectively), which confirms that the nature of gases in the atmosphere of a planet together with its temperature and pressure set the density of the atmosphere.

Greenhouse effect. The key greenhouse gas in the atmosphere of Venus and Earth is CO_2 (see Information Box 2.7). The large proportion of CO_2 in the atmosphere of Venus (96% by volume) creates a very high average temperature at the surface of the planet (+464 °C, T_{atmos} in Table 5.3). In contrast, the low concentration of CO_2 in the Earth's atmosphere, which is presently 0.04% but was higher in the past (Fig. 8.1), contributed to keep the temperature at the surface of the planet at a level compatible with the presence of liquid water during 4 billions of years (see Sect. 3.7.2). However, the temperature of the Earth's environment is now increasing in response to the current increase in atmospheric CO_2 (Fig. 9.5), and the latter is the key driver of the ongoing anthropogenic *global warming* and *ocean acidification* (see Sects. 11.1.1 and 11.1.3).

This analysis confirms the following conclusions from a previous section concerning the phases of water on Earth, Venus and Mars (see Sect. 5.5.3). On Mars where the atmospheric pressure is almost zero, the presence of temporary liquid water could be possible if there were brines in the soil with a melting point above the daytime soil temperatures, whose monthly averages are between −23 and +4 °C. On Venus, the effects of the very high atmospheric pressure on both the vaporization point of water (280 °C) and the temperature at the surface of the planet (average: +464 °C) prevent the existence of water in the liquid phase. On Earth, the effects of the moderate atmospheric pressure on both the freezing and vaporization points of seawater (-2 and $+101$ °C) and the temperature at the surface of the planet (average: $+15$ °C) favour the prevalence of water in the liquid phase.

We had seen in a previous chapter that two important planetary characteristics that allow Earth to have a significant atmosphere are its mass and its distance from the Sun (see Sect. 2.3.4). We saw in this section that the Earth's atmospheric pressure determines a combination of surface temperatures and freezing and vaporization points of seawater that allowed the creation and maintenance of a planet-wide liquid ocean during the course of the planet's history. We also saw that the concentration of atmospheric CO_2 contributed to create a greenhouse effect that kept the temperature of the planet at a level compatible with the presence of liquid water. It follows that the Earth's mass, distance from the Sun, atmospheric pressure and concentration of atmospheric CO_2 are major determinants of the presence of water in mostly liquid phase on Earth..

5.6.3 Three Additional Planetary Characteristics for the Occurrence of Liquid Water on Earth

In the previous section, it was explained that most of the water on Earth is liquid because the atmospheric pressure of the planet determines a favourable combination of surface temperatures and freezing and vaporization points of seawater. However, the atmospheric pressure determines only the average temperature on the planet, and not the seasonal and daily variations in temperature. As a consequence, at some very high latitudes and altitudes on Earth, there is no liquid water at any time of the year or water is liquid only during the warmest season, and the phase of water may even change between liquid and ice within one day. In contrast in areas where water is present in mostly the liquid phase, it generally remains liquid over the whole day-and-night cycle and during most seasons or even all of them. What are the planetary characteristics responsible for the relatively low amplitude of diurnal and seasonal variations in the Earth's temperature, and thus for the continual prevalence of liquid water on the planet?

In order to provide a first answer to this question, let us look at the closest neighbour of Earth, the Moon. There, surface temperature can reach more than $+100$ °C during the day and less than -150 °C during the night. The reason for these huge variations in temperature is the absence of atmosphere on the Moon. In contrast, the lower part of the Earth's atmosphere stores heat during the day and releases part of it during the night, thus reducing temperature extremes between day and night (see Sect. 3.4.2). Indeed, the lower atmosphere is a good reservoir of heat even if the amount of heat absorbed by a unit of mass of air to raise its temperature by 1 °C (or released when it cools by 1 °C) is not as high as that of seawater (see Sect. 4.2.2). It follows that the ability of the lower atmosphere to store and release heat is among the planetary characteristics that make liquid water a permanent feature of most environments on Earth.

However, the Earth's atmosphere is relatively thin, between 500 and 1000 km (see Sect. 2.2.1), and because of this, the heated side of the globe could become very hot during the day and the unheated side very cold during the night. Such large-amplitude daily variations in temperature are not observed on Earth, where the temperature differences between day and night are generally quite small. This is partly explained by the redistribution of heat over the whole planet by movements of air masses in the atmosphere and water masses in the oceans (see Sect. 3.5.4), and partly by the fast rotation of the globe. Indeed, planet Earth rotates on its axis once every 24 h, which is the fastest rotation among the four rocky planets in the Solar System (Table 1.1). This rotation contributes, directly and through resulting movements of air and water masses, to minimize temperature differences between day and night and among different areas (see Sect. 3.3.3). Hence, the fast rotation of Earth is another planetary characteristic that favours the day-and-night and year-round presence of liquid water over most of the planet.

A third planetary characteristic contributes to minimize seasonal temperature variations. This characteristic is the axial tilt (or obliquity) of Earth. The small value of the axial tilt (currently about 23°) prevents the widespread occurrence of extreme seasonal conditions (see Sect. 3.3.2). It is not known if the value of the axial tilt was always near 23°, but this value presently restricts the extreme seasonal variations to the polar regions and subtropical deserts. It follows that the small value of the Earth's axial tilt, which is a consequence of the existence of the Earth-Moon system, is also a planetary characteristic that contributes to explain the long-term year-round occurrence of liquid water on the planet.

Liquid Water on Earth, Improbable but Self-Sustaining In summary, the widespread and long-term occurrence of liquid water at the surface of Earth results from the combination of several characteristics of the planet. The presence of a significant atmosphere on Earth is due to its mass and distance from the Sun, which are therefore two major determinants of the presence of water in mostly liquid phase on Earth. In addition, the atmospheric pressure determines a favourable combination of surface temperatures and freezing and vaporization points of seawater that allows the creation and maintenance of a planet-wide liquid ocean during the course of the planet's history. Also, the concentration of atmospheric CO_2 reached after the formation of the ocean contributed to create a greenhouse effect that has since maintained the temperature of the planet at a level compatible with the presence of liquid water. Three additional planetary characteristics contribute to the presence of liquid water on Earth. These are the storage and release of heat by the lower atmosphere, the fast rotation of Earth, and the small value of the axial tilt stabilized by the existence of the Earth-Moon system.

5.6.4   Liquid Water Throughout the Earth's History

It was explained above how planet Earth acquired huge amounts of liquid water early in its history, and how this water largely determined the later fate of the planet and its ecosystems (see Sect. 5.3). Hence, liquid water was present on Earth since its early beginning, initially as a warm ocean under high atmospheric pressure, followed by a moderate-temperature ocean under decreasing atmospheric pressure (see Sect. 5.5.4). Does this indicate that Earth was never extremely hot or extremely cold, and thus always provided conditions suitable for organisms?

There were periods during which Earth was much warmer than currently, but the temperature of the planet never reached the vaporization point of seawater (101 °C, Table 5.2) after its initial cooling. This was a consequence of the combined formation of $CaCO_3$ on and within the seafloor of the warm-water ocean and the removal of this $CaCO_3$ by plate tectonics (see Sect. 5.5.4). If the Earth's global atmospheric temperatures had reached values below the vapour-liquid line in the phase diagram of water (Fig. 5.3), for example more than 101 °C at an atmospheric pressure of 1 atm, this global warming could have caused massive evaporation of water and the resulting water vapour would have fed the greenhouse effect (see Information Box 2.7), thus initiating a positive feedback loop leading to a runaway warming (see Information Box 6.2). This warming would have caused the planet to lose all its water to the outer space as hypothesized to have occurred on Venus (see Sect. 5.5.4). The continuous presence of an ocean on Earth's since more than 4.4 billion years shows that such a runaway warming did not happen at any time during the Earth's history.

Even if Earth had not been in danger of losing its water to the outer space, it must be remembered that the planet continually loses water to space, and only the steady resupply of water by interplanetary dust and rocks keeps the Earth's water budget even (see Sect. 5.4.2). In the relatively "recent" past, the average temperature of Earth reached 20 °C and perhaps 25 °C on several occasions during the last 500 million years (Figs. 4.8 and 7.1). Average temperatures of 20 °C and more are very high compared to the present 15 °C.

We now turn to the possibility of very low temperatures since 4.4 billion years. Given that the freezing point of seawater is −1.9 °C (Table 5.2), the surfaces of all oceans and freshwater bodies would have frozen if the global temperature had dropped below −2 °C. In such a case, the deeper waters, still liquid, would have been without contact with the atmosphere. A decrease of the global temperature below −2 °C would have initiated a positive feedback loop leading to a runaway cooling, meaning that the initial drop in temperature would have caused growth of the ice and snow cover, which would have increased the Earth's albedo (see Information Box 2.9), which would have caused a further drop in temperature, and so forth until the occurrence of a runaway cooling. This would have caused the freezing of the entire surface of Earth, a state from which the planet could not have easily broken out afterwards.

In fact, Earth has probably frozen at least once and perhaps several times more than 650 million years ago and as early as 2.4 billion years ago. This phenomenon, called *snowball Earth*, is described in Chapter 4 (see Sect. 4.5.4). Planet Earth has not known snowball episodes since 600 million years, but it has experienced several icehouses during this period as well as before (see Sect. 4.5.6). During the latest icehouse, in which we presently live, there have been successive cycles of glacial and interglacial episodes since 2.6 million years ago (beginning of the Quaternary period), known as *Quaternary Ice Age* (see Sect. 4.5.2). The present cycle of glacial-interglacial phases is a geologically very recent phenomenon, but other cycles of glaciation may have existed during the previous icehouses in the distant past (Table 4.3).

The initial question was whether Earth had sometimes been extremely hot or extremely cold? The answer to this question is that there were indeed periods when the planet was much warmer than currently and others when it was largely or partly covered by ice. Despite these large changes in temperature, Earth always provided conditions suitable for the presence of liquid water.

Liquid Water has Always Been Available for Organisms and Ecosystems The history of Earth indicates that its planetary characteristics generally favoured moderate temperatures, and thus the maintenance of liquid water at the surface of the planet. However, temperatures were temporarily out of control during the snowball Earth episodes, but ecosystems could survive and perhaps even thrive in deep waters, away from the iced surface. Biological innovations, such as calcification, contributed to keep temperature variations within limits generally compatible with the presence of vast expanses of liquid water at the Earth's surface (see Sect. 4.7.2). The wide availability of liquid water largely explains the success of organisms and ecosystems on Earth.

5.6.5   The Earth's Water Cycle

The interactions between the numerous processes described in the previous sections determine the fluxes of water among reservoirs located within and among the various components of the Earth System. These reservoirs are: the atmosphere; the oceans and other water bodies; the ice sheets, ice caps, glaciers and sea ice; the continents' surfaces and rocks; the aquifers; the deep crust and the mantle; and the organisms (see Sect. 5.2.1). Together, these reservoirs and fluxes are called the *water cycle*, or the *hydrological cycle*. There are several ways to schematize this complex cycle, of which Fig. 5.7 provides an example.

Figure 5.7 represents the main reservoirs of Earth's water described in Sect. 5.2.1, which are the oceans (94% of the total volume), the underground storage (4%), the ice sheets, caps, glaciers and snow (2%), the freshwater bodies

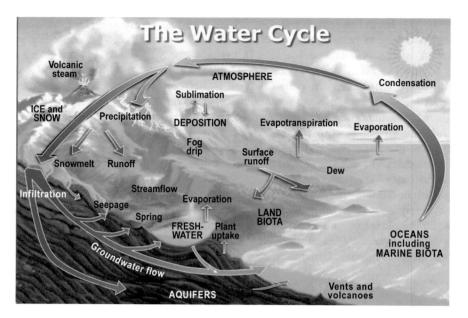

Fig. 5.7 The Earth's water cycle. The main reservoirs of water are written in capital letters, and the arrows represent physical fluxes whose names are written close to the arrows. Credits at the end of the chapter

(lakes, rivers), and the atmospheric vapour. Part of the water is also contained in the oceanic, continental and freshwater biota. Together, the freshwater bodies, the atmosphere and the biota comprise less than 1% of the water on Earth.

Starting the description of the cycle with the atmosphere, Fig. 5.7 shows that evaporation moves water to the atmosphere from the oceans and fresh waters, and evapotranspiration does the same from the continents. In the atmosphere, water vapour condenses onto small solids (called *cloud condensation nuclei* and forms clouds, which eventually return the water to the oceans and continents by precipitations. The latter include rainfalls and snowfalls, and water also leaves the atmosphere in the form of fog and dew. In cold conditions, the accumulation of snow can contribute to the growth of glaciers, and in warmer conditions, rain accelerates the melt of snow and ice. Water can also be exchanged directly between the atmosphere and the ice or snow by sublimation and deposition. The surface runoff from rain and snow melt, including the water flowing in streams and rivers (called *streamflow* in Fig. 5.7), brings water to freshwater bodies and the continental biota (including people), and carries water to the oceans (see Sects. 3.5.2 to 3.5.4, and Information Box 5.7).

The previous paragraph describes only one part of the cycling of water on Earth. Indeed two-thirds of the fresh water on the planet are located under the ground (see Sect. 5.2.1). Water infiltrates from continental surfaces into aquifers, where it can be stored for long periods. For example, water in some aquifers goes back 20,000 years. Some of the water in aquifers flows under the ground and different parts of it seep out, surge in springs, feed freshwater bodies, and sustain the growth of plants and other land biota. In many coastal areas, there is a significant flow of groundwater into the ocean.

Finally, some water circulates in the deep Earth. One of the ways water reaches these areas is penetration deep into the seafloor through porous sediments and rocks. From there, water is later emitted in hydrothermal vents, which are typically located along mid-ocean ridges (see Sect. 7.3.1). Another way by which water reaches the deep Earth is downward transport by subduction deep in the mantle of water trapped in sedimentary layers of the subducting seafloor. This water is later emitted in the atmosphere and the ocean by land and submarine volcanoes, respectively (see Sect. 5.2.1).

The amount of water on the planet is finite, but through its continual cycling, water has irrigated the different components of the Earth System for billions of years. Hence, the cycling of water has made its availability almost infinite. Unfortunately, human interference with the water cycle often causes *water scarcity* (see Sect. 11.1.3). The book of Spellman (2020) provides further reading on the different sources of freshwater on continents, and the effects of human use, misuse and reuse of fresh and waste water on the overall water supply.

5.7   Key Points: Liquid Water and the Living Earth

We conclude this chapter with a brief summary of some of the key points of the above sections, as done in previous chapters (see Sects. 1.6, 2.5, 3.8, and 4.8). Here, we examine interactions between the Solar System, Earth, water, and organisms. The context of this section is the geological history of Earth (see Sect. 1.6, Table 1.3), and more specifically the three phases of the Earth's history illustrated in Fig. 1.11, namely the Mineral, Cradle and Living Earth. These three phases occurred from 4.6 to 4.0, 4.0 to 2.5, and 2.5 to 0 billion years ago, respectively.

The presence of large amounts of water on Earth was initially determined by events that took place during the Mineral Earth phase. There is a debate among specialists concerning the origin of Earth's water, which could have come from the solar nebula when the planet formed or brought later to Earth by water-containing asteroids and/or water-rich comets (see Sect. 5.3). One key factor in the retention of water on Earth may have been the formation of a planet-wide ocean very early in the Earth's history, which moved gaseous water from the atmosphere to the liquid ocean (see Sect. 5.6.4). This ocean could form on a still warm planet because the high atmospheric pressure that then existed allowed water to be in the liquid phase despite the high temperature (see Sect. 5.5.4).

During the Cradle Earth phase, the abundant water was liquid and its temperature became low enough for organisms. This is because during tens of millions of years, most of the atmospheric CO_2 was transformed chemically into $CaCO_3$ in the seafloor, causing a large reduction in the natural greenhouse effect, which brought the temperature down (see Sect. 5.5.4). Further into the Living Earth phase, the Earth's moderate atmospheric pressure determined both the freezing and vaporization points of seawater and the temperature at the surface of the planet, whose values favoured the prevalence of water in the liquid phase (see Sect. 5.6.2). Two major planetary determinants of the presence of water mostly in the liquid phase on Earth are the mass of the planet and its distance from the Sun, which allowed Earth to have a significant atmosphere (see Sect. 5.6.2). In addition, the ability of the lower atmosphere to store and release heat is among the planetary characteristics that make liquid water a permanent feature of most environments on Earth (see Sect. 5.6.3). Also, the fast rotation of Earth favoured the day-and-night presence of liquid water over most of the planet year-round, and the small value of the Earth's axial tilt contributed to the long-term year-round occurrence of liquid water on the planet (see Sect. 5.6.3).

Liquid water is the most powerful natural solvent, and thus plays a key role in transporting most of the substances used or released by the cells (see Sect. 5.1.1). The existence of large amounts of liquid water was favoured by adequate combinations of temperature and atmospheric pressure since the Cradle Earth phase, which allowed organisms to get established and build up the huge biomasses that led them to progressively take over the whole planet (see Sect. 5.5.1).

During the Living Earth phase, the widely diverse aquatic habitats, from salt to fresh waters, were suitable for the diversification of organisms and ecosystems (see Sect. 5.5.2). One major result of biological evolution was the O_2-producing photosynthesis, which requires water and uses the hydrogen atoms from H_2O to synthesize organic matter while releasing O_2 in the environment (see Information Box 5.1). Conversely, O_2 respiration releases as H_2O the hydrogen atoms contained in organic matter combined with the oxygen taken from the environment (see Information Box 5.2). Even though there were periods when Earth was much warmer than it is now and other times when it was largely or partly covered by ice, the planet has always provided conditions conducive to the presence of liquid water, which was essential for organisms (see Sect. 5.6.4).

Figure Credits

Fig. 5.1 Original. Figure 5.1 is licensed under CC BY-SA 4.0 by Philippe Bertrand, Louis Legendre and Mohamed Khamla.

Fig. 5.2 Original. Figure 5.2 is licensed under CC BY-SA 4.0 by Philippe Bertrand, Louis Legendre and Mohamed Khamla.

Fig. 5.3 Modified after http://planet-terre.ens-lyon.fr/article/eau-glace-pression.xml. With permission from Prof. Olivier Dequincey, Ecole Normale Supérieure de Lyon, France.

Fig. 5.4 This work, Fig. 5.4, is a derivative of https://commons.wikimedia.org/wiki/File:3D_model_hydrogen_bonds_in_water.svg by Qwerter at Czech wikipedia https://cs.wikipedia.org/wiki/Wikipedista:Qwerter, in the public domain. I, Mohamed Khamla, release this work in the public domain.

Fig. 5.5 Original. Values calculated using the equation given in https://en.wikipedia.org/wiki/Atmospheric_pressure. Figure 5.5 is licensed under CC BY-SA 4.0 by Philippe Bertrand, Louis Legendre and Mohamed Khamla.

Fig. 5.6 Original. Figure 5.6 is licensed under CC BY-SA 4.0 by Philippe Bertrand, Louis Legendre and Mohamed Khamla.

Fig. 5.7 This work, Fig. 5.7, is a derivative of https://commons.wikimedia.org/wiki/File:Watercyclesummary.jpg by John Evans and Howard Periman, USGS, in the public domain. I, Mohamed Khamla, release this work in the public domain.

Further Reading

Ball P (2001) Life's matrix: a biography of water. University of California Press
Ball P (2017) Water is an active matrix of life for cell and molecular biology. Proc Natl Acad Sci USA 114:13327–13335. https://doi.org/10.1073/pnas.1703781114
Hand K (2020) Alien oceans: the search for life in the depths of space. Princeton University Press, Princeton
Hanslmeier A (2011) Water in the Universe. Astrophysics and Space Science Library, vol 368. Springer, Dordrecht
Kratz RF (2020) Molecular & cell biology for dummies, 2nd edn. Wiley, Hoboden, NJ
Lécuyer C (2013) Water on Earth: Physicochemical and biological properties. Wiley Online Library. https://doi.org/10.1002/9781118574928
Pinti DL, Arndt N (2015) Oceans, Origin of. In: Gargaud M et al (eds) Encyclopedia of astrobiology. Springer, Heidelberg and Berlin
Spellman FR (2020) The science of water, concepts and applications, 4th edn. CRC Press, Boca Raton
Taylor FW (2014) The scientific exploration of Venus. Cambridge University Press, Cambridge

Chapter 6
The Building Blocks of Organisms: Connections with Gravitation

6.1 Organisms: Hardware

6.1.1 Chemical Hardware of Organisms

Two major components of organisms on Earth are their chemical constituents and physical structures (hardware) and their genetic information (software) (Fig. 1.2). This chapter examines successively: the chemical constituents of the cells and their roles in carrying the genetic information (Sects. 6.1 and 6.2); the chemical bricks of organic matter in the outer layers of the planet (Sect. 6.3); the ions in the ocean and the organisms (Sects. 6.4); and the availability of the key biogenic elements in the environment (Sect. 6.5). The next section explains how gravitation contributed to make available in the Earth's crust the chemical elements that would become the building blocks of organisms, and ensures their long-term availability for ecosystems (Sect. 6.6). The chapter ends with a summary of key points concerning the interactions between the Solar System, Earth, the building blocks of the cells, and the organisms and ecosystems (Sect. 6.7).

The chemical constituents of the cells are water and various organic compounds that make up the organic matter (see Information Box 3.5). These compounds are: the carbohydrates and the lipids, which are the main components of the structural elements of the cells; the proteins, which make the cells work; and the nucleic acids, which carry the genetic information. These different groups of substances are described in turn below, showing how they are structurally and functionally interconnected, and how they are chemically connected with the Earth's environment (see Sects. 6.1.4 and 6.3).

© Springer Nature Switzerland AG 2021
P. Bertrand and L. Legendre, *Earth, Our Living Planet*, The Frontiers Collection,
https://doi.org/10.1007/978-3-030-67773-2_6

6.1.2   *Elemental Composition of Organisms*

The organisms on Earth belong to a wide diversity of biological groups, which range from microscopic short-lived unicellular organisms to 5,000-years old trees and 170-ton blue whales. The latter are thought to be the largest animals that ever existed. Despite these huge differences, the chemical elements that compose all organisms are quite similar. This is illustrated in the last four columns of Table 6.1, for an animal (human being *Homo sapiens*) and a leguminous plant (alfalfa, also known as lucerne), where the concentrations are expressed in % atoms (see Table 6.1). The chemical elements listed in Table 6.1 are the primary building blocks of cells and organisms.

Table 6.1   Atom % of the main chemical elements in the Earth's crust, the atmosphere, the seawater, the human body, and alfalfa (a leguminous plant, also called lucerne). Values in each cell of the table: % abundance (number of atoms) of a given element relative to the total abundance (number of atoms) of all the elements present in an Earth's compartment or an organism. The order of the elements in the table reflects their general abundance in organisms

Chemical elements (and symbols)	Earth's crust	Atmosphere[a]	Seawater[b]		Human body		Alfalfa	
	Without flowing H_2O	With 1% water vapour	Including H_2O	Dissolved forms	With 60% H_2O	Without H_2O (dry)	With 75% H_2O	Without H_2O (dry)
Hydrogen (H)	–	1	66	–	61	50	62	45
Oxygen (O)	47	21	33	–	26	11.2	31	21
Carbon (C)	–	0.02	0.001	0.1	11	30	6	31
Nitrogen (N)	–	77	–	–	2	6.9	0.4	2
Calcium (Ca)	3.5	–	0.006	0.6	0.2	0.7	0.1	0.5
Phosphorus (P)	–	–	–	–	0.1	0.4	0.01	0.1
Sulphur (S)	–	–	0.017	1.7	0.1	0.4	0.02	0.1
Sodium (Na)	–	–	0.28	28	0.07	0.2	0.01	0.1
Potassium (K)	–	–	0.006	0.6	0.04	0.1	0.04	0.2
Chlorine (Cl)	–	–	0.33	33	0.03	0.1	0.01	0.1
Magnesium (Mg)	2.2	–	0.033	3.3	0.01	0.03	0.02	0.1
Silicon (Si)	28	–	–	–	–	–	–	–
Aluminium (Al)	7.9	–	–	–	–	–	–	–
Iron (Fe)	4.5	–	–	–	–	–	–	–
Argon (Ar)	–	0.5	–	–	–	–	–	–

[a]In the atmosphere, the main substances containing H, O C and N are, respectively: H_2, H_2O, CH_4; O_2, H_2O, CO_2, N_2O, NO_2; CO_2, CH_4; and N_2, N_2O, NO_2
[b]Seawater with a salinity of 35, that is, 35 g of dissolved substances per kilogram of seawater. The dissolved substances do not include dissolved gases

---

Information Box 6.1 Abundance Expressed in Number of Atoms or in Mass Units The % abundance of a chemical element in an organism or in the environment can be expressed in terms of either its *mass* or its *number of atoms*, or for a gas as the *volume* it occupies in the air (see Information Box 2.1). Because the atomic masses of the various chemical elements are very different, computing the % abundance of an element in term of its mass or its number of atoms leads to very different results. The *atomic mass* used in the following paragraphs is defined in Chap. 5 (see Information Box 5.6).

Taking the water molecule (H_2O) as example, the % abundances of hydrogen and oxygen expressed in numbers of atoms are 2/3 and $1/3 = 67\%$ H and 33% O, respectively. However because the atomic masses of H and O are 1 and 16, respectively (the mass of the H_2O molecule is 18), the % abundances of hydrogen and oxygen expressed in mass are $[(2 \times 1)/18] \times 100$ and $[(1 \times 16)/18] \times 100 = 11\%$ H and 89% O, respectively.

In the generalized formula of sugars (CH_2O), already used in previous chapters (see Information Boxes 3.4, 5.1 and 5.2), the % abundances of carbon, hydrogen and oxygen expressed in number of atoms are 1/4, 2/4 and $1/4 = 25\%$ C, 50% H and 25% O, respectively. Because the atomic masses of C, H and O are 12, 1 and 16, respectively (the mass of the CH_2O molecule is 30), the % abundances of carbon, hydrogen and oxygen expressed in masses are $[(1 \times 12)/30] \times 100, [(2 \times 1)/30] \times 100$ and $[(1 \times 16)/30] \times 100 = 40\%$ C, 7% H and 53% O, respectively.

In the two examples above, the values based on the numbers of atoms are very different from those based on the atomic masses. Hence, one must always check carefully which units are used when consulting the literature. In Table 6.1, each value corresponds to the percentage of the total *number of atoms* of a given chemical element in the Earth's compartment or in the organism considered. For example in the Earth's crust column of Table 6.1, the value of 28 for silicon means that 28% of the atoms of all the elements in the Earth's crust are atoms of silicon.

All organisms contain more than 60% and up to 95% of water (expressed in atom %, as in Table 6.1), and their composition is consequently dominated by the chemical elements hydrogen (H) and oxygen (O), which are the two constituents of water (H_2O). Because H and O are systematically the most abundant chemical elements in all organisms, it is informative to examine the atom % of the main elements in organisms excluding their constitutive water. Indeed, the chemical composition of "dry" organic matter (that is, without water) provides information on the chemical composition of compounds other than water in organisms (Table 6.1). The abundances of elements without considering water can be measured directly on desiccated organisms, or it can be calculated from the composition of "fresh" biomass when its water content is known, in which case it can be subtracted from the total H and O.

In Table 6.1, the atom % of chemical elements—with and without water for humans and alfalfa (last four columns)—are compared to the atom % of elements in the environment (first columns), that is, the Earth's crust, the atmosphere (including water vapour), and the oceans. Given that the H and O atoms of water molecules (H_2O) account for 99% of all atoms in the oceans, the atom % in Table 6.1 are provided for seawater including its H_2O molecules, and also for seawater excluding the H and O atoms of its H_2O molecules, thus only considering the elements that make up the main compounds dissolved in seawater.

We first examine the five top chemical elements in Table 6.1. These are hydrogen, oxygen, carbon, nitrogen and calcium.

Hydrogen is the most abundant chemical element in organisms, whether including or not H_2O in the calculation (Table 6.1, last four columns). This indicates that hydrogen is the key element in organisms not only as part of their constitutive water, but also as a component of their other molecules than water. The source of the hydrogen used to build organic compounds is not the same for the organisms that use O_2-producing or O_2-less photosynthesis or chemosynthesis (see Information Box 5.1). On the one hand, O_2-producing photosynthetic organisms extract hydrogen from their surrounding water by splitting H_2O molecules into H_2 and O, and they release the oxygen in the environment as molecules of free O_2. On the other hand, O_2-less photosynthetic and chemosynthetic organisms derive hydrogen from other H-containing compounds than water, for example hydrogen sulphide (H_2S). Figure 6.1 shows the connections between the cycle of hydrogen and the cycles of carbon and oxygen in O_2-producing photosynthesis and O_2 respiration (see Information Box 3.4).

Figure 6.1 illustrates key steps of O_2-producing photosynthesis and O_2 respiration. The steps of *O_2-producing photosynthesis* are: the splitting of water (H_2O)

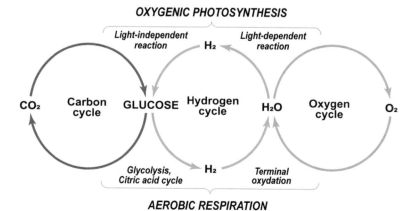

Fig. 6.1 Interconnections between the cycles of carbon, hydrogen and oxygen in O_2-producing photosynthesis and O_2 respiration (see Information Box 3.4). Credits at the end of the chapter

molecules into H_2 and O, using the energy carried by photons of light (light-dependent reaction); the accompanying release of O_2 molecules as by-product; and the synthesis of glucose ($C_6H_{12}O_6$) and H_2O by the combination of H_2 with CO_2 (light-independent reaction). In summary, O_2-producing photosynthesis uses water, carbon dioxide and energy, and produces glucose, water and free oxygen (see Information Box 5.1). The steps of O_2 *respiration* are: the release of energy by the reaction of O_2 with $C_6H_{12}O_6$ (glycolysis); the breakdown of $C_6H_{12}O_6$, which liberates CO_2 (through the citric acid cycle); and the accompanying combination of H atoms from $C_6H_{12}O_6$ with the O_2 forming H_2O molecules (terminal oxidation). In summary, O_2 respiration uses glucose and free oxygen, and produces carbon dioxide, water and energy (see Information Box 5.2).

The combination of the two reactions in Fig. 6.1 shows that the two biological O_2-processes continually cycle H and O atoms from H_2O and back to it, in a complementary manner. The O_2 respiration returns to the environment as H_2O both the *hydrogen* atoms taken from H_2O by O_2-producing photosynthesis and the *oxygen* atoms from H_2O released to the environment by O_2-producing photosynthesis.

Oxygen is the second most abundant chemical element in organisms when including H_2O in the calculation, and the third when H_2O is not included (Table 6.1, last four columns). This indicates that oxygen is a key component of organisms both as part of their constitutive water and also as a constituent of their organic compounds. The inorganic source of the oxygen used by organisms to synthesize organic compounds is the CO_2 gas, whose concentration in the atmosphere and dissolved in natural waters is quite low. The oxygen atoms from H_2O are not used to synthesize organic matter, but are instead released in the environment during the O_2-producing photosynthesis. This explains the abundance of free dioxygen (O_2) both as atmospheric gas and dissolved in natural waters. Figure 6.1 illustrates the connections between the cycle of oxygen and the cycles of hydrogen and carbon in O_2-producing photosynthesis and O_2 respiration (see Information Box 3.4).

The abundant O_2 resource is not used by organisms as a source of oxygen in the synthesis of organic matter, but is employed instead by O_2 respiring organisms in the process of respiration (see Information Boxes 3.4 and 5.2). To synthesize organic matter, the organisms do not use the oxygen available in its most abundant environmental pool, which is H_2O, but they instead use the oxygen from relatively scarce CO_2 (Table 6.1).

Carbon comes third among the chemical elements in organisms when including H_2O in the calculation, and second when H_2O is not included (Table 6.1, last four columns). This stresses the fact that carbon plays a central role in organic matter, not only because of its abundance in the organic compounds, but also for reasons explained in the next paragraphs. Figure 6.1 illustrates the connections between the cycle of carbon and the cycles of hydrogen and oxygen in O_2-producing photosynthesis and O_2 respiration (see Information Box 3.4).

The inorganic source of the carbon used by autotrophic organisms to synthesize organic compounds is the CO_2 gas, which is also the inorganic source of the oxygen as explained just above. The three key chemical components of organisms are H, C and O, and the CO_2 gas provides organisms with C and O, whereas the H used by O_2-producing photosynthetic organisms comes from water.

Conversely O_2 respiration releases energy by the reaction of O_2 with organic molecules, and the process is accompanied by the production of CO_2 and H_2O. More precisely, the reaction of O_2 with reserve compounds of the cells, such as sugars and fats, releases energy and forms CO_2, and the hydrogen contained in the organic matter recombines with the respired O_2 to form H_2O. The respiratory CO_2 is released in the environment, where it is part of the Earth's *fast carbon cycle* (see Sect. 2.1.6 and Information Box 9.1).

Almost all chemical compounds that contain carbon are called *organic*, with the notable exception of CO_2 (see Information Box 3.5). Carbon atoms can form strong links with up to four other atoms, which can be other carbon atoms or atoms of hydrogen, oxygen, nitrogen or sulphur (these links are called *hydrogen bonds*, Fig. 5.4). Because of this property, the atoms in organic molecules can link together in long straight or branching chains, or in rings. This generates a huge range of molecular structures and shapes, which result in the astonishing number of *several millions* of different organic compounds that exist in nature. In addition, humans have created numerous synthetic organic compounds during the last two centuries. Although some natural organic compounds are produced outside organisms by purely chemical reactions (see Information Box 3.5), carbon plays a unique role in living matter because of its ability to form long, complex and very differently shaped chains. This unique characteristic of carbon explains its high abundance in organisms.

Another factor that contributes to explain the special role of carbon in organic matter is the fact that, at the Earth's surface, the most abundant natural solvent is liquid water. This is the case in both the natural environment (oceans and lakes, and moisture in the soil and the atmosphere) and the cells of organisms. The ubiquity of liquid water largely explains the key role of carbon in organic matter. Indeed, carbon readily combines with hydrogen and oxygen, which are the two components of water, with the result that these three chemical elements are the most abundant in living matter (Table 6.1). Also in water solutions, carbon compounds are more stable than compounds based on other chemical elements able to make chains, such as silicon. Finally, almost all the inorganic and organic compounds naturally dissolve in liquid water, given time (see Sect. 5.5.2), and once in solution, the smaller components (ions or molecules) of the dissolved compounds can recombine into larger carbon-containing molecules. Hence, the abundance of liquid water at the surface of Earth (see Sect. 5.2.1) has favoured, since the early history of our planet, the formation of large molecules based on carbon over molecules based on other chemical elements.

The takeover of Earth by carbon-based organisms is largely due to the ubiquitous presence of liquid water at the surface of the planet since billions of years. Indeed, water could be present in the liquid phase under the conditions of temperature and atmospheric pressure that prevailed at the Earth's surface as early as 4.4 billion years ago (see Sect. 5.5.4). Most astronomical bodies inside and outside the Solar System have surface temperature and pressure conditions very different from those of Earth, so that surface water there, when present, would be either solid (ice) or gaseous. As a consequence, the development of life on these astronomical bodies would require the presence, at their surface, of large amounts of other liquids than water, which could act as solvents in which small chemical components (ions or molecules) could recombine.

Specialists of extraterrestrial life have identified a variety of compounds as possible life-sustaining solvents outside Earth, which differ according to the surface temperature and pressure conditions on extraterrestrial planetary bodies. These compounds include ammonia, hydrocyanic acid, and methanol, and under very special conditions, hydrogen sulphide, sulphur dioxide, methane, and ethane. With these solvents, some different forms of organic matter could hypothetically be based on such elements as silicon, boron, nitrogen, phosphorus, sulphur or germanium, although this has never been observed. Given the theoretical diversity of possible life forms in the Universe, the detection of organisms based on other chemical elements than carbon would not be an easy task. The book of Schulze-Makuch and Irwin (2018) provides further reading on the topic of exotic extraterrestrial life forms.

Nitrogen is the fourth most abundant chemical element in organisms (Table 6.1, last four columns). The ultimate inorganic source of nitrogen for organisms is the very abundant N_2 gas (called dinitrogen), which is found both in the atmosphere and dissolved in natural waters, although the solubility of nitrogen in water is lower than that of oxygen. However, most organisms cannot use N_2, and only a few types of bacteria and archaea can directly fix N from the N_2 gas into organic matter. Some, but not all, of the N_2-fixing microorganisms are autotrophs, and some of them live in symbiosis with autotrophs to which they supply N. Because of the limited sources of bioavailable N in aquatic or continental environments, these may be, at times, in short supply of this essential chemical element despite the high ambient concentration of the N_2 gas.

The autotrophic organisms unable to use N_2 find the N they need in N-containing compounds released by the N_2-fixers or by the decomposition of organic matter in natural waters and soils. These compounds are called *nitrogenous nutrients*, and they are used by autotrophs to elaborate their organic matter. Heterotrophic organisms also need N, which they acquire from N-containing organic matter. Leguminous plants, such as alfalfa in Table 6.1, host symbiotic N_2-fixing bacteria in their root nodules, which allow them to produce high-protein organic matter regardless of the bioavailable nitrogen present in the soil. In aquatic environments, N is mostly fixed from N_2 by microscopic cyanobacteria, whose ancestors "invented" O_2-producing photosynthesis (see Information Box 5.1). The N_2-fixing bacteria and archaea in water and soil produce two molecules of ammonia (NH_3) from one molecule of inorganic N_2 gas molecule, this NH_3 being the initial source of the nitrogen used by all organisms for the synthesis of amino acids, proteins, nucleotides and nucleic acids in their organic matter (see Sects. 6.1.3 and 6.2.2, below).

Calcium comes next after nitrogen in the % atom composition of organisms (Table 6.1, last four columns), and its inorganic source for the elaboration of organic matter is the large pool of Ca^{2+} ions dissolved in natural waters. The concentration of calcium is about 10,000 times lower in the cells of eukaryotes than in their surrounding environment, and the mechanisms that maintain the calcium gradient across the cell membrane are called *calcium pumps*. Calcium is a key chemical element for the structure of cell membranes and the building of skeletons and shells (see Sect. 6.4.1), and also for the transmission of electrical signals in organisms.

The above five top chemical elements in Table 6.1 dominate the composition of organisms, but only four of them are also abundant in one or more compartments of the Earth's environment. These are *hydrogen, oxygen, nitrogen*, and *calcium*. Four other elements are abundant in organisms but not in the Earth's outer envelopes. These are *carbon, phosphorus, sulphur* and *potassium*. Finally, many other elements are abundant in the environment but not in organisms, although they may have important biological roles. These include, among others, *sodium, potassium, chlorine, magnesium, silicon*, and *iron*. The differences between the elements of the three groups show that the abundance of chemical elements in the environment was not the only factor that determined the chemical composition of organisms during the course of biological evolution. Indeed, the chemical composition of organic molecules was also determined by their efficiency at fulfilling their functions in organisms.

The biological roles of hydrogen, oxygen, carbon, nitrogen and calcium were explained above. Some of the biological roles of other eight abundant elements in Table 6.1—phosphorus, sulphur, sodium, potassium, chlorine, magnesium, silicon and iron—are briefly described here.

Phosphorus (P) is an essential component of molecules called phospholipids, which are the main components of cell membranes (see Sect. 6.1.3, below). This indicates that phosphorus was involved very early in the process of biological evolution. The organic matter contains less phosphorus than nitrogen (Table 6.1), reflecting the lower availability of phosphate than nitrate in many aquatic and continental environments (see Sect. 9.3.1). This is because N can be fixed from the abundant N_2 by specialized bacteria and archaea (see above), whereas P must be obtained from dissolved phosphate (PO_4^{3-}) and other mineral forms that are not controlled by organisms. As a result, phosphorus limits the production of organic matter by autotrophs in many environments, both aquatic and continental. In addition, vertebrates use phosphorus for bones and teeth, in the form of calcium phosphate, $Ca_3(PO_4)_2$, and this process also often incorporates other elements such as magnesium and iron.

Sulphur (S) is a component of two essential amino acids, cysteine and methionine (see Sect. 6.1.3, below), and is a central chemical element for some organisms and ecosystems. For example, specialized bacteria and archaea in hydrothermal vents on the seafloor use the energy supplied by the reaction between O_2 and hydrogen sulphide (H_2S) to synthesize organic matter using chemosynthesis (see Information Box 5.1). Conversely in absence of O_2, sulphate-reducing bacteria respire sulphate (SO_4^{2-}) to oxidize organic matter and thus obtain energy (see Information Box 5.11, Eq. 5.11). This process takes place in our own guts.

Sodium (Na) plays a key role in regulating the volume of the cell through *osmotic balance* between the cell and its environment. This refers to the balance in the salt and water of the fluids inside and outside the cell, across the cell membrane. As a result, not enough or too much internal Na causes water to leave or enter the cell, respectively (see Sect. 6.4.1, below).

Potassium (K) is more abundant in plants than in animals (see the alfalfa and the humans in Table 6.1). Plants require K for protein synthesis and photosynthesis.

Chlorine (Cl) contributes, together with Na, to regulate osmotic balance between the cell and its environment, and thus the volume of the cell.

Magnesium (Mg) is a major component of many enzymes, including all those that synthesize or utilize ATP, which fuels the chemical reactions in the cells (see Sect. 6.2.2). Furthermore, ATP must bind to Mg to be biologically active.

Silicon (Si) plays a very important biological role, in particular because some ecologically key groups of photosynthetic plankton consist of organisms with a cell wall made of silica SiO_2), whose development largely depends on the availability of dissolved silicon. The most abundant of these organisms, called *diatoms*, form the basis of marine food webs that sustain many commercially important fisheries. Conversely, the lack of silicon in nutrient-rich effluents from agriculture prevents the growth of diatoms, with the result that these effluents cause the production of great masses of inedible phytoplankton whose decomposition in the water column leads to oxygen depletion.

On the continents, silicon is especially used by plants that belong to the group of *monocotyledons* (plants that have only one embryonic leaf in their seeds), which often have high concentrations of Si in their leaves and stems. These plants include grass, whose Si-rich leaves are often as sharp as glass shards, and bamboo and palm trees, whose remarkable strength partly lies in the high Si content of their stems. The popular science book of De La Rocha and Conley (2017) provides further reading on the role of silicon in the Earth System.

Iron (Fe) is essential to the metabolism of cells, being an integral part of the proteins involved in such key biological functions as photosynthesis, respiration, and assimilation of nitrate. Biological evolution selected biochemical mechanisms that enhanced both the efficiency of iron acquisition by the cells and the regulation of its storage and toxicity inside the cells (iron is toxic when present in excessive amounts). Because of its low concentrations in surface waters of different regions of the oceans, iron limits the growth of phytoplankton in about 20% of the World Ocean even if the supply of nitrogenous and phosphorous nutrients is adequate. The major areas affected by iron deficiency are called *High Nutrient*

Low Chlorophyll regions (see Sect. 9.3.5). On the continents, iron is among the abundant elements in soils, but it is largely present in chemical forms that cannot be taken up by plants. An article written by Underwood (2019) provides further reading on the role of iron and on HNLC regions.

A chapter on marine biogeochemical cycles by Legendre (2014) provides further reading on the cycles of carbon, oxygen, nitrogen and phosphorus. The next sections examine structural and functional aspects of the chemical composition of organisms. Readers familiar with the biochemistry of organic molecules and cell biology could go directly to Sect. 6.3, while taking a look at Sects. 6.1.4 and 6.2.5.

6.1.3   From Chemical Elements to Complex Organic Molecules

Cells and organisms chemically consist of water and organic molecules. The "chemical bricks" of water are hydrogen and oxygen, and the main chemical bricks of organic matter are the thirteen chemical elements examined in the previous section. The organisms use the latter elements and several others to build organic compounds, which range from very small to incredibly large. Two extreme examples of organic molecules synthesized by the cells are formaldehyde (CH_2O), which is made of only 4 atoms, and DNA, which contains more than one billion atoms—one *billion* is not a typo, as seen below (see Sect. 6.2.2).

In the cells, the chemical elements are initially combined into small and (relatively) simple molecules, called *monomers* (meaning "one part"), which are themselves combined into larger and more complex molecules called *polymers* (meaning "several parts") or *macromolecules* (meaning "large molecules"). Although water molecules do not form real polymers, groups of water molecules called *water clusters* have been found under experimental conditions. Such groups of water molecules will not be considered here, and all the polymers examined below are made of carbon-based monomers.

In our day-to-day life, we are surrounded by a large number of synthetic monomers and polymers, which chemists initially designed by studying the chemical structure of natural organic compounds. Examples of synthetic organic compounds that are part of our everyday environment are fibres such as nylon, chemical dyes, synthetic pesticides, and 60,000 types of plastics. In this chapter, we do not consider synthetic organic compounds, and we focus exclusively on substances that naturally occur in organisms, called *biopolymers*, and their constitutive monomers. The four main groups of biopolymers are: carbohydrates, lipids, proteins, and nucleic acids (Table 6.2). The following paragraphs briefly describe each group of biopolymers and explain its significance for organisms.

Table 6.2 Main groups of biochemical compounds in organisms. From left to right: biopolymers and their constitutive monomers with examples of compounds, and the main chemical elements (last column) that make up the monomers and polymers

Biopolymers	Examples of biopolymers	Monomers	Examples of monomers	Main chemical elements
Carbohydrates: polysaccharides	Starch; glycogen; cellulose; chitin	Carbohydrates: monosaccharides (sugars)	Glucose; fructose	C, H, O
Lipids	Phospholipids; fats; cholesterol; rubber	Fatty acids	Palmitic acid; arachidic acid	C, H, O, (P)[a]
Peptides and proteins	Collagen; keratin; haemoglobin	Amino acids	Aspartate; glutamate	C, H, O, N, (S)[b]
Nucleic acids	Deoxyribonucleic acid (DNA, with A, C, G, T); ribonucleic acid (RNA, with A, C, G, U)	Nucleotides (sugar, phosphate, and one of the N-containing bases); ATP	N-containing bases: adenine (A), cytosine (C), guanine (G), thymine (T), uracil (U)	C, H, O, N, P

[a]phosphorus is not present in all lipids
[b]sulphur is not present in all peptides or proteins

Carbohydrates include monosaccharides and polysaccharides. The *monosac-charides* are mall molecules of simple *sugars* (monomers), formed of carbon, hydrogen and oxygen. There are a few tens of different types of monosaccharides in the organisms. Chains of a few and up to thousands of monosaccharides form the *polysaccharides* (biopolymers), which either store energy in cells or contribute to their structure. Some sugars are integral parts of the nucleotides, which are the building blocks of nucleic acids (see Sect. 6.2.2, below).

Lipids include fatty acids and various types of biopolymers. The lipids present a wide variety of compounds made of different combinations of carbon, hydrogen and oxygen (and in some cases phosphorus). Some *fatty acids* (monomers) are used by the cells to build different types of polymers, such as the *phospholipids*, which are major constituents of the cell membrane. *Fats* are another kind of lipids, which store energy in the cells for long periods. Other types of lipidic biopolymers include the *steroids* (example: cholesterol) and the *terpenes* (example: natural rubber).

Peptides and proteins are made of amino acids. About 500 different *amino acids* are found in nature. These monomers are formed of carbon, hydrogen, oxygen and nitrogen, and in some cases sulphur. Out of the hundreds of amino acids, only 22 combine into the *peptide* chains (called *polypeptides*, which contain up to about 50 amino acids) that are the building blocks of *proteins* (biopolymers). The hundreds of other amino acids not directly incorporated into polypeptides play various roles in the functioning of cells, and some are secondarily involved in proteins. The two terms polypeptide and protein are often used interchangeably. Small proteins contain from about 50 up to a few hundred amino acid molecules, and the largest have more than 25,000 amino acids. The functions of many amino acids, including some of the most abundant in cells, are not known

Proteins play a multitude of key roles in organisms. For example, some proteins regulate the physiology and behaviour of organisms under the names of *hormones*; others contribute to the structure of the cells, for example collagen in skin and keratin in hairs; others speed up chemical reactions, such as digestion, and are then called *enzymes*; others destroy foreign substances or invaders, under the names of *antibodies*; and still others carry messages or materials inside or among cells, such as haemoglobin which carries oxygen in the blood of most vertebrates and in the tissues of some invertebrates.

Plants can synthesize all the amino acids they need, using bioavailable chemical resources in the natural environment. Contrary to the plants, the animals, including human beings, cannot synthesize themselves all the amino acids they need, and they must thus acquire the amino acids they cannot synthesize by eating plants or other animals. Once eaten, the proteins contained in the food are broken by the process of *digestion* into small peptides and amino acids, which are used by the animals to build their own proteins. The indispensable (or *essential*) amino acids the animals cannot synthesize, and must thus acquire in their food, vary among species.

More generally, the heterotrophs must break down not only the proteins, but also the carbohydrates and the lipids they acquire in their food to build their own biopolymers. The process includes the chemical breakdown of the food biopolymers into simpler molecules by the action of enzymes, and the assemblage of the resulting simple molecules into new, larger biopolymers. During the first step, carbohydrates are cut into simple sugars, lipids are split into free fatty acids and other compounds, and proteins are cut into small peptides and amino acids. The general role of digestion is illustrated in Fig. 5.1. More information on the composition and functioning of cells can be found in the general audience book of Kratz (2020).

6.1.4   Internal Controls: Feedback Inhibition

The huge number of chemical compounds that exist in the cells and the complexity of their synthesis raise the following question: By which mechanisms do cells ensure that there is always the right amounts—that is, neither too little nor too much—of the various compounds (bricks) required to synthesize the amounts of the compounds they need? This feat is achieved through the key cellular control mechanism of *feedback inhibition* (see Information Box 6.2 for the concept of *feedback*). This mechanism chemically inhibits the enzyme that favours the production of a particular substance when that substance accumulates to a certain level in the cell. For example, the synthesis of most amino acids is controlled by enzyme-based feedback inhibition, meaning that the cell stops making a given amino acid when its concentration becomes high enough to satisfy the cell's metabolic needs. This occurs because the end product of the reaction (here, the synthesis of an amino acid) chemically interferes with the enzyme that helped produce it. The synthesis of the amino acid restarts when there is not enough of it in the cell. This mechanism is a *feedback inhibition*.

Information Box 6.2 Feedbacks and Feedback Loops In a *feedback*, a causal process (**A**) is amplified (positive feedback) or mitigated (negative feedback) by the response of another process (**B**) following the direct or indirect action of **A** on **B**. A *feedback loop* is created when the feedback-modified process **A** further influences process **B** (Fig. 9.1).

An example of *natural positive feedback* is provided by the reflection of the solar radiation by large surfaces covered by snow and ice, which causes high albedo (see Information Box 2.9). Here we use as example the alternation between glacial and interglacial episodes during the Quaternary Ice Age (see Sect. 4.5.2). At the beginning of a glacial episode, the cooling climate favours the persistence of high-albedo surfaces in high-latitude areas in summer, which increases the reflection of solar radiation to space and thus leads to further global cooling. In contrast at the beginning of an interglacial episode, the rising temperature causes the shrinking of high-albedo surfaces in high-latitude areas, which decreases the global albedo and favours further global warming. In both cases, the change in albedo occurs in response to a change in temperature, which it amplifies, hence a positive feedback. The change in temperature resulting from the feedback enhances the change in albedo through a *positive feedback loop*, which causes a *runaway* cooling (glacial) or warming (interglacial) (see Sect. 9.1.3; Fig. 9.3b, c and accompanying text).

Another example of *positive feedback* is provided by the global greenhouse effect of water vapour in the atmosphere (see Information Box 2.7). When the climate warms up as a consequence of an increased greenhouse effect, there is stronger evaporation of oceanic and continental waters, and the higher content of water vapour in the atmosphere amplifies the global warming. In contrast when the greenhouse effect decreases, the lower temperature reduces the evaporation, and the lower content of water vapour in the atmosphere amplifies the global cooling. In both cases, the action of water vapour occurs in response a change in the greenhouse effect, which it amplifies, hence a positive feedback. The resulting *positive feedback loop* can contribute to a *continuing* increase or decrease in temperature (see Sect. 9.1.3; Fig. 9.3d and accompanying text).

An example of *negative feedback* is provided by the formation of clouds resulting from the production of aerosols of the chemical substance dimethylsulphide (DMS). Some groups of marine phytoplankton release dimethylsulphoniopropionate (DMSP) in the surrounding water as they grow, and part of this DMSP is transformed by the planktonic ecosystem into DMS gas, which escapes to the atmosphere. There, DMS is oxidized to various compounds, of which some can form aerosols that act as cloud condensation nuclei. This can develop into a *negative feedback loop*, where the formation of clouds increases the albedo of the atmosphere (see Information Box 2.9), which limits the penetration of solar light and thus slows down the development of phytoplankton. It was hypothesized that an increase in sea-surface irradiance could favour the development of DMPS-producing phytoplankton and thus the release of DMS by the ocean, which would cause a decrease in the penetration of solar radiation in the atmosphere and a corresponding decrease in phytoplankton development. This negative feedback loop could help *stabilize* the Earth's temperature by changing the cloud albedo (see Sect. 9.1.3; Fig. 9.4f and accompanying text).

All answers of natural systems are not feedbacks. For example, a mass extinction following the impact of a large meteoroid on Earth or long episodes of volcanic eruptions does not involve any feedback (see Sect. 7.4.6 and Information Box 1.6). In such a case, the extinction is due to a brutal change in the global environmental conditions, and there is no return effect, positive or negative, of the global ecosystem on the environmental conditions that caused the extinction, except perhaps in the very long term.

In living cells, enzyme-based feedback inhibitions control the synthesis of not only amino acids, but also several carbohydrates and lipids, and they can similarly control the synthesis of nucleotides and thus nucleic acids, described below (see Sect. 6.2.2). Enzymes, which are proteins, play a central role in controlling the amounts of most biochemical compounds produced by cells. The feedback inhibition mechanism explains how a tiny cell, which can be less than one thousandth of a millimetre in size, controls the simultaneous synthesis of thousands of highly complex molecules. The control of the chemical composition of the cell constituents is examined below (see Sect. 6.2.5).

6.2   Organisms: Software

6.2.1   Chemical Software of Organisms

Another key component of organisms is their "software", which is contained in the nucleic acids that encode the genetic information (Fig. 1.2). The next five sections explain how nucleic acids control the chemical composition of organic molecules, and how their interactions with the environment, through the hardware and the software of organisms, have driven the evolution and diversification of life forms on Earth since almost 4 billion years.

6.2.2   The Most Complex Organic Molecules: Nucleic Acids

The largest molecules that exist on Earth are the *nucleic acids.* These are formed of *nucleotides*, which combine atoms of carbon, hydrogen, oxygen nitrogen and phosphorus. There are five primary nucleotides, which are each composed

of three chemical units: a sugar, a phosphate group, and one of the five primary *N-containing bases* identified by letters A, G, C, T and U in Table 6.2. The
N-containing bases are flat-shaped organic molecules with a nitrogen atom. In
summary: a N-containing base together with a sugar and a phosphate group form a
nucleotide, and a long chain of nucleotides forms a nucleic acid molecule

There are two types of nucleic acid molecules. *Ribonucleic acid* (RNA) is
formed of nucleotides that each contains one ribose sugar and one of the four
bases A, G, C or U, and the resulting nucleic acid is generally single-stranded.
RNA strands may contain less than 200 nucleotides and up to several thousands,
and they may take a wide variety of shapes (twisted, folded, etc.) *Deoxyribonucleic
acid* (DNA) is formed of nucleotides that each contains one deoxyribose sugar and
one of the four bases A, G, C or T, and the resulting nucleic acid is generally double-stranded, forming the well-known three-dimensional double-helix structure
(Fig. 6.2). A DNA strand may contain up to several hundred million nucleotides,
and given that there are about 25 atoms per nucleotide, a single strand of nucleic
acid with several hundred millions of nucleotides contains more than one billion
atoms! This is a staggering number of atoms, especially remembering that these
highly complex molecules are synthesized by cells that can be very tiny.

In the spiral-helix DNA, the nucleotides are arranged into pairs linked by
their respective bases: base A is always paired with base T, and base C with base
G (Fig. 6.2). The pairs of bases connect the two strands of the helix together, as
rungs connect the two sides of a ladder. The five primary bases are the fundamental
units of the genetic code. Bases T and U, which are only found in DNA and RNA,
respectively, are identical except for the lack of a small chemical structure (a methyl
group) in U. Hence, Nature has been using only four letters—A, G, C, T and/or
U—to write the genetic epic of all organisms that existed on Earth since about 4
billion years. This small number of letters is somewhat similar to the two digits (0
and 1) used in computers. In contrast, the English language uses 26 letters (A to Z).

Some individual nucleotides also have, in addition to being part of nucleic acids,
other essential functions in cells. For example, the N-containing base adenine (A)
is a central component of the molecule *adenosine triphosphate* (ATP), which fuels
the chemical reactions in the cell by liberating its energy where and when needed
(see Information Box 3.3). The ATP molecule is the energy currency of all cells.
Other triphosphate compounds, which contain different nucleotides than adenosine,
are also involved in energy storage and transport. Another group of compounds that
contain nucleotides participates in cell signalling (communication among cell components). Nucleotides thus play key roles in the functioning of cells.

6.2.3   The Code of the Genetic Software

The cells elaborate organic matter by combining individual atoms of chemicals
elements (Table 6.1) into monomers, which are progressively assembled into
increasingly complex biomolecules, some of which contain more than one billion atoms, as explained in the previous section. How do living cells control the

Fig. 6.2 The DNA molecule is made up of two strands of nucleotides organized into a double helix. The two strands are held together by base pairs belonging to the nucleotides that form the two strands: base G always pairs with base C and base A with base T. Credits at the end of the chapter

incredible variety and complexity of the millions of organic compounds that interact in an organism? How do cells pass the information from generation to generation in such a way that, despite the similarity of humans and alfalfa in term of chemical elements (Table 6.1), human beings do not give birth to strands of alfalfa? Trying to answer these complex questions from the bottom up, that is, starting from single-atom chemical elements and progressively building up highly complex billion-atoms biomolecules, would generate more unknowns than answers. Fortunately, there is an unexpectedly simple answer to these questions, which is found by looking from the most complex biomolecules, at the top of the biochemical world, down towards the simpler molecules. Doing so is possible because the chemical composition of the molecules that make up the living cells is controlled from the top down.

It was explained above (see Sect. 6.1.3) that proteins largely control the functioning of living cells. The chemical agents through which proteins act within and among the cells are their constitutive amino acids, whose particular sequence within a given protein determines if the protein functions as a hormone, a structural agent, an enzyme, an antibody, or other. The questions are then: Firstly, which mechanism determines the exact sequence of amino acids within a protein, remembering that some proteins contain more than 25,000 molecules of amino acids? Secondly, which mechanism ensures that this exact sequence is repeated over and over again when copies of the protein are made in a given cell, and from parent cell to daughter cells during thousands and millions of cell generations?

Computer scientists and linguists know that the general answer to questions concerning the transmission of information is the existence of a code, that is, a system of symbols and rules that specify how the symbols are organized to faithfully carry the information. The number of symbols in a code is not very important. In computer science for example, the code uses only the two symbols 0 and 1 (called *bits*, meaning "binary digits"), and two different *sequences* of the same bits have different meanings. For instance, the two four-bit numbers 0110 and 1001 mean 6 and 9, respectively, in the decimal system. Taking language as a second example, English has 26 different symbols (letters), but the meaning of words made up of the same letters depends on the *sequence* of letters. For example the five words POST, POTS, SPOT, STOP and TOPS are composed of the same four letters (O, P, S, and T), but their meanings are totally different depending on the order of the letters. These two simple examples illustrate the importance, in codes, of rules that specify both the *lengths of the sequences* of symbols and the *order* of symbols in the sequences.

For the synthesis of proteins, the (genetic) code uses the four symbols that correspond to the N-containing bases listed Table 6.2, which are A, C, G, and T or U for DNA or RNA, respectively. In nucleic acids, the four bases are grouped into triplets, within which the ordering of the three bases provides the code for synthesizing a specific amino acid. Compared to a written language, the triplets of four bases are the "words" of heredity, and the sequences of triplets are the "sentences". For example as shown in Table 6.3 for RNA, the triplets AGC and GCA code for the amino acids serine and alanine, respectively. In addition, different triplets can code for the same amino acid; for example, AAA and AAG both code for lysine. Finally, the synthesis of a given protein is stopped by any of the three triplets UAA, UAG or UGA, which would correspond to a full stop in an English sentence. Conversely, it is methionine that generally signals the start of protein translation from mRNA (see Sect. 6.2.3, below). Nature uses not only letters and words, but also punctuation marks!

Mathematically, triplets of 4 bases could code for $4^3 = 64$ different amino acids. In practice, however, Table 6.3 shows that some amino acids are coded by a single triplet (for example tryptophan) whereas others are coded by any of several triplets (up to six in the cases of arginine, leucine and serine). As a result, the 64 possible triplets made of three N-containing bases only code for the 20 amino acids listed in Table 6.3. These 20 amino acids make up all the polypeptides that are the building blocks of the millions of different proteins. In the text on peptides and proteins, above, it was indicated that the polypeptides that make up proteins could contain up to 22 different amino acids (see Sect. 6.1.3). The difference between the 22 amino acids cited there and the 20 in Table 6.3 comes from the fact that two amino acids (*selenocysteine* and *pyrrolysine*, which are not listed in Table 6.3) are incorporated into proteins by other mechanisms than those described here.

6.2.4   Genetic Control of the Constituents of Organic Matter

How is the code for the synthesis of proteins actually implemented in living cells? The answer lies in the nucleic acids, DNA and RNA, which control the synthesis of proteins using the codes listed in Table 6.3.

Table 6.3 The genetic code of RNA for 20 of the 22 amino acids that make up proteins. Letters A, C, G and U represent the N-containing bases in the nucleotides (Table 6.2). In DNA, U is replaced by T

Amino acid	RNA triplet codes	Amino acid	RNA triplet codes	Amino acid	RNA triplet codes
Alanine	GCA, GCC, GCG, GCU	Glycine	GGA, GGC, GGG, GGU	Proline	CCA, CCC, CCG, CCU
Arginine	AGA, AGG, CGA, CGC, CGG, CGU	Histidine	CAC, CAU	Serine	AGC, AGU, UCA, UCC, UCG, UCU
Asparagine	AAC, AAU	Isoleucine	AUC, AUU	Threonine	ACA, ACC, ACG, ACU
Aspartic acid	GAC, GAU	Leucine	CUA, CAC, CUG, CUU, UUA, UUG	Tryptophan	UGG
Cysteine	UGC, UGU	Lysine	AAA, AAG	Trosine	UAC, UAU
Glutamic acid	GAA, GAG	Methionine (Start)	AUA, AUG	Valine	GUA, GUC, GUG, GUU
Glutamine	CAA, CAG	Phenylalanine	UUC, UUU	(Stop)	UAA, UAG, UGA

The genetic information of each organism is stored in its DNA molecules in the form of very long sequences of the 4 nucleotides, which contain the bases A, C, G and T (Table 6.2). Let us compare the storage capacity of a computer and of amino acids, using as example a very short sequence of 10 symbols. A sequence of 10 bits (0 or 1) in a computer can store $2^{10} = 1024$ different combinations of the two bits, whereas a sequence of 10 nucleotides (with bases A, C, G or T) on one of the two strands of a DNA molecule can store 4^{10} different combinations of the four bases, which is more than 1 million. This value of more than 1 million possible combinations of the four bases is for a short sequence of only 10 nucleotides, whereas real DNA molecules can contain up to several hundred million nucleotides. It follows that a single DNA molecule can store tens of billions of different combinations of nucleotides, where each nucleotide contains one of the 4 bases. This incredibly large number explains the almost infinite diversity of organisms that exists not only among species but also among the individuals within each species. The role of RNA is explained a few paragraphs below.

It is known since more than one century that some biological characteristics of an offspring come from the parent(s) through units of heredity called *genes*. It was later discovered that the genes were located on *chromosomes*, which are structures found in cells, more precisely in the cell nucleus in the case of eukaryotes. A chromosome consists chiefly of a DNA molecule and some proteins, and a gene is the entire DNA sequence necessary for the synthesis of a given functional polypeptide or protein (see Sect. 6.1.3). The qualifier "functional" means that a gene includes more than the nucleotides that encodes the amino acid sequence required to synthesize a given polypeptide or protein, because to be functional the latter requires additional sequences located on the same DNA molecule.

Prokaryotes (bacteria and archaea) do not have a membrane-bound nucleus. They generally have a single chromosome, which is a circular double-stranded molecule of DNA, and the region of the cell occupied by this chromosome is called the *nucleoid* (meaning "like a nucleus") In some bacteria, the chromosome is linear instead of circular. For simplicity, we only describe below the processes for the eukaryotic cells, which have a nucleus.

The *first step* in the flow of genetic information within a biological system is the *replication* of its genetic material. Replication ensures that, when the cell of an organism divides, the two daughter cells each have an identical copy of the DNA of the parent cell. This is possible because base G in the DNA always pairs with base C, and base A always pairs with base T, hence each of the two strands of the parent DNA molecule makes an exact copy of the other strand when it replicates. DNA replication takes place inside the nucleus of the cell. There, the enzyme *DNA polymerase* synthesizes the new strands by adding nucleotides (present in the cell nucleus) that complement each of the two strands of the replicating DNA molecule (Fig. 6.3).

The *second step* is called *transcription*, and also occurs inside the nucleus. It prepares the transfer of genetic information from DNA, which is located inside the nucleus, towards organelles called *ribosomes* where the synthesis of proteins takes place. The ribosomes are located in the cell cytoplasm outside the nucleus. The genetic information required to synthesize a protein is encoded in a gene, which is a specific sequence of nucleotides present in the double-strand DNA. During transcription, the genetic information on one of the two strands of the DNA, called *template strand*, is copied on a single-strand *messenger RNA* (mRNA), through the progressive addition of nucleotides (present in the cell nucleus) to the growing mRNA by the enzyme *RNA polymerase* (Fig. 6.4). Because the DNA bases C, A, T, G pair with the mRNA bases G, U, A, C, respectively, the sequence of bases in the mRNA is identical to the sequence on the other strand of the DNA, called *coding strand* (base T in the DNA is replaced by base U in the mRNA). Once the gene is transcribed on the mRNA strand, the latter detaches from the DNA and moves out of the nucleus towards a ribosome.

Transcription can make many copies of the same information on several mRNA strands, each of which is used to make a molecule of a given protein. Hence, transcription can result in the production of many copies of a protein. This is different from replication (first step, above), which makes a single copy of the genetic information contained in a DNA molecule.

The *third step*, called *translation* occurs in a ribosome, where the information carried by the mRNA strand is used to assemble a protein (Fig. 6.5). The latter being a chain of amino acids, building a protein consists in assembling amino acids using the information in the mRNA. Indeed, each group of three N-bases in the mRNA strand codes for one specific amino acid (Table 6.3). Following the instructions encoded in the succession of base triplets on the mRNA strand, molecules of *transfer RNA* carry to the ribosome various amino acids present in the cytoplasm, whose bases complement those on the mRNA strand in the ribosome.

Fig. 6.3 Replication of a parent DNA molecule produces two identical daughter molecules. The enzyme DNA polymerase adds free nucleotides, which are present in the cell nucleus, to complement each of the two strands of the parent DNA molecule. Bases A, G, C, T pair with bases T, C, G, A, respectively. Credits at the end of the chapter

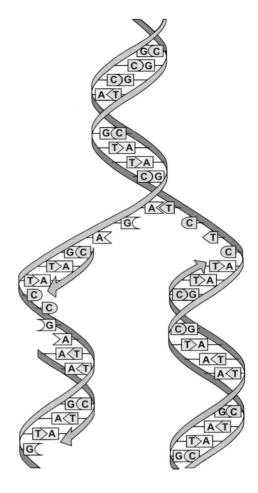

The assembling of the protein chain is initiated when the "start" triplet is identified on the mRNA, and is terminated when any of the three "stop" triplets is encountered (the start and stop triplets are identified in Table 6.3). Once a stop triplet is reached, the ribosome releases the protein in the cytoplasm, where it starts to play its intended role in the functioning of the cell. This way, the nucleic acids, and thus the genes, control the chemical composition of proteins.

Two key components of organisms are their software and hardware (see Information Box 1.1). The first two steps in the flow of genetic information within a biological system, which occur within the cell nucleus, correspond to the "software" of organisms. The third step produces proteins, which are part of the chemical "hardware" of organisms and whose assemblage takes place in ribosomes, outside the nucleus. The three steps make up the *central dogma of molecular biology* (Fig. 6.5), whose consequences are further discussed in the next section.

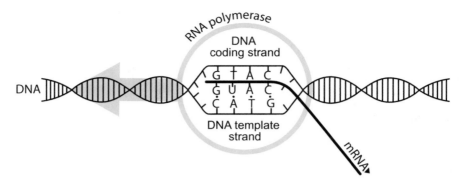

Fig. 6.4 During transcription, the enzyme RNA polymerase pairs, on the growing mRNA strand, free nucleotides present in the cell nucleus with the nucleotides on the template strand of the DNA molecule. Bases C, A, T, G on the DNA template strand pair with bases G, U, A, C on the mRNA strand, respectively. Credits at the end of the chapter

Fig. 6.5 Central dogma of molecular biology: genetic information flows from DNA to RNA (mRNA) and proteins. RNA is the link between DNA and proteins. Three processes are involved: replication and transcription (which occur in the nucleus and involve the enzymes DNA and RNA polymerase, respectively), and translation (which occurs in ribosomes). Credits at the end of the chapter

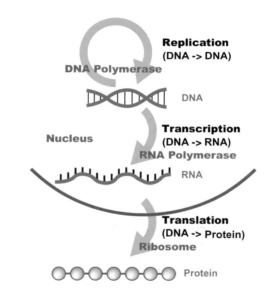

6.2.5   Roles of the Hardware and the Software in Biological Evolution

In theory, the series of processes that link DNA replication inside the cell nucleus to protein translation in a ribosome is seamless (Fig. 6.5). However, "copy errors" do occur, especially during the replication process. These mistakes are called *mutations*, and some may modify a single amino acid in a protein whereas others may affect whole proteins, with different consequences. Indeed, some of the mutations do not influence the body functions of the organism that carry them, and thus go undetected, in which case these mutations are kept in the genetic patrimony of the species and passed to descendants. Other mutations are lethal to the individuals carrying them, and these individuals are removed from the population without reproducing, with the result that the lethal mutations they carry are eliminated and cannot accumulate in the species' genetic patrimony. Still other mutations that are harmful without being lethal can be passed on to descendants, such as hereditary diseases. Other, rare mutations may be beneficial to offspring, and these mutations are then incorporated in the species' genetic patrimony, which is the starting point of biological evolution.

It was explained above how enzyme-based feedback inhibition controls the synthesis of amino acids, several carbohydrates, lipids and nucleotides, which are constituents of nucleic acids (see Sect. 6.1.4). In addition, it was explained in the above paragraphs that DNA controls the chemical composition of proteins through the agency of RNA. Similarly, DNA controls the chemical composition of lipids and carbohydrates through the agency of enzymes, which are proteins whose chemical composition is itself controlled by DNA. It follows that the cells and organisms function as systems, more specifically open systems. A *system* is an entity made of interrelated and interdependent parts that is more than the sum of its parts (see Information Box 3.3). This is indeed the case of cells and organisms, whose biochemical constituents control each other through complex interactions. An *open system* exchanges matter and/or energy with its surroundings, and these exchanges keep organisms alive (see Information Box 1.1). These characteristics of the cells and organisms allow them to maintain a high level of organization among their constituents as long as these stay alive.

The chemical hardware of organisms, described in the above sections, ensures the metabolic functioning of the cells and organisms. The biological hardware also has a structural aspect, which concerns the body, behaviour and reproduction characteristics of unicellular and multicellular organisms. These characteristics include the types of motion (toward the food, away from predators, etc.), the modes of food acquisition (grazing, predation, commensalism, parasitism, etc.), the means of self-protection (transparency, shells, mimicry, camouflage, etc.), and the reproductive strategies (asexual or sexual reproduction; dispersal of large numbers of gametes or offspring, or parental attention to a small number of offspring with feeding and education).

The cells of multicellular organisms are organized in various components (tissues and organs) and structures (body systems) (see Information Box 1.1). Most multicellular organisms have different types of cells, which each have a specific structure and function. Ensembles of similar cells that carry out the same function are called a *tissue*. Examples of tissues in humans are the nervous tissue, which contains neurones and their surrounding neuroglia, and the blood, which contains the red blood cells and a number of other components. In complex multicellular organisms, several tissues form functional units called *organs*. Examples of organs in humans are the brain and the heart. The combination of two or more organs that execute a given function is called a *body system*. Examples of body systems in humans are the nervous system and the circulatory system.

Both the chemical and structural hardware are transient states between the birth, reproduction and death of cells or organisms. Survival and reproduction are continuously threatened by challenges arising from the environment, other species and even congeners, which are agents of natural selection acting on the hardware of individuals. It is also the hardware that conveys, or not, to the next generation the genetic information coded in the DNA (software), which includes the mutations not eliminated by natural selection. Hence, the software is subject to random mutations and controls the composition and functioning of the hardware, and the hardware largely determines the long-term evolution of the software through the mechanism of natural selection. Further reading on the genetic material and the genetic code is provided by the general audience book of Robinson and Spock (2020), and the more theoretical book of Zhegunov (2012).

6.3 The Chemical Bricks of Organic Matter in the Outer Layers of Planet Earth

The establishment of organisms on Earth, their progressive expansion to most of the environments on the planet, and the regulation they have exerted on Earth's biogeochemical cycles and climate for billions of years were possible because the bricks (chemical elements and compounds) of organic matter were readily available in the outer layers of the planet (crust, ocean, and atmosphere). Chemical elements present in all organisms are called *biogenic*, and the six *major biogenic elements* are (in alphabetical order): carbon, hydrogen, nitrogen, oxygen, phosphorus and sulphur (Table 6.2). These chemical elements are present in relatively large quantities in organisms, hence the adjectives *major* and *biogenic* used to qualify them (group A in Fig. 6.6).

Because of their chemical and physical properties, the major biogenic elements offer an incredibly large number of possibilities of combinations into highly complex organic compounds. The existence of these amazingly numerous compounds allowed evolution to create the past and present highly diverse organisms, and will undoubtedly continue to produce new types of organisms in the future. Briefly, the Earth's ecosystems are based on: a wide array of organic molecules, the interactions among these molecules within cells and organisms, and the interactions of cells and organisms with their environment. It is through these interactions that organisms acquire the materials and energy they need to sustain their metabolism and ensure their reproduction (see Information Box 1.1).

In the natural environment, the chemical elements with the lightest atomic nuclei are chemically *volatile* (meaning that they can easily evaporate) below temperatures of less than a few hundreds of degrees (Celsius) or even a few tens of degrees (Table 6.4). Some elements are volatile in the their elementary or molecular states

Periodic Table of the Elements

1 H		A	B	C	D												2 He
3 Li	4 Be											5 B	6 C	7 N	8 O	9 F	10 Ne
11 Na	12 Mg											13 Al	14 Si	15 P	16 S	17 Cl	18 Ar
19 K	20 Ca	21 Sc	22 Ti	23 V	24 Cr	25 Mn	26 Fe	27 Co	28 Ni	29 Cu	30 Zn	31 Ga	32 Ge	33 As	34 Se	35 Br	36 Kr
37 Rb	38 Sr	39 Y	40 Zr	41 Nb	42 Mo	43 Tc	44 Ru	45 Rh	46 Pd	47 Ag	48 Cd	49 In	50 Sn	51 Sb	52 Te	53 I	54 Xe
55 Cs	56 Ba	a	72 Hf	73 Ta	74 W	75 Re	76 Os	77 Ir	78 Pt	79 Au	80 Hg	81 Tl	82 Pb	83 Bi	84 Po	85 At	86 Rn
87 Fr	88 Ra	b	104 Rf	105 Db	106 Sg	107 Bh	108 Hs	109 Mt	110 Ds	111 Rg	112 Cn	113 Nh	114 Fl	115 Mc	116 Lv	117 Ts	118 Og

1 — atomic number
H — symbol

a Lanthanide series	57 La	58 Ca	59 Pr	60 Nd	61 Pm	62 Sm	63 Eu	64 Gd	65 Tb	66 Dy	67 Ho	68 Er	69 Tm	70 Yb	71 Lu
b Actinide series	89 Ac	90 Th	91 Pa	92 U	93 Np	94 Pu	95 Am	96 Cm	97 Bk	98 Cf	99 Es	100 Fm	101 Md	102 No	104 Lr

Fig. 6.6 The main chemical elements used by cells and organisms are represented in the periodic table of elements: (A) six major biogenic elements (see Sect. 6.3), (B) light-mass alkali (see Sect. 6.4.1), (C) two other important biogenic elements (see Sect. 6.5), and (D) essential trace elements (see Sect. 6.5). Credits at the end of the chapter

(for example H, H_2, O, O_2, O_3, N, N_2), and some of their molecular compounds (for example CO_2, CH_4, H_2O, NO_x, SO_x, H_2S) are mostly found as gas on Earth, except H_2O that can occur as solid, liquid or gas in the Earth's environment. Additionally, these compounds have various degrees of *chemical stability* in the environmental conditions of Earth. A substance is said to be chemically stable if it does not react easily with other substances in given environmental conditions.

During the formation of the early Earth, the light chemical elements rose to the surface of the fluid planet (future crust and atmosphere), and the heavier elements sank to the centre under the effect of gravity (see Sect. 6.5.1). This phenomenon is known as *planetary differentiation*. In the outer envelopes of the planet, where the organisms later developed, the chemical elements differentiated in successive steps followed by the formation of chemical compounds. These steps are described in the following paragraphs for the atmosphere and the ocean.

The Sun and the planets were formed from an interstellar cloud of gases, called *solar nebula*, where the major components were the chemical elements hydrogen, helium and lithium, which have 1, 2 and 3 protons, respectively, corresponding to the atomic numbers 1–3 (Fig. 6.6). The reason for this composition is that the nuclei of the atoms of hydrogen and helium, which are the two simplest and lightest chemical elements, were formed early in the history of the Universe (primordial nucleosynthesis, Table 1.2) shortly after the Big Bang, which itself occurred 13.7 billion years ago (see Sect. 1.4.1). The nuclei of the atoms of lithium, the next lightest chemical element after hydrogen and helium, were also created in part during the primordial nucleosynthesis. The solar nebula also contained small amounts of most other chemical elements, including the heavy, radioactive chemical elements created in the stars (see Sect. 1.4.2). It follows that the early atmosphere of Earth and the other planets of the Solar System was predominantly composed of hydrogen and helium. This lasted until the mass of the Sun, which progressively contracted under the force of gravity, became sufficiently concentrated to cause hydrogen atom nuclei to fuse into helium atom nuclei, and this giant nuclear fusion reaction liberated enormous amounts of energy.

When the nuclear fusion reaction of the Sun became self-sustaining, the resulting stream of particles emanating from the Sun, called *solar wind* (see Information Box 2.6), likely swept away the components of the early planetary atmospheres, which were blown to the outer areas of the Solar System. There, they contributed to the formation of the giant, gaseous planets (Jupiter, Saturn, Uranus and Neptune), and smaller bodies that orbit the Sun in the asteroid belt (located between Mars and Jupiter), the Kuiper belt and the Oort cloud (see Sect. 1.2.2). Among the components of the early planetary atmospheres were the hydrogen and oxygen atoms that formed the water of the comets, which may have brought water to Earth from the outer areas of the Solar System later (see Sect. 1.2.2).

After the giant collision that created the Earth-Moon system about 4.5 billion years ago (see Sect. 1.5.1), gaseous compounds made of the light biogenic elements cited above (Table 6.4) escaped from the mass of magma, and current knowledge suggests that the *first phase of the atmosphere* was predominantly composed of hydrogen (H), helium (He), and hydrogen-containing gases such as ammonia (NH_3), methane (CH_4) and water vapour (H_2O). Because Earth was very hot,

Table 6.4 Molar mass and temperature of volatilization (boiling point) of some major biogenic elements and compounds

Element or molecule	Molar mass (g mol^{-1})	Boiling point (°C)	Chemical compound	Molar mass (g mol^{-1})	Boiling point (°C)
H	1.01	Gas only	CO_2 (carbon dioxide)	44.01	−78.5
H_2	2.02	−253	CH_4 (methane)	16.04	−161.5
N	14.01	Gas only	H_2O (water)	18.01	+100
N_2	28.02	−196	NO_x (nitrogen oxide)	NO: 30.01 NO$_2$: 46.01	−152 +21.2
O	16.00	Gas only	SO_x (sulphur oxide)	SO$_2$: 64.06 SO$_3$: 80.06	−10.0 +44.6
O_2	32.00	−183	H_2S (hydrogen sulphide)	34.08	−60.7
O_3	48.00	−112			

atmospheric water was in the form of vapour, and the loss of water by atmospheric escape of hydrogen was probably very high (see Sects. 5.4.2 and 8.1.1). As a result, the initially degassed water was partly or even completely lost (see Sect. 5.3.2).

The proportion of water vapour in the *second phase of the atmosphere* before the formation of the primitive ocean is not known, and depends on the relative importance of the different losses and sources of water at the time. On the one hand, if all or most of the water that had degassed from the hot Earth had been lost from the atmosphere due to escape of hydrogen, it is possible that Earth passed through a phase when the atmosphere was dry and mostly composed of CO_2. This would have been the case until a new stock of water was supplied by the bombardment of comets and/or meteoroids, especially during the Late Heavy Bombardment about 4.1–3.9 billion years ago (see Sect. 5.3.2). On the other hand, some geochemical indices argue for the presence of a very early liquid-water ocean 4.4 billion years ago, in which case at least part of the degassing water would have been retained, and this water would have formed the primitive ocean by condensation (see Sect. 5.3.2). In both cases, the high concentration of CO_2 in the atmosphere before the formation of the ocean, during at least the first 100–150 million years of Earth's existence, caused a considerable greenhouse effect (see Information Box 2.7) and thus temperatures at the Earth's surface of the order of several hundreds of degrees (Celsius).

Despite the very high atmospheric temperature, the cooling of the Earth's surface and the formation of a solid crust together with the considerable atmospheric pressure allowed water vapour to condense and thus form a liquid-water ocean (see Sect. 5.5.4). In the meantime, the biogenic elements cited above (Table 6.4) had combined into a variety of chemical compounds, for example CO_2. The early formation of the liquid-water ocean was the key determinant leading to the creation of the wide array of chemical compounds that organisms would use to firmly establish themselves on the planet. The abundance of liquid water favoured the formation of a wide array of carbon-based molecules, which contributed to establish carbon as the key chemical element in the Earth's organisms (see Sect. 6.1.2).

The suite of chemical reactions in the early ocean could have been as follows. The atmospheric CO_2 gas partly dissolved in water, where it interacted with water molecules to give the three dissolved forms of inorganic CO_2 in water, which are the CO_2 gas and the bicarbonate and carbonate ions (HCO_3^- and CO_3^{2-}, respectively) (see Information Box 6.3). Similarly, the atmospheric gases sulphur dioxide (SO_2) and nitrogen dioxide (NO_2) partly dissolved in water leading to dissolved SO_2 gas and the hydrogen sulphite, sulphate and nitrate ions (HSO_3^-, SO_4^{2-} and NO_3^-, respectively). Chemical interactions between the carbon and nitrogen compounds present in the water led to the formation of ammonia in its simple and hydrogenated forms, NH_3 (ammonia) and NH_4^+ (ammonium), respectively. In other words, the chemical reactions resulting from the exchanges between biogenic compounds in the atmosphere and in the water of the primitive ocean led to the formation of a wide variety of dissolved compounds, on which the organisms could later develop. The complexity of the ocean's solution of chemical compounds progressively increased through continual exchanges between the chemical compounds in the primitive ocean and those in both the atmosphere above the ocean

Information Box 6.3 Carbonate Equilibrium in Natural Waters When the atmospheric carbon dioxide gas, $CO_{2(gas)}$, dissolves in water, $CO_{2(aqueous)}$, it combines with water molecules (H_2O) to make carbonic acid (H_2CO_3):

$$CO_{2(gas)} \rightleftharpoons CO_{2(aqueous)} \tag{6.1}$$

$$CO_{2(aqueous)} + H_2O \rightleftharpoons H_2CO_3 \tag{6.2}$$

The H_2CO_3 molecule dissociates to form a negative bicarbonate ion (HCO_3^-) and a positive hydrogen ion (H^+):

$$H_2CO_3 \rightleftharpoons HCO_3^- + H^+ \tag{6.3}$$

Because H_2CO_3 dissociates in less than one second, it does not exist in any sizeable amount natural water and is thus not considered in this text. Some of the HCO_3^- ions dissociate to form negative carbonate ions (CO_3^{2-}) and positive hydrogen ions (H^+):

$$HCO_3^- \rightleftharpoons CO_3^{2-} + H^+ \tag{6.4}$$

The four dissolved forms of dissolved CO_2—$CO_{2(aqueous)}$, H_2CO_3, HCO_3^-, and CO_3^{2-}—make up the *carbonate system*. In natural waters, H_2CO_3 is generally not considered as explained just above.

The double arrows in above equations indicate that these chemical transformations are all reversible. The directions of the arrows are controlled by equilibrium constants that depend on environmental conditions such as the pH (see Information Box 6.4), temperature and hydrostatic pressure of the solution. Hence according to the conditions, atmospheric $CO_{2(gas)}$ progressively dissolves as $CO_{2(aqueous)}$ in the ocean, where it is initially stored as dissolved HCO_3^- and CO_3^{2-}, or on the contrary, the HCO_3^- and CO_3^{2-} dissolved in the ocean are transferred back to $CO_{2(aqueous)}$, which is degassed to the atmosphere as $CO_{2(gas)}$. Under the average conditions of the current surface ocean (temperature between 0 and 30 °C, pressure of 1 atm, pH of about 8.2), the dissolved inorganic carbon in seawater is mostly in the form of HCO_3^- (Fig. 9.7).

Equations 6.3 and 6.4 show that dissolution of $CO_{2(gas)}$ in water and the reverse processes change the concentration of hydrogen ions (H^+) in water, and thus modify its pH (see Information Box 6.4). Hence the dissolution or degassing of CO_2 both influences the pH of water and are influenced by it (previous paragraph). The ongoing anthropogenic increase in atmospheric CO_2 is accompanied by a CO_2 increase in seawater, which shifts the reaction to the right in Eqs. 6.1–6.4 and thus releases H^+ into seawater, which in turn causes *ocean acidification* (see Sects. 11.1.1 and 11.1.3).

The carbon dissolved in natural waters can form solid carbonates through the combination of dissolved bicarbonate negative ions (HCO_3^-) with positive ions of mostly calcium (Ca^{2+}) and magnesium (Mg^{2+}), which are ultimately transferred to sediments as solid calcium carbonate ($CaCO_3$) and magnesium carbonate ($MgCO_3$) (see Information Box 5.11). Solid carbonates can dissolve back to CO_3^{2-} and Ca^{2+} or Mg^{2+} when the ions become undersaturated in water. These two mechanisms can lead to large exchanges between the dissolved and the solid reservoirs of inorganic carbon on Earth. The book of Middelburg (2019) provides further reading on the chemistry of CO_2 in the ocean.

(by dissolution and evaporation of such compounds as CO_2, N_2, NO_x, and SO_x), and the solid seafloor below (which involved carbonate, phosphate, sulphate or sulphide, and nitrate ions, that is, CO_3^{2-}, PO_4^{3-}, SO_4^{2-} or S^{2-}, and NO_3^-, respectively).

6.4   Chemical Composition of Seawater, Biomineralization, and Calcification

6.4.1   *Electrical Charges of Ions in the Ocean and the Organisms*

The major biogenic elements H, O, C, N and S, identified in the previous section, were present in the primitive ocean in the form of dissolved ions with negative electrical charges (HCO_3^-, CO_3^{2-}, HSO_3^-, SO_4^{2-}, NO_3^-). The overall electrical neutrality of the ocean was maintained by the existence in the ocean's solution of an equivalent amount of positively charged ions. The positive charges were partly in the form of hydrogen ions (H^+), which resulted from the dissociation of water molecules, H_2O, into H^+ and OH^- ions (see Information Box 6.4). The positively charged ions also included major elements present in the solid outer envelope of Earth—rocks and sediments—such as sodium (Na^+), potassium (K^+), calcium (Ca^{2+}), magnesium (Mg^{2+}) and boron (B^{3+}). These light-mass *akali* had dissolved in seawater when the primitive ocean was formed, together with such negative ions as chlorine (Cl^-) and fluorine (F^-).

Information Box 6.4 pH and Alkalinity In pure water, the H_2O molecules are partly dissociated into H^+ and OH^- ions, with an equal number of ions of each type. The balance between the concentrations of H^+ and OH^- ions in a solution is quantified as the *power of hydrogen* or *potential of hydrogen* of the solution (abbreviated to *pH*):

$$pH = \frac{1}{\log[H^+]} = -\log[H^+] \tag{6.5}$$

where $[H^+]$ is the concentration of H^+ ions. The pH is defined in such a way that its value is equal to 7.0 (pH = 7.0) when the concentrations of H^+ and OH^- ions in the solution are equal. The pH of water solutions with a concentration of H^+ ions greater than that of OH^- ions is lower than 7.0 (pH < 7.0), and the solution is said to be *acidic*. Conversely, the pH of water solutions with a concentration of H^+ ions smaller than that of OH^- ions is higher than 7.0 (pH > 7.0), and the solution is said to be *basic*.

The average pH of seawater has been about 8.2 since a long time. At a pH above 7.0, the H^+ ions are less concentrated than the OH^- ions, and the negative charges of the excess OH^- ions are compensated by the positive charges of dissolved ions such as Na^+, K^+, Ca^{2+}, Mg^{2+} and B^{3+}. Because these metals (except boron, B^{3+}) belong to the chemical groups of either *alkali* or *alkaline earths* (first and second columns from the left in the periodic table of elements; group B in Fig. 6.6), the resulting solution is called *alkaline*. The current ocean having a pH higher than 7.0 is thus slightly alkaline.

A related concept is the *alkalinity* of a solution. It is defined as the excess of *bases* (substances that accept H^+) over *acids* (substances that donate H^+) in the solution (see Information Box 1.5). One example is provided by Eq. 9.4, where the dissolution of solid calcium carbonate ($CaCO_3$) in seawater produces the negatively charged HCO_3^- and/or CO_3^{2-} ions that can interact with H^+. The additional HCO_3^- and CO_3^{2-} ions increase the alkalinity of seawater.

The early "mineral soup" was the chemical reservoir from which the first organic molecules were assembled. Together with the organic compounds possibly carried from space by comets and meteoroids (see Sect. 1.2.2), the primitive organic molecules evolved into the more elaborate biochemical components of organisms described in previous sections. The main constituents of modern organisms are chemical elements abundant in the early ocean.

The formation and dissolution of solid calcium carbonate ($CaCO_3$) (see Information Boxes 5.11 and 6.3) are controlled by the pH of seawater, namely acidic conditions favour the dissolution of solid $CaCO_3$, and basic conditions favour its formation. In return, the dissolution of $CaCO_3$ increases the pH of seawater, and its formation decreases it. The reciprocal control of chemical reactions by the pH and of the pH by chemical reactions minimizes changes in the ocean's pH, a property that chemists call *buffer*. This *negative feedback loop* (see Information Box 6.2) has maintained the current pH of seawater close to a value of about 8.2, but the ongoing release of CO_2 to the atmosphere by human activities is causing *ocean acidification* (see Sects. 11.1.1 and 11.1.3).

Bacteria and archaea are able to live under conditions either very acidic (pH of 4, and even 0) or very alkaline (pH higher than 9). For example, cyanobacteria can occupy hyper-acidic environments such as the acid lakes of volcanoes, where pH = 0. Conversely, some bacteria live in highly alkaline environments such as soda lakes (high concentrations of typically sodium carbonate, Na_2CO_3; pH between 9 and 12) or the effluents of cement factories (pH higher than 9). The organisms maintain their internal pH different from that of the external medium thanks to a mechanism that involves exchanges of ions through the cell membrane (Fig. 6.7).

The average salinity of seawater is around 35 (see Information Box 5.9), and it is likely that it was so since comets and/or meteoroids completed their supply of water to Earth about 4.1–3.9 billion years ago (see Sect. 5.3.2). This salinity corresponds to about 500 mmol of both Na^+ and Cl^- per litre of seawater (mmol L^{-1}; or millimolar, mM). Cells belonging to a wide range of organisms (bacteria Escherichia *coli*, yeast Saccharomyces *cerevisiae*, and mammals) maintain their internal concentrations of Na^+ and Cl^- at values of 10–30 and 10–200 mM, respectively, which are lower than the concentrations of Na^+ and Cl^- in seawater. In order to maintain their internal pH close to neutrality (pH = 7), the cells exchange Na^+ and Cl^- ions with the surrounding medium, whose pH is generally higher than theirs. When the internal pH of a cell is too low compared to that of

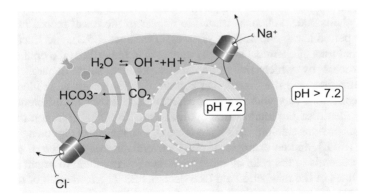

Fig. 6.7 Schematic representation of the ion exchanges through the cell membrane that regulate the internal pH and salinity of the cell. Credits at the end of the chapter

the external medium, the cell expels H^+ and HCO_3^- ions from its internal medium, and the loss of these ions is compensated by the entry of Na^+ and Cl^- ions to maintain the *osmotic balance* between the inside of the cell and the ambient water; and when the internal pH is too high compared to that of the external medium, the reverse exchange of ions takes place (Fig. 6.7). This mechanism, which appeared during the course of biological evolution billions of years ago, continues to be used by organisms.

Terrestrial organisms, including humans, evolved from aquatic organisms, and genetically inherited the cell physiology of their marine ancestors. For example, the regulation of the internal salinity and pH of the cells and blood of humans (salinity and pH of about 9 and 7.3, respectively) is done by the same exchanges of ions through the cell membrane as those found in marine organisms. In multicellular organisms, complementary mechanisms contribute to the regulation of the internal salinity, such as the excretion of urea in urine, or through the skin by sharks.

The presence of light-mass alkali dissolved in seawater led to the evolution of the major biological innovations of mineral tests, shells and skeletons in various groups of organisms. These hard structures result from the fact that some chemical elements—such as calcium, magnesium, silicon and phosphorus—are in excess in the cytoplasm of the cells, and must be excreted to regulate their cellular concentrations. The excretion of these elements outside the cells initially resulted in the formation of hard solid masses, called *concretions*. This process began much before the advent of eukaryotic cells between 2.1 and 1.6 billion years ago (see Sect. 1.5.4).

One example of the early formation of hard calcareous structures is the construction of stromatolites involving prokaryotes in shallow waters, which have transferred dissolved carbon and alkaline elements to solid carbonate since about 3.5 billion years (see Sect. 6.4.2). Calcification began in aquatic environments, and when some of the calcifying aquatic organisms moved to continents, they carried their solid structures with them and adapted them to their new environment (one example, is the shell of land snails). Hard parts of eukaryotes (mineral tests, shells and, skeletons) started to appear in the fossil record in the early Cambrian period about 540 million years ago (see Sect. 1.5.3), and likely evolved as functional uses of chemical elements in excess in the cells of the organisms.

The process by which living organisms produce solid minerals is called biomineralization, and the tissues that contain these minerals are called mineralized tissues. It is a very widespread phenomenon, and over 60 different minerals have been identified in biomineralizing organisms. The most common minerals in these organisms are calcium carbonate ($CaCO_3$), magnesium carbonate ($MgCO_3$) and silica (SiO_2), and an example of other minerals is the presence of crystals of magnetic minerals in the cells of some bacteria allowing them to orient and move along the lines of the magnetic field (see Sect. 8.1.2). The formation of carbonates by organisms is called *calcification*, and that of silica is known as *silicification* (see Sect. 4.7.2).

Although primitive concretions may not necessarily had functional uses for the organisms, biological evolution promoted life forms that took advantage of concretions for buoyancy, motility, protection and other functions. Examples in unicellular planktonic organisms include the calcareous and siliceous tests of such microscopic algae as coccolithophores and diatoms, respectively, and the calcareous tests of foraminifer protists. Colonial organisms such as sponges have calcareous or siliceous spicules in their tissues, and corals developed the spectacular calcareous external structures known as coral reefs (see Figs. 7.8 and 7.9). In multicellular organisms, hard structures or skeletons developed either externally—for example, the calcareous shells of molluscs, and the chitin exosqueleton of arthropods—or internally—for example, the silicified cells of some groups of plants, and the calcium-phosphate bones of vertebrates (see Sect. 6.1.2).

6.4.2   Formation of Solid Carbonates During the Earth's History

The period following the formation of the global ocean, perhaps as early as 4.4 billion years ago, was decisive for the later emergence of organisms on Earth and the takeover of the planet by ecosystems. Indeed during the following hundreds of millions of years, most of the initially very abundant atmospheric CO_2 was sequestered (stored) in carbonate rocks, where more than 99% of the Earth's carbon is still stored today, which brought the global temperature down to levels suitable for organisms (see Sects. 5.5.4 and 8.1.1). The process that led to the early formation of the carbonate rocks—mostly calcium carbonate ($CaCO_3$), but also magnesium, aluminium and iron carbonates ($MgCO_3$, $Al_2(CO_3)_3$ and $FeCO_3$, respectively)—was strictly chemical (see Information Box 5.11). It remained so during at least one billion years, until organisms developed the ability to form solid $CaCO_3$ and other carbonates. This section describes the progressive involvement of organisms in the history of $CaCO_3$ on Earth.

Fossil structures made of carbonates in which microorganisms are locally conspicuous go back about 3.5 billion years. Such structures are the only witnesses of the possible involvement of organisms in the formation of solid $CaCO_3$ until the occurrence in the fossil record of carbonate structures of multicellular organisms, about 500 million years ago. Indeed, multicellular organisms appeared probably on the planet between 1 billion and 500 million years before present, and until that time all organisms on Earth were microbes, which lived in aquatic environments, mostly oceanic. All these microorganisms belonged to the prokaryotes (archaea and bacteria) until about 2.0 and even perhaps 1.5 billion years ago, at which time eukaryotic unicellular organisms appeared in aquatic environments. The presence of microbes in carbonate fossils is very difficult to determine, and their presence of does not necessarily mean that they had a role in the formation of the fossils.

The ancient carbonate structures that involve microorganisms were mostly associated with marine sediments, and the most spectacular of these structures are the *stromatolites* and the *thrombolites* (Fig. 6.8). These structures occurred as domes and columns up to 1.5 m high in shallow waters of lakes and seas. There were similar structures based on silicon, phosphorus, iron, manganese and sulphur, but we focus here on the structures made of $CaCO_3$. Stromatolites and thrombolites are not the only types of carbonate accumulations with locally conspicuous microorganisms, but we only examine these two major groups in the following paragraphs.

Carbonate *stromatolites* show internal lamination, meaning that their carbonate is arranged in successive layers. Stromatolites already existed perhaps as early as 3.4 billion years before present, and they are still being formed in some marine lagoons of Australia and elsewhere. In contrast, carbonate *thrombolites* are not internally laminated, and they exhibit external shapes of individual masses of microbial carbonate, such as clots or small shrub-like masses. They go back about 500 million years, and they still exist in some Australian brackish lakes and elsewhere.

The microbes present in stromatolites and thrombolites were initially bacteria, including O_2-photosynthetic cyanobacteria, and they later included microscopic algae when these appeared during the course of biological evolution. Specialists distinguish three main processes in the formation of the $CaCO_3$ fossils that involve microorganisms: inorganic formation, grain trapping, and microbial formation. We briefly examine each process in turn.

The first process was the *chemical formation of $CaCO_3$*, which mostly occurred in environments where the carbonate saturation state of seawater (Ω) was high (see Sect. 11.1.3). Indeed, solid $CaCO_3$ can form chemically when Ω is high. This was generally the case earlier than 1 billion years ago, when the concentration of atmospheric CO_2 was more than 10 times the present level (Fig. 8.1) and Ω was

Fig. 6.8 Modern representatives of microbial carbonate structures that already existed more than 3 billion years ago (Fig. 6.9): **a** stromatolites at low tide, Highbourne Cay, Exuma Islands, Bahamas; and **b** thrombolites during low water in summer and autumn, Lake Clifton, Australia. Credits at the end of the chapter

thus very high. In this environment, microorganisms did not contribute much to the formation of the carbonate of stromatolites, but they influenced their fabric, especially their lamination, as they grew on top of the forming $CaCO_3$ layers.

The second process was the *trapping of calcareous sediment grains by microbial mats*. Cyanobacteria secrete a *mucilaginous sheath* that makes their mats very efficient at trapping sediment grains. Later on, evolution produced algae whose addition to the mats enhanced their trapping ability. The presence of a mat on top of a calcareous layer favoured the trapping of grains, whose binding created a next layer on which a new mat could grow, and the repetition of the process created a layered structure.

The third process was *microbial calcification*. One mechanism was the *calcification of the cyanobacterial sheath*, meaning the formation of $CaCO_3$ within the sheath. This mechanism was favoured by the drop in the concentration of atmospheric CO_2 to less than 10 times the present level that occurred 1 billion years ago. The lower concentration of CO_2 in seawater led to the evolution of CO_2-concentrating mechanisms in cyanobacteria, which assisted their photosynthetic carbon uptake. The enhanced photosynthesis locally lowered the CO_2 concentration of seawater and thus increased the pH of the cyanobacterial sheath, which promoted its calcification. Another mechanism of microbial calcification was the *calcification of various products secreted by heterotrophic bacteria*. The calcification that occurred in the mat contributed to cement together the grains trapped there.

Figure 6.9 illustrates the roles of the above three mechanisms in the formation of $CaCO_3$ structures since 3.5 million years. It shows that the chemical formation

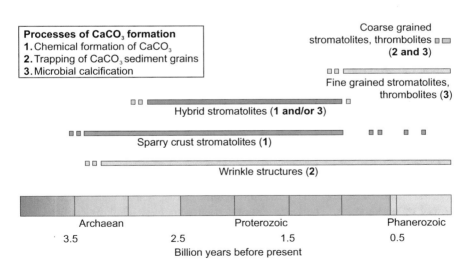

Fig. 6.9 Evolution of carbonate structures that involved microbes since 3.5 million years. For each type of structure: technical name and (in brackets) main processes of $CaCO_3$ formation. Credits at the end of the chapter

of $CaCO_3$ and grain trapping dominated until about 1 billion years before present, with possible contribution of some microbial calcification. Since about 1 billion years, microbial calcification became an important mechanism in the formation of $CaCO_3$.

Microbes were not only involved in the formation of the above carbonate structures in relation with the sediment, but they also contributed to the formation of solid carbonate in the water column. As already explained above for the cyanobaterial mats, phytoplankton photosynthesis causes local increases in pH. This also occurs in the water column, where it can induce the formation of small crystals of $CaCO_3$. The presence of large amounts of $CaCO_3$ crystals in suspension in the water can make the surface of the ocean or lakes milky, a phenomenon known as *whiting*. The whiting phenomenon can also be caused by large blooms of coccolithophores (phytoplankton; Fig. 7.8a) followed by the shedding in the water of the small calcareous plates that cover their cells (called *coccoliths*). Finally, substances secreted by heterotrophic bacteria can serve as support for the formation of $CaCO_3$ crystals in the water column. Some marine blooms of coccolithophores reach almost 1 million square kilometres.

There are regular occurrences of whiting in modern oceans and lakes, and there is clear evidence that the phenomenon occurred as well in the past. Indeed, the sinking over millions of years of $CaCO_3$ crystals and coccoliths formed in the water-column created huge accumulations of solid $CaCO_3$ on the seafloor. One example is provided by the well-known White Cliffs of Dover in England (Fig. 2.1), which are composed of fossilized coccolith aggregates that fell to the bottom of shallow seas about 70 million years ago. There, the coccoliths formed deposits whose thickness reaches up to 500 m, which were mineralized to the chalky form of $CaCO_3$. The formation of solid $CaCO_3$ by microorganisms in the water column is responsible for a large part of the calcareous rocks that exist on Earth.

The mechanisms by which microorganisms are involved in the massive formation of carbonate goes well beyond those cited above for the stromatolites and whiting event. Especially, the microbial formation of solid carbonate can occur as a by-product of other metabolic activities than photosynthesis, which include the following processes: denitrification (see Sect. 9.3.4, Eq. 9.14), ammonification (see Sect. 9.3.2, Eq. 9.9), sulphate reduction (see Information Box 5.11, Eq. 5.11), methane oxidation (see Sect. 9.2.5, Eq. 9.6), and the bacterial hydrolysis of urea.

Finally, the organisms involved in the formation of carbonate also include multicellular eukaryotes since about 500 million years. As a result, today's calcifying organisms include unicellular and multicellular eukaryotes in addition to a wide variety of prokaryotes. The calcifying eukaryotes include planktonic microalgae (example: coccolithophores, Fig. 7.8a), planktonic and benthic protozoa (examples: foraminifers), planktonic molluscs (example: pteropods, Fig. 7.8c), benthic algae (crustose corraline algae), benthic invertebrates (examples: sponges and corals, Fig. 7.8d), and benthic and terrestrial molluscs (examples: shellfish, squids, and snails, Fig. 7.8b). Multicellular eukaryotes create highly visible carbonate structures, such as coral reefs, but these are only the "tip of the iceberg" as most of the biological formation of solid carbonate on Earth continues to be done by microorganisms today, as it was the case in the previous 3 billion years. The article of Poppick (2019) provides further reading on coccolithophores and their role in the climate.

Roles of Organisms in Atmospheric CO_2 Regulation

Carbonate formation is a key process in the global feedback mechanisms that determine the level of atmospheric CO_2, and hence the Earth's greenhouse effect (see Sect. 9.2.3). During the course of the Earth's history, the formation of carbonate from atmospheric CO_2 began as a physical and chemical process about 4.4 billion years before present (see Sect. 5.5.4). It remained that way for at least 1 billion years, until the advent of the first carbonate structures that involved microbes (Fig. 6.9). During the following 2.5 billion years, the chemical production of carbonates and grain trapping initially dominated the formation of *stromatolites* with possible contribution of some microbial calcification, after which microbial calcification became an important mechanism around 1 billion years ago or earlier. Since about 500 million years, the formation of carbonates has also involved multicellular eukaryotes, and today about 90% of the total production of solid carbonate is done by unicellular and multicellular organisms, the remainder being chemical. These values do not include the formation of carbonates within the seafloor by chemical alteration of young basalt near ocean ridges (see Information Box 7.4).

6.5   Availability of the Key Biogenic Elements in the Earth's Environment

The previous sections stressed the importance for organisms of the six major biogenic elements (carbon, hydrogen, nitrogen, oxygen, phosphorus and sulphur) and the light alkaline elements (sodium, potassium, calcium and magnesium). Many species depend on the availability of these elements in solution in the water of aquatic bodies and soils. The next paragraphs examine the forms of various chemical elements used by organisms, and their natural reservoirs. This information is summarized in Table 6.5.

Hydrogen (H) is very abundant in the environment as part of water (H_2O) and other H-containing compounds such as hydrogen sulphide (H_2S). Different autotrophic organisms use various sources of hydrogen to synthesize organic matter (see Information Box 5.1). The bacterial digestion (in guts) or decomposition (in the environment) of organic matter in absence of oxygen produces H_2S, which is accompanied by the well-known "rotten-egg" odour.

Table 6.5 Bioavailable forms of important biogenic elements and their reservoirs in the Earth's environment

Biogenic elements	Main forms in the environment that can be used by organisms	Natural reservoirs
Hydrogen (H)	O_2-producing photosynthetic organisms: water (H_2O) Other autotrophs: H-containing compounds such as H_2S	H_2O: water bodies and soil moisture H_2S: sediments and water bodies without O_2, guts of organisms
Carbon (C)	Carbon dioxide (CO_2), bicarbonate and carbonate (HCO_3^- and CO_3^{2-})	Atmosphere, water bodies, and soil moisture
Oxygen (O)	Carbon dioxide (CO_2), bicarbonate and carbonate (HCO_3^- and CO_3^{2-})	Atmosphere, water bodies, and soil moisture
Nitrogen (N)	Dinitrogen gas (N_2), nitrate (NO_3^-)	Atmosphere, water bodies and soil moisture
Calcium (C)	Dissolved calcium (Ca^{2+})	Water bodies and soil moisture
Phosphorus (P)	Dissolved phosphate (PO_4^{3-})	Water bodies and soil moisture
Sulphur (S)	Dissolved sulphate (SO_4^{2-})	Water bodies and soil moisture
Silicon (Si)	Dissolved silicate (SiO_4^{4-})	Water bodies and soil moisture
Iron (Fe)	Dissolved iron (Fe^{2+})	Water bodies and soil moisture

Bio-available *carbon* (C) does not generally limit the development of organisms because it is abundant in the atmosphere and widely present as dissolved ions in natural waters. In addition, the ocean and inland waters are in contact with carbonate sedimentary rocks, which are by far the largest reservoir of carbon at the Earth's surface. In natural waters, any deficit or excess in dissolved carbon ions (HCO_3^- or CO_3^{2-}) can thus be quickly compensated by the dissolution or formation of solid carbonates. This process contributes to the regulation of the oceanic reservoir of dissolved inorganic carbon (see Sect. 9.2.4). On continents, plants use the CO_2 present in the atmospheric carbon reservoir. Although the CO_2 content of the atmosphere is continually depleted by the chemical alteration of silicate rocks ($CaSiO_3$) (see Sect. 7.3.3), this key greenhouse gas was never completely exhausted—which could have caused a global freezing of the planet and the extinction of organisms—because the atmosphere was continually replenished in CO_2 by volcanic eruptions (see Sect. 7.2.3).

The source of *oxygen* used by cells to build organic molecules is CO_2, which is also their source of carbon (previous paragraph). This is different from the oxygen used for respiration, which is present in the environment under different forms that include free oxygen (O_2), nitrate (NO_3^-), sulphate (SO_4^{2-}) and oxides of metals, in particular iron and manganese. The latter forms of oxygen are used for respiration in absence of O_2 (see Information Box 5.2).

Nitrogen gas (N_2) has been abundant in the Earth's atmosphere since the origin of the planet. However, most organisms cannot assimilate N_2, and they use instead oxidized forms of nitrogen, primarily nitrate (NO_3^-). Nitrate occurs naturally in minerals such as nitratine ($NaNO_3$), which was extensively mined as fertilizer for agriculture before the industrial production of N-fertilizers (see Sect. 2.1.4). Despite the natural occurrence of NO_3^-, this inorganic nutrient is in short supply in most environments, and biological evolution produced bacteria and archaea that can fix N directly from the N_2 gas, and thus supply nitrogen to the oceanic and continental food webs (see Sect. 6.1.2). After circulating in the food webs, N is released into the environment by the decomposition of organic matter, and part of it is lost as N_2 gas (see Sect. 9.3.4) while another part is converted into NO_3^-. This NO_3^- is used by autotrophic organisms to synthesize new organic matter.

Calcium (Ca) is abundant in the Earth's crust, and is thus available as dissolved Ca^{2+} ions in natural waters (Table 6.1). As for carbon (above), any deficit in dissolved calcium ions in natural waters can be compensated by the dissolution of calcareous sediments (mostly composed of $CaCO_3$).

Bioavailable *phosphorus* (P) is mostly present in natural waters in the form of dissolved phosphate ions (PO_4^{3-}). Phosphorus is mostly supplied to the ocean by rivers, which transport dissolved PO_4^{3-} ions resulting from the erosion of continental rocks, for example rocks of the apatite family. Phosphorus is mostly lost through sedimentation of P-containing biological structures. Overall, phosphorus is in short supply in the natural environment, and thus limits the growth of plants in various continental areas and of phytoplankton in many lakes and in some parts of the ocean. Because of this short supply, there is heavy use of phosphorus fertilizers in agriculture to maintain high production of plants. However, excessive P-fertilization causes water pollution and the resulting noxious excess of biological production in natural fresh waters called *eutrophication* (see Sect. 10.3). The book of Butusov and Jernelöv (2013) provides further reading on phosphorus.

Sulphur (S) is found as a free chemical element near volcanoes and hot springs, and is otherwise widely distributed on Earth in many minerals. It is abundant in seawater (Table 6.1) in the form of dissolved sulphate (SO_4^{2-}). It is provided to the ocean by the continental erosion (dissolution) of sedimentary sulphates (example: gypsum, $CaSO_4$) and sulphides (S^{2-}) (example: pyrite, FeS_2). The availability of sulphur in soils depends on both the contribution of dissolved sulphate by inland waters and the composition of the soils and the underlying bedrock, which may contain sulphur in various forms. In addition, marine aerosols contribute significant amounts of sulphates to continental soils (about 5 billion tons per year).

Organisms also depend on the bioavailability of silicon and iron, which are important chemical elements for biological activity (see Sect. 6.1.2). They are identified as group C in Fig. 6.6, and their bioavailability is as follows.

Silicon (Si) is very abundant in the Earth's crust (Table 6.1) and is thus generally abundant in the outer envelopes of the planet. In natural waters including in soils, it is mostly present as dissolved silica (SiO_2), often as hydrated orthosilicic acid ($Si(OH)_4$). The latter is especially important for organisms because it is the only bioavailable form of silicon.

Iron (Fe) is very abundant in the Earth's crust (Table 6.1), but the abundance and/or supply of bioavailable iron is low in several continental and oceanic areas. On continents, most of the iron in soils is in the form of insoluble iron oxide

(Fe_2O_3), which cannot be used by photosynthetic organisms. However, some organic molecules known as *chelators* bind to iron atoms and thus make them bio-available. In contrast, the bioavailability of iron can be reduced by a number of factors that include soil pH higher than 6.5 and waterlogged soil. Also, high levels of some chemical elements in the soil, such as phosphorus, nitrogen, zinc, manganese and molybdenum, can interfere with the uptake of iron by plants.

In environments where the concentration of oxygen or other oxidizers (such as sulphur) is high, elemental iron (Fe) and ferrous iron (Fe^{2+}) tend to be transformed into ferric iron (Fe^{3+}). In the literature, Fe^{2+} and Fe^{3+} are also written Fe(II) and Fe(III), respectively. In dry environments such as soils, the reaction transforms solid Fe and Fe^{2+} into solid Fe^{3+} compounds, although it is favoured by traces of moisture, whereas in water the reaction involves ions. The most *bioavailable* forms of iron are dissolved Fe^{2+}, and to a lesser extent Fe^{3+}. In water, depending on conditions and especially pH, Fe^{2+} exists as dissolved ions, or combine with H_2O (water) to form ferrous hydroxide, $Fe(OH)_2$, whose combination with O_2 produces insoluble Fe_2O_3. Still in water and depending on conditions, Fe^{3+} exists in solution as *chelate complexes* (previous paragraph) or combine with H_2O to form insoluble ferric hydroxide, $Fe(OH)_3$. As a consequence, most of the iron supplied to the ocean is progressively transformed into insoluble compounds, which sink out of the surface waters, thus depleting the upper ocean of iron

The iron found in the oceans mostly comes from emerged lands (continents and islands), where the erosion of continental rocks and soils by water and wind release iron in running waters and in the air. The meltwater running from the Antarctic and Greenland Ice Sheets and the icebergs they calve also supply large amounts of iron to the oceans. The water released by hydrothermal vents often contains iron sulphides (FeS, and other chemical forms), which are used by some organisms in the hydrothermal communities and also contribute iron to the global ocean. The iron from emerged lands is transported to the oceans by atmospheric currents and running waters. Deserts and volcanoes are major sources of the iron transported by atmospheric currents. The supply of iron can be very low in oceanic areas located far from continents and thus not fertilized by iron-rich dust blown by the wind. In these areas, known as *High Nutrient Low Chlorophyll* regions, the growth of phytoplankton is generally limited by the low availability of iron (see Sect. 9.3.5). The article of Raiswell and Canfield (2012) provides further reading on iron in the Earth's environment.

It is interesting to note that bioavailable iron was very abundant in the ocean earlier than 2.4 billion years ago because there was no O_2 in the environment at the time. After the advent of O_2-producing photosynthesis, the O_2 released in the ocean led to the sinking of most of the dissolved iron to the seafloor as explained above for the modern ocean. This went on during hundreds of millions of years, and seawater thus lost most of its bioavailable iron (Sect. 9.4.3). This was a very significant environmental change given that all ecosystems were aquatic, and thus mostly marine, at the time.

The release of O_2 by photosynthesis caused a permanent environmental deficiency in bioavailable iron, and this created a permanent limitation of photosynthesis because iron both plays a role in the formation of the main photosynthetic pigment (chlorophyll) and is essential for a key step in photosynthesis. As a consequence, the cellular process of O_2-producing photosynthesis created a self-limiting environmental condition by enhancing the global oxidation of iron in seawater. This is an example of negative feedback (see Information Box 6.2), where the increasing concentration of photosynthetic O_2 in aquatic environments caused a decrease in the bioavailability of iron, which limited photosynthesis. However, this feedback did not create a feedback loop because the relatively low photosynthesis did not affect, in return, the long-term concentration of O_2 in the atmosphere, which is controlled otherwise (see Sect. 9.4.4).

Unlike the major biogenic elements, the light alkaline elements, and silicon and iron examined above, other elements of biological importance are in very low concentrations in the outer layers of the planet (Table 6.1). There are two different explanations for this scarcity. On the one hand, some of these chemical elements are scarce in the whole Universe, and this explains their scarcity on Earth. On the other hand, some elements abundant in the Universe were concentrated in the Earth's mantle or core by the planetary differentiation billions of years ago (see Sect. 6.3), and this explains their scarcity in the outer layers of the planet. Some of these rare elements are nevertheless essential to the organisms, and are called *essential trace elements* (group D in Fig. 6.6). These include boron (B; cell wall of plants), copper (Cu; respiratory proteins of some invertebrates), fluorine (F; bones and teeth of vertebrates); molybdenum (Mo; N_2 fixation, and enzymes), manganese (Mn; enzymes) and zinc (Zn; enzymes). The book of Ochiai (2008) provides further reading on the bioinorganic chemistry of a wide range of chemical elements.

6.6   Gravitation and the Planetary Differentiation of Chemical Elements

6.6.1   Gravitation

Gravitation (also called *gravity*) describes the phenomenon by which all masses in the Universe are brought toward one another. Gravitation influences all masses, from the smallest components of matter, such as the atoms and their constitutive particles, to the largest components of the Universe, such as the planets, the stars and the galaxies. The mathematical bases of mechanics were established by Sir Isaac Newton in his famous book *Philosophiæ Naturalis Principia Mathematica* (1687), where he described gravitation as a force that causes bodies to be attracted to each other, whereas Albert Einstein's *theory of general relativity* (1915) describes it as a curvature of space-time caused by the uneven distribution of masses. Figure 6.10 illustrates the prediction that space-time around a rotating astronomical body would be warped, and twisted by the body's rotation. In this

Fig. 6.10 Illustration of the prediction of the theory of general relativity concerning gravity. Details in the text. Credits at the end of the chapter

figure, the deformed grid around Earth is a visual analogy of the curvature of the space-time caused by the presence of the planet. The curved geometry of space-time is felt as gravity. By comparison, classical, Newtonian physics considers that space and time are independent of the presence of matter and energy, in which case the grid in Fig. 6.10 would be flat with lines at right angles. The book of MacDougal (2012) provides further reading on Newton's gravity and the mechanics of the Universe, and that of Fischer (2015) presents relativity for everyone.

In 2004–2005, the US scientific satellite Gravity Probe B collected data later interpreted as consistent with Einstein's prediction concerning gravity. This satellite is represented in Fig. 6.10 pointing towards the distant star IM Pegasi, which is located in the Pegasus constellation 329 light-years from Earth, for precise positioning. The relativity theory also predicts that massive astronomical bodies (planets or stars) that orbit each other cause ripples of space that travel at the speed of light, called *gravitational waves*. This prediction was shown to be correct by astronomical observations made in 2015, that is, one century after the formulation of the theory by Albert Einstein in 1915. The book of Petkov (2009) provides further reading on relativity and the nature of space-time.

For most practical purposes, gravity is well described by Newton's law of universal gravitation, where the force that causes two bodies to be attracted to each other is proportional to the product of their masses and inversely proportional to the square of their distance. However, Einstein's relativity must be used in cases of objects that move very fast. For example, the satellites of Global Navigation Satellite Systems (GNSS) are located very high above the ground (approximately 20,200 km for the Global Positioning System, GPS) and move very fast (14,000 km per hour for the GPS) (see Sect. 8.1.3). The GNSS deliver positions at the Earth's surface within a few centimetres, which requires that atomic clocks on board the satellites have an accuracy of 20–30 ns. Because the satellites move constantly relative to the users on Earth, the measurements of their on-board clocks must be corrected for the effects predicted by the theory of relativity. Without Einstein's relativity, there would be no GNSS!

Gravitation is one of the fundamental characteristics of the Universe. It caused the original gaseous matter in the Universe to start coalescing into stars, and the stars to form galaxies (see Sect. 1.4.1). It also causes the nascent stars, including the Sun, to progressively contract, leading to the fusion of their hydrogen atoms into helium and the ensuing nuclear fusion reactions (see Sect. 6.3). At the end of their lives, the stars die through different mechanisms, which all involve gravitation, and the dying stars produce the wide varied of chemical elements that exist in the Universe (see Sect. 1.4.1). Within the Solar System, gravitation caused the matter in the solar nebula to coalesce into the Sun, the planets, and billions of other astronomical bodies (see Sect. 1.3.1). The spatial range of gravitation encompasses the whole Universe, but the effects of gravitation are stronger for closer objects.

6.6.2 Gravitation and the Distribution of Chemical Elements on Earth

One example of the effect of gravitation on Earth is the weight that gravity gives to physical masses, meaning that the same amount of matter (called the *mass*) has different *weights* depending on local gravity. Indeed, an astronaut weighing 70 kg on Earth (without the spacesuit) weighs only 11.5 kg on low-gravity Moon, and has no weight at all in no-gravity space, this without any change of mass. Another example of the effect of gravitation is provided by the Earth's tides, which result from the combined gravitational attraction of the Moon and the Sun.

Gravitation was not only involved in the formation of Earth from the solar nebula, but it also influenced the early development of the planet more than 4.5 billion years ago. Indeed, gravitation largely caused the planetary differentiation of the early Earth, described above (see Sects. 6.3 and 6.4). Because of gravitation, the different chemical elements in the nascent planet had different weights (previous paragraph), with the result that the heavier elements sank towards the centre of the still fluid Earth, and the lighter ones rose to the surface. In addition, gravity, together with the heat released by the radioactive elements in the initial mass of the future Earth, created pressures and temperatures high enough to melt some of the materials, which facilitated their movements. The combination of these

Fig. 6.11 Layered structure of the inner solid Earth: (1) continental crust, (2) oceanic crust, (3) upper mantle, (4) lower mantle, (5) outer core, and (6) inner core. X-axis: depth below ground of the limits of the six layers; Y-axis: thickness of layers. Credits at the end of the chapter

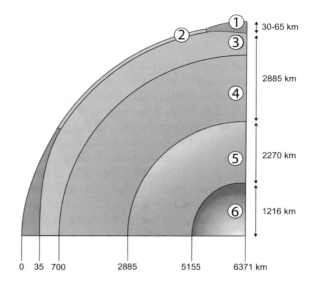

gravitation-related factors caused the concentration of the heavy metals in the centre of the planet by planetary differentiation of the early Earth, and increased the concentrations of lighter chemical elements in the outer envelopes of the planet where they could become the building blocks of organisms (see Sects. 6.3 and 6.4).

The result of planetary differentiation was a structure in stacked layers of decreasing densities, like that of a layered cocktail or an onion (Fig. 6.11 and Table 6.6). At the centre of present Earth is the *core*, which is very hot and very dense, contains most of the heavy elements, especially iron and nickel, and accounts for 17% of the volume of the globe. It has a solid inner part, and a fluid outer part. The core is surrounded by the lighter (less dense), thick *mantle* (81% of the volume of the globe), which is made of hot material called *magma*, with a solid lower part and a fluid upper part (except the most superficial layer, which is solid). The upper mantle is covered by the lighter, thin *crust*, which is solid as a result of its external cooling and acounts for less than 2% of the volume of the globe.

Going down from the surface of the crust into the planet, there is a rapid increase in the density of materials, pressure and temperature (Table 6.6). For example in the very deep TauTona gold mine in South Africa, the temperature of the rocks at 3.9 km depth is 55 °C, which shows how rapidly the temperature increases with depth. At the surface of the globe, there are major differences between the oceanic and continental crusts, which correspond to the seafloor and the continents, respectively. The crust of oceans is largely composed of basalt rocks, which are rich in atoms of iron, calcium and magnesium, and that of continents is dominated by granite rocks, which are rich in atoms of potassium, silicon and sodium. The oceanic crust is denser and thinner than the continental crust, its higher density reflecting the much larger proportion of iron (which has the high density of 7,870 kg m^{-3}) in basaltic than granitic rocks. The pressure inside Earth is thousands of times higher than at the surface of the planet, and the temperature in the upper mantle is so high that the silicate matter there is fluid (or more precisely in the fluid state called "plastic"), whereas the pressure in the lower mantle, below, is so great that the silicate matter cannot be fluid any more. In the deeper outer core, below the lower mantle, the even higher temperature there causes the matter (mostly iron and nickel) to be fluid, whereas at the centre of the planet, the incredibly large pressure prevents the matter to be fluid.

Above the crust are the fluid envelopes of Earth (ocean and atmosphere), which are less dense than the underlying layers because they are mostly composed of light chemical elements and also because of the lower pressure (Table 6.6). About 70% of the surface of the globe is covered by oceans (mostly H_2O, with dissolved Na and Cl and small proportions of all the other chemical elements; Table 6.1), whose average depth is 3.7 km and which is stratified into layers of different densities according to the temperature and the salinity of the seawater. Above the oceans and the continents is the atmosphere, where the main gases are N_2 and O_2 (Table 6.1) and which is stratified in four layers of decreasing densities (see Sect. 2.2.1). The main characteristics of the different layers of the ocean and the atmosphere are summarized in Table 6.6.

Table 6.6 Main characteristics of the successive layers of Earth, from the top of the atmosphere to the centre of the globe

Earth's layer (limits: altitude or depth)	Average density (kg m^{-3})	Average pressure (atm)	Average temperature (°C)	State	Main chemical elements
Atmosphere (altitude: km)					
Thermosphere (85–1000)	–	–	+500	Gas	N, O
Mesosphere (50–85)	–	5×10^{-12}	−50	Gas	N, O
Stratosphere (10–50)	0.01	0.01	−40	Gas	N, O
Troposphere (0–10)	0.7	0.5	−20	Gas	N, O
Liquid Ocean (depth: m)					
Surface (0–100 m)	1,024	5	17	Liquid	H, O, Na, Cl
Intermediate (100–1000 m)	1,026	55	10	Liquid	H, O, Na, Cl
Deep (>1000 m)	1,028	370	2	Liquid	H, O, Na, Cl
Solid Earth (depth: km)					
Oceanic crust (0–5)	3,000	800	10	Solid	O, Si, Mg, Fe, Ca
Continental crust (0–35)	2,700	5,500	55	Solid	O, Si, Al, K, Na, Ca
Upper mantle (35–700)	4,000	115,000	1,200	Fluid	O, Si, Mg, Fe
Lower mantle (700–2885)	5,000	1,000,000	2,100	Solid	O, Ca, Ti
Outer core (2885–5155)	10,500	2,300,000	4,000	Fluid	Fe, Ni, S, Si
Inner core (5155–6371)	12,000	3,300,000	5,500	Solid	Fe, Ni

Hence from the centre of the globe upwards, Earth is structured in successive, increasingly lighter layers, from the solid core to the liquid ocean and the gaseous atmosphere, whose average temperatures range from several thousands of degrees (core) to 3.9 °C (ocean) and to freezing cold (mesosphere). The outer envelopes of the planet are rich in the six major biogenic elements carbon, hydrogen, nitrogen, oxygen, phosphorus and sulphur (see Sect. 6.3). This richness results from the planetary differentiation that took place more than 4.5 billion years ago (above), and also from the supply of materials by the meteoroids that heavily impacted Earth until 3.9 billion years ago and continued to hit the planet afterwards (see Sect. 1.3.6). Within these outer envelopes, ecosystems thrive in the layers where temperatures are compatible with the thermal limits of organisms (see Sect. 3.1.1). These layers are the crust, the ocean, and the lower part of the atmosphere called *troposphere* (see Sect. 2.2.1), whose characteristics are summarized in the coming three paragraphs.

Solid crust. Ecosystems occupy the whole surface of continents, with the exception of the areas covered by glaciers and some deserts. In addition, organisms are abundant in the soil, whose formation largely results from their activity. Ecosystems also thrive at the surface of the oceanic crust and in the underlying sediments. Deep in the solid crust, microbial organisms have been found down to at least 5 km in the subsurface of continents and oceans, where they form a deep biosphere whose study is in progress.

Liquid ocean. In the ocean's waters, ecosystems are found at all depths, from the permanently or seasonally hospitable surface layer down to the generally cold and permanently dark deep waters. Photosynthesis takes place in the well-illuminated surface waters, and chemosynthesis mostly occurs on and inside the seafloor.

Troposphere. Ecosystems are also present in the lowest part of the gaseous troposphere, in the forms of flying and gliding organisms (insects, birds, bats, and gliding ants, squids, fish, frogs, lizards and mammals). In addition, microbes have been observed on particles up to 10 km in the upper troposphere, but their abundance and longevity in this environment is poorly documented.

It follows from the above brief description of the different layers of the planet inhabited by organisms that the latter can withstand a wide range of pressures, so that temperature and liquid water are the factors that determine the habitability of the various environments of Earth. These go from 10 km high in the atmosphere down to 5 km under the ground, and the ranges of pressure and temperature between these two extremes are 0.26 to 1300 atm and −50 to +150 °C, respectively. The domain occupied by ecosystems, between +10 and −5 km, is very thin, and its 15 km represent only 0.25% of the total thickness of the combined solid Earth and gaseous atmosphere (more than 6000 km; Table 6.6). This book examines the conditions that allowed the organisms to take over the whole planet from this thin foothold.

The onion-layered structure of Earth is the source of three major dynamic components of the Earth System. Going from the upper atmosphere downwards, the first major dynamic component is the *climate*, whose main source of energy is the solar radiation, which comes from outside the planet and affects the atmosphere, the oceans and the continental surfaces. The second major dynamic component is the *tectonic activity*, which results from the interactions between the upper fluid mantle and the solid crust and is driven by the flow of heat energy coming from inside the planet. The third major dynamic component is the *magnetic field*, which is linked to interactions between the core and the lower mantle fuelled by internal heat energy acting deep inside the planet. These three dynamic components of the Earth System deeply influence the ecosystems, as explained in Chaps. 3, 7 and 8 for the climate, the tectonic activity and the magnetic field, respectively.,

The climate, tectonic activity and magnetic field are powered by energy flows of different origins, but they all primarily rely on gravitation, which has been at work in the Universe since the Big Bang 13.7 billion years ago. Some of the key effects of gravitation are as follows.

In the Universe, stars started to form by gravitational accretion of the dispersed primordial matter (mainly hydrogen and helium, the lightest nuclei). As these grew and became denser, they released energy that led to the formation of nuclei of heavier elements (nucleosynthesis). As stars died, the resulting chemical elements added up to the primordial elements to feed the next generations of stars and planets, which were also formed by gravitational accretion of dispersed chemical elements. The stellar mechanisms involved in the formation of the wide variety of chemical elements all relied on gravitation (see Sect. 1.4.1). The processes of star formation by gravitational accretion, dispersion of the chemical elements forged during the lifetimes of stars, and incorporation of the whole suite of elements— from hydrogen to uranium and even heavier nuclei—into new star systems have been going on since the Big Bang, continues nowadays, and will likely carry on until the end of the Universe. It follows that gravitation has been central to the formation of the building blocks of organisms.

On Earth, as shown above, the main source of internal energy is the slow decay of the radioactive elements concentrated in the mantle and the crust by gravitational accretion followed by planetary differentiation. The concentrations of radioactive elements inside the globe are low, but on the long run the thermal energy they release sustains the Earth's geothermal heat flow, tectonic movements and volcanic activity (see Sect. 3.2.3). Conversely, the gravitation-driven planetary differentiation concentrated in the Earth's crust, more than 4.5 billion years ago, many of the chemical elements that would later become the building blocks of organisms.

Still on Earth, gravitation plays another important role in favouring the circulation of chemical elements needed by organisms between different layers of the planet. Indeed, the chemical elements naturally tend to accumulate in the crust, mostly as sedimentary rocks, whereas ecosystems need them where they thrive, which is at the surface of continents and in the oceans. Gravitation has a key role in the tectonic mechanisms that keep chemical elements moving slowly among different layers of Earth, and thus ensure their long-term availability for ecosystems (see Sect. 7.3.1).

Gravitation connects the day-to-day dynamics of Earth and its ecosystems to the creation of chemical elements in the stars. In most cases, this creation occurred very far away from the Solar System and billions of years ago, and gravitation links the Earth System and its organisms and ecosystems to the whole Universe, past and present.

6.7 Key Points: The Building Blocks of the Cells and the Living Earth

We conclude this chapter with a brief summary of some of the key points of the above sections, as done in previous chapters (see Sects. 1.6, 2.5, 3.8, 4.8, and 5.7). Here, we examine interactions between Earth, the building blocks of the cells, and the organisms and ecosystems. As in the corresponding sections of previous chapters, the context of this section is provided by the three phases of the Earth's history illustrated in Fig. 1.11, which are the Mineral, Cradle and Living Earth. These three phases occurred from 4.6 to 4.0, 4.0 to 2.5, and 2.5 to 0 billion years ago, respectively.

Since the Big Bang 13.7 billion years ago, gravitation has been central to the formation of the whole suite of chemical elements, which include the building blocks of organisms (see Sect. 6.6.2). During the formation of the early, Mineral Earth, the light chemical elements rose to the surface of the planet (crust and atmosphere), and the heavier elements sank to the centre under the effect of gravity (see Sect. 6.3). Through this mechanism, gravitation-driven planetary differentiation concentrated in the crust many of the chemical elements that would later be the building blocks of organisms (see Sect. 6.6.2). In addition, gravitation has a key role in the tectonic mechanisms that keep chemical elements moving slowly among different layers of the planet, and thus ensure their long-term availability for ecosystems (see Sect. 6.6.2).

Since the Cradle Earth phase, the availability of the bricks (chemical elements and compounds) of organic matter in the outer layers of the planet enabled the establishment of organisms on Earth, their progressive expansion to most of the environments, and their regulation of the Earth's biogeochemical cycles and climate for nearly 4 billion years (see Sect. 6.3). The abundance of liquid water at the surface of Earth favoured large molecules based on carbon (see Sect. 6.6.2). More generally, the early formation of the liquid-water ocean led to the creation of the wide array of chemical compounds that organisms would use to firmly establish themselves on the planet (see Sect. 6.3).

The presence of light-mass alkaline elements dissolved in seawater led to the evolution of the major biological innovations of mineral tests, shells and skeletons in various groups of organisms (see Sect. 6.4.1). During the course of the Earth's history, the formation of carbonate was a chemical process during at least 1 billion years, after which microorganisms were involved in the formation of such mineral structures as stromatolites, and since about 500 million years, the formation of carbonates has also involved multicellular eukaryotes (see Sect. 6.4.2).

Various organic compounds have had complementary roles in cells and organisms since the Cradle Earth phase. At the level of the cellular hardware, polysaccharides store energy in cells and contribute to their structure, lipids are major constituents of the cell membrane and store energy in cells for long periods, proteins have a multitude of key roles in organisms, and enzyme-based feedback inhibition loops control the synthesis of many complex molecules in cells (see

Sects. 6.1.3 and 6.1.4). Concerning the cellular software, nucleic acids control the chemical composition of proteins (see Sect. 6.2.5). Also, the interactions of nucleic acids with the environment have driven the evolution and diversification of life forms on Earth (see Sect. 6.2.1). The large number of different combinations of nucleotides in a single DNA molecule and the mutations that occur during DNA replication explain the almost infinite diversity of organisms that exists among species and among the individuals within each species (see Sects. 6.2.4 and 6.2.5). Also, some nucleotides play key roles in cellular energy storage and transport and in cell signalling (see Sect. 6.2.2). Overall, the numerous, highly complex organic compounds allowed biological evolution to create the past and present high diversity of organisms (see Sect. 6.3).

In the present, Living Earth phase, the hydrogen, carbon and oxygen used to build organic compounds are supplied by photosynthesis and chemosynthesis, and the nitrogen is supplied by the N_2-fixation of bacteria and archaea (see Sect. 6.1.2). Ecosystems have had cumulated effects, over millions of years, on systems that control the cycling of hydrogen, carbon and oxygen on the planet (see Sect. 6.1.2). These examples show that ecosystems are based on a wide array of organic molecules, on the interactions among these molecules within living cells and organisms, and on the interactions of cells and organisms with their environment (see Sect. 6.3). Overall, gravitation links organisms and ecosystems to the whole Universe (see Sect. 6.6.2).

Figure Credits

Fig. 6.1 This work, Fig. 6.1, is a derivative of https://commons.wikimedia.org/wiki/File:CHO-cycles_en.png by Qniemiec https://commons.wikimedia.org/wiki/User:Qniemiec, used under GNU FDL and CC BY-SA 3.0. Figure 6.1 is licensed under GNU FDL and CC BY-SA 3.0 by Mohamed Khamla.

Fig. 6.2 This work, Fig. 6.2, is a derivative of https://commons.wikimedia.org/wiki/File:DNA_replication_split.svg by Madprime https://en.wikipedia.org/wiki/User:Mad_Price_Ball, used under GNU FDL, CC BY-SA 3.0 and CC BY-SA 2.5. Figure 6.2 is licensed under GNU FDL and CC BY-SA 3.0 by Mohamed Khamla.

Fig. 6.3 This work, Fig. 6.3, is a derivative of https://commons.wikimedia.org/wiki/File:DNA_replication_split.svg by Madprime https://en.wikipedia.org/wiki/User:Mad_Price_Ball, used under GNU FDL, CC BY-SA 3.0 and CC BY-SA 2.5. Figure 6.3 is licensed under GNU FDL and CC BY-SA 3.0 by Mohamed Khamla.

Fig. 6.4 This work, Fig. 6.4, is a derivative of https://commons.wikimedia.org/wiki/File:DNA_transcription.jpg by Dovelike, used under CC BY-SA 3.0. Figure 6.4 is licensed under CC BY-SA 3.0 by Mohamed Khamla.

Fig. 6.5 This work, Fig. 6.5, is a derivative of https://commons.wikimedia.org/wiki/File:Central_Dogma_of_Molecular_Biochemistry_with_Enzymes.jpg by Daniel

Horspool, used under CC BY-SA 3.0 and GNU FDL. Figure 6.5 is licensed under CC BY-SA 3.0 and GNU FDL by Mohamed Khamla.

Fig. 6.6 This work, Fig. 6.6, is a derivative of https://commons.wikimedia.org/wiki/File:Discovery_of_chemical_elements.svg by Sandbh, used under CC BY-SA 3.0. Figure 6.6 is licensed under CC BY-SA 3.0 by Mohamed Khamla.

Fig. 6.7 Modified after Figure 12 of http://ressources.unisciel.fr/biocell/chap2/co/module_Chap2_7.html. With permission from Prof. Ijsbrand Kramer, Université de Bordeaux, France.

Fig. 6.8a https://commons.wikimedia.org/wiki/File:20130118-HighbormeCay-Stromatolite-03.JPG by Vincent Poirier https://commons.wikimedia.org/wiki/User:Vfp15, used under CC BY-SA 3.0.

Fig. 6.8b https://commons.wikimedia.org/wiki/File:Lake_Clifton_SMC_2008.jpg by SeanMack https://commons.wikimedia.org/wiki/User:SeanMack, used under GNU FDL and CC BY 3.0.

Fig. 6.9 Modified after Figure 11 of Riding (2011). With permission from Prof. Robert E. Riding, Cardiff University, Wales.

Fig. 6.10 Image 17 in Gravity Probe B • Image Gallery, Artwork by Gravity Probe B, Stanford University, NASA and Lockheed Martin Missiles & Space Company, http://einstein.stanford.edu/gallery. All of the photos and images in the GP-B Image Gallery may be downloaded at no charge and used in news and media stories, publications and for educational purposes, https://einstein.stanford.edu/RESOURCES/press-index.html.

Fig. 6.11 This work, Fig. 6.11, is a derivative of https://commons.wikimedia.org/wiki/File:Slice_earth.svgby Dake, used under CC BY-SA 2.5. Figure 6.11 is licensed under CC BY-SA 3.0 and GNU FDL by Mohamed Khamla.

References

Einstein, A (1915). Die Feldgleichungen der Gravitation [The Field Equations of Gravitation] Sitzungsberichte. Königlich Preussische Akademie der Wissenschaften 1915: 844–847. English translation at https://en.wikisource.org/wiki/Translation:The_Field_Equations_of_Gravitation

Newton I (1687) Philosophiæ naturalis principia mathematica, London. Translated in English by Andrew Motte in 1729 https://books.google.fr/books?id=ySYULc7VEwsC&printsec=frontcover&hl=en&source=gbs_ge_summary_r&redir_esc=y#v=onepage&q&f=false

Riding R (2011) Microbialites, stromatolites, and thrombolites. In: Reitner J, Thiel V (eds) Encyclopedia of geobiology. Encyclopedia of Earth sciences Series, Springer, Amsterdam, pp 635–654

Further Reading

Butusov M, Jernelöv A (2013) Phosphorus. An element that could have been called Lucifer. Springer Briefs in Environmental Science, vol 9. Springer, New York, NY

De La Rocha C, Conley DJ (2017) Silica stories. Springer, Cham

Fischer K (2015) Relativity for everyone. How space-time bends, Second Edition. Undergraduate Lecture Notes in Physics. Springer, Cham

Kratz RF (2020) Molecular & cell biology for dummies, 2nd edn. Wiley, Hoboden, NJ

Legendre L (2014) Marine biogeochemical cycles. In: Monaco A, Prouzet P (eds) Oceans in the Earth System ISTE. Hoboken, London and Wiley, NJ, pp 145–187

MacDougal DW (2012) Newton's gravity. An introductory guide to the mechanics of the universe. Undergraduate Lecture Notes in Physics. Springer, New York, NY

Middelburg JJ (2019) Marine carbon biogeochemistry. A primer for Earth System scientists. Springer Briefs in Earth Sciences. Springer, Cham

Ochiai E (2008) Bioinorganic chemistry: A survey. Elsevier/Academic Press, Amsterdam

Petkov V (2009) Relativity and the nature of spacetime, 2nd ed. The Frontiers Collection. Springer, Berlin, Heidelberg

Poppick L (2019) Tiny living stones of the sea. Know Mag. https://doi.org/10.1146/knowable-062519-1

Raiswell R, Canfield RE (2012) Iron biogeochemical cycle past and present. Geochem Perspect 1:1–220. https://doi.org/10.7185/geochempersp.1.1

Robinson TR, Spock L (2020) Genetics for dummies, 3rd edn. Wiley, Hoboden, NJ

Schulze-Makuch D, Irwin LN (2018) Life in the Universe. Expectations and constraints, 3rd edn. Springer-Praxis Books, Springer, Cham

Underwood E (2019) The iron ocean. Know Mag. https://doi.org/10.1146/knowable-121919-2

Zhegunov G (2012) The dual nature of life. The Frontiers Collection. Springer, Berlin Heidelberg

Chapter 7
The Natural Greenhouse Effect: Connections with Geological Activity

7.1 The Natural Greenhouse Effect and Organisms

7.1.1 Natural Versus Anthropogenic Greenhouse Effect

The expression *greenhouse effect* is often understood nowadays as equivalent to *global warming*, and thus associated with the increasingly negative effects of the ongoing rise in global temperature on organisms and ecosystems. However, the greenhouse effect in itself is a *natural* process of the Earth System (see Information Box 2.7), and the present global climate problem is caused by the alteration of this natural process by the massive release of greenhouse gases into the atmosphere by human activities (see Sect. 2.1.6 and Fig. 9.5). The *anthropogenic* greenhouse effect, which causes the ongoing large and rapid warming of the atmosphere and the ocean, was initiated by the Industrial Revolution in the second half of the eighteenth century and has progressively gained momentum since then. Global warming is one of the characteristics of the latest geological epoch, which is called Anthropocene by reference to the major impacts of human activities on the Earth's environment (see Sect. 1.5.4). Some of the global and regional impacts of the anthropogenic greenhouse effect on the Earth System are examined in Chapter 10 (see Sect. 11.1.1–11.1.3).

In contrast with the two-century-old anthropogenic greenhouse effect, the natural greenhouse effect contributed to keep the Earth's temperature suitable for organisms since their appearance on the planet between 3.8 and 4.0 billion years ago (see Sect. 1.5.4). This allowed biomasses to build up, and organisms to progressively take over the Earth System. The present chapter deals with the connections between the geological activity of Earth and the natural greenhouse effect, and the remainder of this section addresses the habitability of the planet and the long-term natural greenhouse effect. It is followed by descriptions of: the tectonic activity of

P. Bertrand and L. Legendre, *Earth, Our Living Planet*, The Frontiers Collection, https://doi.org/10.1007/978-3-030-67773-2_7

Earth (Sect. 7.2) and its effects on the long-term functioning of the Earth System (Sect. 7.3); the effects of tectonically-driven carbon recycling on climate variations (Sect. 7.4); and the feedback of climate into tectonic activity (Sect. 7.5). The chapter ends with a summary of key points concerning the interactions between the Solar System, Earth, the natural greenhouse effect, and organisms (Sect. 7.6).

7.1.2   The Earth's Habitability

The effects of high and low temperatures on organisms and ecosystems and the mechanisms that determine the range of temperatures on Earth were examined in previous chapters. Some prokaryotes (archaea and bacteria) can survive scorching temperatures above +100 °C, and others can withstand freezing temperatures close to −200 °C. Equivalent extreme temperatures for the eukaryotes are up to +90 °C for some species, and down to almost −200 °C for others.

The present book addresses the takeover of Earth by organisms, and thus does not focus on extreme temperature because only a few organisms or species can survive such conditions. Indeed, the takeover of the planet required the build-up of huge biomasses of organisms, and thus conditions that not only allowed survival but favoured growth. The ranges of temperatures for growth are narrower than those for survival, that is, the limits for growth range from near +100 °C to above −20 °C for prokaryotes and up to +65 °C to above −20 °C for some eukaryotes (Table 3.1). These ranges are very wide because they reflect the combined temperature limits for all known species in each domain of life, whereas the actual ranges of extreme temperatures for given species are much narrower than the overall ranges cited here, and are generally limited to few tens of degrees or even less. The profusion of organisms and ecosystems in aquatic environments and on continents indicates that temperatures within the limits for growth have generally existed in at least some parts of Earth from the time organisms appeared on the planet, between 3.8 and 4.0 billion years ago, until now.

The thermal habitability of Earth is examined in Chapter 3. The average temperature at the surface of Earth is presently +15 °C, and the lowest and highest recorded values are −89 and +57 °C, respectively (Table 3.2). The usual diurnal range of temperatures in a given area is generally 10 to 20 °C, except in some deserts where temperatures can range from less than 0 °C during the night to more than 50 °C during the day, and where there are no or very few organisms. The main source of heat at the surface of Earth is the radiation from the Sun, which progressively increased since the early period of the faint young Sun (see Sect. 3.7.1). Over the last billions of years, the temperatures of the Earth's atmosphere and oceans remained within values suitable for organisms by the combination of four characteristics of the Solar System and the Earth System, which are: the Earth's fast rotation (see Sect. 3.3.3), the small the tilt of the axis of rotation of the planet maintained by the Earth-Moon system (see Sect. 3.3.2), the presence of an atmosphere with the right density and greenhouse gas composition to create a moderate greenhouse effect (see Sect. 3.6.1), and the Earth's distance from the Sun (see Sect. 3.6.1). In return, the presence of large biomasses of organisms in the oceans and on continents influenced the heat budget of Earth (see Sect. 3.5.5).

The overall habitability of Earth, that is, the long-term hospitability of the planet's environments to organisms, is dealt with in Chapter 4. The co-evolution of species and the Earth's environments led ecosystems to thrive in most environments of the planet, even those with extreme temperatures provided that liquid water was present at least temporarily (see Sect. 4.1.4). Chapter 4 examines changes in overall habitability corresponding to periods of variability in the Earth's environment ranging from one day to billions of years (Table 4.1). It also considers the mass extinctions of species of organisms that occurred during the last 542 million years, as witnesses of major crashes in habitability.

7.1.3   The Long-Term Greenhouse Effect

The conditions of temperature that prevailed at the surface of Earth until the advent of the ongoing global warming were regulated by a moderate greenhouse effect. Without it, the average temperature of the lower atmosphere (at ground level) could have been about $-18\,°C$, and perhaps as cold as $-50\,°C$ because of the development of a planet-wide glaciation (see Sect. 2.4.1). At such low global temperatures, all water on Earth would have been frozen, and the lack of liquid water would have prevented the presence of organisms (see Sect. 5.1.1). Conversely, an excessively strong greenhouse effect could have caused high temperatures that would have vaporized all the water (see Sect. 5.5.1), and the lack of liquid water would have had the same drastic consequence for organisms as a very weak greenhouse effect. Even if water can be liquid at temperatures higher than $100\,°C$ when the atmospheric pressure is high, organisms could not take advantage of the liquid water in such circumstances because the temperature would be too hot for their growth or even their survival (see Sect. 3.1.1).

It is thought that over the last 3.8 billion years or even more, the average temperature at the Earth's surface ranged between about 5 and $30\,°C$ (Fig. 7.1). The six cool and six warm periods in Fig. 7.1 correspond to icehouse and greenhouse Earths, respectively (see Sect. 4.5.5), and average temperatures colder than $10\,°C$ have occurred during the episodes of snowball Earth (see Sect. 4.5.4). The major cold periods tended to occur when the movements of continents brought a large landmass over one of the poles, where ice could accumulate. Figure 7.1 also shows that a general cooling has been going on for the last 50 million years, which may have been be related to Antarctica geophysical events that included changes in Southern Ocean currents and the opening of the Drake Passage, which led to a progressive accumulation of ice on the Antarctic continent (see Sect. 4.5.2).

The relatively small range of temperature variations is consistent with the fact that organisms have been abundant on the planet since billions of years (see Sect. 1.5.4), and this abundance suggests that the greenhouse effect was generally

Fig. 7.1 Temperature history of Earth from 4.6 billion years ago to present (note that the time axis is not linear). The dashed line indicates the present average temperature of around 15 °C. The original figure includes additional information on some of the temperature events. Credits at the end of the chapter

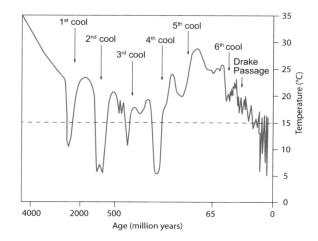

moderate during the whole period. This raises the question of the conditions that have led to the establishment of the Earth's moderate, natural greenhouse effect in the first place, and its regulation in the billions of years that followed. It is thought that the high concentration of atmospheric CO_2 initially caused a very strong greenhouse effect, and the mechanism responsible for the decrease in the CO_2 concentration and the associated greenhouse effect are examined in Chapter 5 (see Sect. 5.5.4). This process lowered the temperature of the initial ocean from perhaps more than 200 °C to temperatures that allowed the establishment and development of organisms and ecosystems, that is, below 100 °C. The present chapter addresses the geological mechanisms that contributed to keep the Earth's greenhouse effect moderate over the last 4 billion years.

7.2   The Tectonic Activity of Earth

7.2.1   Plate Tectonics

During the course of the Earth's billion-year history, the configuration of the surface of the planet has continually changed, that is, continents joined together and broke apart, mountains rose high in the sky and were eroded flat, the ground shook frequently, and volcanoes erupted repeatedly. These large-scale geological events continue to happen today, and it was discovered in the 1960s that they are interconnected. It was then found that all these geological phenomena were linked because the outermost shell of the globe is made of rigid pieces, called *plates*, that continually move relative to each other. Their motions are referred to as *plate tectonics*. The plates include the Earth's crust and the top part of the upper mantle (Fig. 6.11).

The plates float like rafts on the underlying fluid mantle because their density is lower than that of the upper mantle (Table 6.6). They move at the surface of Earth at speeds lower than 10 cm per year, carried by convection movements in the fluid magma of the upper mantle. These movements are fuelled by heat emitted by the Earth's core (see Sect. 3.2.3). Plate tectonics may have started very early on Earth. This could happen because the upper mantle contained some of the initial water of the planet (see Sect. 5.3.2), which reduced its viscosity and thus made movements of the plates possible. The low viscosity of the mantle was maintained subsequently by the injection of water from the overlying ocean, as soon as it formed or perhaps only later. Without water, there would likely be no plate tectonics on Earth (see Sect. 7.2.5, below).

The plates either move away from each other (called *divergence*), or collide with each other (called *convergence*), or slide past each other. The areas where the plates diverge, converge or slide past each other determine their boundaries. Given that the plates are rigid, their relative movements are accompanied by very large friction at the plate boundaries, and this friction causes earthquakes. Mapping the locations of the epicentres of earthquakes on the planet has shown that most epicentres were located in relatively narrow belts (the *epicentre* of an earthquake is the point at the Earth's surface located vertically above the point of origin of the earthquake under the ground) (Fig. 7.2). Researchers understood that these narrow belts delineate the boundaries of plates and this approach, together with the analysis of magnetic reversals in the seafloor (Sect. 8.5.2), led them to identify seven large plates and a number of smaller ones, shown in Fig. 7.2. Although the plates are rigid, their shapes have changed continually over the geological times. Indeed as the plates diverged and converged, some of their parts grew and others were destroyed. The combined rate of growth of all plates is equal to their rate of destruction, with the result that the overall surface of the globe (and thus its volume) does not increase or decrease.

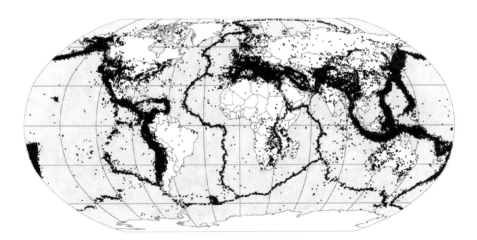

Fig. 7.2 Epicentres of 358,214 earthquakes recorded from 1963 to 1998, delineating the boundaries of the major tectonic plates: African, Antarctic, Eurasian, North American, South American, Pacific, and Indo Australian (sometimes subdivided into Indian and Australian). Credits at the end of the chapter

A given plate can have both continental and oceanic parts (Fig. 7.2).

The *continental parts of plates* have an average thickness of about 200 km. The rocks that form the continental crust are derived from magma from the underlying mantle and the fusion of materials (oceanic crust and sediments, and water) carried down by the subduction of plates (see Sect. 7.2.3, below). These mixed materials cooled and solidified in different ways depending on their relative proportions and depths. Some of the magma was trapped deep into the crust, where it cooled down slowly and generally produced coarse-grained rocks called *granite*, whereas other magma erupted on the surface, mostly through volcanoes, where it cooled down rapidly and generally produced fine-grained rocks called *basalt* (see Sect. 1.5.1). Basalt is denser than granite. In addition to these *igneous* rocks, the upper part of continental plates includes: *sedimentary* rocks, which are formed by the accumulation, at the bottom of water bodies and on land, of small mineral and organic particles; and *metamorphic* rocks, which result from the transformation of sediments (either marine or continental) by high temperature and pressure (see Sect. 1.5.2).

The *oceanic parts of plates* are denser than the continental parts, and they are also thinner with a thickness of typically 100 km. The seafloor is made of rocks formed by the solidification of volcanic magma coming from the mantle (igneous rocks, mostly basalt), on which sediments are deposited. The thickness of the sediment layer generally increases with the age of the seafloor, because older seafloor has received sinking particles during a longer time than younger seafloor, but local conditions can modify this rule (Figs. 7.3 and 8.9a). Indeed relative to the age of the seafloor, the thickness of the sediment layer is reduced in areas of low sedimentation rate or where bottom currents are eroding the accumulated sediments,

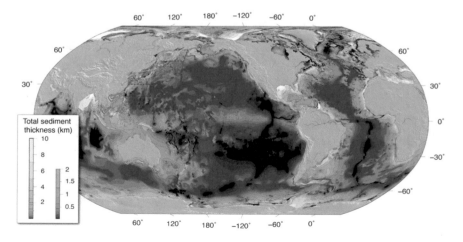

Fig. 7.3 Total thickness of sediments in the world's oceans and marginal seas. Credits at the end of the chapter

whereas the sediment can be very thick in areas of high sedimentation rate, for example near the mouth of large rivers (more than 15 km in some areas of the Bay of Bengal, between the Indian sub-continent and Southeast Asia). The book of Siebold and Berger (2017) provides further reading on the seafloor.

When the combined pressure of water and the accumulated sediments becomes high enough, the unconsolidated sediment at the bottom of the sediment layer consolidates into sedimentary rocks, examples of which are limestone ($CaCO_3$) and shale (stratified rock formed of consolidated mud or clay). The sediments from which sedimentary rocks are formed often include both mineral and organic particles, with the result that these rocks contain organic carbon in their matrix. The organic carbon present in rocks account for about 20% of all the carbon on Earth.

The development of the theory of plate tectonics in the 1960s accounted for a large number of geological phenomena previously thought as being more or less independent. These included: the distribution of earthquakes and volcanoes, which reflect upward and downward movements that occur at plate boundaries (Fig. 7.2); the existence of ocean trenches, where the oldest seafloor is destroyed by subduction; and the building of mountains, which results from plate convergence. These mechanisms are explained in the next sections.

The theory of plate tectonics resulted from the works of many researchers between the beginning of the twentieth century (pioneering book of Alfred Wegener on continental drift in 1915) and the 1960s. It provided a unifying theory for Earth Sciences, as did the publication of the first article on the theory of relativity for physics about fifty years earlier (Einstein 1905) and the publication of the founding work of the theory of evolution for the Life Sciences another fifty years before (Darwin 1859, followed by several theoretical developments in the twentieth century). The book of Frisch et al. (2011) provides further reading on plate tectonics, including the researchers involved in the development of the theory.

7.2.2   Growth of Oceanic Crust

The areas of the seafloor or continents where plates move away from one to another are *divergent plate boundaries*. Plates diverge because convection currents exist in the fluid magma that forms the upper part of the underlying mantle (Fig. 7.4a,b). These currents are caused by the flow of energy from the interior of Earth (see Sect. 3.2.3). The space that opens between the diverging plates is filled with fluid magma that surges upwards through submarine volcanoes, and this magma becomes part of the two plates as it cools down and solidifies into basalt rock. This process forms new plate material at the divergent boundary, which causes the two plates to grow horizontally (Fig. 7.4b).

Most active divergent boundaries are in the oceans, where they form the well-known *mid-ocean ridges* (Fig. 7.4b), whose combined length is about 70,000 km; by comparison, the circumference of Earth around the equator is 40,075 km.

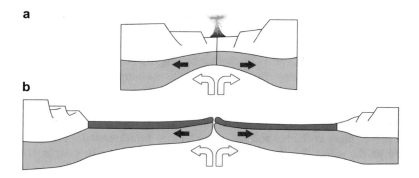

Fig. 7.4 Schematic representation of divergent plates, which move away from each other (solid arrows) under the effect of convection currents in the fluid magma in the upper part of the underlying mantle (open arrows): **a** on a continent, and **b** in an ocean. Credits at the end of the chapter

These submarine volcanic mountain chains typically have a valley, known as a rift, along their spines. As the two plates grow and move away from the divergent boundary, they are progressively covered by sediment whose thickness increases with the distance from the divergent boundary (Fig. 7.3). After a long time, the sediments consolidate into sedimentary rocks, so that the older basaltic rocks are generally overlaid with sedimentary rocks (see Sect. 7.2.1).

Some divergent boundaries are located on continents, where they form *rift valleys* such as the East African Rift Valley, which is located above sea level and is partly occupied by large lakes (Fig. 7.4a). As they progressively widen, continental rift valleys are invaded by the ocean, and an example of a completely developed rift valley is the Red Sea where the two continental parts of the former single plate are fully separated and the central valley is under sea level. A continental divergence often leads the opening of an ocean through the fragmentation of the continental plate, and an oceanic divergence is the underwater time continuation of the process when the ocean is already open.

The earthquakes caused by the combined lateral movements of the rigid plates and vertical movement of the fluid magma generally occur within 60 km of the surface, and are called *shallow earthquakes*. The oceanic and continental divergent boundaries correspond to the long, narrow bands of earthquakes in Fig. 7.2, where their epicentres delineate the locations of divergent boundaries.

7.2.3   Growth of Continental Crust, and Destruction of Oceanic and Continental Crust

Continental crust grows at *convergent plate boundaries*, where the converging plates simultaneously undergo growth and destruction. There are three different types of convergence, with different consequences for the destruction and growth of oceans and continents: ocean-ocean, ocean-continent and continent-continent (Fig. 7.5).

Ocean-ocean convergence. When oceanic parts of two plates meet, one of the two passes under the other, a phenomenon known as *subduction* (Fig. 7.5a). As explained above, the seafloor is made of igneous, basaltic rocks formed by the solidification of volcanic fluid magma from the mantle, overlaid with sedimentary rocks resulting from the consolidation of accumulated sediments (see Sect. 7.2.2). As the subducting part of one plate sinks into the mantle under the second, it becomes increasingly warmer, and at a depth of about 150 km, the temperature and pressure become so high that the plate and the overlying sediments begin to melt into magma. As a consequence of chemical reactions at depth, the resulting magma has a lower density than the basaltic rocks of the ocean floor, and part of this lower-density magma attaches to the base of the overriding plate. Another part of this magma moves up towards the surface, where it erupts in volcanoes and forms new continental crust whose density is lower than that of the oceanic crust. These volcanoes form *volcanic island arcs* above the subducted plate, such as Japan and the Eastern Caribbean Islands.

Fig. 7.5 Schematic representation of convergent plates moving towards each other. There are three types of convergence: **a** ocean-ocean, **b** ocean-continent, and **c** continent-continent. Solid arrows: movements of tectonic plates. Credits at the end of the chapter

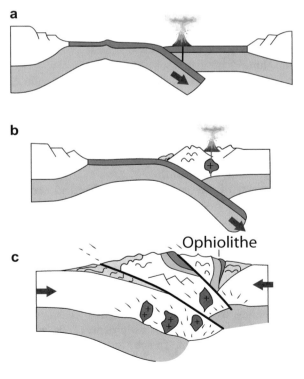

The penetration of the plate into the mantle causes destruction of oceanic crust belonging to the subducting plate, whose downward movement at about 45° under the overriding plate causes *intermediate and deep earthquakes* (60–300 and 300–700 km, respectively) along the subducting plate. Both the material attached to the base of the overriding plate and the volcanic islands on that plate form new continental crust, which contributes to the growth of the overriding plate and thus the growth of the continental crust.

Ocean-continent convergence. When the oceanic part of a plate meets the continental part of another, the first subducts under the second (Fig. 7.5b) because the oceanic crust is denser and thinner than the continental crust (Table 6.6). In addition, huge amounts of unconsolidated and consolidated sediments accumulated over a thickness of several kilometres on the seafloor are scraped off the ocean's floor and plastered onto the continental margin during subduction, a phenomenon called *sediment accretion.* As in the case of the ocean-ocean convergence (previous paragraphs), the basaltic rocks and the overlying sediments not plastered onto the continental margin do melt into magma, which has a lower density than the basaltic rocks of the ocean floor. This leads to both the addition of lighter material to the base of the continental plate and volcanic activity on that plate. The two phenomena create additional continental crust, for example the Cascade Mountain Range and the Andean Volcanic Belt along the west coasts of North and South America, respectively.

The continental margin of a plate, including the material scraped off the subducted seafloor, can be deformed by the *compression* resulting from the convergence. This compression creates chains of non-volcanic high mountains, such as the North-American Rocky Mountains.

The penetration of the plate into the mantle causes destruction of seafloor (basaltic rocks and overlying sediments), and the downward movement of the subducting plate at about 45° under the continental plate causes *intermediate and deep earthquakes* along the subducting plate. The material attached to the base of the continental plate, the accumulation of material from volcanoes on that plate and the accretion of sediments along the continental margin all contribute to the growth of the continent.

Several of the geological and chemical processes described from this section to Sect. 7.3.3 are represented in their environmental setting in Fig. 7.6. The exchanges of water between the atmosphere and the ocean (evaporation and precipitation) illustrated in Fig. 7.6 are described in Chapter 3 (see Sect. 3.4.2).

Consequences of ocean-ocean and ocean-continent convergence. Continental crust is formed in both ocean-ocean and ocean-continent convergence, with the result that the first type of convergence can progressively evolve into the second as the first type creates continental crust. In addition, the plate boundaries corresponding to both ocean-ocean and ocean-continent convergence are characterized by the presence of deep depressions in the seafloor, called *trenches.* For example, the Mariana Trench, located at the subduction boundary of the Pacific

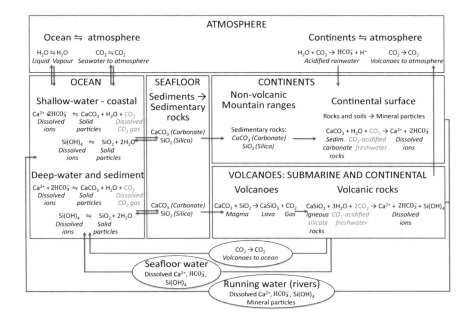

Fig. 7.6 Cycle of inorganic carbon among the atmosphere, the ocean, the seafloor, the continents, and the volcanoes (see Sect. 7.2.3–7.3.3). Blue arrows: exchanges of inorganic carbon between compartments. The CO_2 molecules in green colour highlight the difference between the dissolution of molecules of $CaSiO_3$ and $CaCO_3$ by CO_2-acidified fresh water (details in Fig. 7.7). Each arrow is explained in the text. Credits at the end of the chapter

plate beneath the Philippine Sea plate, is the deepest point in all oceans (10,994 m below sea level; if the tallest mountain on continents, Mount Everest, were moved to the Mariana Trench at its deepest point, the peak of the mountain would be more than 2 km underwater). Some trenches are partially or completely filled with sediments, which had been accumulated there by accretion.

The major mineral components of the seafloor rocks carried into the mantle by a subducting plate are calcium carbonate ($CaCO_3$) and silica (SiO_2) (arrow seafloor-to-volcanoes in Fig. 7.6; the production of these two minerals is explained below) (see Sect. 7.3.3). The mineral components also include such chemical elements as aluminium (Al), magnesium (Mg), potassium (K) and iron (Fe). The melting of the plate frees the chemical elements of $CaCO_3$ and SiO_2 and also CO_2 in the hot magma. The chemical elements and the CO_2 dissolved in the magma are carried to volcanoes located either above the trenches or far away on the seafloor (mostly on mid-ocean ridges). The CO_2 from the first type of volcanoes escapes to the atmosphere (arrow volcanoes-to-atmosphere in Fig. 7.6) and the CO_2 from the second type dissolves in seawater (arrow volcanoes-to-ocean in Fig. 7.6).

Upon cooling and solidification of the lava, the chemicals elements it contains combine to form calcium silicate ($CaSiO_3$). These reactions are represented by the following summary chemical equation:

$$CaCO_3 + SiO_2 \rightarrow CaSiO_3 + CO_2 \qquad (7.1)$$

The formation of $CaSiO_3$ rocks by solidification of volcanic lava occurs both on continents at convergent boundaries, and on the seafloor at divergent boundaries. Silicate minerals are present in volcanic rocks under different forms (examples: $CaSiO_3$, wollastonite; $CaAl_2Si_2O_8$, anorthite), but for simplicity, we use $CaSiO_3$ in this book as generic silicate mineral, and we call the rocks containing $CaSiO_3$ *silicate rocks*. We will see below that the $CaSiO_3$ and CO_2 released by volcanoes both play important roles in the global cycling of carbon (see Sect. 7.3.3). Equation 7.1 and other chemical equations in the present chapter are shown in their environmental setting in Fig. 7.6.

Continent-continent convergence. When continental parts of two plates meet, they collide (Fig. 7.5c). The intense compression can cause the formation of ranges of high mountains, such as the Himalayas and the European Alps (not to be confounded with the Australian Alps, the Japanese Alps, the Southern Alps in New Zealand … or the Montes Alpes on the Moon). The collision between the continental parts of two plates causes compression and thus shortening and thickening of the continental plates. It is accompanied by generally *shallow earthquakes*.

The relative occurrence of the different types of earthquakes on the planet reflects the tectonic mechanisms associated with divergent and convergent boundaries examined in the previous section and this one. Indeed, about 85, 12 and 3% of all the earthquakes on the planet are shallow, intermediate and deep, respectively. The shallow earthquakes occur at divergent boundaries, at most of the continent-continent convergent boundaries, and along the upper part of the subducting plates at ocean-ocean and ocean-continent convergent boundaries. The intermediate and deep earthquakes occur along the intermediate and deep parts of the subducting plates at ocean-ocean and ocean-continent convergent boundaries. In addition, some deep earthquakes are observed under a few mountain ranges.

Marine sedimentary rocks on continents. The sedimentary rocks formed in the deep ocean and in shallow seas have different fates. Most deep-ocean rocks are subducted into the mantle, and their materials are returned to the surface through volcanoes, as explained above in this section. However, tectonic movements in areas of plate convergence sometimes uplift (raise) intact portions of the seafloor at the surface of continents, where these pieces of ancient deep seafloor are known as *ophiolites* (Fig. 7.5c). These typically consist of igneous and metamorphic rocks overlaid with sedimentary rocks, which is the sequence of rocks that exists in the seafloor. The thickness of some ophiolites reflects that of the seafloor of which there were part millions of years before their uplifting to the land. One example is the Semail Ophiolite in Oman (Arabic Peninsula), which has a thickness of about 15 km, consistent with the thickness of the oceanic crust.

The fate of sedimentary rocks formed in shallow seas is generally different from that of their deep-ocean counterparts, and these rocks often form large deposits on continents. Several mechanisms explain the presence of shallow-sea sedimentary rocks on continents (arrow seafloor-to-continents in Fig. 7.6). In some cases, a piece of a carbonate continental shelf was trapped between two convergent plates when these met and was thus uplifted onto the continent. A spectacular example is the summit of Mount Everest in the Himalaya (Nepal), which is the highest point on Earth and is made of limestone. This $CaCO_3$ was formed on the continental shelf of the ancient Tethys Ocean (Fig. 4.9), and uplifted by tectonics to its present location more than 8,800 m above the sea level. In other cases, the limestone was formed at the bottom of a shallow sea, whose present location was under the sea level in the past and is now above the sea level. One example is the limestone of the Paris Basin (France), which was at the bottom of a sea 45 million years ago and is now above sea level, which allowed its use in the last centuries to erect the beautiful Paris buildings. In still other cases, the limestone was formed on an active continental margin then under water, and later uplifted to the continent by the formation of a non-volcanic mountain chain. One example is provided by the Rocky Mountains in western Canada, which are made almost exclusively of sedimentary rocks. The article of Arndt (2013) provides further reading on the formation and evolution of the continental crust.

7.2.4   Ages of the Seafloor and Continents

Because the seafloor is continually created at mid-ocean ridges and destroyed in subduction zones, no parts of the seafloor are very old. In fact, very little of the seafloor is older than 150 million years, and almost none exceeds 200 million years (Fig. 8.9a). The oldest seafloor (280 million years old) is found in the Mediterranean Sea, because that sea is the last remnant of the oldest ocean still represented on Earth, called Tethys (Fig. 4.9), which is progressively disappearing as the African and Eurasian plates collide and the Mediterranean Sea thus shrinks (Fig. 7.2). A piece of seafloor 280 million years old may seem very ancient, but is in fact very young compared with the oldest known continental rocks, whose ages range between about 3.7 and 4.3 billion years.. It may be that the oldest intact Earth rock was found on the Moon. Indeed, the study of one of the rocks collected by Apollo 14 astronauts in 1971 came to the conclusion that this rock had been formed on Earth more than 4.0 billion years ago, and got jettisoned to the Moon by one of the many impacts of meteoroids on the early Earth.

Based on plate movement, it is estimated that the current average age of ocean plates is of the order of 70 to 80 million years. This time period corresponds, for example, to the time it will take for the basalt presently formed on the East Pacific Ocean seafloor near the equator, at a divergent plate boundary located in the vicinity of the Galapagos Islands, to reach the nearest subduction zone, which is the Tonga-Kermadec Trench in the West Pacific Ocean. This means that marine sediment particles remain, on average, 70 to 80 million years on the eseafloor. The sediments deposited near mid-ocean ridges stays on the seafloor longer than this average time, and those deposited near subduction zones stays on the seafloor shorter.

7.2.5   Tectonic Activity in the Solar System

Volcanic activity occurs not only on Earth, but also on other bodies of the Solar System. Indeed, robotic space exploration found active volcanoes on the moons Io of planet Jupiter, Triton of Neptune, and Enceladus of Saturn. In addition, the tallest mountain in the Solar System, Olympus Mons on planet Mars, is a volcano, but it is not active nowadays (see Information Box 5.10).

Despite the presence of active volcanoes on some moons of giant planets, plate tectonics seems to presently occur only on Earth. One reason for this could be that presence of large bodies of surface liquid water is a necessary condition for having plate tectonics, and this condition seems to presently exist only on Earth. However, the possible existence of tectonic activity elsewhere in the Solar System is debated among specialists. Indeed, space exploration has provided signs of possible tectonic activity on the moons Europa of Jupiter and Titan of Saturn. It is interesting to note that these two moons are also suspected to harbour underground liquid oceans (see Information Box 5.4), which would be consistent with the idea that tectonics requires the presence of large bodies of surface liquid water.

7.3   Tectonic Activity and Long-Term Functioning of the Earth System

7.3.1   Tectonic Recycling of Carbon and Other Biogenic Elements

The eruptions of continental and submarine volcanoes not only emit rock-forming lava (see Sect. 7.2.3, above), but also release gases into the atmosphere and the seawater (arrows volcanoes-to-atmosphere and volcanoes-to-deep water in Fig. 7.6). These gases result from the melting of the subducting oceanic crust, including its consolidated sediments, during which the gases are dissolved in the fluid magma of the mantle (see Sect. 7.2.3). For example volcanic CO_2 results from the melting of both sedimentary rocks such as limestone ($CaCO_3$), which is mostly formed in the ocean by planktonic and benthic organisms and by coral reefs, and shales, which contain organic material. The melting of seafloor rocks during subduction injects huge amounts of CO_2 into the fluid magma, where this CO_2 is retained until it is released into the atmosphere.

When the magma moves up towards the ground or the seafloor, the pressure decreases and the gases leave the magma. As a consequence, the emissions of magma are accompanied by the release of gases into the atmosphere by volcanoes, fumaroles (openings in the ground, often near volcanoes) and geysers, or in the seawater by submarine volcanoes (Eq. 7.1). The main atmospheric volcanic gases are water vapour (H_2O), carbon dioxide (CO_2), and according to the type of emission, different proportions of sulphur dioxide (SO_2) and hydrogen sulphide (H_2S). The composition of volcanic gases varies considerably among volcanoes,

---

Information Box 7.1 Erosion and Volcanic Emissions: Recycling of Carbon The decomposition of rocks, soils and their minerals through contact with the atmosphere, water or organisms is termed *weathering*, and the combined process of weathering and transport of the resulting products is called *erosion*. For simplicity, we only use the word *erosion* in this text.

The greenhouse effect was sustained in the long term by the release of carbon contained in continental rocks by erosion and the volcanic emission of CO_2. These two processes represent together a total flux of 0.2 billion tons of carbon per year.

The pre-industrial stocks of carbon in the hydrosphere (mainly the ocean) and the atmosphere were 40,400 and 590 billion tons, respectively. The combined 41,000 billion tons of carbon in these two reservoirs made up the stock of free carbon dioxide in the Earth's fluid envelopes. Assuming that this stock remained constant in pre-industrial times, then the rate of carbon burial in the seafloor and the mantle was equal to the flux of CO_2 by volcanic emissions and rock erosion (driven by tectonic activity), that is, 0.2 billion tons. In the hypothetical case where 0.2 billion tons of carbon would have been buried every year in the seafloor and the mantle and no carbon would have been recycled by tectonically-driven activity, then the reservoir of free carbon dioxide (in the ocean and the atmosphere) would have been exhausted in about 200,000 years (41,000 billion tons divided by 0.2 billion tons per year).

A period of 200,000 years may seem long, but it is only a tiny fraction of the duration of the 4 billion years of the Living Earth. Indeed, 200,000 years out of 4 billions are equivalent to about 4 seconds in a 24-h day. Without a steady supply of CO_2, Earth would have frozen very rapidly. This shows that the recycling of carbon by volcanic emissions and rock erosion had the essential role of sustaining the greenhouse effect needed to maintain the thermal habitability of the planet during the course of the Earth's history.

---

but water vapour is always the most common gas with more than 60% of the total emissions, and CO_2 accounts for 10–40%. The high proportions of water vapour and CO_2 in volcanic emissions is important for the Earth's climate because they are both major greenhouse gases (see Information Box 2.7).

On continents, the recycling of carbon linked to tectonic activity comprises the volcanic emission of CO_2 and the release of carbon contained in rocks by the action of erosion (see Sect. 7.3.3, below), whose resulting carbon is carried to the ocean by running waters. The recycling of carbon by these two mechanisms had an essential role during the course of the Earth's history, as it regulated the long-term greenhouse effect needed to maintain the thermal habitability of the planet (see Information Box 7.1).

In oceans, sediments contain not only carbon but also other biogenic elements required by organisms for their key functions, which are to preserve genetic information, build biomass, sustain activity, and reproduce (see Sect. 6.3). These chemical elements are in the sediments because the biological activity taking place in the ocean and on continents continually produces remains that either sink to the seafloor or are buried in soils. A large part of these remains is rapidly recycled by bacteria, but over thousands and millions of years, most of the biogenic elements contained in biological remains eventually reach the seafloor, where they are buried as organic sediments. The fact that these elements are generally available where organisms need them, that is, in ocean waters and on continents, means that they are efficiently recycled at the global scale. The book of Shikazono (2012) provides further reading on the cycling of carbon and other chemical elements.

Tectonic activity ensures the global recycling of biogenic elements in two major ways. These are the volcanic and hydrothermal activity, and the erosion on continents, as explained in the following two paragraphs

Volcanic and hydrothermal activity. The sedimentary rocks carried into the mantle by the subducting plates are molten in the magma, and their chemical elements are carried to the surface of continents or the seafloor by volcanic activity (see Sect. 7.2.3). Through continental and submarine volcanoes, the biogenic elements are released by both the effusion of lava (Ca, K, Al, Si, Fe, Ni, P) and the emission of volcanic gases (CO_2, SO_2, CO, H_2S). In addition, the circulation of seawater within the seafloor near submarine volcanoes transports deep-crust minerals upwards, and the resulting hydrothermal fluids fuel ecosystems that surround hydrothermal vents and also contribute some essential chemical elements (such as iron) to the global ocean (see Sect. 3.2.3). The hydrothermal vents and most submarine volcanoes are located at mid-ocean ridges. The articles of Staudigel and Clague (2010) and Rubin et al. (2012) provide further reading on submarine volcanoes.

Erosion on continents. The formation of mountains (including volcanoes) on continents creates slopes, which increase the rates of both water erosion of rocks and transport of eroded materials by mechanisms described below (see Sect. 7.3.3). Conversely, erosion affects the relief. Indeed, erosion decreases the relief in the long term, but it can create slopes in the shorter term for example when a river cuts a canyon.

Water erosion releases biogenic elements, which are both used by ecosystems on continents and transported by rivers to the oceans where they are used by marine ecosystems. The recycling of biogenic elements is as important for the functioning of the Earth's ecosystems as that of carbon described above in this section. The book of Bauer and Velde (2014) provides further reading on the physical and chemical processes that affect the surface of continents.

7.3.2   *Tectonic and Chemical Sequestration of Biogenic Element*

The global circulation of carbon and other biogenic elements has two components, that is, recycling, examined in the previous section, and sequestration, described in this section. In the context of Earth sciences, the words *sequester* and *sequestration* refer to the long-term storage of compounds in natural reservoirs by physical, chemical and/or biological processes.

The basic mechanism of sequestration is the transfer of chemical elements from the gaseous to the liquid phase and from the liquid to the solid phase. One example in the current ocean is the following suite of processes: dissolution of atmospheric CO_2 in seawater (Eq. 6.3); fixation of carbon from dissolved CO_2 into organic matter by autotrophic organisms (see Information Box 5.1); circulation of this carbon in food webs (see Information Box 5.3), where most of it is released as CO_2 (see Information Box 5.2); burial of the remaining organic matter in sediments, where some of it sequestered for a very long time. The sequestered chemical elements include hydrogen (H), oxygen (O), carbon (C), nitrogen (N), calcium (Ca), phosphorus (P) and sulphur (S). These elements are sequestered as inorganic and organic compounds, which are continually formed by a large number of natural processes and a small fraction of which is sequestered. Examples of sequestered compounds are water (H_2O), calcium carbonate ($CaCO_3$), and organic matter which contains H, O, C, N, Ca, P, S and a variety of other chemical elements (Table 6.1).

Small fractions of the minerals and organic matter formed in the oceans and on continents accumulate on the seafloor and in soils. There, the minerals and organic compounds may consolidate to form rocks in which organic matter may be dispersed, or organic-rich deposits that include coal and peat. Both dispersed and concentrated organic matter may subsequently give rise to petroleum and natural gas due to the secondary processes of thermal maturation, migration and accumulation in porous sedimentary reservoirs. The fractions deposited at any time are very small and the processes of formation of rocks and fossil organic matter are very slow, but given the long times involved, the amounts of compounds sequestered are very large.

In addition to the above chemical and biological processes, plate tectonics contributes to sequester materials at great depths under the seafloor. For example, a significant quantity of the interstitial water of sediments may be stored between 400 and 650 km deep in the mantle at the transition between the upper and lower mantle, fed by the subduction of the sediments carried by the plates (see Sect. 5.2.1). Similarly, large amounts of the subducted carbon and nitrogen may be trapped deep in the mantle.

There was a period during the history of Earth when chemical elements that would be required later by organisms and ecosystems (carbon, oxygen, nitrogen, sulphur) could have been largely lost to outer space. This occurred when a new atmosphere was created 4.4 billion years before present or earlier by the degassing of the cooling magma (see Sect. 8.1.1). This atmosphere was very rich in the above biogenic elements, and as long as these remained in gaseous forms, they could be lost from the planet (see Sect. 8.4.1). The formation of the ocean about 4.4 billion years ago removed these chemical elements from the atmosphere, and thus prevented their loss from the Earth System.

The sequestration mechanisms that operated more than 4 billion years ago were the transfer of chemical elements to the liquid or solid phase through physical processes that condensed gases into liquids, and the formation of solid compounds from ions dissolved in liquid water by chemical processes. An example of condensation was the formation of liquid water (H_2O) from atmospheric water vapour, which was a key mechanism for the retention of water on Earth (see Sect. 5.4.2). An example of chemical process was the formation of solid $CaCO_3$ from dissolved calcium and bicarbonate ions in water (see Information Box 5.11 and Sect. 6.4.2). These were followed by the storage of compounds as described in above paragraphs for the modern ocean. As a consequence, more than 99% of the main biogenic elements that were in the initial atmosphere 4.4 billion years ago (carbon, oxygen, nitrogen, and sulphur) were sequestered, so that less than 1% has been available for use by organisms since their early occurrence on Earth. We examine in the remainder of this chapter the mechanism by which the rare biogenic elements remained available to the organisms and ecosystems during the following billions of years.

7.3.3   Erosion of Rocks and Carbon Recycling

It was explained above that tectonic activity brings to the continents two types of carbon-containing rocks (see Sect. 7.2.3). On the one hand, carbonate rocks, which are mostly composed of $CaCO_3$ but contain also magnesium carbonate ($MgCO_3$) and other carbonates, are formed on the seafloor and moved later to the continents. On the other hand, silicate rocks, represented by the generic $CaSiO_3$ mineral in this book, are formed from volcanic lava on land and on the seafloor. The minerals that contain silicon are much more abundant in the Earth crust than those without Si. Indeed, more than 90% of the volume of the crust is occupied by a wide variety of minerals that contain Si, and less than 10% by Si-less minerals (which include carbonates). The proportions of the two types of minerals are consistent with the fact that silicon accounts for 28% of all the atoms in the crust and carbon only a very small value (Table 6.1). The Si-containing minerals are continually replenished by large-scale, long-term geophysical processes (underwater formation of oceanic crust at ocean ridges, formation of continental crust in subduction areas, volcanic eruptions on continents) whereas the carbonates are involved in smaller-scale, shorter-term processes of the Earth System. We consider the fate of the two types of rocks in turn in the remainder of this section.

The rocks and soils on the continents are subject to erosion (see Information Box 7.1). The agents of physical breakdown include changes in temperature, the action of winds and waves, and the mechanical breaking of soils and rocks by organisms through such activities as the growth of plant roots and the burrowing of animals. The agents of chemical breakdown include dissolution by CO_2-acidified water, oxidation, and the release by plants and animals of compounds that are acidic or promote chemical dissolution. The products of physical and chemical breakdown (called *weathering* by specialists) are transported away by water, ice, snow, wind, waves, and gravity. In the remainder of this book, we mostly consider erosion by physical and chemical agents, especially the dissolution of solid materials (rocks and sedimentary particles) by CO_2-acidified water and the transport of the resulting compounds by water runoff on continents and water circulation in the seafloor. The organisms both accelerate erosion, as agents of physical and chemical breakdown, and reduce it, especially plants whose roots stabilize the soil on slopes and dunes and along shores. Through their roles in erosion, organisms contribute to the long-term recycling of carbon, which sustains the natural greenhouse effect.

One result of erosion is the formation of mineral particles, which are transported by the atmosphere (dust), running waters (suspended solids) and other agents, and eventually find their way to the ocean where they accumulate in sediments. As they sink in the water or after their burial in sediments, these particles often undergo physical and chemical transformations, thus creating a wide variety of debris called (in order of decreasing size) *sand*, *silt* and *clay*, which eventually consolidate into sedimentary rocks. The latter generally include some fraction of organic matter disseminated throughout the rock, and the sedimentary rocks that contain more than 3% of organic carbon are called *organic-rich sedimentary rocks*.

We examine in this section the chemical erosion of continental and seafloor rocks by CO_2-acidified water. Natural water is acidified by the dissolution of CO_2 gas, which creates dissolved ions: bicarbonate (HCO_3^-), carbonate (CO_3^{2-}) and hydrogen (H^+). Under the average conditions of the current surface ocean, the dissolved inorganic carbon in seawater is mostly in the form of HCO_3^- (Eq. 6.3), and we consequently do not consider CO_3^{2-} below (see Information Box 6.3). Combined Eqs. 6.2 and 6.3 are renumbered here Eq. 7.2:

$$CO_2 + H_2O \rightleftharpoons HCO_3^- + H^+ \tag{7.2}$$

The equilibrium double arrows in the equation indicate a balance between the CO_2 and H_2O molecules (left side) and the HCO_3^- and H^+ ions (right side). The dissolution of one molecule of CO_2 gas in water (left side of the equation) shifts the equilibrium to the right, which increases the concentration of H^+ ions in the water. The higher concentration of H^+ ions makes the water more acidic, which lowers its pH (see Information Box 6.4). The dissolution of CO_2 in water described by Eq. 7.2 is represented by the arrows from the atmosphere to other compartments of the Earth System in Fig. 7.6.

In the following paragraphs we examine in turn the chemical alteration by CO_2-acidified water of continental $CaCO_3$ rocks, continental $CaSiO_3$ rocks, and seafloor $CaSiO_3$ rocks. The chemical equations representing the erosion of $CaCO_3$ (on continents) and $CaSiO_3$ (on continents and in the seafloor) are summarized

Information Box 7.2 Chemical Alteration of Continental $CaCO_3$ Rocks The first step in the chemical alteration of continental $CaCO_3$ rocks is the dissolution of atmospheric CO_2 in rainwater (arrow atmosphere-to-continents in Fig. 7.6):

$$CO_2 + H_2O \rightleftharpoons HCO_3^- + H^+ \qquad (7.2)$$

The sign $\rightleftharpoons$ indicates that this reaction and the following ones are reversible, meaning that according to the environmental conditions, the reaction can go in either direction. The ensuing process is summarized in the following equation (see Information Box 5.11, Eq. 5.10 read from right-to-left):

$$CaCO_{3(solid)} + H_2O + CO_{2(aqueous)} \rightleftharpoons Ca^{2+}_{(aqueous)} + 2HCO^-_{3(aqueous)} \qquad (7.3)$$

This equation does not detail the chemical reactions involved in the process, but shows instead the overall balance of the reactions. An increase in the concentration of dissolved $CO_{2(aqueous)}$ in the water (left side of the equation) shifts the reaction towards the right, which causes the dissolution of solid $CaCO_3$. Hence one molecule of CO_2-acidified water has the power to dissolve one molecule of solid $CaCO_3$, and the reaction forms two molecules of dissolved HCO_3^-.

The HCO_3^- and Ca^{2+} ions dissolved in continental fresh waters are ultimately transported into the ocean by running waters (arrow continents-to-ocean/running-water in Fig. 7.6). Once they are in the ocean, some of the dissolved HCO_3^- and Ca^{2+} ions are used by calcifying organisms to form solid $CaCO_3$ (see Information Box 5.11, Eq. 5.10):

$$Ca^{2+}_{(aqueous)} + 2HCO^-_{3(aqueous)} \rightleftharpoons CaCO_{3(solid)} + H_2O + CO_{2(aqueous)} \qquad (7.4)$$

This equation (read from left-to-right) shows that $CaCO_3$ formation by organisms, called *calcification*, is the reverse of $CaCO_3$ dissolution (Eq. 7.3). The different meanings of *calcification* are explained in Chapter 4 (see Information Box 4.4). Equation 7.4 indicates that the formation of one molecule of solid $CaCO_3$ in the ocean, which requires two molecules of dissolved HCO_3^- (left side of the equation), has two different environmental effects (right side). First, it releases CO_2 in seawater, from where the dissolved CO_2 generally escapes to the atmosphere where it increases the concentration of this gas (arrow ocean-to-atmosphere in Fig. 7.6). Second, the solid $CaCO_3$ sinks to the seafloor or remains there after the death of the calcifying organisms. The formation of $CaCO_3$ occurs in the ocean, lakes and rivers, but the bulk of it takes place in the ocean. Hence for simplicity, we only consider the formation $CaCO_3$ in the ocean in this book.

Combining Eqs. 7.3 and 7.4 summarizes the erosion of continental $CaCO_3$:

$$\textit{Continental } CaCO_{3(solid)} + H_2O + CO_{2(aqueous)} \rightleftharpoons$$
$$\textit{Seafloor } CaCO_{3(solid)} + H_2O + CO_{2(aqueous)} \qquad (7.5)$$

Equation 7.5 shows that chemical alteration of continental $CaCO_3$ transfers this mineral from continents to the seafloor without changing the concentration of aquatic or atmospheric CO_2 (Fig. 7.7).

and compared in Fig. 7.7. The net results and summaries of the chemical reactions in Fig. 7.7 show that the erosion of $CaCO_3$ rocks does not change the concentration of atmospheric CO_2, whereas the erosion of $CaSiO_3$ rocks causes the removal of CO_2 from the atmosphere and its long-term storage in solid $CaCO_3$. We carefully explain this key difference below.

First, we consider the chemical alteration of *continental $CaCO_3$ rocks* (limestone and others; arrow non-volcanic mountain chains-to-continental surface in Fig. 7.6). The chemical equations are given separately in Information Box 7.2, and summarized in the upper part of Fig. 7.7.

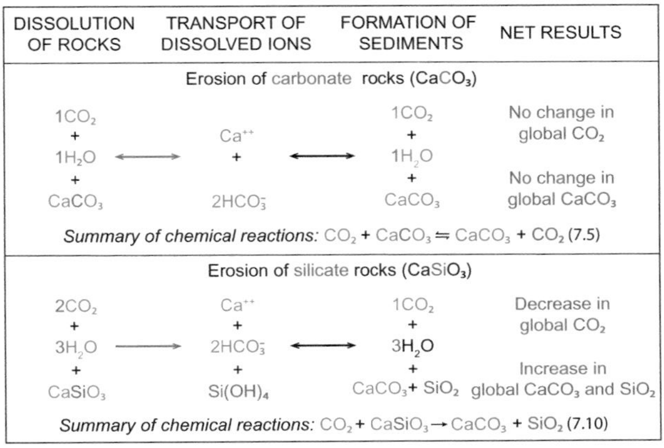

DISSOLUTION OF ROCKS	TRANSPORT OF DISSOLVED IONS	FORMATION OF SEDIMENTS	NET RESULTS
Erosion of carbonate rocks ($CaCO_3$)			
$1CO_2$ + $1H_2O$ + $CaCO_3$	Ca^{++} + $2HCO_3^-$	$1CO_2$ + $1H_2O$ + $CaCO_3$	No change in global CO_2 / No change in global $CaCO_3$
Summary of chemical reactions: $CO_2 + CaCO_3 \rightleftharpoons CaCO_3 + CO_2$ (7.5)			
Erosion of silicate rocks ($CaSiO_3$)			
$2CO_2$ + $3H_2O$ + $CaSiO_3$	Ca^{++} + $2HCO_3^-$ + $Si(OH)_4$	$1CO_2$ + $3H_2O$ + $CaCO_3 + SiO_2$	Decrease in global CO_2 / Increase in global $CaCO_3$ and SiO_2
Summary of chemical reactions: $CO_2 + CaSiO_3 \rightarrow CaCO_3 + SiO_2$ (7.10)			

Fig. 7.7 Processes of chemical alteration of carbonate and silicate rocks: dissolution of rocks by CO_2-acidified water, transport of the resulting ions in water, formation of $CaCO_3$ sediments, and net results of the chemical reactions (see Information Boxes 7.2 and 7.3). Characters in blue and red: similarities and differences, respectively, between the two processes. Double and rightwards arrows: two-way (reversible) and one-way (irreversible) chemical reactions, respectively. The summary chemical equations are numbered as in the text. Credits at the end of the chapter

Information Box 7.3 Chemical Alteration of Continental $CaSiO_3$ Rocks It is shown in Information Box 7.2 that the dissolution of one molecule of $CaCO_3$ in continental waters takes up *one molecule* of atmospheric CO_2. The following equation shows that the dissolution of one molecule of $CaSiO_3$ takes up *two molecules* of CO_2 from the atmosphere:

$$CaSiO_{3(solid)} + 3H_2O + 2CO_{2(aqueous)} \rightarrow Ca^{2+}_{(aqueous)} + 2HCO^-_{3(aqueous)} + Si(OH)_{4(aqueous)} \quad (7.6)$$

This equation does not detail the chemical reactions involved in the process, but shows instead the overall balance of the reactions. An increase in the concentration of dissolved $CO_{2(aqueous)}$ in the water shifts the reaction towards the right, which causes the dissolution of solid $CaSiO_3$. One product of the reaction is orthosilicic acid, $Si(OH)_4$.

It must be noted that the dissolution of $CaSiO_3$ is a one-way reaction, meaning that $CaSiO_3$ cannot be formed by combining Ca^{2+}, HCO^-_3 and $Si(OH)_4$. This is contrary to $CaCO_3$, which can be both dissolved into Ca^{2+} and HCO^-_3 and formed from these two ions as explained above (see Information Boxes 5.11 and 7.2). Because the dissolution of $CaSiO_3$ cannot be reversed, it has the same effect on the suite of processes leading to the formation of $CaCO_3$ in the ocean as a *ratchet* in a mechanical system, that is, it ceaselessly moves carbon from the atmosphere into the ocean.

Running waters carry the dissolved Ca^{2+}, HCO^-_3 and $Si(OH)_4$ ions to the ocean (arrow volcanoes-to-ocean/running-water in Fig. 7.6). There, the two reactions of *calcification* and *silicification* take place. The different meanings of *calcification* and *silicification* are explained in Chapter 4 (see Information Box 4.4).

On the one hand, some of the Ca^{2+} and HCO^-_3 ions resulting from the chemical alteration of $CaSiO_3$ (right side of Eq. 7.6) are used by calcifying organisms to form solid $CaCO_3$ in the ocean through the process of calcification (see Information Box 7.2, Eq. 7.4). The combined reaction of $CaSiO_3$ dissolution and $CaCO_3$ formation is obtained by replacing $[Ca^{2+}_{(aqueous)} + 2HCO^-_{3(aqueous)}]$ by $[CaCO_{3(solid)} + H_2O + CO_{2(aqueous)}]$ (Eq. 7.4) on the right side of Eq. 7.6:

$$CaSiO_{3(solid)} + 3H_2O + 2CO_{2(aqueous)} \rightarrow$$
$$CaCO_{3(solid)} + H_2O + CO_{2(aqueous)} + Si(OH)_{4(aqueous)} \quad (7.7)$$

On the other hand, silicifying organisms use the dissolved $Si(OH)_4$ to build structures of solid silica, SiO_2, through the process of silicification (equation read from left-to-right):

$$Si(OH)_{4(aqueous)} \rightleftharpoons SiO_{2(solid)} + 2H_2O \quad (7.8)$$

Combining Eqs. 7.7 and 7.8 summarizes the alteration of continental $CaSiO_3$:

$$CaSiO_{3(solid)} + 3H_2O + 2CO_{2(aqueous)} \rightarrow$$
$$CaCO_{3(solid)} + SiO_{2(solid)} + 3H_2O + CO_{2(aqueous)} \quad (7.9)$$

and removing the H_2O and CO_2 molecules that appear on the two sides of Eq. 7.9 gives:

$$\begin{aligned} Continental\ CaSiO_{3(solid)} + CO_{2(aqueous)} &\rightarrow \\ Seafloor\ CaCO_{3(solid)} + SiO_{2(solid)} \end{aligned} \qquad (7.10)$$

Equation 7.10 shows that the chemical alteration of continental $CaSiO_3$ rocks stores carbon from atmospheric CO_2 into solid seafloor $CaCO_3$ (Fig. 7.7).

On continents, the dissolution of one molecule of $CaCO_3$ by CO_2-acidified water *takes up one molecule of CO_2* from the atmosphere and releases one Ca^{2+} and two HCO_3^- ions in the water, which flows to the ocean. In the ocean, the combination of one Ca^{2+} and two HCO_3^- ions leads to the formation of one molecule of $CaCO_3$ and *releases one molecule of CO_2* to the seawater and thus the atmosphere. Hence as shown in the summary of chemical reactions in Fig. 7.7 (Eq. 7.5), the chemical alteration of $CaCO_3$ on continents followed by the formation of $CaCO_3$ in the ocean *does not change* the long-term concentration of CO_2 in the ocean or the atmosphere. In other words, it *does not result in a net transfer* of atmospheric CO_2 to marine sediments in the form of $CaCO_3$, and thus does not contribute to the long-term regulation of the CO_2 greenhouse effect.

In the ocean, the $CaCO_3$ structures built by calcifying plankton sink to the seafloor upon the death of the organisms (examples: coccolithophores, planktonic foraminifers, pteropods), and those built by benthic organisms (examples: crustose coralline algae, benthic foraminifers, molluscs, corals) stay on the seafloor (Fig. 7.8). There, they form carbonate sediments (two arrows ocean-to-seafloor in Fig. 7.6). When environmental conditions favour their dissolution (see Sect. 7.4.2), some of these sediments dissolve back to Ca^{2+} and HCO_3^- ions (two arrows seafloor-to-ocean in Fig. 7.6). In the long term, the carbonate sediments become carbonate rocks, some of which are eventually recycled in the mantle and volcanoes by tectonic activity (see Sect. 7.3.1, above).

Second, we examine the chemical alteration of *continental $CaSiO_3$ rocks* (arrow volcanoes-to-volcanic rocks in Fig. 7.6). The chemical equations are given separately in Information Box 7.3, and summarized in the lower part of Fig. 7.7.

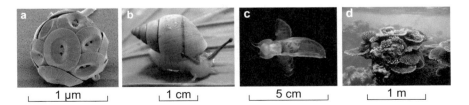

Fig. 7.8 Examples of modern organisms with $CaCO_3$ structures: **a** planktonic coccolithophore *Coccolithus pelagicus*; **b** land gastropod mollusc *Pachnodus praslinus*; **c** planktonic pteropod mollusc *Clione limacina*; and **d** corals on the Great Barrier Reef in Australia. Credits at the end of the chapter

On continents, the dissolution of one molecule of $CaSiO_3$ by CO_2-acidified water takes up *two* molecules of CO_2 from the atmosphere and releases one Ca^{2+} and two HCO_3^- ions and one molecule of silicic acid, $Si(OH)_4$, in the water, which eventually flows to the ocean. In the ocean, the combination of one Ca^{2+} and two HCO_3^- ions leads to the formation of one molecule of $CaCO_3$ and releases *one* molecule of CO_2 to the seawater and thus the atmosphere. Hence, the formation of one molecule of solid $CaCO_3$ in the ocean *takes up two molecules of CO_2* from the atmosphere to dissolve continental $CaSiO_3$ and *releases only one molecule of CO_2* in the ocean and thus the atmosphere (left and right sides, respectively, of Eqs. 7.7 or 7.9). The summary of chemical reactions in Fig. 7.7 (Eq. 7.10) shows that for each molecule of $CaSiO_3$ dissolved on continents, *there is a long-term net transfer of one molecule* of atmospheric CO_2 to marine sediments in the form of $CaCO_3$.

The dissolution of solid $CaSiO_3$ by CO_2-acidified water (Eq. 7.6) is very slow, but because it is a one-way reaction, it leads to the accumulation of $CaCO_3$ on the seafloor and its burial there over millions of years. This causes a long-term reduction of the greenhouse effect produced by atmospheric CO_2. The formation and fate of the $CaCO_3$ sediments resulting from the dissolution of continental $CaSiO_3$ rocks are the same as those described above in this section for the dissolution of continental $CaCO_3$ rocks.

In addition, the dissolution of continental $CaSiO_3$ is accompanied by the formation of solid silica, SiO_2 (Eq. 7.8 read from left-to-right). Examples of biological structures made of silica are the frustules of diatoms (both planktonic and benthic), the internal siliceous skeletons of silicoflagellates (plankton), and the spicules of siliceous sponges (Fig. 7.9). Upon the death of organisms, their SiO_2 remains are buried in the seafloor, where they form silicate sediments (two arrows ocean-to-seafloor in Fig. 7.6). Some of these sediments eventually dissolve back to $Si(OH)_4$ (Eq. 7.8 read from right-to-left), but the dissolution of biogenic silica is very slow (two arrows seafloor-to-ocean in Fig. 7.6).

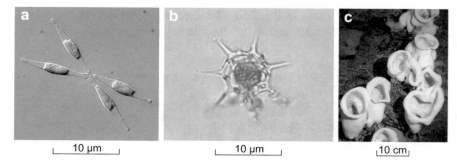

Fig. 7.9 Examples of modern organisms with SiO_2 structures: **a** planktonic diatom *Phaeodactylum tricornutum*, **b** planktonic silicoflagellate *Dictyocha speculum*, and **c** benthic siliceous sponge *Staurocalyptus* sp. Credits at the end of the chapter

The fate of SiO_2 sediments is different depending on water depth (arrows seafloor to volcanoes and seafloor-to-continents in Fig. 7.6). On the one hand, particles that sediment in the deep ocean are recycled in volcanoes, where some of the sedimentary SiO_2 rocks are molten and their chemical elements combine with those of $CaCO_3$ to form $CaSiO_3$ (see Sect. 7.2.3). On the other hand, materials that sediment in shallow waters can be found later on continents, in the form of sedimentary SiO_2 rocks such as diatomite and opal. Continental SiO_2 is not involved in the carbon cycle, and its fate is thus not considered here. The popular science book of De La Rocha and Conley (2017) provides further reading on the fascinating world of silicate.

Like most chemical reactions, the rate of dissolution of $CaSiO_3$ is sensitive to temperature, and this sensitivity is one of the most significant characteristics of carbon recycling on Earth. Indeed, when the global temperature tends to increase, there is a corresponding increase in the rate of dissolution of continental $CaSiO_3$ rocks, which speeds up the uptake of CO_2 from the atmosphere. The increased uptake of atmospheric CO_2 decreases the greenhouse effect, which counteracts the increase in global temperature (see Information Box 2.7). Hence, the increased dissolution of continental $CaSiO_3$ exerts a *negative feedback* into temperature. Conversely when the global temperature tends to decrease, there is a corresponding decrease in the rate of dissolution of continental $CaSiO_3$ rocks, which slows down the uptake of CO_2 from the atmosphere. The decreased uptake of atmospheric CO_2 allows the CO_2 emitted by volcanoes to accumulate in the atmosphere and thus strengthens the greenhouse effect, which counteracts the decrease in global temperature. The decreased dissolution of continental $CaSiO_3$ exerts a *negative feedback* into temperature. The combination of the two reactions creates a *negative feedback loop* (see Information Box 6.2), called *tectonic CO_2 feedback loop* in Chapter 9 (see Sect. 9.2.3).

Negative Feedback Loop Reducing the Long-Term Temperature Variations The above two negative feedbacks into temperature affect the rate of dissolution of continental $CaSiO_3$ rocks, thus creating a *negative feedback loop*. This feedback loop adjusted the atmospheric CO_2 concentrations to changes in global temperature throughout the history of the planet, and generally reduced the magnitude and duration of long-term temperature variations. The timescale of this feedback loop is of the order of 400,000 years (see Sect. 9.2.3). As explained in the next paragraph, the action of this feedback loop is modulated by variations in tectonic activity, which modify the quantity of $CaSiO_3$ rocks in contact with the atmosphere.

It was explained above that the rate of water erosion and transport of eroded materials on continents strongly depends on the steepness of the slopes of surfaces, which itself largely depends on the tectonic formation of mountains (see Sect. 7.3.1). Hence together with climate, the steepness of the slopes of

Information Box 7.4 Alteration of Seafloor CaSiO₃ Rocks The chemical reactions involved in the process of $CaCO_3$ formation from seafloor $CaSiO_3$ rocks are the same as those given in Eq. 7.7 for the chemical alteration of continental $CaSiO_3$ rocks (see Information Box 7.3). These processes are summarized in the following, simplified equation:

$$CaSiO_{3(solid)} + 2H_2O + CO_{2(aqueous)} \rightarrow CaCO_{3(solid)} + Si(OH)_{4(aqueous)} \quad (7.11)$$

This equation is obtained by removing the H_2O and CO_2 molecules that appear on the two sides of Eq. 7.7.

In Information Box 7.3, Eq. 7.7 represents the following suite of processes for *continental CaSiO₃*: dissolution of atmospheric CO_2 in rainwater, which carries the CO_2 from the atmosphere to the $CaSiO_3$ rocks; dissolution of the $CaSiO_3$ rocks by the CO_2-acidified water; transport of the Ca^{2+}, HCO_3^- and $Si(OH)_4$ ions resulting from the dissolution of $CaSiO_3$ to the ocean by running fresh waters; in the ocean, formation of solid $CaCO_3$ by the combination of dissolved Ca^{2+} and HCO_3^- ions.

Here, we consider the alteration of *seafloor CaSiO₃*, and the CO_2-rich seawater that circulates in the crust plays the same roles as those performed by rain and running fresh waters on continents. Indeed, the seawater that circulates in the crust carries dissolved CO_2 toward the seafloor $CaSiO_3$, and also carries the ions resulting from the chemical alteration of $CaSiO_3$ to the sites where $CaCO_3$ accumulates within and on the seafloor. The chemical process of $CaCO_3$ formation in the ocean from the products of dissolution of seafloor $CaSiO_3$ is the same as described in Information Box 7.3 for continental $CaSiO_3$, and this process stores in the same way carbon from dissolved CO_2 into solid seafloor $CaCO_3$.

continental surfaces and thus the formation of mountains affects the chemical alteration of $CaCO_3$ and $CaSiO_3$ rocks on continents, and the transport of the resulting dissolved ions by running waters to the ocean. It follows that tectonic activity affects the long-term climate in two ways. Firstly, it drives the volcanic production of $CaSiO_3$ rocks and CO_2 (see Sect. 7.2.3). Secondly, it generates slopes on continents, which largely drive the water erosion of the $CaSiO_3$ rocks by CO_2-acidified water and the transport of the resulting dissolved Ca^{2+}, HCO_3^- and $Si(OH)_4$ ions to the ocean, where solid $CaCO_3$ and $CaSiO_2$ are formed and buried in the seafloor. These combined processes contribute to regulate the greenhouse effect on Earth.

Third, we examine the fate of *seafloor CaSiO₃ rocks* (two arrow volcanoes-to-volcanic rocks in Fig. 7.6). Submarine volcanoes produce silicate rocks in the same way as their continental counterparts, and the volcanoes on mid-ocean ridges continually produce the $CaSiO_3$ basaltic rocks that form the bulk of the new seafloor. The deep waters generally contain high concentrations of dissolved CO_2 (in the form of HCO_3^-) as a consequence of the decomposition of sinking organic particles, and the concentration of dissolved CO_2 is even higher in the vicinity of submarine volcanoes because of the large amounts of CO_2 released during eruptions. The chemical equations are given separately in Information Box 7.4, and summarized in the lower part of Fig. 7.7.

Around submarine volcanoes, which are mostly located on mid-oceanic ridges, CO_2-rich seawater circulates in cracks and open spaces in the new seafloor, where its interaction with $CaSiO_3$ rocks creates $CaCO_3$ deposits within and on the seafloor (arrow volcanoes-to-ocean/seafloor water in Fig. 7.6). These processes only take place in areas where the new basalt rocks are not already covered by sediments, which are the mid-ocean ridges and their vicinity (Fig. 7.3). These areas being also the location of most of the active submarine volcanoes, there is a correspondence between the bands of shallow sediments in Fig. 7.3 (dark colour) and the narrow bands of seafloor earthquakes in Fig. 7.2. Farther away from active volcanoes, thick layers of sediments prevent the seawater from reaching the seafloor rocks and thus dissolving them. Hence, the alteration of seafloor $CaSiO_3$ rocks essentially takes place on and near the mid-ocean ridges.

It was explained above in this section that the dissolution of $CaSiO_3$ rocks on continents followed by the formation of solid $CaCO_3$ in the oceans caused a long-term reduction of the CO_2 greenhouse effect through the long-term storage of carbon into $CaCO_3$ rocks. Here, the same storage of CO_2-carbon into $CaCO_3$ rocks for very long times is achieved by the dissolution of seafloor $CaSiO_3$ rocks combined with the formation of solid $CaCO_3$ within and on the seafloor.

Carbon sequestration by the agency of chemical alteration of seafloor $CaSiO_3$ played a key role in the early history of Earth, when it led to the storage of huge amounts of carbon in $CaCO_3$ rocks, which had the effect of lowering significantly the initially very high concentrations of atmospheric CO_2. This lessened the corresponding greenhouse effect, and the temperature of the ocean went down from about +200 °C to perhaps +70 °C or lower, which was suitable for the development of organisms (see Sect. 5.5.4). This created huge $CaCO_3$ deposits, in which more than 99% of the Earth's carbon was sequestered. These deposits still exist on today's Earth, but their initial materials were progressively replaced by newly formed $CaCO_3$, which compensated for the seafloor carbonates destroyed by subduction (see Sect. 7.2.3).

It was explained above that silicate minerals are present in volcanic rocks under different forms and that, for simplicity, we use $CaSiO_3$ in this book as the generic silicate mineral. The chemical formula $CaSiO_3$ corresponds to the mineral called *wollastonite*, and we cited above as example of other silicate mineral the *anorthite*, whose chemical formula is $CaAl_2Si_2O$ (see Sect. 7.2.3). The chemical alteration of minerals with compositions other than $CaSiO_3$ produces other compounds than those shown in Eq. 7.6. For example, *feldspar* minerals, which are the main constituents of basalt rocks and make up more than 50% of the volume of the Earth's crust, have the general formula $XAl_{(1-2)}Si_{(3-2)}O_8$, where X can be potassium (K^+), sodium (Na^+) or calcium (Ca^{++}). The chemical alteration of *K-feldspar* produces (right side of Eq. 7.12) the clay mineral *kaolinite*, $Al_4Si_4O_{10}(OH)_8$, potassium (K^+) and bicarbonate (HCO_3^-) ions, and silica (SiO_2):

$$4KAlSi_3O_8 + 6H_2O + 4CO_2 \rightarrow$$
$$Al_4Si_4O_{10}(OH)_8 + 4K^+ + 4HCO_3^- + 8SiO_2 \qquad (7.12)$$

Similar reactions affect *Na-feldspar* and *Ca-feldspar*. Kaolinite has traditionally been the main component of *porcelain*, and has now many industrial uses. Equation 7.12 shows that the reaction of K-feldspar with CO_2-acidified water produces HCO_3^- ions, as did the chemical alteration of $CaSiO_3$ above (Eq. 7.6). When two of these HCO_3^- ions are used to form one mole of $CaCO_3$ in the ocean (Eq. 7.4), the net effect of the chemical alteration of one mole of feldspar on a continent or in the seafloor is to withdraw one mole of CO_2 from the environment and sequester its carbon in solid $CaCO_3$, as is the case of $CaSiO_3$ (Eq. 7.10). Hence, the chemical alteration of all types of silicate rocks leads the sequestration of atmospheric CO_2 in carbonate rocks.

In summary, tectonic activity ensures the long-term recycling of the Earth's carbon through two complementary suites of processes. In the first, continental carbonate and silicate rocks ($CaCO_3$ and $CaSiO_3$) are altered by CO_2-acidified fresh water, and the resulting HCO_3^- and Ca^{2+} are transported by rivers to the ocean; there, part of the carbon sinks to the seafloor as solid $CaCO_3$. Similarly, seafloor silicate rocks are altered by CO_2-rich seawater that circulates in the seafloor, and the resulting HCO_3^- and Ca^{2+} are transported by the same seawater to sites of formation of solid $CaCO_3$ within and on the seafloor. In the second suite of processes, part of the carbon contained in marine $CaCO_3$ is transferred to the atmosphere and the ocean by volcanoes, and the resulting CO_2 acidifies both the fresh water that dissolves the continental carbonate and silicate rocks and the seawater that dissolves the seafloor silicate rocks.

The first suite comprises the following processes:

- continental carbonate and silicate rocks (brought to the continents as part of the second suite of processes) are altered by CO_2-acidified fresh water (from the second suite of processes), which dissolves the solid $CaCO_3$ and $CaSiO_3$;
- running waters transport the resulting dissolved HCO_3^-, Ca^{2+} and $Si(OH)_4$ to the ocean;
- in the ocean, calcifying organisms use dissolved HCO_3^- and Ca^{2+} to form solid $CaCO_3$, which sinks to the seafloor; the formation of $CaCO_3$ is accompanied by the release of CO_2 gas, which escapes to the atmosphere; silicifying organisms use dissolved $Si(OH)_4$ to form solid SiO_2, which also sinks to the seafloor;
- the $CaCO_3$ and SiO_2 sediments are consolidated into rocks on the seafloor;

- similarly, marine silicate rocks (formed on the seafloor as part of the second suite of processes) are eroded by CO_2-rich seawater (from the second suite of processes) that circulates in the seafloor, where it dissolves the solid $CaSiO_3$;
- the same seawater transports the resulting dissolved HCO_3^-, Ca^{2+} and $Si(OH)_4$ to sites of chemical formation of solid $CaCO_3$ rocks within and on the seafloor; the formation of $CaCO_3$ is accompanied by the release of CO_2 gas, which escapes to the ocean
- the chemical alteration of continental and seafloor $CaSiO_3$ withdraws CO_2 from the environment and sequesters its carbon in solid $CaCO_3$.

The second suite of comprises the following processes:

- the $CaCO_3$ sedimentary rocks formed in shallow waters are brought to continents by tectonic activity; the $CaCO_3$ and SiO_2 sedimentary rocks formed in the deep ocean are transferred to subduction zones by horizontal movements of the seafloor;
- in the hot mantle, the melting of $CaCO_3$ and SiO_2 contained in the subducted plate releases the chemical elements Ca, Si and O, and CO_2 gas in the magma;
- the CO_2 dissolved in the magma is carried upwards by volcanic activity, and released to the atmosphere and the ocean as CO_2 gas during volcanic eruptions;
- the Ca, Si and O contained in the magma reach the surface of continents or the seafloor as part of volcanic lava, which solidifies into $CaSiO_3$ rock;
- some atmospheric CO_2 dissolves in fresh water (rain or running water) and the ocean CO_2 near submarine volcanoes dissolves in seawater, which acidifies the CO_2-rich waters;
- the CO_2-acidified fresh water dissolves continental $CaCO_3$ and $CaSiO_3$ rocks, and the CO_2-rich seawater dissolves seafloor $CaSiO_3$ rocks;
- the next steps are part of the first suite of processes.

The average time elapsed between the emission of CO_2 by volcanoes and the return of its carbon into the magma through subduction is estimated to be presently about 70–80 million years. This recycling may have been faster at earlier times in the Earth's history because the average tectonic cycle may have been markedly shorter on the young planet.

7.3.4   Global Effects of the Dissolution of $CaCO_3$

In the above paragraphs on the fate of continental $CaCO_3$ and $CaSiO_3$ rocks, we described the effects of $CaCO_3$ formation in the ocean. We now examine briefly two major environmental effects of the opposite process, that is, the dissolution of $CaCO_3$ accumulated on the seafloor. This occurs when there is an increase in the concentration of CO_2 in seawater close to the seafloor, and the resulting dissolution of the solid $CaCO_3$ causes in return a decrease in the concentration of CO_2 in seawater (see Information Box 5.11). Hence the dissolution of marine $CaCO_3$ can buffer (mitigate) changes in the concentration of CO_2 in the ocean. The decreased concentration of CO_2 in seawater has two major environmental effects.

Effect on atmospheric CO_2. The lower concentration of CO_2 in seawater makes the ocean absorb additional atmospheric CO_2. This effect is described in Chapter 9 within the context of the oceanic carbonate feedback loop (see Sect. 9.2.4).

Effect on the ocean's pH. The lower concentration of CO_2 in seawater is accompanied by an increase in the pH of seawater (see Information Box 6.4). It is expected that this mechanism will eventually buffer the present decrease in seawater pH, that is, *ocean acidification*, which results from the increase in atmospheric anthropogenic CO_2 (Fig. 9.5). It is explained below that it would take hundreds of years for this process to contribute to the buffering of the ongoing ocean acidification (see Sect. 7.4.2).

7.4   Effects of Tectonically-Driven Carbon Recycling on Climate Variations

7.4.1   Exchanges of Carbon Among the Outer Envelopes of Earth

The maintenance of temperatures suitable for the growth of organisms and the development of ecosystems since about 4 billion years required a greenhouse effect neither too weak not too strong in the long term. It was explained above that maintaining a moderate greenhouse effect required that carbon not remain sequestered in sedimentary rocks and the mantle, but be regularly recycled to the outer fluid envelopes of Earth (ocean and atmosphere), from where it could rapidly contribute to the processes controlling the climate (see Sect. 7.3.1). This shows that tectonic activity played a key role in the long-term control of the Earth's climate.

The present section examines how the tectonically driven recycling of carbon is involved, beyond its long-term effect on the global climate, in shorter-term climate variations. The fluxes mentioned in the text are illustrated in the simplified global cycle of inorganic carbon of Fig. 7.10, which does not include the formation and chemical alteration of volcanic $CaSiO_3$ rocks in the seafloor.

We begin by looking at the exchanges of carbon among the outer envelopes of Earth. The physical state of all carbon reservoirs is not the same, that is, some reservoirs are either fluid or solid, and others are in-between. The atmosphere and the ocean are fluid, and the seafloor and the continental rocks are solid. The magma in the upper mantle is semi-fluid, and the marine sediments in contact with seawater have both fluid and solid components as they contain large proportions of

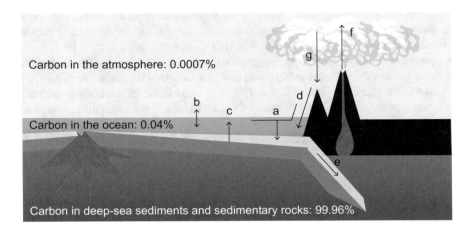

Carbon in the atmosphere: 0.0007%

Carbon in the ocean: 0.04%

Carbon in deep-sea sediments and sedimentary rocks: 99.96%

Fig. 7.10 Main reservoirs of carbon and some of the fluxes of this chemical element in the outer envelopes of the current Earth System. Credits at the end of the chapter

porewater (up to 50% or more in recently-deposited sediments). The latter is also the case for most soils. Organisms also form a reservoir of carbon. They contain between 60 and 95% of water (see Sect. 6.1.2), and they interact directly with the two fluid reservoirs. Most of the fluxes between carbon reservoirs are one-way (fluxes d to g in Fig. 7.10), except the two-way fluxes between the atmosphere, the ocean and marine sediments (fluxes a to c) and between organisms and the two fluid reservoirs (not shown in Fig. 7.10). The residence time of carbon in the solid and magmatic reservoirs is long (up to millions of years), and is much shorter (from a few hours to thousands of years) in the fluid reservoirs and the organisms. According to the environmental conditions in the upper sediments and the overlying bottom water, the sedimented carbon can be recycled rapidly to the overlying water or buried in the sediment for very long periods (up to hundreds of millions of years in the case of petroleum).

The carbon in sediments and sedimentary rocks results from the sinking of biogenic detritus from marine and continental biological activity (flux a). This carbon is recycled through the following three main pathways.

The first pathway (flux c) involves exchanges between sedimentary biogenic carbon (solid carbonates and fossil organic matter) and the dissolved inorganic carbon present in seawater as HCO_3^-, CO_3^{2-} and CO_2 (see Information Box 6.3). One process, whose occurrence depends on the pH of seawater, is the dissolution of solid carbonates on the seafloor (see Sect. 9.2.4). Another process is the oxidation of organic matter in sediments by benthic respiration, which releases the carbon contained in organic matter as CO_2, directly or indirectly (see Information Box 5.2).

The second pathway involves water erosion of continental rocks (fluxes d and g), which are brought to the continents by tectonic activity (see Sect. 7.2.3). These rocks can contain carbon in the forms of solid carbonates and organic matter. As in the first pathway, the alteration of carbonates release dissolved inorganic carbon, whose main form in continental water is HCO_3^-. The HCO_3^- is carried by rivers to the ocean reservoir of dissolved inorganic carbon (see Sect. 7.3.3). The respiration of the organic matter contained in rocks releases CO_2 (not illustrated in Fig. 7.10).

The third pathway (fluxes e and f) is the subduction of carbonate sediment into the mantle and the production of CO_2 and $CaSiO_3$ by volcanic activity (see Sect. 7.2.3).

These three pathways affect carbon recycling in the Earth System and thus the global climate with very different efficiencies and at very different timescales, examined in the next sections. In addition to the fluxes involved in the three pathways, Fig. 7.10 illustrates the exchange of CO_2 between the atmosphere and the ocean (flux b). In the following sections, we examine effects of the tectonically driven recycling of carbon on climate variations related to the three pathways, namely: the formation and dissolution of seafloor carbonates (first pathway; Sect. 7.4.2), the chemical alteration of continental carbonate and silicate rocks (second pathway; Sect. 7.4.3), and different aspects of volcanic activity (third pathway; Sects. 7.4.4–7.4.6).

7.4.2 Formation and Dissolution of Seafloor $CaCO_3$: Centennial Effect

The presence of carbonate sediments and rocks on the seafloor of a given area indicates that the local conditions led to the formation of solid $CaCO_3$ in the past and do not favour the dissolution of $CaCO_3$ today. The solubility of $CaCO_3$ is higher at higher pressure, lower temperature, and higher concentration of dissolved CO_2, the latter causing lower pH (see Information Box 6.3). In addition, the carbonate structures of different organisms are made of one of two mineralogical forms of $CaCO_3$, called *aragonite* and *calcite*, which are chemically the same but have different crystal structures and thus different solubility. The solubility of aragonite is higher than that of calcite, and consequently the aragonite shells of planktonic pteropods (molluscs) are less abundant in marine sediments than the calcite tests of coccolithophorids (phytoplankton), foraminifera (protozoa) and ostracods (crustaceans). In the ocean, pteropod ooze is restricted to water depths shallower than 2,500 m.

With increasing depth in the water column, the hydrostatic pressure increases and the temperature generally decreases, with the consequence that the dissolution of $CaCO_3$ increases. The depth below which the dissolution of solid $CaCO_3$ is more rapid than its formation is called the *carbonate compensation depth* (CCD), and it varies among oceans and regions within oceans. For example, the CCD is shallower in the Pacific than in the Atlantic Ocean, where the values for calcite are 4200 to 4500 and around 5000 m, respectively. This difference mostly reflects the fact that the deep waters of all oceans are formed north and south of the Atlantic Ocean (see Sect. 3.5.3), with the consequence that the deep waters of the Pacific Ocean contain high concentrations of CO_2 (coming from the oxidation of organic matter) accumulated during the long time elapsed since their formation in the Atlantic (Fig. 3.8). The higher concentrations of CO_2 in the deep waters of the Pacific explain the stronger dissolution of $CaCO_3$ there than in the Atlantic.

The CCD is influenced by the concentration of CO_2 in seawater and the temperature of the ocean, which are both partly determined by the concentration of CO_2 in the atmosphere. Hence, the variations in the CCD reflects changes in the concentration of CO_2 in the atmosphere, and specialists use changes in the CCD over time to study ancient climate. To do so, paleoceanographers drill sediment cores in the deep seafloor, whose cumulated length can reach several kilometres and thus provide information on the Earth's environment over tens of millions of years. One type of information is provided by the presence of biological $CaCO_3$ remains at a given depth in a core, which indicates that the CCD was reaching the seafloor at the corresponding time in the past, whereas the absence of $CaCO_3$ at some other depth indicates that the CCD was then located above the seafloor. For example, researchers found, by combining the information from eight sites in the equatorial Pacific Ocean located at depths corresponding to long-term changes in the CCD, that the CCD had deepened from 3.0–3.5 to 4.6 km since 55 million years. They interpreted this deepening of the CCD as corresponding to an overall increase in erosion of silicate rocks over the past 55 million years.

When the CO_2 concentration of deep water increases as a consequence of an increase in atmospheric CO_2, as it is currently occurring, the bottom water and the sediment start exchanging carbon through the dissolution of solid carbonates on the seafloor (which is part of flux c in Fig. 7.10). For example, a decrease in the pH of surface water (as in the case of the present anthropogenic ocean acidification) can be transferred to the deep ocean by the ocean conveyor belt (see Sect. 3.5.3). Once the low-pH seawater reaches the seafloor, it starts dissolving the solid carbonates (previous paragraph), which in return buffers the pH of the seawater (see Sect. 7.3.4). Because it takes centuries for all the acidified surface waters to come in contact with the carbonate sediments, it will also take centuries to buffer the ocean's pH by dissolution of marine carbonates (see Sect. 9.2.4). Ocean acidification is further examined in Chapter 10 (see Sects. 11.1.1 and 11.1.3).

The long-term ability of the ocean to absorb excess atmospheric CO_2 is virtually unlimited because of the possible exchange between the CO_2 dissolved in seawater and the vast reservoir of deep marine carbonates. Indeed, there are at least 1,500 times more atoms of carbon stored in the seafloor carbonates than dissolved in all the oceans. However, the rate at which the ocean actually absorbs the excess atmospheric CO_2 depends on the one hand, on the magnitude and rate of the disturbance causing this excess (such as a large volcanic eruption), and on the other hand, on the conditions of the exchange between the atmosphere and the deep ocean (including the state of the deep ocean circulation, which varies with time).

7.4.3   Chemical Alteration of Continental CaCO$_3$ and CaSiO$_3$: Different Long-Term Effects

The tectonic processes responsible for the presence of carbonate (CaCO$_3$) and silicate (CaSiO$_3$) rocks on continents are explained above (see Sect. 7.2.3), and the chemical alteration of these rocks is part of the tectonic recycling of carbon (flux d in Fig. 7.10). The alteration of these two types of rocks on continents varies according to the composition of the atmosphere and the climatic conditions.

Concerning the *composition of the atmosphere*, higher concentration of atmospheric CO$_2$ lowers the pH of rainwater, and the dissolution in the rain of additional compounds such as sulphur and nitrogen oxides further lowers its pH. The acidified running water resulting from this rain dissolves the continental CaCO$_3$ and CaSiO$_3$ rocks. This type of erosion does not only dissolves rocks at the surface of the ground, but it also creates natural caves under the ground. A condition for the formation of caves is an annual rainfall greater than half a meter. Some caves are magnificent, with their natural decorations of stalagmites and stalactites. In some remarkable cases, people have added *rock art* to the natural decorations thousands of years ago (see Sect. 4.5.1).

Concerning the *climatic conditions*, higher temperature increases the rate of chemical reactions of the dissolution process, and higher rainfall increases the rate of contact of the resulting running water with the surface rocks. The combined effect of temperature and rain explains why the chemical alteration of rocks is generally higher in wet tropical and subtropical areas than at higher latitudes. In addition, water erosion increases with steeper slopes, which are often created by the tectonic processes of mountain formation (see Sect. 7.3.1).

The acidified fresh water on continents alters carbonate rocks more rapidly than silicate rocks and, as mentioned above, the carbonates rocks are much less abundant in the crust than the silicate rocks. Indeed, the carbonate rocks account for less than 10% of the volume of the crust, whereas the silicate rocks (feldspars) account for more than 50% (see Sects. 7.3.2 and 7.3.3, above). Given that carbonate rocks account for less than 10% of the volume of the crust and are altered quite rapidly, their maintenance since more than 4 billion years has required a fast rate of carbonate formation during the whole history of Earth. This was possible because carbonates are formed by chemical and biological processes, which have the characteristic of responding rapidly to changes in environmental conditions (see Sect. 6.4.2). More than 99% of all the carbon on Earth is sequestered in carbonate rocks, and this key characteristic explains why the ground temperature of the planet has generally been mild since more than 4 billion years. In contrast, all the CO$_2$ of planet Venus is in its atmosphere, and the ground temperature of that planet is scorching hot (see Sect. 5.5.4).

Unlike their carbonate counterparts, silicate rocks are altered slowly and the long-term maintenance of their very large volume has been achieved by tectonic mechanisms (see Sect. 7.2.3). The rate of alteration of silicate rocks is regulated by temperature, and in return the alteration of silicate rocks regulates the long-term changes in the Earth's temperature (see Sects. 7.3.3 and 9.2.3).

7.4.4   Types of Volcanic Activity

Volcanic activity can contribute significantly to either temporary or longer-term changes in the greenhouse effect and thus the climate. We examine in this section some general characteristics of volcanoes, and consider in the following two sections the short- and longer-term climatic effects of volcanic CO_2 gas, ash, aerosols, and lava.

The present volcanic activity in the oceans and on the continents produces annually a global amount of about 4 km³ of material in the form of lava and ash. This present value is small compared to the volumes produced during intense phases of volcanic activity in the past. Table 7.1 reports selected examples of extreme volcanic events, whose intensity increases with their distance in time. Eruptions larger than 1000 km³ are sometimes qualified of *supereruptions*. The global volume of material presently emitted by all volcanoes on the planet (4 km³ per year) is small compared to the values in Table 7.1.

Volcanic eruptions can be explosive or effusive, and volcanoes emit three main types of materials, which are gases, ash and aerosols, and lava. In some eruption, masses of molten lava larger than 64 mm are ejected in the air and cool into solid *volcanic bombs* before they reach the ground; some bombs can be much larger

Table 7.1 Selected examples of very large volcanic eruptions on Earth

Year	Volcano	Location	Volume emitted (km³)	Significance
1912	Novarupta	Alaska, USA	30	Largest eruption of the twentieth century
1815	Mount Tambora	Indonesia	150	Largest eruption in at least 1300 years
−73,000	Present location of Lake Toba	Indonesia	2,000 to 3,000	One of the two largest Quaternary eruptions
−630,000	Yellowstone	USA	1,000	Third largest Quaternary eruption
−2 millions	Yellowstone	USA	2,500	One of the two largest Quaternary eruptions
−68 to −60 millions	Deccan Traps volcanic province	India	500,000	Intense regional volcanism over thousands of years

than 64 mm, and reach up to 5–6 m. *Explosive* volcanoes erupt violently and send materials high into the air, whereas *effusive* volcanoes ooze red-hot lava that flows down their sides. In the ocean, the increase of pressure with increasing depth restricts the release of volatile gases, resulting in a decrease in the occurrence of explosive eruptions at depths greater than 500 m.

The main *volcanic gases* are water vapour, CO_2, and depending on the characteristics of the emission, different proportions of SO_2 and H_2S (see Sect. 7.3.1). *Volcanic ash* consists of the solid products of explosive volcanic eruptions. It is formed when the gases dissolved in the volcanic magma escape violently into the atmosphere, propelling the magma to an altitude that can reach 50 km. In the stratosphere, the ejected magma solidifies into fragments of volcanic rock and glass, which make up the volcanic ash. The latter is transported by wind, up to thousands of kilometres away. *Lava* is the molten rock expelled during non-explosive, effusive volcanic eruptions, or more generally through fractures in the crust of a planet, whose temperatures range between 700 and 1200 °C. Ancient lava has also been observed on Mars. When lava cools down and solidifies, it forms igneous rocks, typically basalt. Basaltic rocks are mostly composed of silicate minerals, including $CaSiO_3$ and various feldspars (see Sect. 7.2.3).

Volcanic rocks are produced in different proportions in three areas of tectonic activity: 73% are produced at divergent plate boundaries (mid-ocean ridges, and rift valleys), 15% at convergent plate boundaries (subduction zones), and 12% at hotspots. Divergent and convergent boundaries are explained above (see Sects. 7.2.2 and 7.2.3, respectively). A *hotspot* is a region of the crust where a volcano is fed by magma coming from a hypothetical abnormally hot area in the Earth's mantle, and the volcanoes found at locations outside plate boundaries, both in oceans and on continents, correspond to hotspots in the underlying mantle. A hotspot has a fixed location, and it creates volcanoes as the plate moves above it. One example of hotspot is provided by the chain of Hawaiian volcanoes, which extends 6,000 km across the northwest Pacific Ocean and whose hotspot, which has created the Hawaiian islands one after the other, has been active for at least 70 million years. A second example is the Yellowstone chain of calderas and volcanic features, which extends 650 km in northwest USA and whose hotspot has been active for at least 15 million years. A third example is the Afar hotspot, which is located under northeastern Ethiopia and may have contributed to the opening of the East African rift (see Sect. 7.2.2).

7.4.5   Volcanic Activity: Temporary Effects on the Global Climate

The different types of outbreaks of volcanic fever can have different, and even opposite effects on the climate. In the following paragraphs, we examine the temporary climate effects of volcanic CO_2 gas and ash.

Emission of CO_2 gas. In the case of moderate volcanic activity, the CO_2 flux from eruptions (flux f in Fig. 7.10) can be compensated rapidly by the dissolution of the excess atmospheric CO_2 in surface seawater, through the process summarized in above Eq. 7.2 (see Sect. 7.3.3):

$$CO_{2(aqueous)} + H_2O \rightleftharpoons HCO_3^- + H^+ \qquad (7.2)$$

This corresponds to the current situation of moderate volcanic activity, and even to most of the larger eruptions. In such eruptions, the CO_2 emissions by volcanoes do no cause significant increases in the global temperature.

In some other cases, massive and/or sudden increases in emissions of volcanic CO_2 may disrupt the global climate, as these are too large to be rapidly compensated by CO_2 dissolution in surface seawater. Examples are described in Sect. 7.4.6.

Emission of ash and aerosols. The emission of ash and aerosols in the upper atmosphere can have an immediate, but relatively brief effect on the climate, by decreasing the global solar radiation at ground level. The effect of ash is generally of short duration because the airborne ash falls to the ground within weeks, whereas that of aerosols may last up to several years. Volcanic ash is made of very fine particles of volcanic rocks created by the solidification of magma in the atmosphere into volcanic crystals and glass (previous section). Volcanic aerosols generally result from the combination of volcanic gases with atmospheric water, which forms fine, solid particles. For example, the combination of sulphur dioxide (SO_2) with with H_2O forms sulphuric acid (H_2SO_4).

Volcanic ash and aerosols can have large-scale effects other than climatic, as they can disrupt or damage infrastructures necessary to the functioning of modern societies. The most dramatic example to date is the eruption of the Eyjafjallajökull volcano (Iceland) in 2010, which was a small event in term of the volume of erupted materials (0.14 km^3), but the accompanying airborne plume of ash reached an altitude higher than 8 km, and the spreading of ash in the atmosphere caused the largest air-traffic shutdown across western and northern Europe since World War II and before the Covid-19 epidemic. Indeed, the fear of the damage that ash could cause to the turbine engines of aircrafts caused about 20 countries to close their airspace to commercial jet traffic during several weeks, which affected approximately 10 million travellers. The Eyjafjallajökull experience shows that the airborne ash resulting from a large volcanic eruption could cause a global disruption of air traffic. In some other eruptions, the fallout of volcanic ash affected key infrastructures of urban areas such as transportation, electricity and water supply, sometimes very far away from the sites of the events. The book of Wilson et al. (2021) provides further reading on the effects of volcanic ash on infrastructures.

Overall, volcanic CO_2 emissions tend to increase the natural greenhouse effect, and thus temporarily warm the global climate. The effect of ash emissions is opposite, as these tend to cool the atmosphere.

7.4.6 Volcanic Activity: Long-Term Effects on the Global Climate

In the following paragraphs, we examine long-term climate effects of volcanic CO_2 and SO_2 gases, ash, and lava. These effects are either global warming or global cooling.

Emission of CO_2 gas: global warming. The emission of CO_2 resulting from very large volcanic eruptions, regional-scale eruptions, or long eruptive phases can lead to large increases in atmospheric CO_2. The uptake of the excess CO_2 may take several centuries, during which the greenhouse effect and the global climate are durably perturbed.

The mechanism by which increases in atmospheric CO_2 resulting from volcanic eruptions are buffered in the long-term is the uptake of the excess CO_2 by the dissolution of deep-sea carbonate sediments and rocks, mostly solid $CaCO_3$ (see Sects. 7.4.2 and 9.2.4). The carbonate materials dissolve once the surface waters loaded with CO_2, in the forms of dissolved CO_2, HCO_3^- and HCO_3^{2-} (see Information Box 6.3), are transported to depth, where the CO_2-rich deep waters are distributed throughout the World Ocean by the deep ocean circulation (see Sect. 3.5.3). The time required for the downward transport and bottom distribution of the CO_2-rich surface seawater is that of the deep ocean circulation, which is hundreds of years. The dissolution of sedimentary carbonates occurs when these are in contact with the CO_2-containing deep seawater. It follows that the timescale for buffering large amounts of volcanic atmospheric CO_2 by the ocean, and the resulting increase in the greenhouse effect, is hundreds of years.

Emission of SO_2 gas: global cooling. Volcanoes emit other gases than CO_2 during eruptions, especially sulphur dioxide, SO_2 (see Sect. 7.3.1). This gas causes the formation, between 10 and 50 km high in the atmosphere, of tiny droplets of sulphuric acid (H_2SO_4) and other sulphur-rich aerosol particles that can stay there for two to three years. There, these droplets reflect part of the incoming solar radiation back into space, and the increased albedo causes the cooling of the lower atmosphere and the Earth's surface (see Information Box 2.9). Hence, the cooling is not caused directly by SO_2, but instead by sulphur-rich aerosols. This climate effect is the reverse of the effect of CO_2, whose increase contributes to warm the lower atmosphere through the greenhouse effect. Table 7.2 lists some of the volcanic eruptions followed by a decrease of at least 2 °C in the global temperature at ground level. The values in Table 7.2 indicate that the emissions high in the atmosphere of 100 million tons of sulphur or more generally caused a global cooling of at least 0.5 °C.

One example of the effect of SO_2 emission on the climate is the global cooling of about 0.5 °C that lasted about 6 months following the explosive eruption of Mount Pinatubo (Philippines) in 1991. Similarly but on a longer timescale, the eruption of Mount Tambora (Indonesia) in 1815, caused a global cooling of 0.4–0.7 °C that lasted several years. The year after this cool period, 1816, was known

Table 7.2 Selected examples of volcanic eruptions whose emissions of sulphur high in the atmosphere caused a global cooling of at least 0.2 °C

Year	Volcano	Location	Emission (millions of tonnes of sulphur)	Cooling at ground level, °C
1991	Pinatubo	Philippines	20	0.1 to 0.5
1963	Agung	Indonesia	10 to 20	0.3
1912	Novarupta or Katmai	Alaska, USA	20	0.2
1902	Santa Maria	Guatemala	20	0.4
1883	Krakatoa	Indonesia	50	0.3
1815	Tambora	Indonesia	200	0.4 to 0.7
1784	Laki or Lakagigar	Iceland	80 to 100	1
1600	Huaynaputina	Peru	100	0.8
1452	Kuwae	Vanuatu	150 (?)	0.5
−75,000	Location of present Lake Toba	Indonesia	1000	3 to 5

as the "year without a summer", during which crops failed and livestock died in the Northern Hemisphere, resulting in the worst famine of the nineteenth century in Europe. At a still longer timescale, it was proposed that the Toba supereruption (Indonesia) about 75,000 years ago (Table 7.1) caused a global "volcanic winter" that lasted several years and was followed by several centuries of global cooling. It was hypothesized that this long period of cool weather could have been responsible for a sharp reduction in the size of the human population, which was then quite small, and this controversial hypothesis is debated among volcanologists, geologists, climatologists and archaeologists.

At a very long timescale, it seems that the cooling caused by the release of SO_2 in the Deccan Traps volcanic province (India) during thousands of years between 60 and 68 million years ago (Table 7.1) contributed significantly to the fifth mass extinction, during which about three-quarters of the plant and animal species disappeared, including the dinosaurs although the dinosaurian ancestors of modern birds survived (see Sect. 4.6.2). An alternative or complementary explanation to this mass extinction is the impact of an asteroid 10–15 km in diameter near the present town of Chicxulub, in Yucatan (Mexico), which would have produced a sunlight-blocking dust cloud that lowered the global temperature and reduced plant photosynthesis during a decade or more (see Information Box 1.6). The Deccan Traps are not listed in Table 7.2 because the amount of sulphur released in the atmosphere during this very long volcanic period is not known, although it was likely very large.

Emission of ash and SO₂ gas: large-scale cooling. The emission of ash in the upper atmosphere can cause global cooling by decreasing the solar radiation at ground level (see Sect. 7.4.5). However, this effect is generally of short duration because the airborne ash falls to the ground within weeks. Nevertheless, there were events of large-scale cooling caused by successive eruptions that combined the emissions of ash and SO_2 gas. One example is provided by two consecutive eruptions in years 535–536 and 539–540, which occurred at sites that have not yet been determined and were likely responsible for a major cooling in the Northern Hemisphere that lasted a decade.

Another example may be the initiation of the Little Ice Age, which is the period of relative cooling—possible global—that followed the Medieval Warm Period (between about 950 and 1250) in the North Atlantic region (see Sect. 4.5.3). The cooling of the Little Ice Age, which ended around 1850, may have begun with the expansion of the ice pack in the North Atlantic around 1250. This expansion was possibly triggered or enhanced by the massive eruption of the now-extinct Mount Samalas on the island of Lombock (Indonesia) in 1257, followed by three smaller eruptions of the same volcano during the following decades. A second pulse of cooling may have been triggered by the eruption of the submarine volcano Kuwae (Vanuatu; western South Pacific Ocean) in 1452–1453.

Emission of lava and CO_2 gas: snowball Earth. A number of geological observations led researchers in the Earth Sciences to hypothesize that the Earth's surface became entirely or nearly entirely covered by ice during millions of years, a phenomenon called *snowball Earth* or *ice ball Earth*. This occurred at least three times, around 2,200, 710 and 640 million years ago (see Sect. 4.5.4). The mechanisms responsible for the initiation of the global cooling and for breaking out of the global glaciation millions of years later are not known for sure, but some of the hypotheses invoke tectonic activity related to volcanoes. The explanation involves the volcanic production of both lava and CO_2, which may have affected the global climate during the initiation and the termination of the snowball Earth as explained in the next paragraphs. Snowball Earth is a remarkable case of the effect of volcanic activity on the global climate over millions of years.

The last snowball Earth episode (around 640 million years ago) occurred in the geographical setting of the supercontinent Rodinia, which was then centred on the equator, and extended between 60 °N and 60 °S (Fig. 4.7). It started to break up about 800 million years ago, under the action of volcanic hotspots. The breakup of Rodinia was accompanied by very extensive zones of volcanic activity, which produced huge amounts of continental lava, accompanied by the formation of silicate rocks ($CaSiO_3$). These and other volcanic rocks were exposed to the atmosphere by the fragmentation of Rodinia. Following the suite of geochemical reactions described above (see Sect. 7.3.3), the continental $CaSiO_3$ rocks were progressively altered by CO_2-acidified continental waters, leading to a major reduction of atmospheric CO_2 whose carbon was sequestered in solid carbonates on the seafloor. The generally warm climate of Rodinia, resulting from the location of lands at low and mid latitudes, favoured the chemical alteration of continental rocks.

The decrease in atmospheric CO_2, and thus the greenhouse effect, caused a global cooling of the Earth's climate. The progress of glaciers on land and of the ice pack at sea increased the albedo of the planet (see Information Box 2.9), which amplified the cooling, which in return favoured the progress of ice, this process finally leading to a very large ice cover on the planet. This runaway cooling was the result of a *positive feedback loop* (see Sect. 9.1.3; Fig. 9.3e and accompanying text). The expansion of the ice cover also had the effect of reducing the surfaces of $CaSiO_3$ rocks exposed to erosion, with the consequence that the progress of the glaciation inhibited its triggering mechanism, which corresponded to a *negative feedback* (see Sect. 9.1.3; Fig. 9.4b and accompanying text). It is thought that once Earth reached this frozen state, it could not easily break out of it. However after millions of years in the snow/ice ball state, Earth finally managed to break out. How did the planet do it?

During the period when the Earth's surface was covered by ice, volcanic activity continued to emit CO_2 and other greenhouses gases into the atmosphere. The presence of the global ice cover prevented, or at least strongly impeded the flux of atmospheric gases into the ocean. As a consequence, the CO_2 and other greenhouse gases released by volcanoes were not, or were only very slowly absorbed by the ocean, with the result that they accumulated in the atmosphere. The accumulation of greenhouse volcanic gases in the atmosphere generated a greenhouse effect large enough to trigger the melting of the ice pack and glaciers, and this allowed Earth to break out of the global glaciation.

The snowball Earth example illustrates two key roles that volcanoes generally have in the regulation of the Earth's climate. First, volcanoes supply to the atmosphere and the ocean the CO_2 that both largely causes the greenhouse effect and acidifies the water that alters the silicate and carbonate rocks. Second, volcanoes bring to the continents and the seafloor the lava from which the silicate rocks are formed. The CO_2 and the silicate rocks are both central to the negative feedback regulation mechanisms of the Earth's temperature (see Sects. 9.2.3 and 9.2.4).

As in the case of temporary effects of volcanic activity on the global climate (previous section), volcanic emissions of CO_2 tend to increase the natural greenhouse effect and thus warm the global climate in the long term, whereas those of ash and SO_2 are opposite. In addition, the emission of lava has the longest-lasting effect on the global climate.

7.4.7 Long-Term Effects of Tectonic Processes on Ecosystems

The case of the snowball Earth in the previous section stresses the importance of the coupling between the different components of the carbon recycling cycle for the global climate. These components are: the production of $CaSiO_3$ rocks and the release of CO_2 by continental and volcanic volcanoes; on continents, the chemical alteration of $CaSiO_3$ rocks by CO_2-acidified water, followed by the transport of the resulting Ca^{2+}, HCO_3^- and $Si(OH)_4$ by running water to the ocean, which continental processes depend on the maintenance of slopes by mountain building; in the ocean, formation of $CaCO_3$, and calcareous sediments and rocks on the seafloor;

in the seafloor, the chemical alteration of $CaSiO_3$ rocks by CO_2-acidified seawater, followed by the transport of the resulting Ca^{2+}, HCO_3^- and $Si(OH)_4$ by water that circulates in the young crust to sites of formation of $CaCO_3$ rocks; the tectonic transport of seafloor rocks to convergent tectonic plate boundaries; the melting of subducted plates in the mantle, which releases Ca, Si, O and CO_2 in the magma; and the emission of magma by volcanoes, from which CO_2 escapes and whose lava forms $CaSiO_3$ upon solidification.

Even when the above processes became uncoupled and the planet froze up during the millions of years of each snowball Earth, ecosystems continued to exist and probably thrive in liquid seawater. Over the last 4 billion years, the overall coupling of the Earth's carbon-recycling processes maintained a long-term moderate greenhouse effect. This created globally suitable temperatures for organisms, which allowed them to build huge biomasses and thus take over the Earth System. Chapter 4 in the book of Ruddiman (2008) provides further reading on the relations between plate tectonics and the long-term climate of Earth.

7.5   Feedbacks of the Climate on Tectonic Activity

7.5.1   Effects of the Earth's Fluid Envelopes on Tectonics

The previous sections explained that tectonic activity is a necessary condition for the continued existence of ecosystems on Earth as it ensures the recycling of carbon and other biogenic elements, and they also described short- and longer-term effects of tectonic processes on climate. Plate tectonics is driven by energy coming from the interior of Earth, and we examined in previous sections the effects of tectonic mechanisms on processes that occur in the outer fluid envelopes of Earth (ocean and atmosphere), especially but not exclusively the greenhouse effect and the climate.

The next section looks at the possibility of reverse effects, that is, feedbacks (see Information Box 6.2) of climatic processes that occur in the outer fluid envelopes of Earth on tectonics. Researchers have not investigated such feedbacks as widely as the effects of tectonics on the climate, but some specialists have looked into them, and we describe below some interesting possibilities published in the scientific literature.

7.5.2   Feedbacks Between Erosion and Mountain Building

Here we examine feedbacks that concern possible effects of the climate on tectonic mechanisms involved in the building of mountains. We briefly describe such feedbacks for four mountain ranges, namely the St. Elias Range in North

America, the Alps in Europe, the Himalayas in Asia, and the Andes in South America (Fig. 7.11). We have seen above that a mountain range forms as the result of a collision between either the continental parts of two plates, which produces continental mountain ranges such as the European Alps or the Himalayas, or the oceanic part of a plate and the continental part of another, which produces coastal mountain ranges such as the Rocky Mountains or the Andes (see Sect. 7.2.3).

In a first step, we consider together the St. Elias Range and the European Alps. These two mountain ranges share a similar feedback loop between erosion and tectonics.

The coastal *St. Elias Range* extends over 400 km between southeastern Alaska (USA) and southwestern Yukon (Canada). It is the highest *coastal* mountain range on Earth, with many peaks higher than 5,000 m, the highest being Mount Logan (5,959 m), and active volcanoes. This mountain range is formed by the northwestward subduction of a small oceanic plate under the continental North American Plate (Fig. 7.2). The present subduction began about 26 million years ago, and glaciers are major agents of erosion there.

The continental *European Alps* (do not confound with the Australian Alps, the Japanese Alps, or the Southern Alps in New Zealand, as already indicated in Sect. 7.2.3) stretch 1,200 km between eastern France and western Austria.

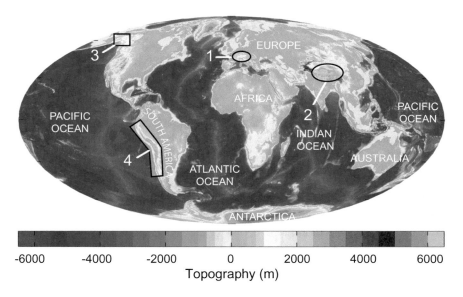

Fig. 7.11 Topographic map of Earth showing the locations of the four mountain ranges examined in the text: two continental mountain ranges (ellipses), (1) the Alps and (2) the Himalayas; and two coastal ranges (rectangles), (3) the St. Elias Range and (4) the Andes. Credits at the end of the chapter

Their formation began about 35 million years ago by the collision of the African plate with small plates arrayed along the southern margin of the Eurasian plate (Fig. 7.2). The highest point of the Alps is Mont Blanc (4,810 m), and there are about 100 peaks higher than 4,000 m in the whole mountain range.

During the building of the St. Elias Range and the Alps, the sediments and rocks at the margin of the incoming plate were pushed against the continental mass of the other plate, which created slow movements of huge masses of materials both upwards (uplift) and forwards. These movements progressively created the mountain ranges. The balance between the upward and forward movements depended on several factors, which included the morphology of the growing mountains, the physical properties of the uplifted materials, and the nature of the interface between these materials and the underlying continental rocks. As they grew, the mountains were continually subjected to erosion by running water, ice and other agents (see Sect. 7.3.3), which changed the distribution of materials within the uplifted mass, and thus modified the balance between the vertical and horizontal movements. This modified in return some of the characteristics of the mountain ranges, such as their rate of growth and their width.

It follows that erosion exerted a direct influence on the tectonic evolution of the St. Elias Range and the Alps. It was shown above that the erosion of continental rocks is determined by such characteristics of the global and regional climate as temperature and rainfall (see Sect. 7.4.3), and also by the steepness of the slopes created by the growth of mountains (see Sect. 7.3.1). The tectonic evolution of the St. Elias Range and the Alps directly influenced the agents of erosion, and these directly influenced the tectonics of mountain building, hence a feedback loop between erosion and tectonics.

In a second step, we examine successively the Himalayas and the Andes. Each of these two mountain ranges illustrates a different relationship between erosion and tectonics.

The continental mountain range of the *Himalayas* forms a 2,400 m-long arc that runs between Pakistan to the west and Buthan-Tibet to the east, with a central southern curve into Nepal. The Himalayas include over 50 mountains peaking at more than 7,200 m, of which ten reach 8,000 m including Mount Everest at 8,848 m. The highest peaks (more than 6,000 m) are located in the central part of the arc, in Nepal, and the mountains on the western and eastern sides average 3,000 and 2,000 m, respectively. The Himalayas started to form about 55 million years ago when the continental part of the Indian plate collided with the Eurasian Plate (Fig. 7.2). About 10 million years ago, the Indian plate began to undergo a counter-clockwise rotation, during which its eastern margins moved faster toward the Eurasian plate than its western margins. This rotation largely explains the overall morphology of the present Himalayas.

The growth of the Himalayas caused an intensification of the monsoon, which strengthened between 9 and 6 million years ago or perhaps earlier, when the Tibetan plateau attained sufficient elevation to act as a heat source and to block northward air fluxes. There has been stronger rainfall on the region and thus stronger erosion since then. Some researchers think that intensification of the monsoon induced sufficient erosion to decrease the elevation of the eastern Himalayas, which caused a change in the gravity (see Sect. 6.6.2) in the area that was sufficient to trigger the counter-clockwise rotation of the Indian plate. The long-term effect of the building of the Himalayas on the monsoon, and of the latter of the tectonic processes of Himalaya building are another example of a feedback between erosion and tectonics.

The *Andes* are the longest coastal mountain range in the world. They form a continuous highland (200 to 700-km wide) over 7,000 km along the western edge of South America. The mountain range extends from Venezuela in the North to Chile and Argentina in the south, and its average height is about 4,000 m, the highest peak being Mount Aconcagua in Argentina (6,961 m). The whole Andes are the site of active volcanism and earthquakes, and the most powerful earthquake ever measured by a seismometer (magnitude 9.5) occurred in Chile in 1960. The Andes are the result of the subduction of the oceanic crust of the Nazca plate (located between South America and the Pacific plate) and the Antarctic plate under the western continental margin of the South American plate (Fig. 7.2). The building of mountains is thought to have begun about 50 million years ago, but this is debated among specialists.

The building of the Andes corresponds to the general cooling of Earth that began about 50 million years ago with the closing of the Drake Passage between South America and Antarctica (see Sect. 4.5.2), suggesting that the two phenomena could have had the same tectonic causes. During the building of the Andes, there were two pulses of rapid widening of the highland away from the continental margin, from about 40 to 30 million years ago and since about 10 million years ago, which corresponded to pulses of increased aridity. The decreased rainfall caused lower erosion, and it was proposed that this allowed the Andes' highlands to grow horizontally eastwards during the long periods of low rainfall. This effect of the climate on the tectonic processes of mountain building is different from those described above for the three other mountain ranges. The article of Witze (2018) provides further reading on the mechanisms involved in the rise of the Andes.

The four cases of the St. Regis Range, European Alps, Himalayas and Andes show that the building of major mountain ranges is not only determined by tectonics, but also by the climatic action of erosion on some of the tectonic mechanisms of mountain building. In most cases, this leads to feedback interactions between erosion and tectonics. The above interactions and feedbacks are summarized in Fig. 7.12, which shows that erosion modifies the way mountains grow by acting on some effects of the tectonic processes of mountain building. In return, the way mountains grow affects the erosion processes.

The growth of mountains is primarily determined by the tectonic processes described previously (see Sect. 7.2.3). However as shown for the above four mountain ranges, the building of mountains is accompanied by increased erosion, which removes materials from the growing mountains. The latter feeds back into the growth of mountains, which is slowed down. The slower growth of mountains

Fig. 7.12 Summary of interactions and feedbacks between erosion and tectonic mechanisms of mountain building in the four mountain ranges identified in Fig. 7.11. Credits at the end of the chapter

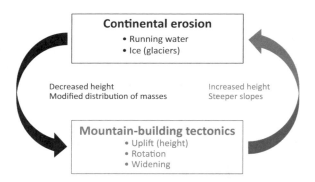

causes slower erosion, which can initiate a *negative feedback loop* (see Sect. 9.1.3; Fig. 9.4e and accompanying text). Hence the removal of huge amounts of materials by erosion and the subsequent redistribution of these materials affect the tectonic processes of mountain building in ways that vary depending on the tectonic environment, and this may reorient the growth of mountains.

The interactions between erosion and mountain building are part of the wider network of processes that regulate the natural greenhouse effect, which were described in this chapter. Given that organisms play a role in continental erosion (see Sect. 7.3.3), their activity likely contributes to the effects of erosion on the tectonics of mountain building. Indeed, organisms became agents of continental erosion after their occupation of continents, which began during the Silurian and Devonian periods 444 to 359 million years ago (see Sect. 1.5.4).

In the very long-term, variations in the natural greenhouse effect were dampened during the history of Earth by the negative feedback of the rate of dissolution of $CaSiO_3$ on the global temperature mediated by the uptake of atmospheric CO_2 (see Sects. 7.3.3 and 9.2.3). In the shorter term, the interplay between erosion and tectonics could have contributed to finer adjustments of the climate.

7.6 Key Points: The Natural Greenhouse Effect and the Living Earth

We conclude this chapter with a brief summary of some of the key points of the above sections, as done in previous chapters (see Sects. 1.6, 2.5, 3.8, 4.8, 5.7, and 6.7). Here, we examine interactions between the Solar System, Earth, the natural greenhouse effect, and organisms. As in the corresponding sections of previous chapters, the context of this section is provided by the three phases of the Earth's history illustrated in Fig. 1.11, which are the Mineral, Cradle and Living Earth. These three phases occurred from 4.6 to 4.0, 4.0 to 2.5, and 2.5 to 0 billion years ago, respectively.

In the Mineral Earth stage, before continents started to form, the storage of carbon from the highly abundant CO_2 into $CaCO_3$ rocks for very long times was driven by the chemical alteration of seafloor $CaSiO_3$ rocks combined with the formation of solid $CaCO_3$ within and on the seafloor (see Sect. 7.3.3). This process caused the storage of huge amounts of carbon in seafloor $CaCO_3$ rocks in the early history of Earth, which sequestered more than 99% of the Earth's carbon in $CaCO_3$ This had the effect of lowering significantly the initially very high concentrations of atmospheric CO_2 and the correspondingly very strong greenhouse effect. As a result, the ocean's temperature went down from about +200 °C to perhaps +70 °C or lower, which was suitable for the development of organisms (see Sect. 7.3.3). This mechanism continues today, as $CaSiO_3$ rocks of the new basaltic seafloor, formed from the lava of submarine volcanoes, are eroded by CO_2-acidified seawater and the resulting carbon is stored into seafloor carbonate rocks (see Sect. 7.3.3).

Since the Cradle Earth stage, the same chemical processes have been taking place on continents, where the chemical alteration of $CaSiO_3$ rocks by CO_2-acidified rain is causing a long-term net transfer of atmospheric CO_2 to marine sediments in the form of $CaCO_3$ (see Sect. 7.3.3). On continents, there is also chemical alteration of $CaCO_3$ followed by formation of $CaCO_3$ in the ocean, but contrary to the chemical alteration of $CaSO_3$, this process does not change the long-term concentration of CO_2 in the atmosphere (see Sect. 7.3.3). Tectonic activity affects the long-term climate by driving the volcanic production of $CaSiO_3$ rocks and CO_2, and by generating slopes on continents that largely drive the water erosion of $CaSiO_3$ (see Sect. 7.3.3). Overall, the recycling of carbon by volcanic emissions and rock erosion regulated the long-term greenhouse effect needed to maintain the thermal habitability of the planet.

Some processes are specific to continents, and others to oceans. On continents, chemical erosion modifies the way the mountains grow, by acting on some effects of the tectonic processes of mountain building, and the interactions between erosion and mountain building are part of the wider network of processes that regulate the natural greenhouse effect (see Sect. 7.5.2). Also, volcanic emissions of CO_2 tend to increase the natural greenhouse effect and thus temporarily warm the global climate, whereas the effect of the emissions of volcanic ash is opposite (see Sect. 7.4.5). After the organisms had occupied the continents, which began during the Silurian and Devonian periods (444 to 359 million years ago), they played a role in continental erosion, and their activity thus likely contributed to the effects of erosion on the tectonics of mountain building (see Sect. 7.5.2). In the ocean, the $CaCO_3$ produced by calcifying plankton sinks to the seafloor upon the death of the organisms, or stays on the seafloor in the case of benthic organisms; there, they form carbonate sediments (see Sect. 7.3.3). In the two environments, the organisms contribute to recycle carbon in the long term, and thus sustain the natural greenhouse effect (see Sect. 7.3.3).

In the very long-term, variations in the natural greenhouse effect were dampened during the history of Earth by the negative feedback of global temperature into the rate of dissolution of $Si(OH)_4$, and the negative feedback of the resulting change in atmospheric CO_2 into the global temperature (see Sects. 7.3.3 and 7.5.2). Over the last 4 billion years, the overall coupling of the Earth's carbon-recycling processes maintained a long-term moderate greenhouse effect. This created globally suitable temperatures for organisms, which allowed them to build huge biomasses and thus take over the Earth System (see Sect. 7.4.7).

Figure Credits

Fig. 7.1 Modified after Fig. 1 of Douglass (2006). With permission from Prof. David H. Douglass, University of Rochester, USA.

Fig. 7.2   https://commons.wikimedia.org/wiki/File:Quake_epicenters_1963-98_ notitle.png by NASA, in the public domain.

Fig. 7.3 Figure 2 of Straume et al. (2019) by NOAA, https://ngdc.noaa.gov/mgg/ sedthick,   https://www.ngdc.noaa.gov/mgg/sedthick/data/version3/fig_2_new_ press.png, in the public domain.

Fig. 7.4 This work, Fig. 7.4, is a derivative of https://commons.wikimedia.org/ wiki/File:Cycle_orogénique.png by Anthony Saphon, used under GNU FDL, CC BY-SA 3.0 and CC BY 2.5. Figure 7.4 is licensed under GNU FDL and CC BY-SA 3.0 by Mohamed Khamla.

Fig. 7.5 This work, Fig. 7.5, is a derivative of https://commons.wikimedia.org/ wiki/File:Cycle_orogénique.png by Anthony Saphon, used under GNU FDL, CC BY-SA 3.0 and CC BY 2.5. Figure 7.5 is licensed under GNU FDL and CC BY-SA 3.0 by Mohamed Khamla.

Fig. 7.6 Original. Figure 7.6 is licensed under CC BY-SA 4.0 by Philippe Bertrand, Louis Legendre and Mohamed Khamla.

Fig. 7.7 Original. Figure 7.7 is licensed under CC BY-SA 4.0 by Philippe Bertrand, Louis Legendre and Mohamed Khamla.

Fig. 7.8a   https://commons.wikimedia.org/wiki/File:Coccolithus_pelagicus.jpg by Richard Lampitt, Jeremy Young, The Natural History Museum, London, used under CC BY 2.5. Size added under the photo.

Fig. 7.8b https://commons.wikimedia.org/wiki/File:Pachnodus_praslinus.JPG by Dreizung https://de.wikipedia.org/wiki/Benutzer:Dreizung, used under CC0 1.0 (Public Domain Dedication). Size added under the photo.

Fig. 7.8c https://commons.wikimedia.org/wiki/File:Sea_angel.jpg by Matt Wilson/ Jay Clark, NOAA NMFS AFSC, in the public domain. Size added under the photo.

Fig. 7.8d https://commons.wikimedia.org/wiki/File:Coral_Outcrop_Flynn_Reef. jpg by Toby Hudson https://commons.wikimedia.org/wiki/User:99of9, used under CC BY-SA 3.0. Size added under the photo.

Fig. 7.9a https://commons.wikimedia.org/wiki/File:Phaeodactylum_tricornutum. png by Alessandra de Martino and Chris Bowler, used under CC BY 2.5. Size added under the photo.

Fig. 7.9b https://commons.wikimedia.org/wiki/File:Dictyocha_speculum.jpg by Minami Himemiya https://commons.wikimedia.org/wiki/User:姫宮南, used under CC BY-SA 3.0. Size added under the photo.

Fig. 7.9c https://commons.wikimedia.org/wiki/File:Staurocalyptus-_noaa_photo_ expl0951.jpg by NOAA/Monterey Bay Aquarium Research Institute, in the public domain. Size added under the photo.

Fig. 7.10 Original. Figure 7.10 is licensed under CC BY-SA 4.0 by Philippe Bertrand, Louis Legendre and Mohamed Khamla.

Fig. 7.11 This work, Fig. 7.11, is a derivative of https://commons.wikimedia. org/wiki/File:AYool_topography_15min.png by Plumbago https://en.wikipe- dia.org/wiki/User:Plumbago, used under GNU FDL, CC BY-SA 3.0 and CC BY 2.5. Figure 7.11 is licensed under GNU FDL and CC BY-SA 3.0 by Mohamed Khamla.

Fig. 7.12 Original. Figure 7.12 is licensed under CC BY-SA 4.0 by Philippe Bertrand, Louis Legendre and Mohamed Khamla.

References

Darwin C (1859) On the origin of species by means of natural selection, or the preservation of favoured races in the struggle for life. John Murray, London

Douglass DH (2006) Observational climate data and comparison with models. In: Ragaini R (ed) International Seminar on Nuclear War and Planetary Emergencies—34th Session. The Science and Culture Series—Nuclear Strategy and Peace Technology. World Scientific Publishing Co. Ltd. https://doi.org/10.1142/6076 [Draft copy via http://www.pas.rochester. edu/~douglass/papers/Erice_final%20draft.pdf]

Einstein A (1905) Zur Elektrodynamik bewegter Körper (On the electrodynamics of moving bodies). Annalen der Physik 322:891–921 [English translation at https://www.fourmilab.ch/ etexts/einstein/specrel/www/]

Straume EO et al (2019) GlobSed: Updated total sediment thickness in the world's oceans. Geochem Geophys Geosyst 20:1756–1772. https://doi.org/10.1029/2018GC008115

Wegener A (1915) Die Entstehung der Kontinente und Ozeane. Friedr. Vieweg & Sohn Akt. Ges, Braunschweig. English edition: Wegener A (2011) The origin of continents and oceans, Revised edition (trans: Biram J). Dover, New York

338   7   The Natural Greenhouse Effect: Connections …

Further Reading

Arndt NT (2013) Formation and evolution of the continental crust. Geochem Perspect 2:405–533. https://doi.org/10.7185/geochempersp.2.3

Bauer A, Velde BD (2014) Geochemistry at the Earth's surface. Movement of chemical elements. Springer, Berlin, Heidelberg

De La Rocha C, Conley DJ (2017) Silica stories. Springer, Cham

Frisch W, Meschede M, Blakey R (2011) Plate tectonics. Continental drift and mountain building. Springer, Berlin, Heidelberg

Rubin KH, Soule SA, Chadwick WW Jr, Fornari DJ, Clague DA, Embley RW, Baker ET, Perfit MR, Caress DW, Dziak RP (2012) Volcanic eruptions in the deep sea. Oceanography 25(1):142–157. https://doi.org/10.5670/oceanog.2012.12

Ruddiman, WF (2008) Earth's climate. Past and future, second Edition, W. H. Freeman and Company, New York

Shikazono N (2012) Introduction to Earth and planetary system science. A new view of Earth, planets, and humans. Springer, Tokyo

Siebold E, Berger W (2017) The sea floor. An introduction to marine geology, 4th edn. Springer Textbooks in Earth Sciences, Geography and Environment. Springer, Cham

Staudigel H, Clague DA (2010) The geological history of deep-sea volcanoes: Biosphere, hydrosphere, and lithosphere interactions. Oceanography 23(1):58–71. https://doi.org/10.5670/oceanog.2010.62

Wilson T, Cole JW, Stewart C (2021) Effects of volcanic ash on infrastructure. Springer Briefs in Earth System Sciences, Springer, Berlin, Heidelberg

Witze A (2018) How to build a mountain range. Knowable Magazine. https://doi.org/10.1146/knowable-071218-080201

Chapter 8
The Long-Term Atmosphere: Connections with the Earth's Magnetic Field

8.1 The Earth's Atmospheres and the Earth System

8.1.1 The Phases of the Earth's Atmosphere

The history of the Earth's atmosphere began shortly after the formation of the Earth-Moon system about 4.5 billion years. During these billions of years, the composition of the atmosphere underwent profound changes, which are described in this book as a succession of four phases. In addition to these four phases, we also examine at the end of this section the characteristics of the present atmosphere, because these are becoming rapidly different from those of the fourth phase. The changes that occurred in the concentrations of the main atmospheric gases since 4.5 billion years before present are not known for certain, especially for the most ancient times. Figure 8.1 summarizes available data and current hypotheses about the changing atmosphere since 4.5 billion years (see Information Box 8.1).

This chapter first examines the role played by the magnetic field in the phases of the atmosphere. The present section describes the succession of these phases, and the corresponding Earth System conditions. The remainder of the chapter focuses in turn on: the use of the magnetic field by wildlife (that is, any organism except humans) and humans (Sect. 8.2); the magnetic fields in the Solar System (Sect. 8.3); the contribution of the magnetic field to the takeover of Earth by organisms in aquatic environments and on continents (Sect. 8.4); and the use by researchers of shifts in polarity of magnetic fields fossilized in rocks to explore ancient Earth (Sect. 8.5). The chapter ends with a summary of key points concerning the interactions between the Solar System, Earth, the long-term atmosphere, and organisms (Sect. 8.6).

© Springer Nature Switzerland AG 2021
P. Bertrand and L. Legendre, *Earth, Our Living Planet*, The Frontiers Collection,
https://doi.org/10.1007/978-3-030-67773-2_8

Information Box 8.1: Evolution of the Earth's Atmosphere: Facts and Assumptions Figure 8.1 summarizes facts and assumptions concerning the evolution of the Earth's atmosphere since 4.5 billion years. Given the uncertainties concerning values in the past and especially the distant past, several values in the figure are only semi-quantitative.

The unit of total atmospheric pressure in this book is the *atmosphere* (atm), which corresponds to the present average atmospheric pressure at sea level (see Information Box 2.1). The *partial pressure* of any gas can be expressed in absolute value (atm), or as the volume fraction of total gases as in Fig. 8.1 (%; see Information Box 2.1). Figure 8.1 considers the following gases: nitrogen (N_2), oxygen (O_2), carbon dioxide (CO_2), methane (CH_4), water vapour (H_2O), and various trace gases. At present, the volume fractions of these gases for dry atmosphere are: N_2 (78.1%), O_2 (20.9%), CO_2 (0.04%), and CH_4 (0.0002%), The sum of the partial pressures of all the gases is the total pressure, which is presently set at 1 atm, but was different in the past (Fig. 8.1a).

Water vapour was always present in the Earth's atmosphere. In the initially hot atmosphere (above 2,000 °C), all water was in the form of vapour, and when temperature went down, some water was also in the forms of liquid droplets and ice crystals in clouds. At present, the partial (vapour) pressure of water in the atmosphere ranges between 0 and 3%. It varies in Fig. 8.1 as a constant ratio of 10% of the partial pressure of CO_2.

During the *first phase of the atmosphere* (before 4.4 billion years ago or even earlier), it is assumed that all the Earth's carbon was in the form of CO_2, whose amount was equivalent to that of the present-day sedimentary carbon, and then made up 98% by volume of all atmospheric gases (Fig. 8.1b). In Fig. 8.1, the partial pressure of N_2 is taken as constant at 0.78 atm during the whole history of the planet. The remainder of the nitrogen is located in the crust and the ocean, which account for less than 20 and 0.5%, respectively, of the Earth's nitrogen.

During the *second phase of the atmosphere* (between 4.4 or earlier and 2.4 billion years ago), there was a rapid decrease in atmospheric CO_2 following the formation of a liquid ocean (Fig. 8.1b). In that ocean, CO_2 was incorporated into solid $CaCO_3$ on and in the seafloor (see Sect. 5.5.4). This way CO_2, which was initially the dominant gas in the atmosphere, became a trace gas. The concentration of CO_2 reached a very low value 2.3 billion years ago (Fig. 8.1c), at the beginning of an episode of snowball Earth whose intensity may have been maximum 2.2 billion years ago (see Sect. 4.5.4).

Atmospheric CH_4 increased with the rising production of this gas by anaerobic organisms in the oceans, up to a maximum value of 4.5% 2.5 billion years ago (Fig. 8.1c). Afterwards, CH_4 declined to its present trace concentration as a consequence of its oxidation to CO_2 by the growing concentration of photosynthetically produced O_2 in the atmosphere (Fig. 8.1b).

The *third phase of the atmosphere* (between 2.4 and 0.5 billion years ago) was characterized by the slow accumulation of photosynthetic O_2, from 0.5 to 15% during a long period, and later up to 35% (Fig. 8.1b). The two CO_2 minima 0.75 and 0.66 billion years ago corresponded to the initiation of the snowball Earth episodes that occurred at 0.71 and 0.64 billion years before present (Fig. 8.1c).

During the *fourth phase of the atmosphere* (since 0.5 billion years ago), the values of the partial pressures of CO_2 and O_2 in Fig. 8.1 reproduce those found in the literature.

The *first phase of the atmosphere* followed the giant collision that created the Earth-Moon system about 4.5 billion years ago (see Sect. 1.5.1). An atmosphere had likely started to form around the proto-Earth before the giant collision, but we begin the history of the Earth's atmosphere with this event. The composition of the atmosphere during its first phase reflects the degassing of the volatile compounds released during the cooling of the hot magma ensued from the giant collision. It is thought that these were mostly hydrogen (H), helium (He), and hydrogen-containing gases such as ammonia (NH_3), methane (CH_4) and water vapour (H_2O). Most of these gases were lost by atmospheric escape to space (see Sect. 8.4.1, below) within about 100 million years.

The *second phase of the atmosphere* likely began 4.4 billion years before present or earlier. In the following hundreds of millions of years, Earth continued to be impacted by meteoroids, comets and small planetary bodies, which added materials to the planet. The degassing of the cooling magma created a new atmosphere, where the concentration of CO_2 was very high. Degassing released a suite of volatile compounds in the atmosphere, which included not only H and He but also heavier volatile compounds that contained such elements as carbon, nitrogen, oxygen and sulphur. The lighter gases escaped to space, leaving behind molecular nitrogen (N_2), methane (CH_4), sulphur dioxide (SO_2), carbon monoxide (CO), water vapour, and especially carbon dioxide (CO_2). The most generally accepted origin of Earth's water is the heavy bombardment of the young planet

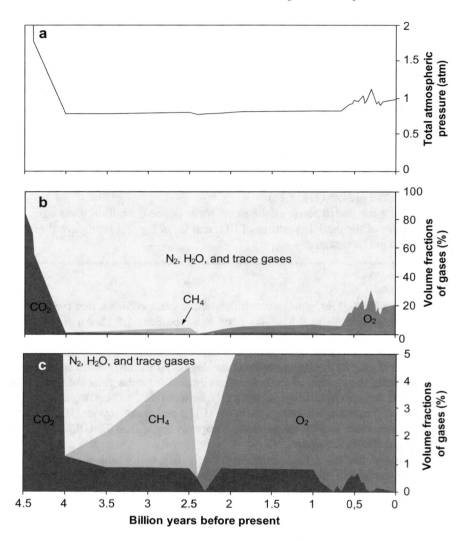

Fig. 8.1 Earth's atmosphere since 4.5 million years before present: **a** total atmospheric pressure, **b** hypothesized proportions of the main gases, and **c** details of the smallest volume fractions below 5%. Note that the Y-axes in the three panels are different. Details are provided in Information Box 8.1. Credits at the end of the chapter

by water-containing asteroids and/or water-rich comets, between 4.1 and 3.9 billion years ago, but that there are alternative explanations (see Sect. 5.3.2). All the explanations ascribe the origin of the abundant water that presently exists on Earth to the original solar nebula, directly or indirectly.

The very high concentration of CO_2 created a very high atmospheric pressure of 60 atmospheres. As Earth cooled down from its initially very high temperature, it reached a point (about 200 °C or higher) where the very high atmospheric pressure allowed the hot atmospheric water vapour to condense into liquid form, leading to the creation of an ocean perhaps 4.4 billion years ago or earlier. Atmospheric CO_2 progressively dissolved in the ocean's waters, where chemical reactions with seafloor silicate rocks ($CaSiO_3$) stored the CO_2 into solid calcium carbonate ($CaCO_3$), which accumulated in the seafloor (Sect. 5.5.4). More than 99% of the initial atmospheric CO_2 gas of the planet was thus sequestered in carbonate rocks, in which most of the Earth's carbon is still stored (see Sect. 5.4.4).

The above chemical processes reduced the initially huge concentration of CO_2 in the atmosphere, with was accompanied by a parallel reduction in the atmospheric pressure and temperature. The more temperate environmental conditions allowed the development of organisms and ecosystems in aquatic habitats, where microorganisms progressively contributed to the transformation of CO_2 into solid $CaCO_3$, which became subsequently stored for the long term (see Sect. 6.4.2). The chemical and biological formation of $CaCO_3$ further decreased the atmospheric pressure and temperature. At the end of its second phase, the atmosphere comprised about 15% CO_2 and 75% nitrogen gases (N_2).

The *third phase of the atmosphere* began with the appearance of free oxygen (O_2) about 2.4 billion years ago, or perhaps more recently. This change took place hundreds of millions of years after marine cyanobacteria had invented the O_2-producing photosynthesis, much earlier than 2.5 billion years ago. Indeed, before photosynthetic production of O_2 by cyanobacteria had any effect on the atmosphere, these organisms had to spread over the global aquatic environments and build up very large biomasses. In addition, all the O_2 produced by photosynthesis combined, during two hundreds million years, with iron and other elements such as sulphur dissolved in water bodies, thus forming oxides, which continued until all these dissolved elements had been oxidized.

Once oxygen escaped from water bodies to the atmosphere, it initially reacted with metals in continental rocks, and it is only when these were completely oxidized that O_2 began to accumulate in the atmosphere. The atmospheric O_2 concentration remained below 1% (per volume) for a very long time (see Sect. 1.5.4). The concentration of oxygen in the upper waters of the ocean was low, and there was no oxygen in the ocean's deep waters. Indeed, the decay of sinking organic matter removed all oxygen from the deeper waters. In shallow seas, microorganisms increasingly contributed to the transformation of CO_2 into solid $CaCO_3$, part of which became sequestered on the seafloor.

The *fourth phase of the atmosphere* was characterized by the presence of the ozone (O_3) layer, which had begun to build up around 500 million years before present. Indeed, O_2 had to accumulate in the upper atmosphere before enough could be transformed into O_3 by the action of the solar the UV solar radiation to form the O_3 layer. The latter is presently located between 15 and 35 km above the Earth's surface (see Information Box 8.2). The concentration of oxygen increased in the upper waters of the ocean, and the deep waters became progressively well oxygenated.

Information Box 8.2 Formation and Destruction of the Ozone (O_3) Layer in the Atmosphere The action of solar ultraviolet (UV) radiation on molecules of free oxygen (O_2) in the upper atmosphere produces the ozone gas, O_3 (see Sect. 2.1.5). This process takes place at an altitude between 10 and 50 km, and the reaction is activated by UV radiation with wavelengths shorter than 240 nm:

$$O_2 + h\nu \rightarrow 2O \qquad (8.1)$$

Each atom of oxygen (O) in Eq. 8.1 combines with one O_2 molecule to form one molecule of O_3:

$$O + O_2 \rightarrow O_3 \qquad (8.2)$$

Through this process, three O_2 molecules produce two O_3 molecules. In physics, the symbol $h\nu$ stands for *energy*. It combines Plank's constant (h) and a frequency (represented by Greek letter "nu", ν). Because the units of Plank's constant are (energy/frequency), those of $h\nu$ are: (energy/frequency) x frequency = *energy*. In the present case, $h\nu$ is the energy of the photons that activate the chemical reaction, and the frequency of the radiation is calculated as follows: ν = speed of light/wavelength of the UV radiation.

In the upper atmosphere, the above process creates the *ozone layer*. This layer is mostly found at an altitude between 15 and 35 km, and its content in O_3 is very low. Indeed, if the thickness of the 20 km O_3 layer were measured at the ground-level atmospheric pressure, it would only be 3 mm. Despite its low concentration in O_3, this layer absorbs 97 to 99% of the solar UV radiation with wavelengths between about 200 to 315 nm. This is crucial because these UV radiations damage the DNA of organisms and, if they reached the Earth's surface, they would prevent the existence of complex organisms on emerged lands (see Sect. 2.1.5).

The O_3 molecule is destroyed below the altitude of 50 km by two main mechanisms. The first is activated by solar UV radiation with wavelengths shorter than 325 nm:

$$O_3 + h\nu \rightarrow O + O_2 \qquad (8.3)$$

Each atom of oxygen (O) in Eq. 8.3 combines with one O_3 molecule to form two molecules of O_2:

$$O + O_3 \rightarrow 2O_2 \qquad (8.4)$$

Through this process, the destruction of two O_3 molecules forms three O_2 molecules. The second mechanism of O_3 destruction is a reaction that involves *free radicals* X present in the atmosphere, where X can be any of NO, Cl^-, Br^-, OH^-, or H^+. This reaction has two steps:

$$X + O_3 \rightarrow XO + O_2 \qquad (8.5)$$

$$XO + O \rightarrow X + O_2 \qquad (8.6)$$

In this reaction, substance X is re-formed in the second step of each cycle (Eq. 8.6), so that its concentration does not change. Such a substance is called a *catalyst*, and because the catalyst does not undergo any permanent chemical change, a small amount is enough to promote a *catalytic reaction*. It follows that, although the upper atmosphere contains only small concentrations of catalysts, the above reaction (Eqs. 8.5 and 8.6) proceeds naturally.

Over 300,000 tons of ozone are made and destroyed in the atmosphere each day. The book of Fabian and Dameris (2014) provides further reading on O_3 in the atmosphere.

The concentration of atmospheric O_2 increased until it reached the extremely high value of 35% during the Carboniferous period (between 359 and 299 million years ago), after which it decreased and finally stabilized, after some fluctuations, at the present 21%. The increase in atmospheric O_2 was accompanied by a corresponding decrease in CO_2, which was caused by the burial of massive amounts of organic carbon in the ground as coal (the name *Carboniferous* means "coal-bearing"), and in the seafloor (part of the latter organic carbon formed petroleum and natural gas). Indeed, if the organic matter buried under the ground and in the seafloor had instead remained in contact with the environmental oxygen, the oxidation of this organic matter would have consumed all the available O_2, whose concentration would consequently not have increased in the atmosphere.

The *present-day atmosphere* is characterized by an ongoing sharp increase in atmospheric CO_2, which is a reversal of the decreasing trend (with ups and downs) that began 4.4 billion years ago. The current increase in atmospheric CO_2 comes from the combustion of fossil organic matter (mostly coal, petroleum, and natural gas) by human societies, and their calcination of $CaCO_3$ rocks to make cement (Fig. 9.5). Given that this CO_2 had been stored away by natural processes over hundreds of millions of years, the present situation could be summarized as "onwards to the past". This new chapter in the history of the Earth's atmosphere is marked by the anthropogenic *climate change* (see Information Box 1.4 and Sects. 11.1.1 and 11.1.3). The book of Catling and Kasting (2017) provides further reading on the Earth's atmosphere in the past, its origins, and how it evolved to its present state.

8.1.2   The Earth System

There are close connections between the evolution of the atmosphere during the course of the Earth's history and that of the Earth System. The latter consists of the atmosphere, the hydrosphere, the solid Earth and the biosphere, and their interacting physical, chemical and biological processes (see Sect. 1.1.1). The following brief review of these connections for the different phases of the Earth's atmosphere leads us to consider the roles of the magnetic field, which is the focus of this chapter.

The *atmosphere in its first phase* (4.4 billion years before present or earlier) was created by the cooling of the initially very hot magma of the young planet, and the Earth System began its development only after the planet had cooled Hence, there was no Earth System as such during the first hundreds of millions of years of the first atmosphere.

The *atmosphere in its second phase* (from before 4.4 to about 2.4 billion years before present) saw the formation of an ocean when the planet had cooled enough to allow the atmospheric water vapour to condense in the liquid form. In this ocean, most of the atmospheric CO_2, whose concentration had been initially very high, was rapidly sequestered in solid $CaCO_3$ (see Sect. 5.5.4). From then onwards, the Earth's atmosphere had the right characteristics and greenhouse gas composition for maintaining a moderate greenhouse effect, which generally kept the temperature of Earth within a range that allowed most of its surface water to be liquid and was suitable for organisms (see Sects. 3.6.1 and 5.6.1). In the vast expanses of liquid water at the surface of the planet, the organisms diversified into archaea and bacteria (see Information Box 3.1), which successively invented chemosynthesis and the two types of photosynthesis (see Information Box 5.1). The organisms, which were then all unicellular prokaryotes, built up very large biomasses in sediments and shallow waters, where they began to modify the chemistry of their environment. This was a first step towards taking over the planet.

The *third phase of the atmosphere* (from about 2.4 to 0.5 billion years before present) saw the first appearance of atmospheric O_2 as a consequence of the invention of O_2-producing photosynthesis by marine cyanobacteria, hundreds of millions of years before. The progressively increasing O_2 in shallow seas caused the *Great Oxidation Event* or *Great Oxygenation Event*, during which the O_2 concentration, although quite low, was high enough to harm the existing organisms given that most of them did not tolerate O_2. The increasing concentrations of environmental O_2 favoured the evolution and development of O_2-respiring bacteria and archaea (see Sect. 10.1.6). During the same, long period, organisms evolved in aquatic environments from unicellular prokaryotes to unicellular and, later, multicellular eukaryotes. The energy-efficient O_2 respiration and the large masses of prokaryotes and multicellular eukaryotes contributed to the takeover of the shallow oceans by organisms.

The *atmosphere in its fourth phase* (from about 500 million years before present) was, and continues to be characterized by the presence of the ozone (O_3) layer, which protected organisms from the DNA-damaging UV radiation from the Sun (see Sect. 2.1.5 and Information Box 8.2). This protection allowed complex multicellular organisms to begin the occupation of emerged lands more than 400 million years ago. Over tens of millions of years, land organisms diversified and developed very large biomasses, and together with marine organisms produced so much organic matter that their physiological activities drastically changed the O_2 and CO_2 composition of the atmosphere. Simultaneously, huge amounts of the organic carbon produced by land and water ecosystems were buried under the ground and in the seafloor, respectively, resulting in increases in atmospheric O_2 that were particularly spectacular during the Carboniferous period (see Sect. 8.1.1, above). This way, organisms completed their takeover of the planet, which led the Earth System to reach the high complexity that currently exists (see Sect. 1.1.1).

The *present-day atmosphere* is still part of the fourth phase, but its characteristics gradually diverge from those described in the previous paragraph. This divergence is the result of the Industrial Age, which began in the second part of the eighteenth century. Indeed, industrial activity was then, and is still largely, fuelled by fossil organic matter (fuel and raw materials) and $CaCO_3$ (cement), whose carbon had been slowly withdrawn from the atmosphere and sequestered by natural processes over hundreds of millions of years. In addition, humans have harnessed electricity and most wavelengths of the electromagnetic spectrum to sustain their activities (see Information Box 2.8). The latter include the use of gamma and X rays for industrial and medical purposes, the application of optics to almost all aspects of life, and the use of the longer wavelengths of the spectrum for radio communications, radar and other applications. As a consequence of technological developments, human activities have become a geophysical force, which is changing the global climate as deeply as a long volcanic eruptive phase (see Sect. 7.4.6) or the fall of a large asteroid (see Sect. 4.6.2). Human societies are profoundly modifying the Earth System from within.

The *magnetic field* was deeply involved in the above succession of atmospheric phases and Earth System conditions, through mechanisms that will be explained below (see Sects. 8.4.1 and 8.4.2). Briefly, the planet did not have a magnetic field during the first phase of its atmosphere, and it lost its initial gases. The establishment of the Earth's magnetic field during the second phase of the atmosphere contributed to the retention of the atmosphere, which protected the expanses of liquid water where organisms diversified and prospered. Later on, the presence of the magnetic field protected the ozone layer, which allowed organisms to populate the continents and complete their takeover of the planet. Finally, the Industrial Age was largely the conquest of electromagnetic processes and the electromagnetic spectrum by humans.

More than 99% of the chemical elements initially present in the gaseous constituents of the Earth's atmosphere during its second phase were progressively sequestered in the crust and in the mantle by a combination of global chemical and tectonic processes (see Sect. 7.3.2). These chemical elements include hydrogen (H), oxygen (O), carbon (C), nitrogen (N), calcium (Ca), phosphorus (P) and sulphur (S). Sequestration protected these elements from being possibly lost to space, but only less than 1% remained available in the outer envelopes of Earth (crust, ocean and atmosphere) for use by ecosystems. This small amount was crucial for the development of organisms and ecosystems on the planet, and it will be explained below how the Earth's magnetic field contributed to protect these chemical elements from being lost to space (see Sect. 8.4.1).

8.2   The Earth's Magnetic Field and Organisms

8.2.1   What Is the Earth's Magnetic Field?

Magnets are fascinating objects that attract or repel some substances by an invisible mechanism. The substances attracted by magnets, or that can be magnetized to become themselves permanent magnets are called *ferromagnetic* There are only a few ferromagnetic substances on Earth, and these include iron, nickel and cobalt. A magnet is surrounded by an invisible *magnetic field* that attracts other ferromagnetic substances, or attracts or repels other magnets. The shape of the magnetic field around a magnet can be evidenced by laying iron filings in its vicinity (Fig. 8.2a). The lines formed by the iron filings show that the magnet has two poles, around which a magnetic field is organized.

At the surface of Earth, magnetized objects, such as the needle of a compass, point toward a direction called *magnetic pole*. In the same way as the layout of iron filing show the existence of a magnetic field around a magnet, the pointing of the compass' needle towards the magnetic pole anywhere on Earth indicates the existence of a magnetic field around the planet. It is as if a gigantic magnet was located at the centre of the globe, generating a magnetic field extending from the interior of the globe into space (Fig. 8.2b). We will examine later what, inside Earth, causes the planet's magnetic field (see Sect. 8.3.1, below).

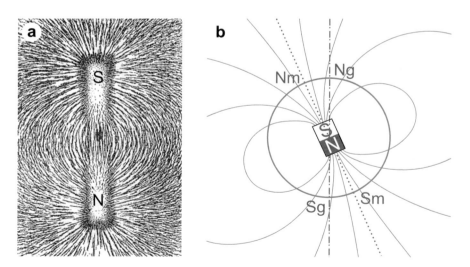

Fig. 8.2 Magnetic field **a** around a rectangular magnet, evidenced by iron filings on a piece of paper placed on top of the magnet bar (S and N: south and north poles of the magnet, respectively), and **b** around Earth, showing the positions of the north and south magnetic poles (Nm and Sm, respectively) and north and south geographic poles (Ng and Sg, respectively). Purple dotted and blue dashed lines: see the text. Credits at the end of the chapter

The Earth's magnetic field has different characteristics, which are its polarity, intensity, inclination, and declination. We will see below that some organisms, including human beings, take advantage of these characteristics (see Sects. 8.2.2 and 8.2.3). Here, we examine the nature of these four characteristics.

Polarity. All magnetic objects have *two poles*, one where the magnetic lines enter the object and the other where they emerge from it, which are called *south magnetic pole* and *north magnetic pole*, respectively. The name *north magnetic pole* comes from the fact that the first manifestation of the Earth's magnetic field known to people was the attraction of the needle of the compass in the direction of the geographic North Pole, which led them to call *north pole* the end of the needle pointing towards the geographic North. However, we now know that the opposite poles of two magnets attract each other, and the similar poles repel. Hence, the north magnetic pole of the compass' needle is attracted by a south magnetic pole. It follows that the magnetic pole of Earth located in the vicinity of the geographic North Pole is physically a south magnetic pole. However, in order to avoid additional confusion, the magnetic pole located in the Northern Hemisphere is called *north magnetic pole*, and that in the Southern Hemisphere *south magnetic pole*. The relative orientation of the magnetic north and south poles is called *polarity*. The Earth's north and south *magnetic poles* are illustrated in Fig. 8.2b (Nm and Sm, respectively), where the north and south *geographic poles* are also represented (Ng and Sg, respectively).

Intensity. The i*ntensity* of the Earth's magnetic field is very low compared to that of the magnets we encounter in day-to-day life. We take here as example the ornaments attached to a small magnet that people affix to refrigerators called *refrigerator magnets*. A strong refrigerator magnet has a magnetic field that 100's of times more powerful than the Earth's magnetic field. Although the latter is not intense, it extends very far from the planet, much beyond the limit of the Earth's atmosphere. The latter is illustrated in Fig. 8.2b.

Inclination. The *inclination* is the angle between the lines of the magnetic field and the Earth's surface. At the magnetic poles, the magnetic lines enter the globe and emerge from it, and the magnetic field is thus vertical (angle of 90° with the surface of the planet). At the magnetic equator, the magnetic lines are parallel to the surface of the globe (angle of 0°). Hence, from one of the poles toward the equator, the inclination of the magnetic field progressively changes from the vertical to the horizontal (that is, from 90 to 0°; Fig. 8.2b).

Declination. At a given location at the surface of the planet, the direction of the Earth's magnetic pole (given by the compass) is not the same as that of the true (geographic) pole. The angle between the two directions is the *declination*. The value of the magnetic declination is not the same everywhere on the globe, and it changes over the years. In some regions, it increases polewards, and in some areas close to the north and south magnetic poles, it can be as much as 80°. The value of the magnetic declination is indicated on navigation maps, as this information is essential to use the magnetic compass efficiently. In Fig. 8.2b, the angle between the purple dotted line and the blue dashed line illustrates the global magnetic

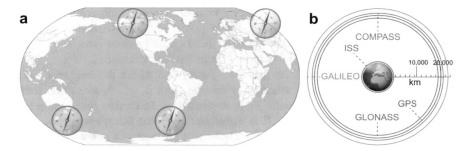

Fig. 8.3   a Map of the world with small images of magnetic compasses showing a magnetic declination of 20° East. **b** Orbits around Earth of the satellites of the four Global Navigation Satellite System (GPS, Galileo, GLONASS, and COMPASS or BeiDoo) and the International Space Station (ISS) for comparison. Credits at the end of the chapter

declination. The map in Fig. 8.3a shows four sites in the world where the declination is 20° East.

8.2.2   Wildlife and the Magnetic Field

Many groups of animals can sense the Earth's magnetic field, and use it for orientation during navigation. In fact, not only animals, but also some bacteria can sense the magnetic field. Indeed, some bacteria have been observed to orient and move along the lines of the magnetic field, and this seems to be related to the presence of crystals of magnetic minerals (iron oxide, Fe_3O_4) in their cells.

The ability of animals to detect the magnetic field has been observed among the arthropods, the molluscs, and many groups of vertebrates. Some animals are sensitive to the north-south polarity of the magnetic field, which provides them with a reference direction during migrations. These include bees, ants, butterflies, lobsters, fish, bats and mole-rats. Other animals seem to be more sensitive to the inclination of the magnetic field, which they compare to the vertical reference provided by gravity (that is, their weight). These include sea turtles and migratory birds, which use the information provided by the changes in the inclination of the magnetic with latitude as they migrate to guide their long-distance navigation. Some migratory animals, such as lobsters, appear to be able to sense the polarity, the inclination and the intensity of the magnetic field, and to combine these three types of information to navigate.

The organs that animals use to sense properties of the Earth's magnetic field remain largely unknown. This is also the case for the ways in which these animals use the magnetic-field information to orient themselves and navigate over often very long distances. The book of Gould and Grant Gould (2012) provides further reading on the methods by which animals find their way both near home and around the globe.

8.2.3 *Humans and the Magnetic Field*

Human beings do not have the biological ability to sense the Earth's magnetic field directly. However, some people claim to suffer from electromagnetic hypersensitivity to the artificial magnetic fields generated by the numerous electric or electronic devices they encounter in the modern environment. Even if humans do not sense the Earth's magnetic field, they have discovered its existence empirically in the past, and used its properties for navigation later on.

The first mention of people taking advantage of the polarity of the magnetic field is in Chinese texts written before the first century of the current era. Chinese people later used the direction indicated by a magnetized needle for determining the orientation of buildings according to the principles of *feng shui* (Chinese geomancy). The first clear mention of the compass being used as a means of navigation is in Chinese texts dating back to the eleventh century. The compass was then a simple magnetic object floating in a bowl of water. It seems that it was initially used in China for military operations on land during the night, and was used for navigation at sea only one century later. The *wet compass* and its dry version (the *dry compass* consists of a freely pivoting magnetized needle mounted on a pin) continued to be used for navigation in Asia in the following centuries.

After the compass became known in Europe in the late twelfth century, it underwent increasingly sophisticated improvements that all aimed at making its use easier for navigation, especially at sea. During several centuries, mariners used dry compasses, and it is only in the nineteenth century that modern versions of the liquid compass progressively took over. One problem of the magnetic compass is that it indicates the *magnetic north*, which is not the same as the *true (geographic) north* as explained above (see Sect. 9.2.1). In fact, the compass aligns itself with the local geomagnetic field, which varies depending on the location at the surface of the globe, and also over the years. Another problem is that the direction indicated by the compass is influenced by the presence of ferromagnetic masses in the vicinity of the compass, such as the steel hull of ships. Hence on ships, the effects of the ferromagnetic disturbances must be nullified by placing *correctors* (magnets and soft iron) adjacent to the compass to create equal but opposing magnetic fields.

Nowadays, the direction of the magnetic north can be determined with a *fluxgate compass*, which was developed for aircrafts in the 1940s and is often used on small boats. It senses the direction of the horizontal component of the magnetic field, and can be connected to other navigational equipment such as an autopilot. It is sensitive to the presence of ferromagnetic masses in its vicinity.

Given the various problems associated with the use of the magnetic compass, a different type of navigation aid was developed in the late nineteenth century, called *gyrocompass*. This navigational instrument indicates the true north, and is not influenced by the presence of ferromagnetic masses. The gyrocompass is based on a fast-spinning disc whose axis sets itself in a position parallel to the Earth's axis of rotation, and thus indicates the true (geographic) north. It became the main instrument for professional navigation in the twentieth century, whereas the magnetic compass continued to be the main orientation device for leisure activities such as boating, hiking and caving (Fig. 8.3a).

A compass, either magnetic or gyro−, provides the direction in which the craft's bow or nose is pointed (called *bearing*), and this information is used to steer the craft in the chosen direction (called *course*). However, a compass does not provide the geographic position of the craft. In order to determine the position of a craft at sea, various radio-navigation systems were developed during the second part of the twentieth century. Each of these was based on a network of transmitters on the ground and a receiver on the craft (examples of such systems are the LORAN and the Decca Navigator). They were replaced by satellite navigation at the turn of the twenty-first century.

At the beginning of the twenty-first century, ships, aircrafts, cars and even hikers rely on one of the various Global Navigation Satellite Systems (GNSS), among which are the American GPS, European Galileo, Russian GLONASS, and Chinese BeiDoo also called COMPASS. Figure 8.3b shows that the orbits of these satellites are very high above ground (between 19,130 and 23,222 km), much higher than the International Space Station (350–400 km). It is explained in Chapter 6 that the functioning of the GNSS involves Einstein's relativity (see Sect. 6.6.1). The different GNSS provide very precise location in the world geodetic system, whose coordinate origin is located at the Earth's centre of mass, and the location provided by a GNSS is stated in terms of longitude, latitude and altitude. This information is used not only for navigation, but also for a number of activities that require precise positioning such as mapping, surveying and satellite communications.

In order to have a back-up system in case of electrical power failure, vessels have on board a means of determining their direction and heading independent of any supply of power, that is, a correctly installed and adjusted magnetic compass. Similarly, experienced hikers carry a magnetic compass in addition to their GNSS-receiving cell phone in case of power or communication failure. Over the centuries, humans made up for their lack of organs to sense the magnetic field by inventing instruments that take advantage of this invisible planetary frame of reference. The book of Gurney (2005) provides further reading on the development of the compass in navigation, essentially in Great Britain.

8.3   Magnetic Fields in the Solar System

8.3.1   The Magnetic Field of Earth

After considering the importance of the magnetic field for the wildlife and the humans, we now turn to the generation of the Earth's magnetic field.). In order to understand the mechanisms by which the Earth's magnetic field is generated, we need to go back to the structure of the globe described in Chapter 6 (see Sect. 6.6.2).

Early planet Earth was a mass of fluid magma. The young planet was very hot because the numerous collisions with other astronomical bodies released much energy and thus generated heat. The last giant collision of proto-Earth with a large body created the Earth-Moon system 4.5 billion years ago, and was followed by smaller hits during hundreds of millions of years (see Sect. 1.5.1). Each of the large and smaller impacts released heat (from gravitational energy), whose accumulation provided the initial heat of Earth. The ensuing fluidity of the magma allowed planetary differentiation to take place, which concentrated the heavy metals (iron and nickel) in the centre of the planet, and increased the concentrations of lighter chemical elements in the outer envelopes of Earth. This resulted in a stacked-layered structure, with the Earth's core at the centre of the globe (see Sect. 6.6.2).

Due to the very high pressure and temperature conditions deep inside Earth, the core has two parts (Fig. 6.11). The first is the *inner core*, with a diameter of 2,430 km and an estimated temperature of more than 5,000 °C (Table 6.6). The size of the inner core is almost as large as that of the Moon (diameter of 3,474 km), and its temperature is close to that at the surface of the Sun (5,500 °C). The inner core consists almost exclusively of iron, together with some nickel. Because of the enormous pressure at the centre of Earth (more than 3 million atmospheres), the iron there is solid, likely in the form of large crystals. The second part of the core, called the *outer core*, is a concentric layer that surrounds the central inner core. Its thickness is 2,270 km, and it mostly consists of iron, nickel, sulphur and silicon. Its temperature ranges between about 5,000 °C close to the inner core and 3,500 °C in the outer part. Given the high temperature of the outer core and its lower pressure than that of the inner core (by 1 million atmospheres), its material is a low-viscosity fluid.

Some researchers think that the heat of the inner core is left over from the initial heat of Earth (see Sect. 1.5.4), and may also include the latent heat of crystallization of the core. The concept of *latent energy* is explained in Chapter 3 for the evaporation of water (see Sect. 3.4.2). In the same way as the passage of water from liquid to vapour generates latent heat, the passage of iron from liquid to solid at the boundary between the outer and the inner core—which causes the inner core to grow at the expense of the outer core—generates latent heat. Other researchers think that the initial heat of Earth is not an important component of the heat of the inner core and the major sources of this heat include instead, in addition to the latent heat of crystallization, the following sources: gravitational energy released by the compression of the core, gravitational energy released by the rejection of light elements at the inner-outer core boundary as it grows, and radioactive decay of potassium, uranium and thorium.

How is this related to the Earth's magnetic field (Fig. 8.4a)? Before answering this question, let us first consider the physical nature of magnetic fields, called *electromagnetism*. This branch of physics explains that electric and magnetic fields are part of the same phenomenon. This means that the magnetic fields of Earth and other bodies of the Solar System are related to electric mechanisms or properties.

The accepted explanation today is the existence, in Solar System bodies with a magnetic field including Earth, of a *dynamo* in which the magnetic field is induced by the rotation of an electrically conductive fluid. Furthermore in order to maintain the magnetic field in the long term, there must be convection currents in the fluid.

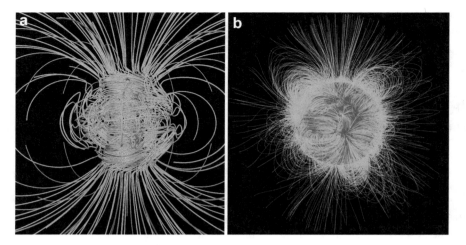

Fig. 8.4   Magnetic fields of **a** Earth and **b** the Sun. Credits at the end of the chapter

These three conditions—electrically conductive fluid, rotation of this fluid, and convection in it—are observed on Earth. Firstly, given that iron conducts electricity, the iron-rich low-viscosity material that makes the outer core is an electrically conductive fluid. Secondly, the fluid in the outer core circulates around the inner core because of the Coriolis effect caused by the rotation of Earth (see Sect. 3.5.2). Thirdly, there is convection in the fluid of the outer core, caused by the heat that escapes from the inner core into the outer core, similar to the convection currents observed between the bottom of a pan of heated water and its surface. The Earth's dynamo (also called *geodynamo*) transforms the rotation of the planet and the heat of the inner core into an electric current in the outer core, which induces the magnetic field. The book of Turner (2011) provides further reading on the study of magnetism and the Earth's magnetic field from over 2,500 years to the present time.

8.3.2   Other Solar System Bodies with Magnetic Fields

Robotic and crewed space missions in the Solar System have shown that many astronomical bodies have a magnetic field. These include: six of the eight planets, namely Mercury, Earth, Jupiter, Saturn, Uranus and Neptune; the largest moon of Jupiter, called Ganymede; and the Sun. Among the eight planets, only Venus and Mars do not have a magnetic field.

It is interesting to note that the first Earth observation satellite, Russian *Sputnik 3* launched in 1958 within the framework of the *International Geophysical Year*, already carried a magnetometer (see Information Box 8.3). Similarly, magnetometers were part of the scientific instruments placed by the astronauts at the

---

Information Box 8.3 Magnetic Compass and Other Magneto-meters Different characteristics of the magnetic field are measured using devices called *magnetometers*. Depending on the type of magnetometer, the instrument measures the direction and/or the intensity of the magnetic field in two dimensions (horizontal) or three (plus vertical), and/or the relative change in the magnetic field at a given location.

The earliest and simplest type of magnetometer is the *magnetic compass*, which measures the direction of the ambient magnetic field, and started to be used for navigation in the eleventh and twelfth centuries (see Sect. 8.2.3). It was only five centuries later, in 1833, that the first magnetometer capable of measuring the magnetic intensity was invented, by the German mathe-matician and physicist Carl Friedrich Gauss (1777–1855). In the following century, many types of magnetometers were developed, based on differ-ent physical properties influenced by the magnetic field. Nowadays, many smartphones contain miniaturized magnetometers used as compasses.

There are many applications of magnetometers, both in the laboratory and in the field, which include mining, medicine, archaeology, and Earth sciences. In some of the field applications, magnetometers are carried by people in backpacks, deployed in planes or spacecraft, or towed behind ships. In this chapter, we refer to various measurements of the magnetic field including magnetometers deployed in spacecraft and towed behind ships.

---

landing site of each of the five American *Apollo* missions that reached the surface of the Moon between November 1969 and December 1972 (namely *Apollo* 12, and 14 to 17; there was no magnetometer in the instrument package of the first lunar landing, during the Apollo 11 mission of July 1969). Magnetometers are now part of the instruments carried by almost all exploration spacecraft and scientific obser-vation satellites.

We examined above the magnetic field of Earth (see Sect. 8.3.1). We now turn to the *three other rocky planets*, Mercury, Venus and Mars.

Similar to Earth, Mercury has a magnetic field, but it is quite weak. This weak-ness is possibly explained by the fact that Mercury is the planet with the second slowest rotation period after Venus (59 Earth days; Table 1.1). Indeed, the rotation of the electrically conductive fluid is a key aspect of a dynamo.

The lack of magnetic field on Venus may be related to its very long rotation period of 243 Earth days, which is the longest of all planets (Table 1.1). The very slow rotation of Venus would not allow a powerful electric current to develop inside the planet, and thus induce a planetary magnetic field. It was also proposed that the entire core of Venus might be molten, which would explain the lack of magnetic field on this planet.

On Mars, there are local magnetic fields in some areas of the planet, whose origins are magnetized crustal rocks. These are interpreted as remnants of an ancient planet-wide magnetic field, which would have existed when the young planet had a hot core. The activity of this core would have created a magnetic field similar to that presently observed on Earth, which would have magnetized rocks in the crust. As the core of Mars cooled down, its outer part would have solidified, shutting down the outer core's electric current and thus the planet's magnetic field 4 billion years ago.

The *four giant planets* are divided in two groups (see Sect. 1.3.1).

The largest two giant planets, Jupiter and Saturn, called *gas giants*, are mostly composed of hydrogen and helium, and they each exhibit a strong magnetic field. Their masses are 316 and 95 times that of Earth, respectively (Table 1.1). It is hypothesized that near the core of these planets, the very high pressure resulting from their huge mass makes the hydrogen to display properties of a metal, including electric conductivity, and the very high temperature makes the metallic hydrogen to be in the liquid phase. Furthermore, their rotation period is the shortest of all the planets (both 0.4 Earth day; Table 1.1). The fast-rotating liquid metallic hydrogen around the cores of Jupiter and Saturn would play the same role as the liquid iron circulating in the outer core of Earth, and the resulting dynamo would cause the observed strong magnetic field.

The other two giant planets, Uranus and Neptune, called *ice giants*, are mostly composed of oxygen, carbon, nitrogen and sulphur, and they each exhibit a strong magnetic field. Their masses are 15 and 17 times that of Earth, respectively, and they both have a rotation period of 0.7 Earth day (Table 1.1). On the two planets, the magnetic field is strongly tilted relative to the axis of rotation, called declination (see Sect. 8.2.1, above), and is not centred on the planet's centre. The masses of these planets are not large enough to generate the pressures required for the existence of metallic hydrogen, and the peculiarities of their magnetic fields are explained by a dynamo generated by convective movements in a thin spherical cell of electrically conducting liquid, which could be a mix of water, ammonia, ammonia hydrosulphide, methane, and others. In this case and contrary to the other planets with a magnetic field, the conducting fluid would not be metallic.

Two of the 79 *moons of Jupiter,* Ganymede and Europa, are known to have a magnetic field. So far, these are the only moons in the Solar System with this characteristic, and their magnetic fields are buried within the much larger magnetic field of Jupiter.

Ganymede is the largest moon in the Solar System. It possesses a metallic core, and it is thought that the dynamo that generates its magnetic field is similar to that of Earth. In addition, Ganymede has a thick liquid, saltwater ocean (see Information Box 5.4), which is located between two or several layers of ice and may be the largest ocean in the Solar System (it may contain more water than all Earth's oceans together). This large saltwater ocean would conduct electricity and thus affect the magnetic field.

The mass of *Europa* is one third that of Ganymede, but Europa is still the sixth most massive moon in the Solar System. Its magnetic field, which is perhaps one fourth that of Ganymede, varies periodically as Europa passes through Jupiter's massive magnetic field. These variations provide evidence that there is a conductive material beneath Europa's frozen surface, which could be the salty liquid ocean proposed to exist under the icy surface of this moon (see Information Box 5.4).

Stars, including the Sun, generate a magnetic field (Fig. 8.4b). Their dynamo is fuelled by convection within their plasma, which is highly conductive of electricity (see Sect. 2.1.2). In the Sun, the convective zone extends down from close to the surface to about 30% of the radius of the star. At the surface, there are dark patches, called *sunspots*, whose numbers vary in space and time. The sunspots correspond to concentrations of magnetic field, where the convective transport of heat is inhibited and temperature is thus lower than the surroundings. Because of this, the sunspots appear dark. There is an 11-year cycle in the number and size of sunspots of the Sun, which corresponds to a cycle in solar magnetic activity. Within this cycle, *solar maxima* can affect human activities through electric power outages and disruptions of satellite communications. At the scale of hundreds of years, there are periods of *grand solar maxima and minima*, which show some correlations with global and regional climate changes (as those shown in Fig. 4.4). However, a correlation does not necessarily indicate that one phenomenon causes the other, because they may be both influenced by a third, unidentified phenomenon.

The solar magnetic field extends far away from the Sun, as it is carried by the *solar wind* (see Information Box 2.6), which is the stream of charged particles released from the upper atmosphere of the Sun. The solar wind is among the factors that explain the rarefied atmosphere (or exosphere) of planet Mercury, which is located close to the Sun (see Sect. 2.3.3). In fact, the solar wind carries solar plasma over the whole Solar System, and thus drags the solar magnetic field to its outer reaches, forming the *interplanetary* or *heliospheric magnetic field.*). In the interplanetary space, electric currents generate magnetic fields much greater than the magnetic field generated by the solar dynamo, for example about a hundred times greater at Earth's location. It will be explained below that the interaction between this magnetic field and that of Earth has a great significance for land organisms (see Sect. 8.4.2).

The region of space around an astronomical body within which charged particles are affected by its magnetic field is called its *magnetosphere*, and more precisely its *intrinsic magnetosphere*. Hence, the Sun, Earth, and Ganymede each have an intrinsic magnetosphere. Even the astronomical bodies without an internal dynamo can have a magnetosphere, which is not intrinsic but is induced. Such an *induced magnetosphere* can occur around planets or other Solar System bodies, including moons, which possess a conducting ionized upper atmosphere but no intrinsic magnetic field, such as Venus. The interaction between the solar wind and the ionosphere of such planets and moons creates an induced magnetosphere, which partly shields them from the solar wind. The book of Lühr et al. (2018) provides further reading on the magnetic fields that exist in the Solar System.

8.4   The Magnetic Field and the Takeover of Earth by Organisms

8.4.1   Organisms and Ecosystems in Aquatic Environments

During the long period of time that extended from 4.6 billion to about 500 million years before present, Earth had known the three successive phases of its atmosphere described above (see Sect. 8.1.1). The fourth phase began 500 million years ago with the creation of the O_3 layer. In this section, we first examine the general mechanisms by which Earth lost its atmosphere during its first phase and retained it since its second phase, different from Venus and Mars, and the consequences for liquid water and the aquatic organisms and ecosystems. In a second part, we focus on the role of the magnetic field in the retention of the Earth's atmosphere and liquid water.

Until about 400 million years ago, all Earth's organisms lived in water, and thus did not have direct contact with the atmosphere. Yet, the characteristics of the atmosphere were important for organisms and ecosystems because these characteristics determined the existence of liquid water at the Earth's surface. Indeed, the conditions of atmospheric temperature and pressure that prevailed more than 4.4 billion years ago led to the formation of an ocean, and later combinations of atmospheric temperature and pressure ensured the continuing existence of vast expanses of liquid water at the surface of the planet (see Sect. 2.1.3). The fact of being in the liquid phase protected the Earth's water from being lost to the outer space through hydrogen escape (see Information Box 8.4). In the resulting ocean and fresh waters, organisms diversified and produced organic matter during billions of years, and eventually built up large biomasses (see Sect. 8.1.2). Which factors contributed to prevent the loss of the Earth's atmosphere, and thus its water, to the outer space?
The key mechanism for the loss of water from a planet is the *atmospheric escape* of its gases to space. In order to escape the gravitational attraction of a planet, gas molecules must have a higher velocity than the planet's *escape velocity*. The latter is determined by the planet's gravity, which depends on its mass (see Sect. 6.6.1). Hence, a planet with a larger mass has a higher escape velocity because of its larger gravity, and consequently this planet loses less gas molecules from its atmosphere to the outer space. Because the mass of Earth is relatively large, its *escape velocity* is 11.2 km s^{-1} at the surface of the planet and 10.8 km s^{-1} at an altitude of 500–600 km in the atmosphere. This relatively high escape velocity contributed to retain most of the Earth's atmosphere (see Sect. 2.3.4).

There are several mechanisms of atmospheric escape (see Information Box 8.4). We examined previously the general mechanisms by which a planet loses its water (see Sects. 2.1.3 and 5.4.2, respectively). In the coming paragraphs, we examine how these mechanisms operated on Earth and its two sister planets, Venus and Mars.

Information Box 8.4 Atmospheric Escape There are several mechanisms by which atmospheric gases escape a planetary body, such as Earth and its sister planets Venus and Mars. These mechanisms are grouped in three broad categories, which we describe here for the rocky planets of the Solar System.

Thermal escape occurs when high-energy solar radiation (mostly soft X rays and extreme UV radiation; Fig. 2.6b) heats up the atmosphere, thus making gas molecules to reach the escape velocity and be lost to space, en masse or individually. Heating can also be caused by a very high heat flux from the interior of a planet.

Nonthermal escape occurs when chemical reactions or interactions between atmospheric ions (see Information Box 1.5) and electrically neutral molecules accelerate individual atoms or molecules to the escape velocity. In the high atmosphere, the chemical reactions are often photochemical, that is, driven by the UV radiation of the Sun, and particles are ionized by the effect of cosmic rays transported by the solar wind.

Impact erosion is the expulsion of huge amounts of atmospheric gases as a result of impacts of large bodies, for example a large asteroid, or the cumulative effect of several asteroid hits.

The first gas to escape the atmosphere of a planet is hydrogen, because it is the lightest chemical element. The escaping hydrogen derives from the dissociation in the atmosphere of H-containing compounds, such as H_2O, by solar UV radiation. However atmospheric escape concerns a wider range of gases than hydrogen because, when hydrogen escapes, it often *drags along* heavier molecules and atoms. This happens when collisions with hydrogen push atoms of heavier gases upwards faster than gravity pulls them downwards.

The loss of hydrogen leaves free oxygen (O_2) behind, whose combination with metals causes the oxidation of the planet. Hence when hydrogen escapes, some matter at the surface or inside the planet is irreversibly oxidized. This explains the red colour of Mars, which lost its hydrogen early in its history and where the subsequent oxidation of volcanic minerals produced red iron oxides. The article of Catling and Zahnle (2009) provides further reading on the escape of planetary atmospheres.

We begin by examining the fate of the *early atmospheres* of Earth, Venus and Mars. We will see that the mechanisms of atmospheric escape in the early histories of the three planets were not the same.

Earth. It is thought that Earth lost its atmosphere in its first phase as a consequence of the high UV radiation from the young Sun (see Sect. 3.7.1). Indeed, the radiation of the young Sun was fainter than today in the visible portion of the electromagnetic spectrum, but brighter in the UV. The UV radiation could dissociate such molecules as water (H_2O) into atomic (H^+) or molecular (H_2) hydrogen, and atomic oxygen (O^-) which recombined rapidly with other elements to form gaseous or solid oxides. The H^+ or H_2 could easily escape the gravitational pull of Earth because of their low masses.

Hydrogen escape from the Earth's atmosphere in its first phase was mostly through a thermal mechanism called *hydrodynamic escape and drag*. In this mechanism, the heating of the atmosphere below 500–600 km above ground by the UV radiation of the Sun caused large upward flows of heated gas and en-masse escape of hydrogen. During the first phase of the atmosphere, the escaping hydrogen dragged along heavier molecules and atoms that included Ne, N_2 and CO_2 (see Information Box 8.4). Contrary to later times when Earth had a *magnetic field* (below), the atmosphere in its first phase was not protected from the solar wind, which could thus strip away the upper atmosphere (see Information Box 2.6).

Venus. It is thought that Venus lost its water early in its planetary history. The mechanism was likely *hydrodynamic escape and drag*. On young, hot Venus, the solar UV radiation would have decomposed the water vapour into H+ and O^- atoms at high altitude, and the escaping hydrogen may have dragged the oxygen with it (see Information Box 8.4). This left behind the CO_2, which has made up the bulk of Venus' atmosphere since then.

Mars. The most important factor responsible for the early loss of most of Mars' atmosphere may have been *impact erosion*, which occurred because the environment of young Mars contained a large number of asteroids. It is thought that many of these asteroids impacted the planet, and the result of these impacts was to strip most of its atmosphere within less than 100 million years. *Thermal escape* also played a role in the loss of the early Martian atmosphere. In addition, Mars lost its *magnetic field* about 4 billion years ago, and without this protection (see Sect. 8.4.2, below), the solar wind progressively eroded what was left of the original atmosphere (see Information Box 2.6). Without the protection of a substantial atmosphere, part of Mars' liquid water evaporated and was lost to space, and part of it froze under the ground. The loss of water by hydrogen escape continues on Mars nowadays.

During the *second and third phases* of the Earth's atmosphere, the atmospheric escape became progressively dominated by *non-thermal mechanisms* (see Information Box 8.4). Nevertheless, thermal escape continued to have a significant role, and the most important thermal mechanisms on today's Earth is *Jeans' escape* (named after the English scientist James Jeans, 1877–1946). This escape mechanism, which currently accounts for about 30% of the loss of hydrogen on Earth, concerns individual atoms because most of the hydrogen is trapped below the altitude of 500–600 km from which it cannot escape. However, some atoms nevertheless escape, as follows. At the high altitude of 500–600 km, the high temperature (above 700 °C) makes the atoms of hydrogen move fast, but not fast enough *on average* to break free from the gravitational pull of the planet (the

escape velocity at this altitude is 10.8 km s^{-1}, see above). However, a small fraction of the atoms move much *faster than average*, as fast as 10.8 km s^{-1} or more,
and thus escape the atmosphere. Although atmospheric gases continuously escape
Earth, the planet also gains hydrogenated materials from the continuous fall of
interplanetary dust and rocks, which partly but not fully compensate the losses
(see Sect. 5.4.2).

The formation of a liquid-water ocean, perhaps as early as 4.4 billion years
before present (that is, about 100 million years after the creation of the Earth-
Moon system), moved water from the gaseous phase in the atmosphere, from
where its hydrogen could escape to the outer space, to the liquid phase in the
ocean, where it was sheltered from escape (see Sect. 5.4.2). This occurred because
the conditions of surface atmospheric pressure and temperature allowed water
vapour to condense into liquid water, and form an ocean (see Sect. 5.5.1). The
atmospheric pressure and temperature are largely determined by the concentration of CO_2 in the atmosphere (see Sect. 5.5.4). As this concentration decreased,
the atmospheric pressure and temperature both decreased, but their combination
always allowed the presence of large amounts of surface liquid water through the
succession of greenhouse and icehouse conditions that marked the Earth's history
(see Sect. 4.5.6).

The mechanisms involved in the loss of the Earth's atmosphere during its *first
phase* were explained above. These mechanisms caused losses from the atmosphere because the planet did not have a *magnetic field* at the time. Indeed, the
earliest record of the magnetic field in Earth's rocks goes back 3.5 billion
years, meaning that the planet did not have a magnetic field during the first
phase of its atmosphere 4.5–4.4 billion years ago. The lack of a magnetic field
is often invoked to explain the rapid loss to space of most of the Earth's initial
atmosphere.

Contrary to the situation during the first phase of the atmosphere, Earth had a
magnetic field during the *second and later phases* of its atmosphere. The next paragraphs explain how the presence of the magnetic field was likely one of the major
factors that protected the Earth's atmosphere, and thus its liquid water.

The key role of the magnetic field in the protection of the Earth's atmosphere.
was demonstrated a few years ago by direct measurements of the atmospheric loss
of oxygen by planets Mars and Earth, knowing that Mars has no global magnetic
field and Earth has a strong one (see Sect. 8.3.2, above). These measurements were
made simultaneously at a time when the two planets were hit by the same stream
of solar wind. They showed that, while the pressure of the solar wind increased at
each planet by similar amounts, the increase in the rate of loss of Martian oxygen
was ten times that of the Earth's increase. This shows that the Earth's magnetic field
efficiently deflects the solar wind, and thus protects the atmosphere. How is it done?

The region of space around Earth in which its magnetic field is more important
than the magnetic field of the Sun is called the *Earth's magnetosphere* (Fig. 8.5).
The outer limit of the Earth's magnetosphere is constantly distorted by the force
of the solar wind (see Information Boxes 2.3 and 8.3), so that on the side facing

Fig. 8.5 Artist's rendition of the Earth's magnetosphere (violet outer envelope, and blue lines of the Earth's magnetic field) deflecting the solar wind (yellow) and being deformed by it. The magnetic lines enter Earth (very small in this figure) and emerge from it at its two magnetic poles. Credits at the end of the chapter

the Sun, it is flattened at an altitude of about 65,000 km, whereas on the opposite side, it extends over several thousand kilometres forming a tail (Fig. 8.5). Although the magnetosphere generally decreases atmospheric escape, gas ions are lost along the open magnetic field lines at the two magnetic poles, and some researchers think that this loss is so large it cancels out the shielding effect of the magnetosphere. On planets without a significant magnetic field, such as present-day Mars and Venus, solar wind particles interact with the atmosphere close to the ground.

The magnetosphere deflects most of the cosmic rays transported in the solar wind, and thus acts as a shield that protects the atmosphere. It is thought that the Earth's magnetosphere has prevented the atmosphere from being stripped away by the solar wind since the second phase of the atmosphere. The Earth's magnetosphere shielded the atmosphere, including its water vapour, from the solar wind, which in turn protected the vast liquid oceans. Indeed, without the protection of the atmosphere by its magnetosphere, Earth would have likely lost its liquid water, as happened on Mars (see above). The presence of the liquid ocean over billions of years was a key factor in the development of organisms and ecosystems and the build-up of large biomasses.

The Earth's Magnetic Field Shields its Atmospheric Shield It follows from the above that the magnetic field protects the Earth's atmosphere, and the latter shields the liquid water from loss to space. Other factors that contribute to the long-term stability of the Earth's atmosphere are the relatively large mass of the planet and its location at a safe distance from the Sun (Sect. 2.3.4). Hence, the long-term occurrence of liquid water and thus organisms on Earth was favoured by three characteristics of the planet, namely its magnetic field, its relatively large mass, and its appropriate distance from the Sun.

8.4.2 Organisms and Ecosystems on Continents

Complex multicellular organisms began to occupy the emerged lands earlier than 400 million years before present, and unicellular organisms may have been present on land much before. The formation of the O_3 layer about 500 million years ago marked the beginning of the *fourth phase* of the atmosphere.

Since the beginning of space exploration, it is known that the greatest danger to the health and even survival of people in space is exposure to high-energy *cosmic rays* or *cosmic radiation*. Because of the ionizing character of cosmic radiation and its prevalence outside the Earth's atmosphere, space agencies try to develop materials that could shield space travellers from this hazard. However, the organisms on *spaceship Earth* are obviously protected from most of the harmful cosmic radiation. What are the mechanisms of their protection?

Planet Earth is continually exposed to cosmic radiation that comes from both the Sun and deep space outside the Solar System (see Information Box 8.5). Fortunately, organisms at the surface of the planet, including humans, are protected from most of the cosmic radiation by two successive magnetic shields and by the atmosphere, which is itself protected by the Earth's magnetic field. The two magnetic shields are the Earth's *magnetosphere* (see Sect. 8.4.1, above) and, surrounding it, the Sun's *heliosphere*, which is the region of space influenced by the solar wind. The heliosphere deflects a large part of the cosmic rays coming from deep space, and the Earth's magnetosphere deflects most of the cosmic rays that pass the heliosphere shield (see Sect. 8.4.1, above). Despite these two magnetic shields, some cosmic rays manage to enter the Earth's atmosphere, but the latter is thick enough to absorb most of the ionizing radiation (see Information Box 8.5). Hence the *thickness of the atmosphere* provides a third shield against cosmic radiation, and the atmosphere is itself protected from loss by the Earth's magnetic field (see Sect. 8.4.1). It follows that organisms at the surface of Earth are largely protected from direct hits by cosmic rays from deep space because the atmosphere, which is protected by the Earth's magnetic field, absorbs most of the cosmic radiation that passes the magnetosphere and heliosphere magnetic shields.

Information Box 8.5 Ionizing Radiation, Cosmic Rays, and Organisms The spectrum of electromagnetic radiation includes: short-wavelength, high-energy ionizing radiation (gamma rays, X rays, and extreme UV); intermediate-wavelength, intermediate-energy visible light (from violet to red); and long-wavelength, low-energy microwaves and radio waves (Fig. 2.4b). The electromagnetic spectrum radiated by the Sun is described with the greenhouse effect in Chapter 2 (see Information Box 2.7).

The *ionizing radiation* carries enough energy to detach some of the electrons of the atoms or molecules with which it interacts, and thus transform them into ions (see Information Box 1.5). Ionizing radiation damages biological molecules by breaking bonds or removing electrons, thus disrupting their structures and functions. Another type of damage results from the ionization of water (the most abundant substance in organisms; see Sect. 6.1.2), which forms a hydronium ion (H_3O^+) and a hydroxyl radical (OH^-). Free radicals, including OH^-, can react with biological molecules such as DNA and proteins, damaging them and disrupting physiological processes.

The two sources of ionizing radiation on Earth are radioactive decay (see Information Box 5.6) and cosmic rays. We focus here on *cosmic rays*, which are high-energy radiation coming from the Sun and from outside the Solar System. Given their ionizing power, cosmic rays could potentially damage Earth's organisms, but the atmosphere of the planet is thick enough to absorb most of the ionizing radiation. This way, the atmosphere protects organisms from damaging effects of cosmic rays.

Cosmic rays that enter the Earth's atmosphere collide with atoms and molecules, mostly N_2 and O_2, which produces an air shower of secondary radiation. Some of this radiation reaches the ground, where it contributes about one-tenth of the background radiation. This fraction increases with altitude, and accounts for one-quarter of the background radiation in high-altitude cities. In addition, chemical reactions that follow the ionizing dissociation of N_2 and O_2 lead to the formation of nitrogen oxides (NO and NO_2) in the high atmosphere, which contribute to deplete the ozone layer (see Information Box 8.2). This could be significant in cases of very high gamma ray flux, such as radiation bursts caused by catastrophic events occurring far away from the Solar System.

The other source of ionizing radiation, namely *radioactive decay*, is not totally independent of cosmic rays. Indeed, the ionization of gas molecules by cosmic rays in the atmosphere produces a number of radioactive isotopes, which in turn produce ionizing radiation during their decay. The best known of these isotopes is carbon-14 (see Information Box 5.6), which is formed naturally in the upper atmosphere (mostly between 9 and 15 km) trough the absorption of neutrons by nitrogen gas (N):

$$n + {}^{14}_{7}N \rightarrow {}^{14}_{6}C + p \tag{8.7}$$

> The reaction in Eq. 8.7 replaces one of the 7 protons (p) of the N atom with one neutron (n). The loss of one proton transforms N (atomic number 7) into C (atomic number 6). The $^{14}_{7}N$ atom has 7 protons (p) and 7 neutrons (n), and its atomic mass is thus $7+7=14$ (see Information Box 5.6). The $^{14}_{6}C$ resulting from the reaction has one proton less and one neutron more than $^{14}_{7}N$, that is, 6 protons and 8 neutrons, and its atomic mass is thus $6+8=14$.

Another hazard that threatens organisms at the surface of Earth is the exposure to solar UV radiation, which damages the genetic material contained in DNA (see Sect. 6.2.2). This is different from the cosmic radiation discussed in the previous paragraph, as the UV radiation comes from the Sun whereas cosmic radiation comes from deep space. Surface dwelling organisms have evolved mechanisms to cope with the harmful effects of UV, but they could not survive on continents or in surface waters if most of the solar UV radiation was not absorbed in the upper atmosphere by the ozone (O_3) layer (see Sect. 2.1.5 and Information Box 8.2). The O_3 layer began to build up about 500 million years ago, after enough O_2 had accumulated in the upper atmosphere where some of the O_2 molecules were transformed into O_3 by the action of the solar UV radiation. The O_3 layer exists because of the presence of a thick atmosphere, which is itself preserved by the effect of the Earth's magnetic field as explained in the previous paragraph. It follows that organisms at the surface of Earth largely avoid critical UV damage to their DNA because the magnetic field protects the atmosphere, where the shielding O_3 layer is continually re-formed by the action of solar UV.

Table 8.1 summarizes changes in the Earth's atmospheric gases and magnetic field, and the evolution of key organisms and ecosystems that occurred since 4.5 billion years before present. This table shows that each phase of the atmosphere was characterized by major changes in the composition of atmospheric gases, which were as follows: rapid loss of the early atmosphere; sequestration of the initially very abundant CO_2 in solid $CaCO_3$; progressive increase in the concentration of O_2; creation of the O_3 layer when the concentration of O_2 in the upper atmosphere became high enough; and current increase in anthropogenic CO_2. Table 8.1 also shows that the magnetic field has protected the atmosphere since its second phase. Finally, Table 8.1 illustrates some of the interactions between the atmosphere and the organisms and ecosystems during each phase of the atmosphere, namely: during the *first phase*, the environmental conditions did not allow the existence of organisms; during the *second phase*, diversification of archaea and bacteria in the aquatic environments created by liquid water, leading to the advent of the O_2-producing photosynthesizers whose O_2 would cause the Great Oxygenation Event that marked the beginning of the next phase; during the *third phase*, advent of the eukaryotes and development of O_2 respiration among the three domains of life (bacteria, archaea and eukaryotes), which consumed most of the O_2 released

Table 8.1 Changes in the Earth's atmospheric gases and magnetic field, and evolution of key organisms and ecosystems during the successive phases of the atmosphere since 4.5 billion years before present (see Sects. 8.1.1, 8.4.1 and 8.1.2, respectively)

Phase of atmosphere (billion years ago)	Changes in atmospheric gases	Magnetic field	Evolution of organisms and ecosystems
First phase (4.5–4.4 or earlier)	Light gases, lost	Absent	None
Second phase (4.4 or earlier to 2.5)	CO_2: initially high, and rapidly sequestered in solid $CaCO_3$	Appeared 3.5 billion years ago or earlier	Archaea and bacteria; O_2-producing photosynthetic organisms
Third phase (2.5 to 0.5)	Slowly increasing O_2	Existing	Eukaryotes; O_2-respiring bacteria, archaea and eukaryotes
Fourth phase (0.5 to the Industrial Revolution[a])	Ozone layer, and rapidly increasing O_2	Existing	Complex eukaryotes on emerged lands
Industrial and technological era[a]	Rapidly increasing CO_2	Existing	Human societies

[a]The Industrial Revolution took place in the second part of the eighteenth century. It marked the beginning of the present industrial and technological era

by photosynthesis; during the *fourth phase*, occupation of emerged lands by complex eukaryotes under the protection provided by the O_3 layer, which led to a major increase in atmospheric O_2; and during the *present industrial and technological era*, ongoing large-scale modification of the atmosphere by human societies.

In summary during the course of the Earth's history, the magnetic field protected the atmosphere from the solar wind, and the atmosphere protected the vast expanses of liquid water that were key to the development of organisms and ecosystems. In addition, the magnetosphere and the heliosphere shielded organisms, especially those on emerged lands, from dangerous cosmic radiation, and the thick atmosphere absorbed most of the cosmic rays that went through the shields. Furthermore, the atmospheric ozone layer protected land-dwelling organisms from DNA-damaging solar UV radiation. In return, ecosystems had major effects on the atmosphere, through the sequestration of atmospheric CO_2 in solid $CaCO_3$, and the production of all the atmospheric O_2. Finally, industrial and technological activities became a geophysical force, which is presently forcing the long-term evolution of the atmosphere backwards, that is, toward an increasing proportion of CO_2 (Fig. 9.8); this causes anthropogenic *climate change* and *ocean acidification* (see Sects. 11.1.1 and 11.1.3).

It was mentioned above that human beings do not have the biological ability to sense the Earth's magnetic field directly (see Sect. 8.2.3). However, this statement was not entirely correct since people at high latitudes can perceive with their eyes the spectacular manifestations of the magnetic field called *polar lights* or *auroras*, more specifically *aurora borealis* and *aurora australis* in the Northern and Southern Hemispheres, respectively. Polar lights are a consequence of interactions between charged particles from the Sun and magnetic lines in the high atmosphere

Fig. 8.6 Polar lights above Bear Lake, Alaska, U.S.A., in January 2005. This phenomenon results from interactions between the solar wind and the Earth's magnetic field over polar areas. Credits at the end of the chapter

over polar areas. There, the charged particles ionize atmospheric constituents, which in turn emit light of varying colours and patterns (Fig. 8.6). These colours and patterns are a consequence of the disturbance of the Earth's magnetic field by the solar wind, which makes visible the presence of the magnetic field otherwise invisible to people.

The waxing and waning of polar lights had fascinated people for hundreds of thousands of years before the Norwegian physicist Kristian Birkeland explained them by invoking the Earth' magnetic field at the beginning of the twentieth century. We now understand that the auroras we see with our eyes high in the atmosphere are caused by the same planetary process that directs the needle of the magnetic compass we hold in our hand. The book of Windridge (2016) provides further reading on the visual beauty, legends and science of northern lights.

8.5   Fossilized Magnetic Fields in Earth's Rocks

8.5.1   Imprints of Magnetic Reversals in Rocks

We know that the Earth's magnetic field existed in the past because, like the organisms, it left fossil traces in some rocks. It may seem strange that something we cannot see or even sense could leave fossil evidence in solid rocks, but this is nevertheless the case for the magnetic field. How do rocks acquire the imprint of the ambient magnetic field? The answer lies in the fact that rocks that exhibit a magnetic signal contain ferromagnetic minerals. As long as a rock is still fluid lava or is in the form of particles suspended in water, its ferromagnetic minerals respond to the Earth's magnetic field like the needle of a compass and align with the Earth's magnetic field. When the rock solidifies, it "freezes" the ferromagnetic minerals in their aligned positions, and thus records information about the ambient magnetic field at that time. This occurs in basaltic and granitic and also sedimentary rocks, as these come from the solidification of fluid magma or the deposition of mineral particles followed by their consolidation, respectively (see Sect. 7.2.1). The magnetic properties acquired by rocks in the past (called *remanent magnetization*) tell researchers what were the main characteristics of the local geomagnetic field at the time the rocks were formed.

The fossilized magnetic evidence recorded in rocks shows that the *polarity* of the Earth's magnetic field has changed on several occasions during the Earth's history. In other words, the positions of the *magnetic* north and south alternated a number of times in the past, that is, the position of the magnetic north switched to that of the magnetic south, and vice versa. Hence in the past, the predominant direction of the magnetic field was either the same as present, a situation called *normal polarity*, or opposite to the present direction (the compass needle would have pointed towards the geographic south), which is called *reverse polarity*. Some volcanic rocks, among others, recorded these *geomagnetic reversals* in their ferromagnetic components as they solidified, thus preserving fossil evidence of the past magnetic fields.

The phenomenon of geomagnetic reversals was discovered by the French geophysicist Bernard Brunhes (1867–1910) in the early twentieth century, but this discovery did not interest more than a few specialists during the following years. Half a century later, geomagnetic reversals where among the key measurements that led to the theory of plate tectonics as explained below (see Sect. 8.5.2). Researchers studying geomagnetic reversals found that their periodicity varies considerably over time. Indeed, some intervals between reversals were as short as 200 years, whereas others lasted more than 10 million years. There were 184 intervals (and thus 183 reversals) in the last 83 million years, and about 300 reversals in the last 200 million years.

Major geomagnetic periods characterized by a dominant polarity (that is, normal or reverse polarity) are called *chrons*, and the periods of reversals within chrons are called *subchrons*. The sequence of chrons, subchrons and geomagnetic reversals during the last 5.25 million years is shown in Fig. 8.7a. Different chrons are named after pioneering scientists in geomagnetism, for example Bernard Brunhes, whose contribution is explained in the previous paragraph, and Carl Friedrich Gauss, who invented the first modern *magnetometer* (see Information Box 8.3). Different subchrons are named after the locations where the rocks leading to their discovery were collected, for instance the Jaramillo Creek in New

Fig. 8.7 Earth's geomagnetic reversals. Dark and light areas: periods of normal and reverse polarity, respectively. Left axes of the two diagrams: millions of years before present (Ma). **a** Chronology of reversals back to 5 million years. Left axis: names of chrons; right axis: names and times of subchrons. **b** Chronology of reversals back to 170 million years, called *geomagnetic polarity timescale*. Right axis: geological eras, periods and epochs (Table 1.3). Credits at the end of the chapter

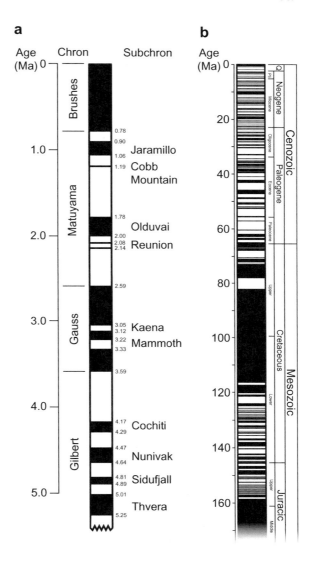

Mexico, U.S.A., and the Olduvai Gorge in Tanzania. For example in Fig. 8.7a, the Brunhes and Matuyama Chrons are periods of normal and reverse polarity, respectively, and the Olduvai Subchron is a period of normal polarity within the reverse-polarity Matuyama Chron.

During a reversal, the intensity of the magnetic field decreases during thousands of years, and many magnetic poles may form chaotically in different places, until the position of the poles and the intensity of the magnetic field stabilize.

Some reversals occurred as quickly as a human lifetime, but most estimates of *polarity transitions* range between 1,000 and 10,000 years. There were situations when the magnetic field decreased but came back to its previous value without a change in polarity, these disruptions of the magnetic field without reversals being called *geomagnetic excursions.*

If we had been able to observe in real time the geomagnetic reversals during the last million years, we would have seen the magnetic needle of the compass frequently switching between the general direction of the geographic North and that of the geographic South. The reversals recorded in seafloor rocks go back about 180 million years, which corresponds to the age of the oldest seafloor in most oceans (see Sect. 7.2.4). More ancient geomagnetic reversals are recorded in volcanic rocks on continents, which are much older than the oldest seafloor. One example of very old magnetic field reversals was recorded in basalt formed 1.1 billion years ago in the mid-continent Keweenawan Rift of supercontinent Rodinia (Fig. 4.7). These rocks are now located near Lake Superior in North America, and their analysis has shown that a rift valley opened in the Rodinia supercontinent, and closed afterwards. When a rift valley forms an ocean, its rocks are eventually destroyed by the subduction of the seafloor (see Sects. 7.2.2 and 7.2.3). Contrary to this usual case, the Keweenawan Rift failed to create a new ocean basin and closed, which preserved its rocks and their magnetic reversals as witnesses of the failed opening of an ocean basin more than 1 billion years ago. Closer in time, the last magnetic reversal occurred 780,000 years ago (Fig. 8.7a). Since then, Earth has been in the Brunhes Chron, during which about fifteen excursions took place.

The above paragraphs described the temporal succession of magnetic reversals. However, they did not explain how researchers determined the ages of the reversals (left sides of the two chronologies in Fig. 8.7a and b). In order to date the reversals, researchers use magnetized continental volcanic rocks, and they assess the orientation of the magnetic field and the age of each rock. The magnetic orientation of each rock is determined with a magnetometer (see Information Box 8.3) and its age using a radiometric dating technique (see Information Box 8.6).

The ages of rocks are determined on samples from all over the world because no volcano produces lava continuously, and the only way to get a time series of reversals going back millions of years is to combine the information provided from as many sites as possible. Even then, the time series is discontinuous, and the measurements are complemented by determining magnetic reversals and ages along sediment cores from the deep ocean (see Information Box 8.7).

Information Box 8.6 Dating Volcanic Rocks The ages of *volcanic rocks* are determined on rock samples brought to the laboratory using *radiometric dating*. This type of method is also called *radioactive* or *radioisotope dating*.

The dating method most generally used for *volcanic rocks* is based on the presence of the chemical element potassium (K) in most of these rocks, and the fact that its long-lived radioactive isotope ^{40}K naturally decays over time into the stable argon isotope ^{40}Ar. Potassium is solid below 759 °C, and is thus a solid at the Earth's surface; it has 24 known isotopes, of which two are stable (^{39}K and ^{41}K) and the other 22 are radioactive, but only ^{40}K has a radioactive period of more than one day (for the radioactive decay and the radioactive period, see Information Box 5.6). Argon is gaseous above -186 °C, and is thus a gas at the Earth's surface; it has 24 known isotopes, the most abundant on Earth being ^{40}Ar, which is stable. When lava solidifies into rocks, the ambient ^{40}Ar gas is trapped inside the rocks. After lava solidification and because ^{40}K decays into ^{40}Ar, the amount of ^{40}K in a rock progressively decreases with time and that of ^{40}Ar increases. Since researchers know the radioactive period of ^{40}K (which is 1.25 billion years), they use the amounts of ^{40}K and ^{40}Ar in a rock sample to compute the time elapsed since the rock solidified. The method is called *potassium-argon dating* or *K-Ar dating*.

Because K and Ar are two different chemical elements, the above dating method requires making two separate isotopic measurements, one for K and the other for Ar. In order to simplify the technique and improve its precision, researchers in the laboratory first transform stable ^{39}K into radioactive ^{39}Ar by irradiation of the rock sample in a nuclear reactor. Once this is done, they determine the amounts of the two argon isotopes ^{40}Ar and ^{39}Ar during a single measurement. They use the amounts of the two isotopes (^{39}Ar represents ^{40}K) to compute the time elapsed since the rock solidified, as in K-Ar dating. The method is called *argon-argon dating* or *^{40}Ar/^{39}Ar dating*.

Information Box 8.7 Dating Marine Sediments The age of *marine sediments* is often determined on a sample in the laboratory using an indirect method based on the measurement of the ratio of two stable isotopes of oxygen, ^{18}O and ^{16}O (see Information Box 5.6), after which the results are compared to a standard chronology of $\partial^{18}O$ (below) based on the Milankovitch cycles of orbital forcing (see Sect. 1.3.4). The resulting age is sometimes called the *astrochronologic age*.

Sediments in ocean and lakes result from the deposition and vertical accumulation of particles initially suspended in the overlying water. There are also sediments on land, called *loess*, which are formed by the deposition of mineral particles transported by the wind. The thickness of loess deposits can reach tens of metres in Midwestern United States and hundreds of metres in China. In all sediments, the most recent part is at the top, and the oldest at the bottom, except when geological accidents (such as earthquakes) have perturbed the vertical structure.

Marine sediment samples are generally in the form of cores (long cylindrical samples), which are collected by ships using various coring or drilling devices. Depending on the sampling instrument, the length of a core ranges between about 1 m and up to hundreds of metres. As a general rule, longer cores provide information that goes farther back in the past, but how far in the past depends not only on the length of the core but also on the local sedimentation rate, whose value can range from 1 to 10 millimetres per 1,000 years. The ages of the sediments at the bottom of a 10-m core in areas with sedimentation rates of 1 and 10 mm per 1,000 year are 10 and 100 million years, respectively.

One technique to determine the ages of sediments down a core is to cut thin slices of sediment at various depths along the core, and pick out of each slice microscopic calcite tests of fossil foraminifers, which are single-cell organisms living in the plankton or on the seafloor (see Information Box 3.1 and Sect. 7.3.3). Several calcite tests from each sediment slice are pooled together, and the relative abundances of the two stable isotopes of oxygen ^{18}O and ^{16}O in the $CaCO_3$ tests are used to compute the *oxygen isotope ratio* $\partial^{18}O$:

$$\partial^{18}O = \left[\left(^{18}O/^{16}O_{sample} \ / \ ^{18}O/^{16}O_{standard} \right) - 1 \right] \times 1000 \qquad (8.8)$$

Variations in the $\partial^{18}O$ of $CaCO_3$ tests with time reflect changes in the size of continental ice sheets during the succession of glacial-interglacial cycles of the Quaternary period (see Sect. 4.5.2). Using models of the times when the glacial and interglacial episodes took place, researchers transform the $\partial^{18}O$ values measured down a marine sediment core into sediment ages. This way, the sediments are dated, from youngest at the top of the core to oldest at the bottom.

The causes of magnetic reversals are not known, but hypotheses try to explain the reversals by changes in the geodynamo, which was described previously (see Sect. 8.3.1). According to one hypothesis, small fluctuations in the convective flow of the outer core could push the system (the geodynamo) out of equilibrium, causing either a geomagnetic excursion (with no reversal) followed by a return of the system to its initial equilibrium, or a reversal that switches the system to a new equilibrium (that is, a different polarity). The triggering of reversals may be of chaotic origin, meaning that very small initial variations within the geodynamo could push it out of equilibrium (this phenomenon is known in meteorology as the *butterfly effect*: "Does the flap of a butterfly's wings in Brazil set off a tornado in Texas?"). According to a different hypothesis, the movements of the fluid in the outer core would be sensitive to changes happening in the mantle, which would be reflected in the distribution of tectonic plates at the surface of the globe. This way, plate tectonics would control magnetic reversals.

The magnetic field affects the magnitude of the cosmic and UV radiations reaching the Earth's surface (Sect. 8.4.2). During a reversal, there is a period of weak or null magnetic field, at which time the intensity of the cosmic and UV flux is expected to be maximum. Given that these radiations damage the DNA of organisms, some researchers have proposed that magnetic reversals were involved in both mass extinctions and biological evolution. According to these researchers, magnetic reversals over the past hundreds of thousands of years have affected continental organisms, especially mammals and even humans, and may have caused the extinction of the human species *Homo neanderthalensis* for example. However, there are many other hypotheses concerning the extinction of the Neanderthals. In fact, these humans have not completely disappeared given that part of the genes of modern humans, *H. sapiens*, come from *H. neanderthalensis*. Similarly, accelerations in evolution, resulting from increased rates of radiation-caused mutations, were ascribed to magnetic reversals. Other researchers hold the view that the observed correlations between some magnetic reversals and mass extinctions or events of biological evolution are fortuitous. These hypotheses are actively debated in the scientific literature.

8.5.2 Magnetic Reversals and Plate Tectonics

In the ocean, new seafloor is formed by the magma that rises through volcanoes located in or near the rift of mid-ocean ridges (Fig. 8.8). When the magma cools, it forms new ocean crust, which is made of basalt near the surface and gabbro deeper. The two types of rocks are both formed from the magnesium-rich and iron-rich volcanic lava, the difference between them being that *basalt* results from rapid cooling at or near the surface of the seafloor, and the *gabbro* from slower cooling deeper beneath the surface. As the magma solidifies near the ridge, the

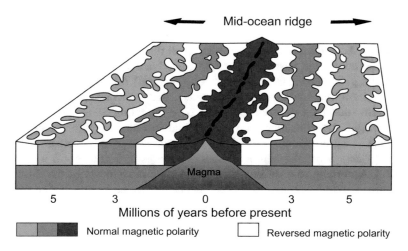

Fig. 8.8 Modelled magnetic anomalies on the seafloor on both sides of a mid-ocean ridge (see Information Box 8.8). Coloured stripes represent normal magnetic polarity (that is, same as today), and white stripes reversed polarity. Credits at the end of the chapter

two types of rocks record the polarity of the ambient magnetic field, as explained above (see Sect. 8.5.1). The seafloor formed at a ridge progressively moves away under the effect of plate tectonics (see Sect. 7.2.2), carrying away from the ridge the magnetic signal imprinted in the basalt and the gabbro in symmetric stripes on both sides of the ridge. Hence, the magnetic information recorded in the seafloor at a ridge and observed at increasing distances from it corresponds to increasingly ancient times.

The magnetic signal recorded in the seafloor is measured by towing a magneto-meter behind a ship (see Information Box 8.8). After removing from the measure-ments the mean values of the magnetic field in the area, the resulting *magnetic anomalies* may look like those illustrated in Fig. 8.8. The next step is to assign ages to the anomalies. One would think that this could be done in the same way as for volcanic rocks collected on continents (see Information Box 8.6). However, such dating cannot be generally done because, except at or near mid-ocean ridges, the seafloor is covered by a thick layer of sediments that prevents access to the underlying rocks. Given this difficulty, the approach used by researchers to date the magnetic anomalies mapped from ships is to compare the magnetic reversals on the seafloor (Fig. 8.8) to the suite of magnetic reversals established for con-tinental volcanic rocks (previous section). For example, if the middle stripes of normal magnetic anomalies in Fig. 8.8 were identified as corresponding to the Oiduval Subchron in Fig. 8.7a, these anomalies would be dated 2 million years before present (Fig. 8.7a).

Information Box 8.8 Magnetic Anomalies on the Seafloor The magnetic characteristics of the seafloor are mapped by towing a magnetometer (see Information Box 8.3) behind a ship at a depth a few metres below the surface. Towed marine magnetometers have the general shape of a fish or torpedo. They are towed far behind the ship to avoid effects of the magnetic field arising from the ship's ferromagnetic mass. The survey generally consists of a series of parallel runs spaced by constant intervals. In the case of a survey relative to a mid-ocean ridge, the runs are perpendicular to the ridge and extend on both sides of it.

The purpose of scientific measurements of the magnetic field at sea is usually to study the magnetic reversals recorded in the seafloor. For this purpose, researchers cannot use directly the magnetic signal recorded by the towed magnetometer, because this signal is dominated by the present Earth's magnetic field. The latter may be hundreds of times stronger than the values of the magnetic field induced by the permanently magnetized seafloor rocks. To obtain the latter values, researchers subtract the value of the present magnetic field from the values measured at sea, and thus get *magnetic anomalies* such as those illustrated in Fig. 8.8. In day-to-day language, the word *anomaly* refers to something "abnormal" or "odd", whereas in the scientific vocabulary, an *anomaly* is a difference between an observed and a mean value, the latter being here the mean value of the present Earth's magnetic field in the area of the measurement.

The values of magnetic anomalies on both sides of a ridge can be positive (normal polarity, like today) or negative (reverse polarity). The values the Earth's magnetic field range between 25,000 and 65,000 nanoteslas, whereas those of the anomalies are only a few hundred nanoteslas. The magnetometers must thus be very sensitive to detect such small values compared to those of the Earth's magnetic field.

In the last decades, the international community has conducted an ambitious program of deep-sea drilling (presently known as the *International Ocean Discovery Program*), which has produced long sediment cores, and some of these cores have reached the crustal rocks under the sediments. The direct dating of these crustal rocks confirmed the ages derived from the comparison of seafloor and continental magnetic reversals. The analysis of seafloor magnetic anomalies was combined with known and dated magnetic polarity reversals on land to construct a long-term time series of reversals called *geomagnetic polarity time scale*, which goes back 175 million years in the past (Fig. 8.7b).

Given the mechanism by which the seafloor is formed, its magnetic stripes are often symmetric about the axis of a ridge (Fig. 8.8). In other words, the magnetic

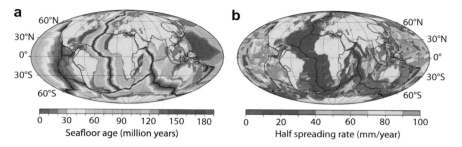

Fig. 8.9 Tectonic information derived from seafloor magnetic anomalies: **a** seafloor age, and **b** half-spreading rates (millimetres per year). Black lines: limits of plates. Credits at the end of the chapter

stripes located on both sides of ridges are generally mirror images of each other. This was also the case in the past as shown by the symmetry of three magnetic reversals going back 1.1 billion years that have been preserved on both sides of the mid-continent Keweenawan Rift, described in the previous section. After researchers had dated the seafloor magnetic anomalies as explained in the previous paragraph, they combined the resulting information in a global map of seafloor ages. The resulting map (Fig. 8.9a) shows that the youngest ages (red) are along ocean ridges, where new seafloor is created and from where it spreads, and the ages generally increase (from yellow to green and to blue) with the distance from the ridges. The oldest seafloor (violet colour in Fig. 8.9a, 280 million years) is in the Mediterranean Sea, where it is the last remnant of the former Tethys Ocean (see Sect. 7.2.4).

The horizontal distributions of age values in the ocean (Fig. 8.7a) are used to estimate the rate (speed) of seafloor spreading as follows: the distance from the ridge is divided by the corresponding age, which provides the distance covered by the seafloor per unit time (millimetres per year in Fig. 8.9b). This value is a *half-spreading rate*, as it provides the speed at which the seafloor moves away from one side of the ridge. It is multiplied by 2 to obtain the *full spreading rate* for both sides of the ridge. The full rate of seafloor spreading at slow, intermediate and fast ridges is less than 50, from 50 to 90 and more than 90 millimetres per year, respectively (see Sect. 7.2.1). For comparison, human fingernails grow at an average rate of about 35 millimetres per year. In Fig. 8.9b, the values are generally low in the Atlantic, intermediate in the Northwestern Pacific and Indian Oceans, and high in the South and East Pacific.

The distribution of magnetic anomalies in the oceans show that the seafloor is spreading, which observation is the cornerstone of plate tectonics. Indeed, the formation of new seafloor at ridges and its general spreading away from ridges explains the following geological characteristics of Earth: the progressive deepening of the seafloor with distance from the ridges, which follows from the contraction of seafloor rocks as they cool down, and thus occupy less volume; the increasing thickness of sediments away from ridges, which results from the accumulation of sediments on the seafloor with time; and the movements of continents, which are determined by the combined action of spreading at different ridges and subduction in trenches.

8.5.3 Plate Tectonics and the Earth System in the Past

By running seafloor spreading backwards in time, researchers can determine what where the positions of continents in the past. This is like running a movie backwards. By reversing the direction of seafloor spreading, which thus becomes *seafloor closing*, the plates that presently diverge reverse their movements and instead converge, until their continental parts get eventually together. This is illustrated in Fig. 8.10 for the positions of continents 65, 150, 200 and 225 million years ago. The figure shows that the present continents formed a single, supercontinent 225 million years ago, this supercontinent being named *Pangaea*, which means "whole Earth".

The landmass of supercontinent Pangaea was distributed about equally between the two hemispheres, differently from the present distribution with 68% of the continental masses in the Northern Hemisphere. It is thought that Pangaea was formed around 335 million years ago by the assemblage of the pieces of a previous supercontinent, and began to break apart about 200 million years ago. In the longer term of 750 million years (Fig. 8.11), the relative proportion of emerged lands has not changed much as it always ranged between 10 and 30%, but there was a trend for continents to change their positions from about 90% in the Southern Hemisphere 400 million years ago to almost 70% in the Northern Hemisphere nowadays. The chapter of Wessel and Müller (2015) provides further reading on the reconstruction of plate movements.

It is also interesting to move forward in time, and look at the predicted positions of landmasses in the future based on the present rates of seafloor spreading. Figure 8.12 shows a representation of one possible configuration of the Earth's landmasses 200 million years from now. The future supercontinent has been named *Pangaea ultima* or *Pangaea proxima* … but if intelligent beings are present on Earth at this distant time in the future, they will likely use a different name!

The initiation of the Quaternary Ice Age, which began about 2.6 million years ago and extends to the present day, may have been related to a succession of tectonically-driven changes in the positions of the continents that occurred during the last 35 million years (see Information Box 4.3). These changes led to the opening of two seaways between large continental masses (the Tasman Strait and the Drake Passage) about 33 million years ago, and the complete closing of two other seaways (the Panama and Indonesian seaways) 3 to 4 million years ago (Fig. 4.5). The opening of the two seaways at high latitudes in the Southern Hemisphere favoured the development of the massive Antarctic ice sheet that still exists today, and the closing of the two seaways may have readied the planet for the succession of glacial and interglacial episodes of the Quaternary period.

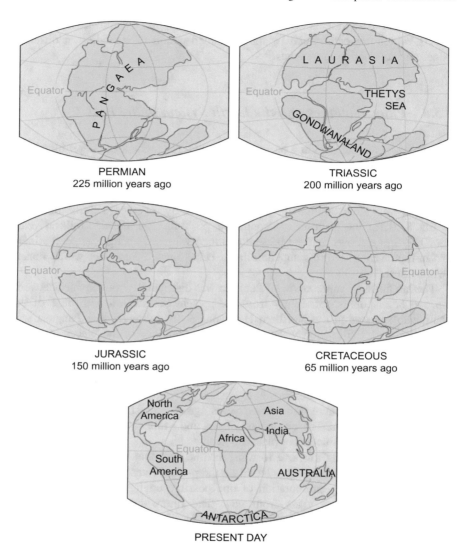

Fig. 8.10 Supercontinent Pangaea, which existed 250 million years ago, broke up from 225 million years ago until now. Credits at the end of the chapter

The above changes took place in the context of a long-term cooling of Earth that has been progressing since about 50 million years. This cooling occurred after a torrid age, which had taken place between 65 and 55 million years ago and during which the average global temperature was about 15 °C warmer than at present (see Sect. 4.5.5). Some researchers think that the opening and closing of seaways played a major role in the long-term cooling trend, whereas others

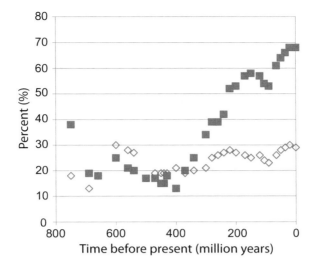

Fig. 8.11   Distribution of the surface areas of emerged lands on Earth during the last 750 million years. Open blue diamonds: global proportion of the Earth's surface occupied by emerged lands; solid red squares: proportion of emerged lands located in the Northern Hemisphere. Data from http://phl.upr.edu/library/notes/distributionoflandmassesofthepaleo-earth. Credits at the end of the chapter

Fig. 8.12   Relative positions of the Earth's landmasses (excluding Antarctica) **a** today and **b** predicted in 200 million years. Credits at the end of the chapter

consider that the cooling was mostly caused by the observed long-term decrease in the concentration of atmospheric CO_2. In any case, the opening and closing of major seaways during the 35 million years before the Quaternary Ice Age are examples of tectonically driven changes in the configuration of continents. These may have caused major changes in the Earth System when the ice sheets began to periodically grow and retreat in the Northern Hemisphere (see Sect. 4.5.2).

At a much longer time scale, Earth went through a succession of at least 16 greenhouses and icehouses during the course of its 4.6 billion years history (Table 4.3). Some of these changing conditions are illustrated in Fig. 8.13 for the

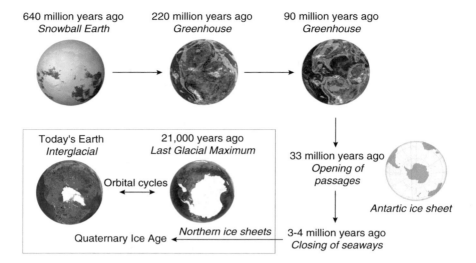

Fig. 8.13 Some of the conditions experienced by Earth during the last 640 million years. Views of the globe: equatorial from 640 to 90 million years ago; south polar 33 million years ago; and north polar 21,000 years ago and today. Credits at the end of the chapter

last 840 million years: icehouse condition (snowball Earth) 640 million years ago; greenhouse condition 220 and 90 million years ago (see Sect. 4.5.4); transition toward the present icehouse with the opening of the Tasman Strait and the Drake Passage 33 million years ago, leading to the development of the Antarctic ice sheet (Fig. 4.4); complete closing of the Panama and Indonesian seaways 3 to 4 million years ago, perhaps causing the beginning of the Quaternary Ice Age 2.6 million years ago (see Information Box 4.3); Last Glacial Maximum 21,000 years ago (see Sect. 4.5.2); and today's interglacial Earth. The latter two conditions correspond to the maximum and minimum extent of the ice sheets in the Northern Hemisphere. These are representative of the succession of glacial-interglacial cycles experienced by Earth since the initiation of the Quaternary Ice Age, which were driven by periodic variations in the Earth's orbit (see Sect. 1.3.4). As explained in the previous paragraph for the Quaternary Ice Age, the tectonically-driven changes in the configuration of continents at longer time scales also had major effects on the Earth System (see Sects. 4.5.4–4.5.6).

8.6   Key: Points: The Long-Term Atmosphere and the Living Earth

We conclude this chapter with a brief summary of some of the key points of the above sections, as done in previous chapters (see Sects. 1.6, 2.5, 3.8, 4.8, 5.7, 6.7, and 7.6). Here, we examine interactions between the Solar System, Earth, the long-term atmosphere, and organisms. As in the corresponding sections of previous chapters, the context of this section is provided by the three phases of the Earth's history illustrated in Fig. 1.11, which are the Mineral, Cradle and Living Earth. These three phases occurred from 4.6 to 4.0, 4.0 to 2.5, and 2.5 to 0 billion years ago, respectively.

After the giant collision that formed the Earth-Moon System 4.5 billion years before present, there were no organisms on Earth until about 3.6 billion years ago, or perhaps earlier. During this Mineral Earth phase, the Earth's atmosphere went through its first phase, which ended 4.4 billion years before present or earlier, and entered its second phase, which lasted until about 2.4 billion years before present. The composition of the atmosphere during its first phase reflected the degassing of the volatile compounds released from the cooling of the hot magma ensued from the collision. Most of these gases were lost by atmospheric escape to space (see Sect. 8.1.1). The hydrogen escaping from the Earth's atmosphere in its first phase dragged along heavier molecules and atoms that included Ne, N_2 and CO_2. This occurred because Earth did not have a magnetic field at the time, which thus could not protect the atmosphere from the solar wind that stripped the upper atmosphere away (see Sect. 8.4.1).

The atmosphere of the second phase had a very high concentration of CO_2 as it was largely created by degassing of the cooling magma. With the progressive cooling of the planet, water vapour condensed into liquid formed at 200 °C or higher because of the very high atmospheric pressure. This led to the creation of the ocean perhaps 4.4 billion years ago or earlier (see Sect. 8.1.1).

Atmospheric CO_2 progressively dissolved in the ocean's waters, where chemical reactions with seafloor silicate rocks ($CaSiO_3$) stored the CO_2 into solid calcium carbonate ($CaCO_3$), which accumulated in the seafloor, thus sequestering more than 99% of the initial atmospheric CO_2 gas in carbonate rocks. This reduction in CO_2 was accompanied by a parallel reduction in the atmospheric pressure and temperature (Sect. 8.1.1). The more temperate environmental conditions allowed the development of organisms and ecosystems in aquatic environments, which marked the passage to the Cradle Earth phase. Microorganisms progressively contributed to the long-term storage of CO_2 into solid $CaCO_3$, and the chemical and biological formation of $CaCO_3$ further decreased the atmospheric pressure and temperature. From then onwards, the Earth's atmosphere had the right density and greenhouse gas composition for maintaining a moderate greenhouse effect, which generally kept the temperature of Earth within a range that allowed most of its surface water to be liquid and was suitable for organisms (Sect. 8.1.2).

During the second and following phases of the atmosphere, the conditions of atmospheric temperature and pressure ensured the maintenance of vast expanses of liquid water at the surface of the planet, which protected water from the loss of its hydrogen to the outer space. In the ocean, organisms diversified and built up the large biomasses during billions of years (see Sect. 8.4.1). In addition, more than 99% of the chemical elements initially in the gaseous constituents of the Earth's atmosphere were progressively sequestered in the sediments and sedimentary rocks, the continental crust and the upper mantle. This sequestration protected the chemical elements from being possibly lost to space, but only less than 1% remained available in the outer envelopes of Earth (crust, ocean and atmosphere) for use by ecosystems (see Sect. 8.1.2).

The third phase of the atmosphere was marked by the appearance of free oxygen (O_2) about 2.4 billion years ago, or perhaps more recently. This change took place hundreds of millions of years after marine cyanobacteria had invented the O_2-producing photosynthesis. With the appearance of atmospheric O_2, Cradle Earth became the Living Earth, which saw the evolution of unicellular prokaryotes to unicellular and, later, multicellular eukaryotes. The energy-efficient O_2 respiration and the large biomasses of prokaryotes and multicellular eukaryotes contributed to the takeover of shallow oceans by organisms (see Sect. 8.1.2). The O_2 progressively accumulated in the upper atmosphere, where some of the O_2 molecules were transformed into ozone (O_3) by the action of the UV radiation of the Sun. The creation of the O_3 layer about 500 million years before present led to the fourth phase of the atmosphere (see Sect. 8.1.1).

The atmosphere in its fourth phase was characterized by the presence of the ozone layer, which provided protection to organisms from the DNA-damaging UV radiation from the Sun. This protection allowed complex multicellular organisms to begin the occupation of emerged lands more than 400 million years ago. The land organisms diversified and developed very large biomasses on the continents, and their physiological activities drastically changed the O_2 and CO_2 composition of the atmosphere. Simultaneously, enormous amounts of the organic carbon produced by land and water ecosystems were buried under the ground and in the seafloor, respectively, which resulted in increases in atmospheric O_2 (see Sect. 8.1.2).

In the present atmosphere, the current increase in atmospheric CO_2 comes from the combustion of fossil organic matter (mostly coal, petroleum and natural gas) by human societies, and the calcination of $CaCO_3$ rocks for making cement. Given that this CO_2 had been stored away by natural processes over hundreds of millions of years, the present situation could be summarized as "onwards to the past", and this new chapter in the history of the Earth's atmosphere is marked by the anthropogenic climate change (see Sect. 8.1.1).

The above changes in the atmosphere of the planet were deeply influenced by the magnetic fields of Earth and the Sun, which generate a magnetosphere in space around each of them. The Earth's magnetic field was involved in the above succession of atmospheric phases and Earth System conditions (see Sect. 8.4.1). During the first phase of the atmosphere, the planet did not have a magnetic field and lost its initial atmosphere. During the second and later phases of its atmosphere, Earth had a magnetic field, and this likely was one of the major factors that protected its atmosphere, and thus its liquid water.

It is thought that the Earth's magnetic field prevented the atmosphere from being stripped away by the solar wind since the second phase of the atmosphere. The Earth's magnetosphere shielded the atmosphere, including its water vapour, from the solar wind, which in turn protected the vast liquid oceans. The presence of the liquid ocean over billions of years was a key factor in the development of organisms and ecosystems and the build-up of large biomasses.

Later on, the presence of the magnetic field protected the ozone layer, which allowed organisms to populate the continents and complete their takeover of the planet (see Sect. 8.1.2). In addition, organisms at the surface of the planet are largely protected from direct hits by cosmic rays because the magnetic field of the planet both shields the atmosphere from most cosmic rays, and protects the atmosphere whose thickness absorbs most of the cosmic radiation that passes the shield (see Sect. 8.4.2).

In summary, during the course of the Earth's history, the magnetic field protected the atmosphere from the solar wind, and the atmosphere protected in turn the vast expanses of liquid water that were key to the development of organisms and ecosystems. In addition, the magnetosphere and the heliosphere (the region of space influenced by the solar wind) shielded the organisms, especially those on emerged lands, from dangerous cosmic radiation, and the thick atmosphere absorbed most of the cosmic rays that went through the shields. Furthermore, the atmospheric ozone layer protected land-dwelling organisms from DNA-damaging solar UV radiation (see Sect. 8.4.2).

Figure Credits

Fig. 8.4a https://commons.wikimedia.org/wiki/File:Geodynamo_Between_Reversals. gif by Dr. Gary A. Glatzmaier, Los Alamos National Laboratory, U.S. Department of Energy, NASA, in the public domain.

Fig. 8.4b https://www.nasa.gov/topics/solarsystem/features/sdo_cycle.html by NASA/ Goddard/SVS, in the public domain.

Fig. 8.5 https://commons.wikimedia.org/wiki/File:Magnetosphere_rendition.jpg by NASA, in the public domain.

Fig. 8.6 https://commons.wikimedia.org/wiki/File:Polarlicht_2.jpg by United States Air Force Senior Airman Joshua Strang, in the public domain.

Fig. 8.7a https://commons.wikimedia.org/wiki/File:Geomagnetic_polarity_late_ Cenozoic.svg by United States Geological Survey, hand-traced to vector by intgr https://en.wikipedia.org/wiki/User:Intgr, in the public domain. Lettering redrawn.

Fig. 8.7b https://commons.wikimedia.org/wiki/File:Geomagnetic_polarity_0-169_ Ma.svg by Anomie https://commons.wikimedia.org/wiki/User_talk:Anomie, in the public domain. Numbers redrawn.

Fig. 8.8 This work, Fig. 8.8, is a derivative of https://commons.wikimedia.org/ wiki/File:Oceanic.Stripe.Magnetic.Anomalies.Scheme.gif by United States Geological Survey, in the public domain. I, Mohamed Khamla, release this work in the public domain.

Fig. 8.9a https://www.ngdc.noaa.gov/mgg/ocean_age/ocean_age_2008.html by Müller et al. (2008), Fig. 1, NOAA, in the public domain.

Fig. 8.9b https://www.ngdc.noaa.gov/mgg/ocean_age/ocean_age_2008.html by Müller et al. (2008), Fig. 2, NOAA, in the public domain.

Fig. 8.10 https://commons.wikimedia.org/wiki/File:Pangaea_to_present.gif by Kious and Tilling (1996), United States Geological Survey, in the public domain.

Fig. 8.11 Original. The data are from http://phl.upr.edu/library/notes/distribution-oflandmassesofthepaleo-earth. Figure 8.11 is licensed under CC BY-SA 4.0 by Philippe Bertrand, Louis Legendre and Mohamed Khamla.

Fig. 8.12a https://commons.wikimedia.org/wiki/File:BlankMap-World-noborders. png by E. Pluribus Anthony at English Wikipedia, in the public domain.

Fig. 8.12b https://commons.wikimedia.org/wiki/File:PangeaUltimaRoughEstimation. png by Pokéfan 95, used under CC0 1.0 (Public Domain Dedication).

Fig. 8.13 Original. Figure 8.13 is licensed under CC BY-SA 4.0 by Philippe Bertrand, Louis Legendre and Mohamed Khamla.

References

Black HH, Davis HN (1913) Practical physics. The MacMillan Co, New York City, NY, USA

Kious WJ, Tilling RI (1996) This dynamic Earth: the story of plate tectonics. U.S. Government Printing Office, Washington, DC. Free access: https://books.google.fr/books?hl=fr&lr=&id=T-5mZVQeigpMC&oi=fnd&pg=PR2&dq=kious+this+dynamic+1966&ots=E-ZnWLM5O-Q&sig=Cw5lYzZX1y9eG2j7tago0BqMha8&redir_esc=y#v=onepage&q=kious%20this%20dynamic%201966&f=false

Müller RD et al (2008) Age, spreading rates and spreading symmetry of the world's ocean crust. Geochem Geophys Geosyst 9:Q04006. https://doi.org/10.1029/2007GC001743

Further Reading

Catling DC, Kasting JF (2017) Atmospheric evolution on inhabited and lifeless worlds. Cambridge University Press, Cambridge

Catling DC, Zahnle KJ (2009) The planetary air leak. Sci Am 300:36–43. http://faculty.washington.edu/dcatling/Catling2009_SciAm.pdf. May 2009

Fabian P, Dameris M (2014) Ozone in the atmosphere. Basic principles, natural and human impacts. Springer, Berlin Heidelberg

Gould JL, Gould CG (2012) Nature's compass. The mystery of animal navigation. Princeton University Press, Princeton

Gurney A (2005) Compass. A story of exploration and innovation. W. W. Norton, New York, NY

Lühr H et al (eds) (2018) Magnetic fields in the Solar System, Astrophysics and Space Science Library, vol 448. Springer, Cham

Turner G (2011) North Pole, South Pole: The epic quest to solve the great mystery of Earth's magnetism. The Experiment, New York

Wessel P, Müller RD (2015) Plate tectonics. In: Schubert G (ed) Treatise on geophysics, 2nd edn, vol 6. Elsevier, Amsterdam, pp. 45–93. https://www.researchgate.net/publication/263551020_Plate_Tectonics

Windridge M (2016) Aurora: In Search of the northern lights. William Collins

Chapter 9
Feedbacks in the Earth System

9.1 The Earth System: Feedbacks and Feedback Loops

9.1.1 Feedbacks and Feedback Loops

The previous chapters described several instances of feedbacks and feedback loops (see Information Box 6.2) that had deeply affected the Earth System at some crucial steps of its long history, and continue to affect it today. In this section and the next three, we revisit the feedbacks and feedback loops from previous chapters, as well as their effects on various aspects of the Earth System. Table 9.1 summarizes ten feedbacks and feedback loops described in previous chapters, which illustrate the involvement of such mechanisms in a wide variety of natural situations.

In Table 9.1, the ten instances of feedbacks and feedback loops from previous chapters are organized in a general framework of positive and negative mechanisms, which are briefly described in Sect. 9.1.3. The general effects of the two types of feedback loops on the Earth System are explained in Sect. 9.1.4. The bulk of the chapter describes how major negative feedback loops contribute to regulate three major characteristics of the Earth System, namely the greenhouse effect (Sect. 9.2), the ocean's nitrogen-to-phosphorus ratio (Sect. 9.3), and the level of atmospheric oxygen (Sect. 9.4). We conclude the chapter with a section on interactions among the major negative feedback loops, and on various effects of forcing factors, the climate and biological evolution on large-scale processes of the Earth System (Sect. 9.5).

It is explained in Information Box 6.2 that, in a *feedback*, the response of a process (B) to a causal process (A) can amplify (positive feedback) or mitigate (negative feedback) the action of the causal process (A). A *feedback loop* is created when the feedback-modified process A further influences process B. In the present text, we consider interactions between macroscopic processes of the Earth System, which often involve organisms and ecosystems. Figure 9.1 illustrates the

P. Bertrand and L. Legendre, *Earth, Our Living Planet*, The Frontiers Collection,
https://doi.org/10.1007/978-3-030-67773-2_9

Table 9.1 Examples of positive and negative feedbacks and feedback loops described in previous chapters. These are briefly described in the accompanying text, and illustrated in Figs. 6.3 and 6.4

Positive feedbacks and feedback loops	Negative feedbacks and feedback loops
Quaternary Ice Age—initiation of a glacial episode (Sect. 4.5.2 and Information Box 6.2)	Seawater pH and $CaCO_3$ (Sect. 6.4)
Quaternary Ice Age—initiation of an interglacial episode (Information Box 6.2)	Snowball Earth—stabilization (Sects. 4.5.4 and 7.4.6)
Snowball Earth—initiation of a global glacial episode (Sects. 4.5.4 and 7.4.6)	Interaction between erosion and mountain building (Sect. 7.5.2)
Greenhouse effect of water vapour (Information Boxes 2.7 and 6.2)	Dissolution of $CaSiO_3$ and the greenhouse effect (Sect. 7.3.3)
Permafrost thaw and release of greenhouse gases (GHG) (Sects. 2.1.7 and 11.1.3)	Marine dimethylsulphide (DMS) and cloud nucleation (Information Box 6.2)

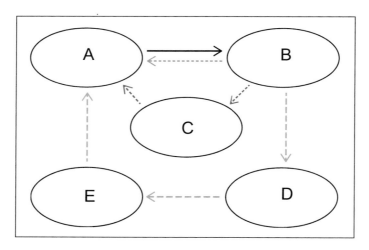

Fig. 9.1 Schematic representation of the mechanism of a *feedback*: state or a process A affects state or process B, which in turn affects A directly, or indirectly through C, or D and E. The diagram also illustrates a *feedback loop*, which takes place when feedback-modified state or process A further influences state or process B, thus initiating a repetition of the loop. Arrows: action of a state or process on another

mechanism of a feedback into process A, mediated by a single process (B), two processes (B and C), or more (B, D, and E). The figure also illustrates the ensuing feedback loops. Examples of feedbacks and feedback loops in organisms and ecosystems are provided in Sect. 9.1.2.

In modern societies, mechanical, electrical and electronic systems regulated by negative feedbacks or feedback loops are very usual. One example is the *thermostat*, which keeps the temperature of a heat-regulated system at a preset value.

Such systems include air-conditioned rooms or buildings, refrigerators, ovens and aquaria, where the temperature is kept at the preset value by the activation of a heating or cooling device when the temperature departs from this value. Another example of device controlled by a negative feedback loop is the *cruise control* of a vehicle, which keeps the vehicle moving at a selected constant speed by accelerating or decelerating it when the speed diverges from the selected value. A third example is the *control of electrical lighting* by a photosensor, which senses the ambient light level and adjust the electrical light up or down to keep the ambient light constant. More generally, a system that regulates a variable at a setpoint is called a *feedback control system*. The following two books provide further reading, the book of Albertos and Mareels (2010) on basic ideas of feedbacks, control, signals and systems for readers interested in the engineering and science of feedbacks, and that of Åström and Murray (2019) on the theory of feedbacks and their use in science and engineering.

9.1.2 Feedbacks and Feedback Loops in Organisms and Ecosystems

The most complex systems regulated by feedbacks and feedback loops are biological. For example in living cells, enzyme-based feedback inhibition loops control the synthesis of amino acids and several carbohydrates and lipids, and similarly control the synthesis of nucleotides and thus nucleic acids (see Sect. 6.1.4). The feedback inhibition mechanism allows tiny cell, which can be less than one thousandth of a millimetre in size, to control the simultaneous synthesis of thousands of highly complex molecules.

Feedbacks and feedback loops are also involved in the regulation of populations and ecosystems. One example is provided by the fur catch records of the Hudson's Bay Company, which extends over 90 years from 1845 to 1935. This company, which was incorporated in England in 1670, operated posts in Northern Canada where agents bought furs of animals caught by First Nations and European trappers. The fur catch records provide unique long-term information on fluctuations of wild animal populations.

Ecologists have used the data series of catches of lynx (*Lynx canadensis*) and snowshoe hare (*Lepus americanus*) from the Hudson's Bay Company (Fig. 9.2a) to explore possible effects of a predator (lynx) on its prey (hare) and the converse effect of the prey on its predator. These reciprocal effects could correspond to a *negative feedback loop*, given that a sustained increase in the hare population would create more food for the lynx, whose population would become larger, so that after some time the predation exerted by the more abundant lynx would cause a decrease in the hare population. This way, the lynx would exert a negative feedback (Fig. 9.1) into its hare prey. In return, the decrease in hare abundance would bring a decrease in lynx numbers, which would allow the hare population to start

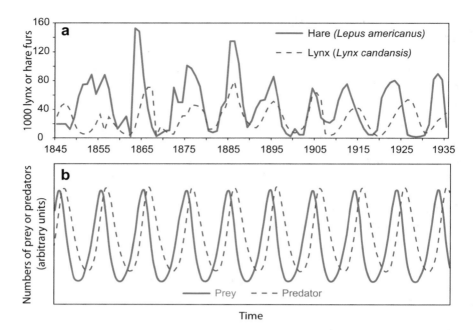

Fig. 9.2 Cyclical variations in predator and prey abundance: **a** numbers of Canada lynx and snowshoe hare furs acquired from trappers by the Hudson's Bay Company between 1845 and 1935, and **b** theoretical Lotka-Volterra predator-prey relationship with a 10-year periodicity

growing again. This would create a negative feedback loop (Fig. 9.1), whose result would be to maintain the abundances of both the lynx and hare populations within limited ranges of values.

The successive peaks in lynx and hare catches show a periodicity of about 10 years (Fig. 9.2a). These fluctuations are quite similar to the curves resulting from Lotka-Volterra equations with a 10-year periodicity (Fig. 9.2b). These equations were initially proposed within the context of the theory of autocatalytic chemical reactions in 1910 by the American researcher Alfred J. Lotka, who later applied it to fluctuations in the abundances of two interrelated species. The same equations were used independently in 1925 by the Italian mathematician Vito Volterra to explain the fluctuations of animal species living together. Since then, these equations have been known as the *Lotka-Volterra equations*, and have been used extensively in theoretical biology and ecology as well as in economic theory to describe *negative feedback loops*.

The Lotka-Volterra curves illustrated in Fig. 9.2b represent the interaction between a prey and a predator in an ideal world without the following constraints: food limitation for the prey, sickness, climate events, and human intervention. In contrast, the fur catches in Fig. 9.2a come from the real world, and they reflect not only the interaction between the populations of lynx and hare, but also the activity of trappers. Indeed, the latter may have selectively targeted one species or the other on different years, depending on such factors as the abundances of either species and the commercial values of their furs. Hence, the intent of the theoretical curves with a 10-year periodicity in Fig. 9.2b is not to reproduce the fluctuations observed in the numbers of lynx and hare furs (Fig. 9.2a), but instead to show that these fluctuations could have corresponded to a Lotka-Volterra predator-prey relationship. The observed periodicity of about 10 years in the fluctuations of lynx and hare Fig. 9.2a is often used as an illustration of a long-term interaction between a prey and its predator corresponding to a negative feedback loop.

9.1.3   Feedbacks and Feedback Loops in the Earth System

Table 9.1 provides ten examples of feedback loops in the Earth System taken from previous chapters and summarized in the following paragraphs. The five positive and five negative feedback loops are illustrated in Figs. 9.3 and 9.4, respectively.

The five *positive feedbacks* in the first column of Table 9.1 are schematized in Fig. 9.3. Diagram 9.3a illustrates the general format of the five other diagrams, which comprises four components: the chain of processes, linked by straight solid arrows; the feedback of the last process on the first, represented by a curved solid arrow; the feedback loop, which repeats the chain of processes, and is represented by dotted curved arrows; and the *runaway* effect resulting from the feedback loop, which is represented by a dashed curved arrow. Summaries of the five positive feedbacks follow.

Figure 9.3b *Quaternary Ice Age—initiation of a glacial episode* (see Sect. 4.5.2 and Information Box 6.2). Each of the 30 to 50 glacial episodes during the Quaternary Ice Age likely began with the incomplete melting in the Northern Hemisphere summer of the snow that had accumulated the previous winter, caused by a general cooling of the climate. This cooling favoured the persistence of snow and ice surfaces with high albedo (see Information Box 2.9) at high latitudes in summer, which increased the reflection of solar radiation to space and thus led to further global cooling. The enhanced global cooling occurred in response to the increase in albedo, hence a *positive feedback*. The cooling resulting from the feedback further increased the albedo, thus initiating a *positive feedback loop* that caused a runaway cooling.

Figure 9.3c *Quaternary Ice Age—initiation of an interglacial episode* (see Information Box 6.2). Each of the 30 to 50 interglacial episodes during the Quaternary Ice Age likely began with some melt of the ice sheets, caused by a general warming of the climate. The rising temperature caused the shrinking of the ice sheets with high albedo (see Information Box 2.9) at high latitudes, which

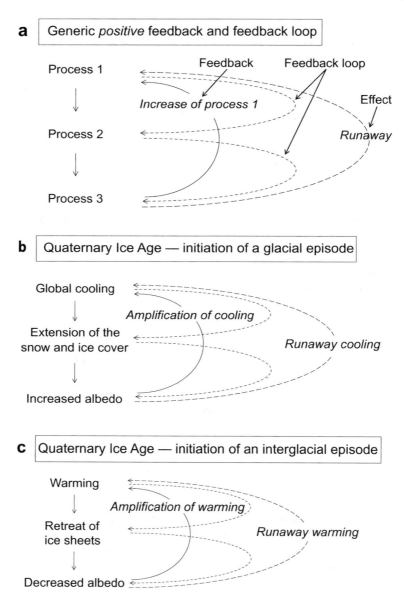

Fig. 9.3a–c Examples of five positive feedback loops described in previous chapters. Diagram 9.3a illustrates the general format of the five other diagrams (diagrams 9.3d–9.3f are on the opposite page). Explanations are given in the text. Credits at the end of the chapter

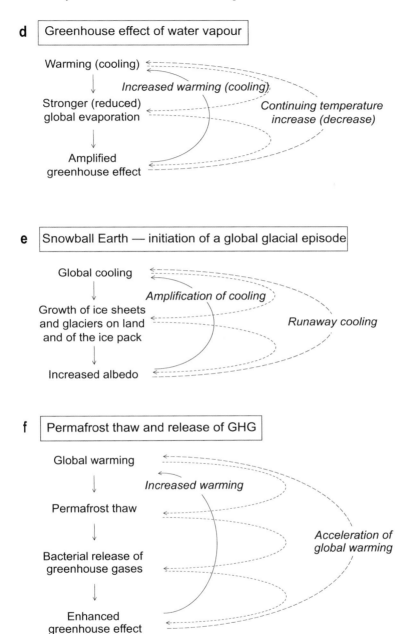

d │ Greenhouse effect of water vapour

Warming (cooling)

Increased warming (cooling)

Stronger (reduced)
global evaporation

*Continuing temperature
increase (decrease)*

Amplified
greenhouse effect

e │ Snowball Earth — initiation of a global glacial episode

Global cooling

Amplification of cooling

Growth of ice sheets
and glaciers on land
and of the ice pack

Runaway cooling

Increased albedo

f │ Permafrost thaw and release of GHG

Global warming

Increased warming

Permafrost thaw

*Acceleration of
global warming*

Bacterial release of
greenhouse gases

Enhanced
greenhouse effect

Fig. 9.3d–f Diagrams 9.3d–9.3f (the general format of the diagrams is given in diagram 9.3a on the opposite page)

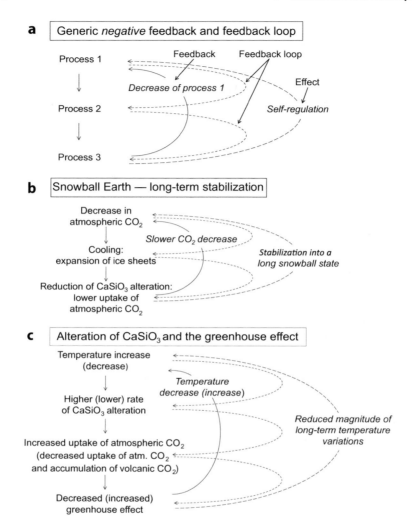

Fig. 9.4a–c Examples of five negative feedback loops described in previous chapters. Diagram 9.4a illustrates the general format of the five other diagrams (diagrams 9.4d–9.4f are on the opposite page). Explanations are given in the text. Credits at the end of the chapter

decreased the global albedo of the planet and favoured further global warming. The rise in the global temperature occurred in response to the decrease in albedo, hence a *positive feedback*. The rise in temperature resulting from the feedback further decreased the albedo, thus initiating a *positive feedback loop* that caused a runaway warming.

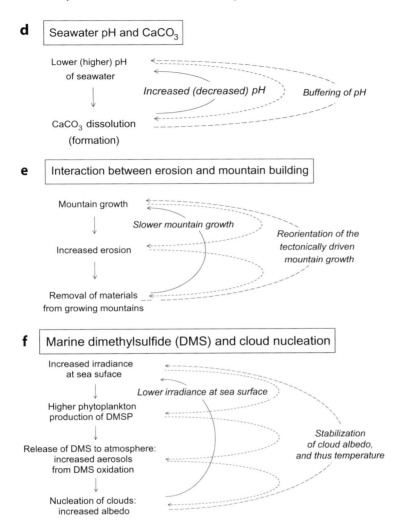

Fig. 9.4d–f Diagrams 9.4d–9.4f (the general format of the diagrams is given in diagram 9.4a on the opposite page)

Figure 9.3d *Greenhouse effect of water vapour* (see Information Boxes 2.7 and 6.2). When the climate warms as a consequence of an increased concentration of atmospheric CO_2 and other greenhouse gases (the greenhouse effect), there is stronger evaporation of oceanic and continental waters, and the higher content of water vapour in the atmosphere amplifies the global warming. In contrast when the greenhouse effect decreases, lower mean temperature reduces evaporation, and the reduced content of water vapour in the atmosphere amplifies the global cooling.

In both cases, the change in the greenhouse effect amplifies the change in global temperature, hence a *positive feedback*. The change in temperature resulting from the positive feedback affects the vapour content of the atmosphere, which initiates a *positive feedback loop* that sustains the increase or decrease in temperature.

Figure 9.3e *Snowball Earth—initiation of a global glacial episode* (see Sects. 4.5.4 and 7.4.6). The last snowball Earth episode (around 640 million years ago) occurred in the geographical setting of the break up of supercontinent Rodinia. This was accompanied by very extensive zones of volcanic activity, which led to the formation of huge amounts of silicate rocks ($CaSiO_3$). The continental $CaSiO_3$ rocks were progressively eroded by CO_2-acidified continental waters, causing a major reduction of atmospheric CO_2 that resulted in a global cooling. The progress of the ice sheets on land and of the ice pack at sea increased the albedo of the planet, which amplified the cooling, hence a *positive feedback*. The additional cooling resulting from the positive feedback favoured the progress of the ice, which initiated a *positive feedback loop* leading to the almost complete glaciation of the planet.

Figure 9.3f *Permafrost thaw and release of greenhouse gases* (GHG) (see Sects. 2.1.7 and 11.1.3). About one fourth of the Northern Hemisphere is covered in permafrost, which is one of the largest natural reservoirs of organic carbon on Earth and is currently thawing at an increasing rate. When permafrost thaws, microbial decomposition can turn the organic carbon into greenhouse gases (GHG), and thus enhance the greenhouse effect, which is a *positive feedback*. The increased warming resulting from the positive feedback would cause additional permafrost thaw, which would initiate *positive feedback loop*. It is feared that as the planet continues to warm, large amounts of GHG will be released from the permafrost, which would create the above positive feedback loop whose effect would be to accelerate the ongoing global warming.

The five *negative feedbacks* in the second column of Table 9.1 are schematized in Fig. 9.4. Diagram 9.4a illustrates the general format of the five other diagrams in the same way as explained above for panel 9.3a, except that the effect of each feedback loop here corresponds to *self-regulation*. Summaries of the five negative feedbacks follow.

Figure 9.4b *Snowball Earth—long-term stabilization* (see Sects. 5.5.4 and 7.4.6). The last snowball Earth episode (around 640 million years ago) started with a period of deep cooling, leading to the almost complete glaciation of Earth (see Fig. 9.3e). Later on, the expansion of the ice sheets on continents had the effect of reducing the surfaces of exposed $CaSiO_3$ rocks, which lowered the uptake of atmospheric CO_2 by the chemical alteration of $CaSiO_3$ (see Fig. 9.4c). Hence, the reduced chemical alteration of $CaSiO_3$ created a *negative* feedback into the concentration of atmospheric CO_2. The lower uptake of atmospheric CO_2 slowed down the expansion of the ice sheets on continents, which initiated a *negative feedback loop* whereby the progress of the ice sheets inhibited the mechanism causing the cooling. The planet then entered in the long frozen period known as *snowball Earth*.

Figure 9.4c *Alteration of CaSiO$_3$ and the greenhouse effect* (see Sect. 7.3.3). When the global temperature tends to increase, there is a corresponding increase in the rate of chemical alteration of continental CaSiO$_3$ rocks. This speeds up the uptake of CO$_2$ from the atmosphere, which decreases the greenhouse effect and thus counteracts the increase in global temperature. The decreased greenhouse effect thus creates a *negative feedback* into the global temperature. Conversely when the global temperature tends to decrease, there is a corresponding decrease in the rate of chemical alteration of continental CaSiO$_3$ rocks. This slows down the uptake of CO$_2$ from the atmosphere, and the decreased uptake of atmospheric CO$_2$ allows the CO$_2$ emitted by volcanoes to accumulate in the atmosphere and thus strengthen the greenhouse effect. This counteracts the decrease in global temperature, which is again a *negative feedback*. The slower change in temperature resulting from either negative feedback affects the rate of dissolution of continental CaSiO$_3$ rocks, which creates a *negative feedback loop*. The latter generally reduced the magnitude and duration of the long-term temperature variations throughout the history of the planet.

Figure 9.4d *Seawater pH and CaCO$_3$* (see Sect. 6.4.1). The pH of seawater controls the formation and dissolution of solid calcium carbonate, CaCO$_3$, that is, acidic conditions favour the dissolution of CaCO$_3$, and basic conditions favour its formation. In return, the dissolution of solid CaCO$_3$ increases the pH of seawater, and its formation decreases it. The two sets of reactions are *negative feedbacks*. The reciprocal control of chemical reactions by the pH and of the pH by chemical reactions dampens changes in the ocean's pH, a property that chemists call *buffer*. This *negative feedback loop* keeps the pH of seawater close to a constant value.

Figure 9.4e *Interaction between erosion and mountain building* (see Sect. 7.5.2). The growth of mountains is primarily determined by tectonic processes. The growth of mountain ranges such as the St. Regis, the European Alps, the Himalayas and the Andes (Fig. 7.11) is accompanied by increased erosion, which removes materials from the growing mountains and thus slows down their growth. The removal of materials thus creates a *negative feedback* into mountain growth. Slower mountain growth causes slower erosion, which may initiate a *negative feedback loop* that modifies the way the mountains grow by acting on some effects of the tectonic processes of mountain building.

Figure 9.4f *Marine dimethylsulphide (DMS) and cloud nucleation* (see Information Box 6.2). Some groups of marine phytoplankton release dimethylsulphoniopropionate (DMSP) in the surrounding water when they grow, and part of this DMSP is transformed by the planktonic ecosystem into dimethylsulphide (DMS) gas, which escapes to the atmosphere. There, DMS is oxidized to various compounds, some of which can form aerosols that act as cloud condensation nuclei. It was hypothesized that an increase in sea-surface irradiance could favour the development of DMSP-producing phytoplankton and thus increase the release of DMS by the ocean. This would enhance the formation of clouds, which would cause an increase in albedo and a decrease in the penetration of solar radiation in the atmosphere, thus creating a *negative feedback* into sea-surface irradiance. The lower irradiance would be accompanied by a decrease in phytoplankton development, and the resulting *negative feedback loop* could help stabilizing the Earth's temperature by changing the cloud albedo.

9.1.4   Long-Term Effects of Feedback Loops, and Self-Regulation

The above examination of the ten cases in Table 9.1 and Figs. 9.3 and 9.4 showed that the long-term effects of positive and negative feedback loops are very different. In a *positive feedback*, a disturbance of process A is *amplified* by the action of process B (or processes B and C, or B, D and E; Fig. 9.1) on process A. When the feedback develops into a positive feedback loop, each successive cycle of the loop further *amplifies* the effect of the disturbance, which leads to an unstable situation and often a *runaway*. In the opposite situation of a *negative feedback*, a disturbance of process A is *mitigated* by the action of process B (or processes B and C, or B, D and E; Fig. 9.1) on process A. When the feedback develops into a negative feedback loop, each successive cycle of the loop progressively *dampens* the disturbance, which reduces the effect of the perturbation and tends to promote a settling to *equilibrium or self-regulation* as explained in the next paragraphs.

Cells and organisms exhibit the most powerful negative feedbacks, and the resulting dynamic state of equilibrium is called *homeostasis*. A more general term for the ability to maintain dynamic stability is *self-regulation*, which applies to both living and non-living systems. The ability to self-regulate allows cells and organisms to maintain a relatively constant environment, while continuously interacting with and adjusting to changes whose origin may be internal or external. Self-regulation is achieved through negative feedback loops, which act at the levels of the cells, tissues, organs, systems, or whole organisms (see Sect. 6.2.5). Examples of characteristics that all cells keep within narrow ranges are their internal pH and chemical composition (see Sect. 6.4.1).

In order to regulate their internal properties, cells and organisms exchange energy and materials with their environment. One example of this process is the regulation of the internal pH of cells by exchanges of ions with the external medium (see Sect. 6.4.1). On a larger scale, warm-blooded animals regulate their temperature (an ability called *homeothermy*) by, for example, releasing heat from their bodies through evaporation of water by sweating (humans), panting (dogs) or flapping ears (elephants). Evaporation removes energy from the liquid water, which cools water down, and this energy is transferred to the water vapour (see Sect. 3.4.2). Self-regulation is generally done by individuals, but there are cases of collective self-regulation where a group of organisms acts as a single entity. An example is provided by honeybees when they need to cool down their hive in case of excessive temperature. One way by which they do this is by evaporating water, which they achieve by bringing water from the outside into the hive and collectively fanning their wings to set up air movements that enhance evaporation and thus cooling.

It was explained previously that cells and organisms are *open systems*, and thus exchange materials and energy with their surroundings, which allows them to maintain a high level of organization among their components as long as these stay alive (see Sect. 6.2.5). It was also explained that part of the energy that fuels cellular processes is released as heat into the environment (see Sect. 5.1.2). In addition, it was explained that in physical systems including the organisms, the materials are continually recycled, whereas the energy is progressively dissipated into heat, which cannot be recovered for further use (see Sect. 3.2.2). Hence, the building and maintenance of biological complexity (diversity of molecules, cells, organisms, species and ecosystems) requires exchanges of materials and energy. The materials are recycled, but the input energy (solar or chemical) is mainly returned to the environment as heat or chemical energy (by-products and waste). The maintenance of the internal properties of biological systems by negative feedback loops thus contributes to the irreversible process of degradation of energy.

Self-regulation is crucial for the maintenance of not only individuals but also species. Indeed, the individuals of many unicellular and multicellular species, including humans, cannot self-regulate forever, and their failure to carry on homeostasis at some point in time (sickness, senescence) causes their death. However, the death of individuals belonging to sexually reproducing species does not generally cause the extinction of their species. Indeed, these species survive beyond the lifetimes of individuals by passing genetic information from generation to generation using sexual reproduction. Unicellular and few multicellular species transfer their genetic information among generations through asexual reproduction, using a number of mechanisms that include cell fission, vegetative propagation, fragmentation, and parthenogenesis.

There is some similarity between the *technological* feedback control systems described above in Sect. 9.1.1 and *biological* self-regulation described here. Examples of of technological and biological temperature regulation are the thermostat and homeothermy by warm-blooded animals, respectively. One major difference between the two types of regulation is the fact that the target temperature of the thermostat is set by the operator, whereas the range of internal temperatures of a warm-blooded animal is a physiological characteristic of the species. More generally, the properties regulated using technological feedback systems can be set arbitrarily at any value by the operator, whereas the range of values of self-regulated biological characteristics is inherent to the species, and keeping within this range is critical for the survival of its individuals.

Self-regulation (or homeostasis) is a vital property of organisms. In the last decades, some researchers proposed that this property could be applied to the

Earth System, and especially the climate system. The origin of this idea goes back to the *Gaia hypothesis* (Lovelock 1995), which posited that the different components of the Earth System—atmosphere, hydrosphere, cryosphere, lithosphere, and biosphere (see Sect. 1.1.1)—formed a self-regulating complex system that seeked an optimal physical and chemical environment for organisms and ecosystems. This proposal was received with enthusiasm by part of the scientific community, and rejected by another part. Gaia remains an important, albeit controversial, hypothesis. Several researchers now study the self-regulating aspects of the climate system (see Information Box 1.4), often without reference to the Gaia hypothesis as in this chapter, but nevertheless stimulated by James Lovelock's original idea.

9.2   Regulation of the Greenhouse Effect: The Carbon Feedback Loops

9.2.1   Cycling of Carbon in the Earth System

For 3.6 billion years or more, the temperature of Earth has remained within a relatively narrow range, which favoured the development of organisms and ecosystems. How did this happen? The answer lies in the functioning of the global carbon cycle, whose relevant aspects, including the roles of carbon feedback loops, are explored in this and subsequent sections.

More than 99% of the Earth's carbon is stored in rocks, where around 80% of the carbon is contained in sedimentary carbonates and the remainder in organic carbon, mainly dispersed in sedimentary rocks or concentrated in deposits (peat, coal, anthracite, graphite, petroleum, natural gas, and oil shale and sands). Some organic carbon is also stored on continents in the organic matter frozen in the permafrost, and in the seafloor as solid methane hydrate (see Sect. 2.1.7). Much smaller amounts of carbon are found (in decreasing order of abundance) in the oceans (mostly dissolved inorganic carbon, and also dissolved organic carbon), in the biomasses of terrestrial and aquatic organisms, and in the atmosphere where CO_2 accounts for less than 0.05% of the total volume of gases (Fig. 7.10). Carbon continually flows between the components of the Earth System—atmosphere, liquid water, rocks and sediments, and organisms (the ice compartment is not much involved in the carbon cycle)—in an exchange called the *carbon cycle*, which has slow and fast components (see Information Box 9.1).

The fast and slow carbon cycles have maintained relatively steady concentrations of carbon in the different components of the Earth System during the last million years, with some variability around the average values. One example of such variability is provided by the cyclical changes in the concentration of atmospheric CO_2 during the last 800,000 years illustrated in Fig. 4.10a. Because carbon

Fig. 9.5 Global anthropogenic CO_2 emissions from 1800 to 2014: total, and resulting from the combustion of four types of fossil organic carbon (petroleum; coal; natural gas; and gas flaring, that is, the burning of natural gas associated with oil extraction processes) and the calcination of $CaCO_3$ for cement production. Credits at the end of the chapter

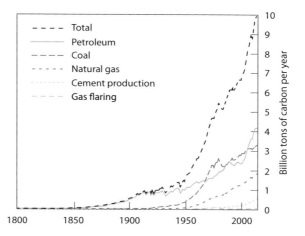

continually flows among the compartments, when a change occurs in the amount of carbon in one of them, the effect ripples through others. A striking example is provided by the progressively increasing emissions of CO_2 in the atmosphere by human societies in the last two centuries. Figure 9.5 shows that the main anthropogenic sources of CO_2 since 1800 were the burning of fossil fuels (mostly coal, petroleum, and natural gas) for energy production, gas flaring (burning of natural gas associated with petroleum extraction processes) and the calcination of $CaCO_3$ rocks for cement production. The global emissions of CO_2 were multiplied by a factor of almost 20 between the years 1900 and 2019.

In the natural situation, that is, without human activities, the carbon contained in fossil fuels would have been discharged slowly as CO_2 into the atmosphere through volcanic activity, over the millions of years of the slow carbon cycle. By burning these compounds, humans are rapidly releasing in the atmosphere increasing amounts of the carbon that had accumulated over millions of years. In other words, humans are moving this carbon from the slow cycle to the fast cycle, with major environmental consequences.

About half the anthropogenic carbon emitted since the beginning of the Industrial Age in the second part of the eighteenth century has accumulated in the atmosphere, and the other half has been absorbed about equally by the continents (plants and uncultivated soils) and the oceans. The resulting rise in the concentration of CO_2 in the atmosphere is mostly responsible for the increase in the greenhouse effect, which causes the ongoing *global warming*. The resulting dissolution of large amounts of CO_2 in seawater is the direct cause of *ocean acidification*. These two and other major negative consequences of the perturbation of the natural cycling of carbon and their effects on human societies are examined in Chapter 11 (see Sects. 11.1.1–11.1.3).

Information Box 9.1 Fast and Slow Carbon Cycles Carbon compounds are continually exchanged among compartments on the Earth System, in various inorganic and organic forms. The atoms of carbon flow between different compartments at different rates, and for conceptual simplicity researchers have grouped these flows in two cycles, with vastly different timescales, which were already mentioned in Chapter 2 (see Sect. 2.1.6).

The *fast* (or *biological*) *carbon cycle* operates on timescales of days to thousands of years. It moves about 1,000–100,000 million tons of carbon every year.

This cycle is focussed on the movement of carbon through the biomasses of organisms. The starting point is the fixation of CO_2 carbon into organic matter by autotrophic organisms (see Information Box 5.1). From then on, this carbon circulates through the food webs (see Information Box 5.3).

During their lives, the organisms respire back to CO_2 most of the carbon contained in the organic matter initially synthesized by autotrophs (see Information Box 5.2). Organisms also release dissolved and particulate organic matter, the latter including the carcasses of dead plants and animals. Most of this organic matter is either consumed by detritivores or used by heterotrophic microbes, and the respiration of these organisms produces CO_2. The total amount of CO_2 released by all the organisms in a biological community is called *community respiration*. In addition, part of the biomass is burned by wildfires and people, a process that recycles carbon back to CO_2 in the same way as O_2 respiration (see Information Box 3.4). As a result, most of the carbon fixed in organic matter by autotrophic organisms is recycled back to CO_2 relatively rapidly, hence the expression *fast carbon cycle*. However, a small part of the biogenic carbon escapes short-term recycling, and enters the slow carbon cycle.

The *slow (geological) carbon cycle* operates on long timescales, and it takes between 100 and 200 million years for carbon to go through it. This cycle moves about 10–100 million tons of carbon every year, that is, 100–1000 times less than the fast carbon cycle, but its effects are very large because its processes are repeated millennia after millennia over very long times.

The slow cycle combines movements of carbon through mostly geological compartments of the Earth System, but it also involves the biota. Most of the atmospheric and oceanic carbon enters the cycle through the chemical alteration of $CaSiO_3$ rocks, which leads to the burial of solid $CaCO_3$ in the seafloor (see Sect. 7.3.3). In the modern ocean, the carbonate minerals ($CaCO_3$ and others) are mostly formed by calcifying organisms (see Sect. 6.4.2). Some of the buried carbonate minerals are transported to the continents, where they form rocks such as limestone and marble. Carbonate minerals make up, on average, about 80% of the carbon-containing rocks, the remaining 20% being organic carbon originating from biological remains embedded in layers of mud on the seafloor.

The mineral and organic components of sediments are compressed and heated over millions of years, forming sedimentary rocks such as shale and coal, within which organic carbon is dispersed (shale) or highly concentrated (coal). This process turns peat into coal, anthracite and even graphite, and the organic carbon from shales into petroleum (the resulting rock is called *oil shale*) and natural gas. Petroleum and natural gas, known as *hydrocarbons*, are expelled from the source rock, and may accumulate in porous geological reservoirs (sand or carbonates) when the latter are covered by impermeable sediments. Coal can also be a source rock for petroleum and natural gas.

The slow cycle returns carbon to the surface through volcanoes. These release CO_2 gas in the atmosphere, and their lava solidifies in part as $CaSiO_3$-bearing rocks (see Sect. 7.2.3).

9.2.2 The Strong Room of the Earth's Carbon and the Greenhouse Effect

Following the formation of the Earth-Moon system 4.5 billion years before present, the planet was very hot, and although it progressively cooled down, its surface temperature could not go below 200 °C or slightly higher because of the high concentration of atmospheric CO_2. Indeed, at the beginning of its second phase 4.4 billion years ago, the Earth's atmosphere mostly comprised CO_2, like the atmosphere of Venus today, and it exerted the very high surface pressure of about 60 atm (see Sect. 8.1.1). Fortunately, a liquid-water ocean formed as early as 4.4 billion years ago, because water could be in the liquid phase at a temperature up to 280 °C under the prevailing very high atmospheric pressure. The 200 °C ocean was too hot for the emergence of organisms, but not to allow the key chemical reaction of solid carbonate formation. The burial of carbonates in the seafloor sequestered more than 99% of the carbon initially present in the CO_2-rich atmosphere. Hence, the early formation of the ocean had the major result of reducing the enormous greenhouse effet caused by the very high initial concentration of atmospheric CO_2 (see Sect. 5.5.4).

Following the dissolution of atmospheric CO_2 in the ocean, inorganic carbon was present in seawater under the three dissolved forms of CO_2 gas and bicarbonate (HCO_3^-) and carbonate (CO_3^{2-}) ions, which are related as follows (see Information Box 6.3, whose Eqs. 6.3 and 6.4 are provided here for convenience):

$$CO_{2(aqueous)} + H_2O \rightleftharpoons HCO_3^- + H^+ \tag{9.1}$$

$$HCO_3^- \rightleftharpoons CO_3^{2-} + H^+ \tag{9.2}$$

The double arrows indicate that the chemical transformations are reversible. The directions of the arrows are controlled by equilibrium constants that depend on environmental conditions such as the temperature, hydrostatic pressure, and pH of the solution. Since the amount of CO_2 in the atmosphere was very large, the concentrations of HCO_3^- and CO_3^{2-} ions in seawater were also very large. These ions combined with the positive ions present in the ocean—such as Ca^{2+}, Mg^{2+} and Fe^{2+}—resulting in the formation of solid calcium carbonate ($CaCO_3$), magnesium carbonate ($MgCO_3$) and iron carbonate ($FeCO_3$), which sedimented on the seafloor. This way, more than 99% of the initial atmospheric CO_2 was stored into solid carbonate sediments and rocks at the bottom of the oceans.

It is not known how long the above process lasted, but it was likely complete by 4 billion years before present. At the end of it, the atmospheric pressure was much lower than initially, and the small concentration of CO_2 in the atmosphere created a moderate greenhouse effect. The cradle for organisms was ready! We will see in the next section how Earth self regulated the concentration of atmospheric CO_2 in the following billions of years, which kept the Earth's temperature within a relatively narrow range appropriate for the organisms.

The solid carbonates in the seafloor are continually transported by subduction in the mantle, where they become part of the magma in which they release CO_2. Part of the magma and dissolved CO_2 reach the surface of the crust through volcanoes, which release the CO_2 in the air and seawater and whose lava forms $CaSiO_3$-bearing rocks. The chemical alteration of continental and seafloor $CaSiO_3$ rocks followed by the chemical and biological formation of carbonates in the ocean sequesters environmental CO_2 in the seafloor in the form of solid carbonates. These are eventually recycled by subduction (see Sects. 7.3.1 and 7.3.3). Because of this continual geological recycling, the stock of solid carbonates formed more than 4 billion years ago does not exit anymore, but was continually replenished by the solid carbonates formed chemically and by organisms (see Information Box 5.11 and Sect. 6.4.2). The Earth's calcareous sediments and sedimentary rocks—which are found from the bottom of the oceans to coral reefs in shallow seas, to sedimentary rocks on continents, and to the top of Mount Everest in the Himalayas—store in solid form most of the carbon present in the Earth's atmosphere 4.4 billion years ago.

The storage of carbon in calcareous sediments and sedimentary rocks protects the Earth's ecosystems from the present trend of human societies to extract and burn or calcine accessible fossil carbon, which releases in the atmosphere increasingly large amounts of the greenhouse gas CO_2 (Fig. 9.5). The resulting increase in atmospheric CO_2 fuels the ongoing anthropogenic global warming. Fortunately, the largest fraction of the carbon present in the early atmosphere is stored in carbonates, whose calcination for producing cement makes only a small dent in this enormous pool of inorganic carbon, and most of the fossil organic carbon is not concentrated in exploitable deposits but is dispersed in the crust. Global warming will create very bad environmental conditions within the coming few decades if human societies do not stop burning or calcining fossil carbon. However, this could not bring the planet back to the Hadean inferno (see Sect. 1.5.4) because humans do not have access to all the carbon present in the atmosphere 4.4 billion years ago, whose release would wipe out all organisms on the planet. Indeed, most of this carbon is stored in carbonate sediments and rocks, which are the strong room of the Earth's carbon. This strong room is largely impervious to human entry.

Over the past billions of years, too much natural greenhouse effect would have been detrimental to ecosystems, but too little would also have been bad. Indeed, without the natural greenhouse effect, the average temperature at the Earth's surface temperature would have been about $-18\,°C$, which would have produced a planet-wide glaciation that would have brought the temperature down to about $-50\,°C$ (see Sect. 2.4.1). There would have been no liquid water except in a few environments, such as the vicinity of volcanoes. Such conditions would not have been favourable for the existence of organisms on Earth, or if some organisms had been present, their small numbers and collective biomass would not have allowed them to have significant effects on the planet.

The temperature of Earth is maintained within a range appropriate for organisms by two global mechanisms that act on two very different timescales, that is, around 400,000 years and 1000 years or more, respectively. This means that the first mechanism responds to slower disturbances with a time constant of around 400,000 years, and the second mechanism responds to relatively rapid disturbances in the concentration of atmospheric CO_2 within 1000 years or more. Figure 9.6 summarizes the main aspects of the two mechanisms: the left panels provide the relative abundances of carbon in compartments of the Earth System; the right panels indicate the timescales of the exchanges of carbon between compartments; and the central panels illustrate the carbon flows within and among compartments. The functioning of the two mechanisms (negative feedback loops) is described in the following two sections. The book of Williams and Follows (2011) provides further reading on the cycling of carbon in the ocean and its connections with the global carbon cycle.

9.2.3   Regulation of Long-Term Variations in the Greenhouse Effect: The Tectonic CO_2 Feedback Loop

Long-term variations in the concentration of atmospheric CO_2 are regulated by interactions between volcanic activity and the chemical alteration of $CaSiO_3$ rocks on continents and on the recent seafloor. This causes the storage of carbon from CO_2 into solid $CaCO_3$ in the seafloor (see Information Box 7.3).

$$CaSiO_{3(solid)} + CO_{2(aqueous)} \rightarrow CaCO_{3(solid)} + SiO_{2(solid)} \tag{9.3}$$

The resulting $CaCO_3$ sediments are progressively consolidated into rocks, which are transported by tectonic processes and recycled in volcanoes (bottom and left arrows in Fig. 9.6a). The whole tectonic cycle takes about 70 to 80 million years (see Sect. 7.3.3), but the timescale of the regulation of long-term variations in the concentration of atmospheric CO_2 by the chemical alteration of $CaSiO_3$ rocks and burial of $CaCO_3$ (Eq. 9.3) is about 400,000 years (top and right arrows in Fig. 9.6a), as explained three paragraphs below.

The rate of dissolution of $CaSiO_3$ is sensitive to temperature, and as a consequence when the global temperature tends to increase, there is a corresponding

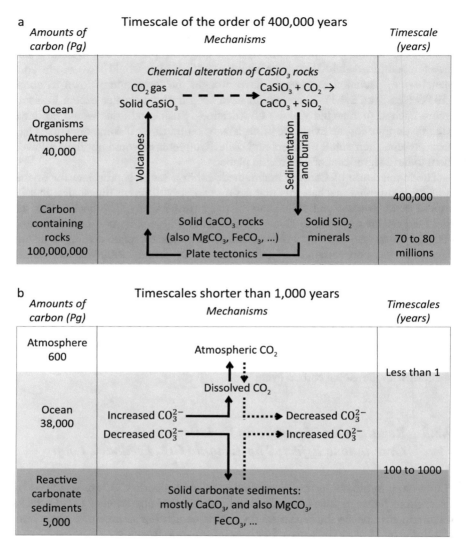

Fig. 9.6 Regulation of the greenhouse effect. **a** At long timescales, regulation by the negative tectonic feedback loop. Solid arrows: physical transport of compounds; dashed arrow: chemical flow. **b** At shorter timescales, partial regulation by the negative oceanic carbonate feedback loop. Solid and dotted arrows: carbon flows corresponding to a deficit or an excess of atmospheric CO_2, respectively. The amounts of carbon are expressed in petagrammes (Pg): $1\ Pg = 10^{15}\ g = 10^9$ t (one billion tons). Details are given in the text (see Sects. 9.2.3 and 9.2.4). Credits at the end of the chapter

increase in the rate of chemical alteration of $CaSiO_3$ rocks. This speeds up the uptake of CO_2, which accelerates the net storage of carbon into $CaCO_3$ on the seafloor (Eq. 9.3), thus causing a decrease in the greenhouse effect. Hence a negative feedback that counteracts the increase in global temperature (Fig. 9.6a).

Conversely when the global temperature tends to decrease, there is a corresponding decrease in the rate of chemical alteration of $CaSiO_3$ rocks, which slows down the uptake of CO_2 (Eq. 9.3). The decreased CO_2 uptake allows the CO_2 emitted by volcanoes to accumulate in the atmosphere, and thus strengthen the greenhouse effect. Hence a negative feedback that counteracts the decrease in global temperature (Fig. 9.6a).

The processes directly involved in the regulation of the greenhouse effect are the interaction between the environmental CO_2 (atmospheric and dissolved in seawater), the $CaSiO_3$ rocks (continents and seafloor), and the seafloor sink of $CaCO_3$ (dashed and right arrows in Fig. 9.6a). The timescale at which changes in the rate of dissolution of $CaSiO_3$ rocks modify the global concentration of atmospheric CO_2 is of the order of 400,000 years. The existence of this negative feedback loop generally prevented the occurrence of global runaway increases or decreases in temperature on the planet. Major exceptions were the three snowball Earth episodes, when the tectonic feedback loop could not prevent the occurrence of runaway cooling. In the case of the last snowball Earth 640 million years ago, the cause of the runaway cooling may have been the exceptionally large amount of $CaSiO_3$ rocks then exposed to chemical alteration (see Sect. 4.5.4).

There is already much anthropogenic CO_2 accumulated in the atmosphere and in order to remove some of the atmospheric CO_2, it was proposed by some authors to enhance the uptake of this greenhouse gas by $CaSiO_3$ rocks. The proposal, known as enhanced silicate rock weathering (ERW), is to amend large areas of agricultural soils in many countries with crushed $CaSiO_3$ rocks (such as basalt), which could take up large amounts of atmospheric CO_2. However, the mining, crushing, transport, and spreading of basalt may release much CO_2 to the atmosphere. If ERW proved effective, the timescale of the atmospheric effect of CO_2 uptake by $CaSiO_3$ could be shortened from the natural value of 400,000 years to a few years only. A benefit for farmers would be that rock dust would improve crop yields. It is explained in Chapter 11 that resolving the anthropogenic CO_2 problem requires a drastic reduction in the extraction and use of fossil carbon, and how negative emission technologies such as ERW could complement this necessary step (see Sect. 112.2). It remains to be seen whether ERW, which would be a new massive industrial sector that would use billions of tons of basalt, could actually be implemented and would be without harmful side effects such as large-scale perturbations of natural environments and the pollution of soils and underground waters.

The global negative feedback loop described here has played a major role throughout the history of the planet as it adjusted the atmospheric CO_2 concentrations to the slow changes that occurred in global temperature. Its timescales was of the order of 400,000 years.

9.2.4 Regulation of Shorter-Term Variations in the Greenhouse Effect: The Oceanic Carbonate Feedback Loop

A regulation of shorter-term variations in the greenhouse effect is provided by exchanges of carbon between the atmospheric CO_2 pool and the much larger pool of $CaCO_3$ on the seafloor. This exchange is mediated by chemical reactions that take place in the sea, as explained in the following paragraphs.

In seawater, inorganic carbon is present under three dissolved forms: CO_2, and HCO_3^- and CO_3^{2-} ions (Eqs. 9.1 and 9.2). *Dissolved inorganic carbon* (DIC) is the sum of the concentrations of these forms ($DIC = [CO_2] + [HCO_3^-] + [CO_3^{2-}]$). The oceanic DIC pool is very large compared to the atmospheric CO_2 pool, that is, for one atom of carbon present in the atmosphere (mostly in the form of CO_2), there are about 50 atoms of inorganic carbon dissolved in the ocean (Fig. 9.6b). The relative abundances of the three dissolved forms of CO_2 in seawater depend on the pH of seawater, which currently ranges, in general, between about 8.1 and 8.2, which is slightly alkaline (see Information Box 6.4). At this pH, the Bjerrum plot of the carbonate system shows that HCO_3^- is by far the most abundant form (91%), followed by CO_3^{2-} (8%) and dissolved CO_2 (1%) (Fig. 9.7).

The three dissolved forms of dissolved CO_2 together with H_2CO_3, which is not included in Fig. 9.7 for the reason explained in Chapter 6, make up the *carbonate system* (see Information Box 6.3). One should not to confuse the *carbonate system* with the *carbonate ion* (CO_3^{2-}) or the *carbonate minerals* (such $CaCO_3$). Figure 9.7 represents the carbonate system as a function of pH, for seawater with a temperature of 20 °C, a salinity of 35 and a DIC concentration of 2.1 mmol kg^{-1},

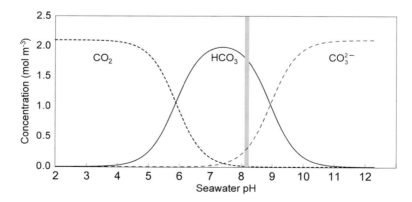

Fig. 9.7 Bjerrum plot of the carbonate system: concentrations of the chemical forms of dissolved inorganic carbon in water as a function of pH (for seawater with a temperature of 20 °C, and a salinity of 35). Shaded band: current pH of seawater, between 8.1 and 8.2. Credits at the end of the chapter

for a pressure of 1 atm. In fact, the positions of the three curves on the pH axis and their shapes (curvatures) depend on the temperature and salinity of the water and also on the pressure. These conditions vary among regions, where water may be colder or warmer, and fresher or saltier, and they also vary with the depth, as the hydrostatic pressure increases with increasing depth. Hence the diagrams for different temperatures, salinities and pressure would look generally similar to Fig. 9.7, but the actual values of CO_2, HCO_3^- and CO_3^{2-} for a given pH would be different from those in this figure.

We consider in the following paragraphs what happens in the two cases of an excess and a deficit of CO_2 in the atmosphere. We examine successively the two-way exchanges of CO_2 between the atmosphere and the ocean, the chemical reactions in the surface ocean, the effect of the downward transport of the chemically modified surface waters, the chemical reactions in the deep ocean, the effect of the upward transport of the chemically modified deep waters, and the effect on atmospheric CO_2. These are schematized in Fig. 9.6b.

Exchanges of CO_2. When there is an *excess* of CO_2 in the atmosphere, this gas dissolves in the surface ocean where it is stored in seawater as dissolved CO_2, HCO_3^- and CO_3^{2-}. Conversely when there is a *deficit* in atmospheric CO_2, some of the CO_2 dissolved in seawater, $CO_{2(diss)}$, is degassed to the atmosphere as $CO_{2(gas)}$, and is replaced by the transformation of dissolved HCO_3^- and CO_3^{2-} ions into $CO_{2(diss)}$ (Eqs. 9.1 and 9.2). In the case of a change in the concentration of either atmospheric or seawater CO_2, it takes on average less than a year for the ocean surface layer and the overlying atmosphere to equilibrate. Because CO_2 is continually exchanged between the atmosphere and the surface ocean, there is a strong link between the greenhouse effect and the pH of the ocean (see Sect. 9.2.6).

Surface ocean. In the case of an *excess* of atmospheric CO_2, there is an increase in the concentration of dissolved CO_2 in surface waters of the ocean (dotted atmosphere-to-ocean arrow in Fig. 9.6b), which increases the concentration of H^+ ions in seawater (Eqs. 9.1 and 9.2). The latter causes a drop in the pH of seawater (see Information Box 6.4), which is accompanied by a decrease in the concentration of CO_3^{2-} ions (Fig. 9.7). Conversely in the case of a *deficit* in atmospheric CO_2, there is a decrease in the concentration of dissolved CO_2 in surface waters (solid ocean-to-atmosphere arrow Fig. 9.6b), which decreases the concentration of H^+ ions in the seawater (Eqs. 9.1 and 9.2). The latter causes a raise in the pH (see Information Box 6.4), which is accompanied by an increase in the concentration of CO_3^{2-} ions (Fig. 9.7).

Hence, an excess of atmospheric CO_2 causes a decrease in the concentration of CO_3^{2-} ions in seawater, and a deficit in atmospheric CO_2 causes an increase in the concentration of CO_3^{2-} ions. This process takes place within less than a year, and is illustrated by the two within-ocean arrows in Fig. 9.6b.

Downward transport. Surface waters are brought in contact with the $CaCO_3$ sediments at the timescale of the deep oceanic circulation, which is of the order of 1,000 years (see Sect. 3.5.3), meaning that most of the water of the ocean comes in contact with $CaCO_3$ sediments within 1,000 years. However, in regions of formation of deep waters, such as the northern North Atlantic Ocean, surface waters would come in contact with the seafloor much more rapidly than 1000 years, that is, within a few years.

Seafloor. In the case of an excess of atmospheric CO_2, the deep oceanic circulation carries the CO_3^{2-} deficit from surface to depth, where this deficit can be compensated by the dissolution of seafloor $CaCO_3$. The dotted sediment-to ocean arrow in Fig. 9.6b illustrates the dissolution of $CaCO_3$ on the seafloor. The following equation (read from right-to-left) summarizes the process:

$$CaCO_{3(solid)} + H^+ \rightleftharpoons$$
$$Ca^{2+}_{(aqueous)} + xCO_{2(aqueous)} + yHCO_{3(aqueous)}^- + zCO_{3(aqueous)}^{2-} \qquad (9.4)$$

where $(x+y+z)=1$; the actual values of x, y and z depend on the hydrostatic pressure (depth), the initial pH and temperature of the water, and on the quantity of $CaCO_3$ dissolved for a given volume of seawater because this dissolution modifies the pH by consuming H^+. Equation 9.4 read from left-to-right shows that the dissolution of one mole of $CaCO_3$ adds to the bottom seawater one mole of CO_2 and/or HCO_3^- and/or CO_3^{2-}.

Given that the mole of HCO_3^- (right side of Eq. 9.4) carries one negative charge, the dissolution of $CaCO_3$ adds one mole of negative charges, and thus one mole of *alkalinity* to the seawater (see Information Box 6.4). The dissolution of $CaCO_3$ also increases the DIC of seawater by one mole ($xCO_2 + yHCO_3^- + zCO_3^{2-}$; right side of Eq. 9.4). Hence the dissolution of $CaCO_3$ increases both the alkalinity and the DIC of seawater.

Conversely when there is a *deficit* of atmospheric CO_2, the resulting CO_3^{2-} surplus can be compensated by the formation of $CaCO_3$. The solid ocean-to-sediment arrow in Fig. 9.6b illustrates the formation of $CaCO_3$ (Eq. 9.4 read from right-to-left), which shows that the formation of one mole of $CaCO_3$ removes from the bottom seawater one mole of CO_2 and/or HCO_3^- and/or CO_3^{2-}. The effects on seawater are opposite to those described in the previous paragraph, that is, the formation of $CaCO_3$ lowers both the alkalinity and the DIC of seawater.

The above reaction in the case of an excess of atmospheric CO_2 is not only theoretical, for it is currently taking place in the ocean. Indeed, some dissolution of $CaCO_3$ sediments has recently been observed in the western North Atlantic Ocean (a region in contact with newly formed deep waters), and the phenomenon has also been reported at various sites in the Southern Atlantic, Indian and Pacific Oceans. Hence the compensation of the CO_3^{2-} deficit caused by anthropogenic CO_2 is already going on in some parts of the ocean.

Upward transport. When the oceanic circulation brings deep waters with lower concentration of CO_2 back to the surface, atmospheric CO_2 dissolves again in the surface ocean, and the cycle is repeated. Conversely when the deep water brought back to the surface is rich in CO_2, this gas is released from surface waters into the atmosphere and the cycle is repeated.

Effect on atmospheric CO_2. In the case of an *excess* of atmospheric CO_2, the invasion of the ocean by atmospheric CO_2 initially increases the DIC of seawater without changing its alkalinity because CO_2 is not electrically charged, whereas the later dissolution of seafloor $CaCO_3$ progressively increases both the DIC and the alkalinity of seawater. Some models have shown that the combination of these two steps would cause the atmosphere to retain more than 20% of its excess CO_2 after it has reached a new equilibrium with the ocean. Such equilibrium would be achieved thousands of years after the end of the emission of the additional CO_2. This retention of an excess of CO_2 in the atmosphere would cause a net warming of the planet, although much smaller than it would have been without the negative feedback resulting from the dissolution of $CaCO_3$. Overall, the above mechanism provides a powerful negative feedback into the greenhouse effect.

The excess of CO_2 remaining in the atmosphere means that global warming will likely continue for many millennia after the end of the present CO_2 emissions. Furthermore, based on our current knowledge, the subsequent cooling would be incomplete, meaning that the Earth's atmosphere and oceans would not cool down all the way back to pre-industrial temperatures before the tectonic feedback involving $CaSiO_3$ erosion remove the excess CO_2 (see Sect. 9.2.3). The return to pre-industrial conditions will probably take hundreds of thousands of years, corresponding to the 400,000-year timescale of the tectonic feedback mechanism. Chapter 11 provides additional information on the possible long-term climate of Earth (see Sects. 11.3.2–11.3.4).

The oceanic carbonate and tectonic CO_2 negative feedback loops described in this section and the previous one, respectively, are essentially abiotic mechanisms, as they they do not directly involve exclusively biological processes. Indeed, although the production of $CaCO_3$ in the ocean is a central component of the two feedback loops and organisms are responsible for most of it in the modern ocean, all $CaCO_3$ was formed abiotically in the early ocean, microorganisms contributed only progressively to the formation of $CaCO_3$ in the following billions of years, and about 10% of the $CaCO_3$ is still formed chemically today (see Sect. 6.4.2). In addition, most of the $CaCO_3$ dissolution is abiotic (see Information Box 5.11). However in areas with a strong downward flux of organic matter, the degradation of this material in the sediment can increase the concentration of CO_2 so much that the resulting local acidification enhances the dissolution of solid carbonates. The chapter of Bertrand (2014) provides further reading on this short-term regulation of the global greenhouse effect by the ocean.

9.2.5   *Variations in the Concentration of Atmospheric CO_2 During the Last 600 Million Years*

The two feedback loops described in the previous two sections have been active in regulating the greenhouse effect for billions of years. They contributed to keep the variations of atmospheric CO_2 within relatively narrow ranges, in the general context of a long-term decrease in the concentration of this gas (Figs. 8.1b,c and 9.8a).

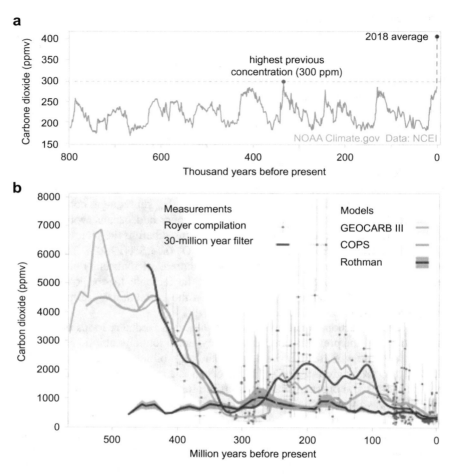

Fig. 9.8 Estimates of changes in atmospheric CO_2 concentrations in the past. **a** Changes during the last 800,000 years. From air bubbles trapped in a 3,200 m-long Antarctic ice core (see also Fig. 11.1). **b** Changes during the last 600 million years: three estimates based on geochemical modelling are compared to a compilation of CO_2 data and the filtered average of these data. Error envelopes are shown when available. Note that there is a factor of one thousand between the X-axes of the two diagrams. Credits at the end of the chapter

The air bubbles trapped in a 3,200 m-long ice core from Antarctica show that the concentration of atmospheric CO_2 varied between 180 and 280 ppmv during the last 800,000 years (Fig. 9.8a), when the low and high CO_2 concentrations corresponded to glacial and interglacial temperatures, respectively (Fig. 11.1). It is debated whether the variations in atmospheric CO_2 were among the proximal causes or instead the effects of the glacial-interglacial cycles, and what were the sources and sinks of atmospheric CO_2, which likely included the continents (vegetation, soils) and the oceans (water, biota). In any case, the observed long-term relationship between the variations in atmospheric CO_2 (Fig. 9.8a) and the

Fig. 9.9 Solubility of CO_2 in water as a function of temperature. Credits at the end of the chapter

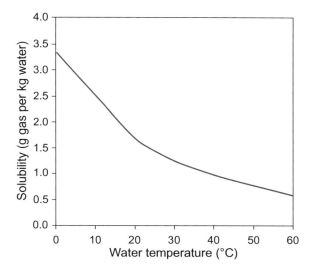

succession of glacial and interglacial episodes (Fig. 4.5a)—that is, low and high atmospheric CO_2 during the glacial and interglacial episodes, respectively—is consistent with the inverse relationship that exists between the solubility of CO_2 and seawater temperature (Fig. 9.9).

Using ice cores, it is possible to go back 2.7 million years in the history of atmospheric O_2, and researchers must use other approaches to explore longer-term variations. One set of such approaches is to make reconstructions based on *proxies*, which are physical, chemical or biological characteristics of the past that stand in for direct measurements, and enable to reconstruct the ancient conditions. Such proxies for dating of volcanic rocks and marine sediments are described in Chapter 8 (see Information Boxes 8.6 and 8.7, respectively). The proxies for past CO_2 concentrations include ratios of carbon isotopes ($^{13}C:^{12}C$) in fossilized soils or the shells of plankton organisms, and the interpretation of the density of stomata in fossil plants (*stomata* are minute pores in the epidermis of leaves or stems). Each of these methods is subject to substantial uncertainties. Another set of approaches is to use geochemical modelling, based on quantifying the geological CO_2 sources and sinks over long time periods. These include volcanic inputs, erosion, and carbonate deposition, such numerical models being largely independent of direct CO_2 measurements.

Long-term variations in atmospheric CO_2 concentrations obtained with the two types of approaches are illustrated in Fig. 9.8b for the last 600 million years. This figure present CO_2 data derived from proxies (Royer et al. 2004) and values from three geochemical models: GEOCARB III (Berner and Kothavala 2001), Carbon-Oxygen-Phosphorus-Sulphur-Evolution COPSE (Bergmann et al. 2004), and Rothman (2001). The values from the two approaches show considerable uncertainty and variability, but they all point to CO_2 levels in the past significantly higher than at present. Overall, atmospheric CO_2 levels have decreased from 500 million years ago, when the concentration was at least 4,000 ppmv, down to the pre-industrial level of 270 ppmv.

The overall pattern of atmospheric CO_2 concentrations in Fig. 9.8b is as follows: high values earlier than 400 million years ago, a large drop from 400 to 300 million years ago, higher values until about 170 million years ago, and a gradual decrease to the present low values. One interpretation of these variations is as follows. Much of the drop from the high CO_2 concentrations 400 million years ago was due to the spread of vascular plants on continents, the deep roots of which enhanced the chemical alteration of silicate rocks and the net uptake of atmospheric CO_2, and thus the sequestration of $CaCO_3$ in the seafloor (see Sect. 7.3.3). The low CO_2 values from 360 to 260 million years ago correspond to cold, icehouse conditions, and the following period to a warmer, greenhouse Earth (Table 4.3). During the last greenhouse, the concentrations of atmospheric CO_2 were several times the pre-industrial value of 270 ppmv.

At an even longer timescale, major variations in the concentration of atmospheric CO_2 and the greenhouse effect are known to have occurred on some occasions during the course of Earth's history. Examples are the three episodes of snowball Earth that likely occurred around 2,200, 710 and 640 million years ago (see Sect. 4.4.4). The initiation of the second and third snowball Earth episodes could have been triggered by the break-up of supercontinents that exposed continental $CaSiO_3$ rocks to chemical alteration by CO_2-acidified running water, which sequestered atmospheric CO_2 into $CaCO_3$ on the seafloor (see Sect. 7.3.3).

The initiation of the first snowball Earth episode could have been caused by the oxidation of atmospheric methane (CH_4), a very powerful greenhouse gas, by the oxygen (O_2) released into the atmosphere by photosynthetic organisms (the concentration of atmospheric O_2 became significant at about this period, and is called *Great Oxygenation Event*).

The oxidation of atmospheric CH_4 occurs through a suite of chemical reactions whose first step is the combination of CH_4 with a highly reactive hydroxyl radical (•OH, which tends to capture one electron by recombining with other elements or ions), which is itself formed by the reaction of water vapour with excited atomic oxygen in the atmosphere:

$$O + H_2O \rightarrow 2 \ \bullet OH \tag{9.5}$$

The suite of reactions ultimately forms formaldehyde (CH_2O), whose combination with hydroxyl radicals (Eq. 9.5) forms CO_2 and H_2O. The net result of the reactions in the atmosphere is:

$$CH_4 + 2O_2 \rightarrow CO_2 + 2H_2O \tag{9.6}$$

During the initiation of the first snowball Earth, the concentration of the CO_2 resulting from the oxidation of CH_4 (Eq. 9.6) would have been drawn down by the chemical alteration of $CaSiO_3$ rocks (see Sect. 7.3.3).

It is thought that in the three snowball Earth episodes, the drawdown of environmental CO_2 by the chemical alteration of $CaSiO_3$ rocks and the sequestration of the resulting $CaCO_3$ were instrumental in causing the initial cooling of the planet. This would have activated the positive feedback loop leading to runaway cooling summarized above (see Sect. 9.1.3, Fig. 9.3e).

9.2.6   Seawater pH and the Greenhouse Effect

The pH of seawater is a very stable characteristic of the modern ocean, which varies very little between the average values of 8.1 and 8.2. One of the main sources of variability is the alternation of community respiration and photosynthesis, which causes short-term changes in the concentration of dissolved CO_2 that increase the pH at night and decrease it during daytime, respectively. The resulting diurnal fluctuations in pH range from 0.1 to 0.15 units in open waters, and up to 0.5 units in areas of dense submarine vegetation such as kelp forests. In coral reefs, the formation and dissolution of $CaCO_3$ cause pH changes that range from 0.16 to 0.78 units.

As long as the pH of seawater remains above 8.0, dissolved CO_2 makes up a very small fraction of the dissolved inorganic carbon (Fig. 9.7), so that seawater can take up atmospheric CO_2 when its partial pressure in the atmosphere is higher than in the ocean. At lower pH values, dissolved CO_2 makes up a larger fraction of the dissolved inorganic carbon, with the consequence that the ocean degases CO_2 into the atmosphere.

The current value of seawater pH is very stable, but how has pH varied since the formation of the ocean around 4.5 billion years ago? The answer to this question is important because many current biological processes depend on pH. Hence, variations in pH may have affected the emergence and development of organisms in the Earth's aquatic environments, although the biological responses to the pH may have changed through the course of biological evolution. In addition, the pH of seawater largely controls the bioavailability of many important chemical elements, such as iron and zinc (see Sect. 6.5), as it determines the chemical species and dissolved state or not of these elements. Furthermore, as explained just above, the pH of seawater is involved in the regulation of CO_2 exchanges between the atmosphere and the ocean, and thus contributes to setting the magnitude of the Earth's greenhouse effect.

Concerning the pH of the ancient ocean, the poor preservation of carbonates deposited 4.5 billion years ago precludes the use of the direct chemical proxies they could contain to estimate past pH values. Similarly, empirical constraints on historical pH values, such as the abundance of old deposits of gypsum (hydrated calcium sulphate), are not really useful as they are interpreted differently in different studies. Because of this, researchers have used numerical models to reconstruct the pH of seawater over the last 4.5 billion years. Different models use different input variables that include: the long-term decrease in atmospheric CO_2 (Fig. 8.1); the cycles of different components of seawater; the erosion of continental and oceanic $CaSiO_3$; the growth of continents; and the biological enhancement of continental erosion by terrestrial plants during the last half billion years (see Sect. 7.3.3).

The results of various models generally agree on a progressive increase in the ocean's pH from about 6.3–7.2 up to as far as 4.0 billion years ago, to around 6.5–7.7 around 2.5 billion years ago, and to the value of 8.2 in the last several millions of years. Models suggest that this evolution was driven by the long-term decrease in atmospheric CO_2, and was influenced by the effects of the continental and seafloor erosion of $CaSiO_3$ on the chemistry of seawater. The negative feedbacks of continental and seafloor erosion of $CaSiO_3$ stabilized the pH near the neutral value of 7.0 (with some variability) during the first 4 billion years of the Earth's history, which contributed to create aquatic environments favourable for organisms.

The long-term decline in atmospheric CO_2 is now being reversed by human activities (see Sect. 9.2.1). Indeed, the ocean has taken up about 30% of the anthropogenic CO_2 released into the atmosphere since the beginning of the Industrial Age in the second part of the eighteenth century, which has caused a significant increase in the concentration of H^+ ions in seawater (Eqs. 9.1 and 9.2). This resulted in a decrease in the average pH of seawater from the pre-industrial value of 8.25 to the current value of 8.14. This decrease is called *ocean acidification*. Because the pH is expressed on a logarithmic scale, the difference of 0.11 pH units represents an increase of almost 30% in the concentration of H^+ ions in seawater. The problem of ocean acidification is further examined in Chapter 11 (see Sects. 11.1. and 11.1.3).

> **The Regulation of Atmospheric CO_2 Contributed to Moderate the Greenhouse Effect and the Variations in Water pH** In summary, the oceanic carbonate and tectonic CO_2 negative feedback loops have generally prevented the occurrence of runaway cooling or warming of the planet. In this way, they have been key mechanisms for the emergence and maintenance of organisms on Earth throughout its entire history. In addition, the concentration of atmospheric CO_2 has largely controlled the pH of aquatic environments, which remained near the neutral value of 7.0 (with some variabiity) during the first 4 billion years of the Earth's history and thus contributed to create environmental conditions favourable for organisms.

9.3   Regulation of Nitrogen and Phosphorus in the Ocean: The N:P Feedback Loops

9.3.1   Nitrogen and Phosphorus in the Ocean

In this section, we discover that the average ratio of nitrogen to phosphorus in phytoplankton biomass is generally nearly constant, that is, N:P = 16:1, and furthermore that the concentrations of dissolved inorganic N and P in seawater are in the same N:P ratio of 16:1. Thousands of observations show that this is not a simple coincidence, but a fundamental property of the Earth System, which can only be explained by the action of powerful feedback loops.

We examined above two carbon feedback loops that contributed to the regulation of variations in the greenhouse effect at two different timescales during the course of the Earth's history, that is, hundreds of thousands and thousands of years (see Sects. 9.2.3 and 9.2.4). The processes involved in these feedback loops were essentially abiotic although carbonates, which are central to the two mechanisms, have been increasingly produced by organisms over the course of the Earth's history (see Sect. 6.4.2). We will see below that the feedback loops involved in the regulation of the balance between nitrogen and phosphorus in the ocean include biological processes related to the dynamics of marine phytoplankton (see Sect. 9.3.6).

The most abundant chemical elements in organisms are hydrogen (H), oxygen (O) and carbon (C), and the other essential elements for the synthesis of key biological molecules include the less abundant nitrogen (N) and phosphorus (P) (see Sect. 6.1.2). The key biomolecules requiring N and/or P include: DNA, which carries the genetic code and passes it from generation to generation; RNA, which mediates the synthesis of proteins; ATP, which is the energy currency of all cells; proteins, enzymes, hormones and neurotransmitters, which directly support the intracellular and intercellular metabolism; pheromones, which are involved in sexual relations between organisms; and phospholipids, which are major constituents of cell membranes (see Sect. 6.1.3).

Biomolecules in the ocean are synthesized either by phytoplankton using photosynthesis, which exploits the energy of light to convert CO_2 into organic matter, or autotrophic bacteria or archaea using one form of chemosynthesis, which exploits chemical energy to produce organic matter from carbon-containing molecules, mostly CO_2 or methane (CH_4) (see Information Box 5.1). The resulting phytoplankton has a relatively uniform average elemental composition across a wide range of taxonomic groups, which is usually cited as $C_{106}H_{263}O_{110}N_{16}P_1$. This means that, for the synthesis of biomass containing 1 unit of P atoms, phytoplankton requires 16, 110, 263 and 106 units of N, O, H and C atoms, respectively. A more recent estimate of the chemical composition of phytoplankton biomass is $C_{106}H_{179}O_{68}N_{16}P_1$ (different proportions of H and O atoms than determined previously), but the ratio C:N:P ratio remains 106:16:1. Some elements that are not abundant in organic matter, and are thus negligible in term of its elemental chemical composition, may nevertheless have significant functional roles in organisms, for example the *trace elements* sulphur (S) and iron (Fe) (see Sect. 6.1.2).

The existence of an almost constant ratio between the key chemical elements in the biomass of phytoplankton means that production of new biomass by these organisms is co-limited by the bioavailability of each of these elements in the environment, including the bioavailability of some trace elements. In other words, even if bioavailable forms of C, O and H were abundant in seawater, a scarcity of bioavailable forms of N, P or/and Fe (which are called *nutrients*) would limit the production of biomass by phytoplankton. Specifically, for N and P, the production of phytoplankton biomass with N:P $= 16$ is impeded when the concentration of bioavailable N is lower than 16 times that of P, or the concentration of bioavailable P is lower than 1/16 times that of N.

It must be noted that phytoplankton photosynthesis can occur in the absence of key nutrients (such as N or P) as long as there are sources of C, O and H, but in such cases the sole products of photosynthesis are carbohydrates, which contain only C, O and H (their generic chemical formula is CH_2O; see also Table 6.2). These carbohydrates are then not used by phytoplankton, which exude them in the environment as dissolved organic carbon (DOC). Hence, photosynthesis in this case does not produce new phytoplankton biomass, but the dissolved organic exudates from phytoplankton can be used by heterotrophic bacteria and archaea to build up their own biomasses. This occurs as long as the latter organisms have themselves access to sources of P, N and other essential elements, for which they often successfully compete with phytoplankton.

The observation that there is a relatively constant C:N:P ratio of 106:16:1 in marine phytoplankton was initially made by the American oceanographer Alfred C. Redfield in 1934, and the value of 106:16:1 is thus called the *Redfield ratio*. In the same study, Redfield also reported the atomic ratio of nitrate (NO_3^-) to phosphate (PO_4^{3-}) in seawater from the Atlantic, Pacific and Indian Oceans and the Barents Sea (Arctic Ocean), which he then found to be N:P$=20:1$. This value was later corrected to 16:1. Redfield noted: "It appears to mean that the relative quantities of nitrate and phosphate occurring in the oceans of the world are just those which are required for the composition of the animals and plants which live in the sea. That two compounds of such great importance in the synthesis of living matter are so exactly balanced in the marine environment is a unique fact and one which calls for some explanation, if it is not to be regarded as a mere coincidence." We will show in following sections that Redfield was right to conclude that the existence of the same C:N$=16:1$ ratio in the plankton and in the seawater was no coincidence.

The remarkable ratio N:P$=16:1$ is illustrated in Fig. 9.10, which displays both the original values of dissolved inorganic NO_3^- and PO_4^{3-} in seawater used by Redfield and additional, modern values from the World Ocean. In this figure, there is a clear, approximate 16:1 relationship between NO_3^- and PO_4^{3-}, with minimal deviation.

The equations of O_2-producing photosynthesis and O_2 respiration (see Information Box 3.4; Eqs. 3.1 and 3.2) can be summarized as follows:

$$CO_2 + 2H_2O + energy \leftrightharpoons [CH_2O] + O_2 + H_2O \qquad (9.7)$$

where $[CH_2O]$ represents generic organic matter (carbohydrate). Equation 9.7 represents photosynthesis when read from left-to-right, and respiration when read from right-to-left. The chemical elements N and P can be incorporated in Eq. 9.7 by adding the N and P nutrients to the left side of the equation, and replacing $[CH_2O]$ by the average composition of phytoplankton on the right side of the equation:

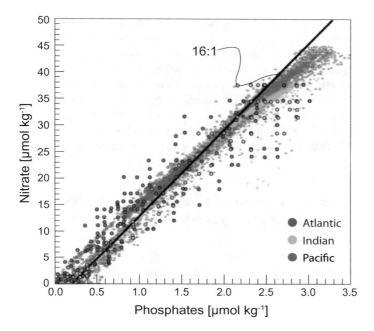

Fig. 9.10 Relationship between nitrate and phosphate concentrations in the ocean: observations used by Redfield in his original 1934 study (large dots) and selected data from the global World Ocean Circulation Experiment survey (WOCE; small symbols). Straight line: 16 NO_3:1 PO_4. Credits at the end of the chapter

$$106CO_2 + 16HNO_3 + H_3PO_4 + 120H_2O \rightleftharpoons$$
$$(CH_2O)_{106}(NH_3)_{16}(H_3PO_4)_1 + 137 O_2 \qquad (9.8)$$

On the left side of Eq. 9.8, HNO_3 and H_3PO_4 represent the inorganic N- and P-nutrients, respectively, and on the right side of the equation, $(CH_2O)_{106}(NH_3)_{16}(H_3PO_4)_1$ represents the average chemical composition of phytoplankton, where C:N:P=106:16:1. Equation 9.8 shows that the building of organic matter requires not only CO_2, water and energy, but also inorganic N- and P-nutrients. More generally and as indicated on different occasions in this book, the photosynthetic and chemosynthetic production of biomass requires the combination of energy, water, CO_2, and inorganic nutrients, and is not possible when one of these requirements is missing.

The discovery of Redfield (1934) launched a *chicken-and-egg causality controversy* (this expression refers to the logical causality dilemma of which came first: the chicken or the egg?) For the Redfield ratio, the question was: Does N:P in plankton reflects the N:P composition of seawater, or does N:P in seawater reflect the N:P composition of plankton? Redfield concluded: "Whatever its explanation, the correspondence between the quantities of biologically available nitrogen and phosphorus in the sea and the proportions in which they are utilized by the plankton is a phenomenon of the greatest interest." In other words, the similarity between the N:P ratios of phytoplankton and seawater in the World Ocean indicates the existence of a strong interaction between seawater composition and the activity of phytoplankton. In fact, this similarity is explained by the existence of negative feedback loops that involve organisms and stabilize the N and P content of the ocean against perturbations, as explained below (see Sect. 9.3.4). The book of Middelburg (2019) considers different aspects of the chemical composition of marine organic matter. The article of Matsumoto et al. (2020) provides further reading the consequences of observed variations in the C:N:P ratio of phytoplankton and particulate organic matter at the scales of ocean basins.

9.3.2 New and Regenerated Phytoplankton Production, Ammonification, and Nitrification

Before looking into the maintenance of the N:P ratio in the ocean, it is important to understand the roles of the different key chemical elements in the production of phytoplankton biomass. The three most abundant chemical elements in organisms are, in order of decreasing numbers of atoms, H, O and C (Table 6.1). Their values for phytoplankton are (relative to 1 P): 263 H, 110 O and 106 C (see Sect. 9.3.1). These three elements do not generally limit the production of phytoplankton biomass because their sources are abundant in the aquatic environment even when other important chemical elements have been exhausted. Indeed, the H atoms used by phytoplankton photosynthesis (Eqs. 9.7 and 9.8) come from the surrounding water (H_2O), and the C and O atoms from the very abundant dissolved CO_2 (Fig. 9.6b). The main chemical elements that limit the production of phytoplankton biomass in the ocean are N and P (Fig. 9.11), and in some areas Fe. The concentrations of N and P in seawater are much higher than those of Fe, and as a consequence, the N- and P-containing nutrients are called *macronutrients* and the Fe-containing nutrients are termed *micronutrients*.

A large part of the organic matter synthesized by phytoplankton is consumed and recycled by the food web in surface waters, and the nutrients released there are rapidly used again by phytoplankton to produce more organic matter. This rapid recycling of organic matter is a major difference between the oceans and the continents. Indeed, the whole phytoplankton biomass in the World Ocean turns over on average every 3–4 days, whereas on continents, the biomass of plants builds up on average during 10 years before being recycled. Hence, the biomass of marine phytoplankton turns over 1,000 times faster than that of terrestrial plants.

Even if most of the organic matter synthesized by phytoplankton is recycled within the surface waters, about 15% on average sinks down or is otherwise transported to depth. The loss of organic matter from surface waters is often called *export*. Because of the C:N:P relationship between the chemical constituents of the organic matter, the amount of organic carbon that can be exported downwards is limited by the external supply of the most limiting nutrient to surface waters. For example, in areas of the ocean where phytoplankton photosynthesis is limited

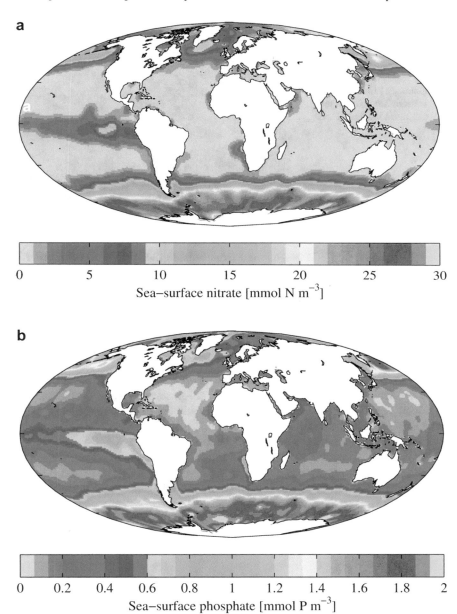

Fig. 9.11 Concentrations of macronutrients in the surface ocean: **a** nitrate, and **b** phosphate. The absolute values of the two nutrients are different, but the same colours in the two maps range from scarcity (violet) to abundance (red). Credits at the end of the chapter

Fig. 9.12 Schematic representation of the cycling of nitrogen in surface and deeper waters, determining new and regenerated phytoplankton production. Explanations are given in the text. Credits at the end of the chapter

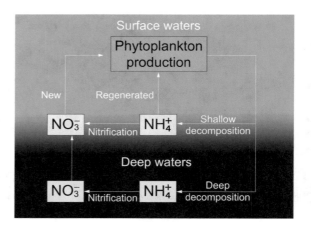

by N, 16 atoms of N must be supplied from outside the surface layer in order to export downwards 106 atoms of carbon in the form of organic matter (Redfield ratio of $C_{106}N_{16}P_1$). The exported organic matter includes sinking phytoplankton cells and organic debris of organisms feeding on phytoplankton, as well as dissolved organic matter. The nutrients supplied to surface waters from outside are called *new nutrients*, and the resulting phytoplankton production is called *new phytoplankton production* (NPP). Conversely, the fraction of the phytoplankton production supported by nutrients recycled in surface waters (called *recycled* or *regenerated nutrients*) is termed *regenerated phytoplankton production* (Fig. 9.12).

At the scale of ocean basins and over one or several years, the amount of organic matter that can be exported downwards without destroying the surface-water ecosystem cannot exceed NPP. Hence at this spatio-temporal scale, the amount of carbon exported is equal to new phytoplankton production.

The transfer of organic matter from surface to depth progressively strips surface waters of their nutrients. This explains the very low levels of N and P in surface waters in most of the World Ocean, where the values of nitrate and phosphate (Fig. 9.11) are generally much lower than the average deep-water concentrations of around 32 and 2 mmol m^{-3}, respectively. At depth, bacterial activity decomposes the organic matter exported from the surface, using a number of metabolic pathways summarized in the following equation, which is not chemically balanced:

$$\text{Organic matter(C:N:P)} + O_2 \rightarrow$$
$$CO_2 + CH_4 + NH_4^+ + NO_x + PO_4^{3-} + H_2O \tag{9.9}$$

where CO_2 results from respiration (Eq. 7 read from right-to-left), CH_4 from fermentation (Eqs. 5.7 and 5.8), NH_4^+ from *ammonification* (explained three paragraphs below), NO_x from nitrification (Eqs. 9.10 and 9.11), and PO_4^{3-} from the decomposition organic matter. The deep decomposition of organic matter builds up concentrations of nutrients in deep waters (Fig. 9.12).

The transfer of organic matter from surface to deep waters and its decomposition there create the following general situation in the ocean: scarcity of nutrients at surface, where irradiance is high enough for phytoplankton photosynthesis; and abundance of nutrients in deep waters, where there is not enough light for photosynthesis. In areas where nutrient-rich deep waters are not efficiently supplied to the surface, NPP and the downward export of organic matter are very low. Conversely, in areas where nutrient-rich deep waters come to the surface, over a few years in upwelling areas or over hundreds of years by the global ocean circulation, NPP and the export of organic matter are high (see Sects. 3.5.3 and 3.5.5).

In marine waters, the production of phytoplankton biomass is most often limited by the availability of N, at least in the short term, so that the concepts of new and regenerated production are exclusively used by reference to the N-nutrients. Figure 9.12 shows that regenerated production in surface waters is fuelled by ammonium (NH_4^+), which is produced by the local rapid bacterial decomposition of organic matter, and new production is fuelled by the upward import of nitrate (NO_3^-) from deeper waters.

At depth, bacterial decomposition of organic matter produces ammonia (NH_3) or NH_4^+ (Eq. 9.9), a process called *ammonification*. The NH_3 or NH_4^+ ions are oxidized to NO_3^- by *nitrifying bacteria* (Fig. 9.12). The process of *nitrification* is performed in two steps, by different bacteria that use different enzymes. Depending on the bacteria performing each step, the chemical reactions may be slightly different. Here is one pair of such reactions:

$$2NH_4^+ + 3O_2 \rightarrow 2NO_2^- + 4H^+ + 2H_2O \tag{9.10}$$

$$2NO_2^- + O_2 \rightarrow 2NO_3^- \tag{9.11}$$

In Eq. 9.10 the NH_4^+ is oxidized to nitrite (NO_2^-), and in Eq. 9.11 NO_2^- is oxidized to nitrate (NO_3^-). The combined two steps can be summarized as follows:

$$NH_4^+ + 2O_2 \rightarrow NO_3^- + 2H^+ + H_2O \tag{9.12}$$

Equation 9.12 shows that the nitrification of one NH_4^+ ion to one NO_3^- ion consumes two molecules of free oxygen (O_2). Nitrification thus consumes large amounts of O_2, which contributes together with O_2 respiration to create the *oxygen minimum* (*OM*) generally observed below the surface layer (see Information Box 9.2).

Nitrate is consumed by phytoplankton in surface waters, and N is exported to depth in the form of organic matter. In the open ocean, the supply of nitrate to surface waters comes from depth, where *ammonification* (production of NH_4^+ by decomposition of organic matter) followed by nitrification takes place (Fig. 9.12). The supply of NO_3^- is different in many coastal areas where a large part of this nutrient is naturally carried from continents by running waters. Nowadays some coastal waters are polluted by large amounts of continental NO_3^- resulting from the overuse of nitrogenous fertilizers by intensive agriculture. In most of the ocean (that is, outside the coastal areas), the amount of organic carbon that can be exported downwards (called *export production*) is equivalent to the supply of NO_3^- to surface waters, in the ratio of 106 atoms of C for 16 atoms of N, or 6.6 atoms of C for 1 atom of N. There, export production is equal to NPP. The article of Ducklow et al. (2001) and the book of Middelburg (2019) provide further reading on the export of organic carbon in the ocean.

Information Box 9.2 Oxygen Minimum, and Ocean Deoxygenation The vertical distribution of dissolved O_2 in the ocean generally shows a minimum at mid-depth, between the well-oxygenated surface waters and deeper waters with higher O_2 concentrations. This phenomenon is called *oxygen minimum*. Oxygen minima can be seen in the north-south vertical distribution of dissolved O_2 in the Pacific Ocean (Fig. 9.13), which shows that there are two types of oxygen minima: on the one hand, the minimum O_2 values in the South Pacific are not very low (left side of Fig. 9.13; located within the upper 1,000 m), whereas on the other hand, the minimum O_2 values in the North Pacific are extremely low (right side of Fig. 9.13; extending down to 2,000 m). What accounts for this difference?

The moderate O_2 minimum observed in the South Pacific Ocean is a general feature of all oceans, called *Oxygen Minimum* or sometimes *Oxygen Minimum Layer*. To avoid any confusion, we call it *oxygen minimum (OM)* in this book. In the ocean, surface waters are generally well oxygenated because they continually exchange O_2 with the atmosphere, and they are the sites of production of organic matter by O_2-producing photosynthetic phytoplankton. Some of this organic matter is exported, directly or through various mechanisms that include the food web, to deeper waters where it is decomposed by bacteria, which consumes O_2 (Eqs. 9.9–9.12). As a consequence, the concentration of dissolved O_2 generally decreases with increasing depth in the surface layer (Fig. 9.13). This decrease continues until a depth of less than 1,000 m, below which O_2 often increases. The decrease in dissolved O_2 with depth followed by an increase creates the OM.

What is the source of the deep O_2 below the OM? The answer to this question lies in the deep ocean circulation (see Sect. 3.5.3). Indeed, the waters that flow deep in the oceans were formed at surface in winter in cold areas where they absorbed large amounts of atmospheric O_2. Cold waters being dense, they sink and carry their absorbed O_2 content with them during their 1,000-year deep journey through ocean basins (Fig. 3.8). Hence the high concentration of dissolved O_2 in deep waters corresponds to events that occurred in faraway areas of deep-water formation centuries in the past, whereas O_2 concentrations in the upper water column reflect local events occurring at timescales of a few months.

The strong O_2 minimum observed in the North Pacific Ocean is called *Oxygen Minimum Zone* or *Oxygen Minimum Layer*. To avoid any confusion, we call it *oxygen minimum zone (OMZ)* in this book. The OMZs are especially developed on the eastern sides of oceans basins (Fig. 9.14; blue dots). These zones of often very low O_2 concentrations are located in highly productive oceanic regions, where the surface layer produces much organic matter whose debris sinks in the underlying waters. There, the combination of enhanced O_2 consumption by respiration of the abundant organic matter and (often) sluggish ocean circulation at mid-depth leads to strong depletion of the dissolved O_2.

The OMZs currently occupy at least 7% of the volume of the oceans, and they are growing fast. This global phenomenon, called *ocean deoxygenation*, is illustrated in Fig. 9.14 (blue dots). The loss of O_2 in open ocean waters, which adds up to the observed decline in oxygen concentrations in coastal waters due to increased anthropogenic nutrient loadings (Fig. 9.14, red dots), is likely related to climate change. If ocean deoxygenation is indeed related to climate change, a number of important factors may be involved, including decreased solubility of oxygen as waters warm up, decreased supply of well-oxygenated surface waters at depth at high latitudes associated with increased ocean stratification, and changes in respiration in the ocean interior. The article of Ramin (2018) provides further reading on the effects of ocean deoxygenation on marine organisms, and the book of Pauly (2019) on the theory of gill-limitation oxygen for fish.

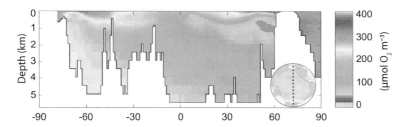

Fig. 9.13 Vertical distribution of the mean annual dissolved O_2 concentration along a north-south transect (right-to-left) at the 180° meridian in the centre of the Pacific Ocean (dotted line in the inset figure). The bottom line corresponds to the depth of the seafloor (down to 5,500 m). Credits at the end of the chapter

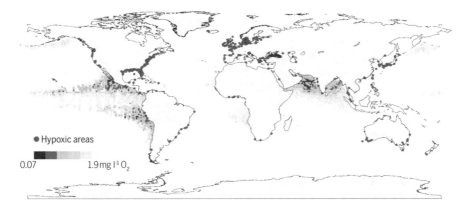

Fig. 9.14 Low and declining oxygen levels in the open ocean and coastal waters. Blue dots: ocean oxygen-minimum zones at 300 m depth (0.07 to 1.9 mg litre^{-1} = 0.002 to 0.05 µmol m^{-3}); red dots: coastal sites where anthropogenic nutrients have exacerbated or caused O_2 declines to less than 2 mg litre^{-1} (less than 0.06 µmol m^{-3}). Credits at the end of the chapter

9.3.3 Limitation of Phytoplankton Production by P

Phytoplankton use phosphorus in the form of dissolved inorganic phosphate (PO_4^{3-}). Because of its generally low concentration in seawater (Fig. 9.11b), the PO_4^{3-} available in surface waters is often taken up rapidly by phytoplankton. Furthermore, the heterotrophic bacteria that use the carbohydrates exuded by phytoplankton to build up their own biomass also need P, and they often successfully compete with phytoplankton for the scarce seawater PO_4^{3-} (see Sect. 9.3.1, above). This further reduces the availability of the PO_4^{3-} macronutrient for phytoplankton.

Dissolved phosphate in the sea has three sources: the erosion and dissolution of sedimentary phosphate rocks, whose P is transported to the ocean by running waters; the remineralization of organic debris leading to decomposition of P-containing organic molecules such as ATP, which releases phosphate ions; and the volcanic recycling of phosphorus-containing rocks. Volcanic recycling may create transient unstable forms of phosphorus, such as the phosphine gas (PH_3) and elemental phosphorus (P_4), but these oxidize rapidly to the PO_4^{3-} form under which they are found in seawater. It follows that the recycling of phosphorus by erosion and volcanoes is very slow, and takes place at the geological timescales of several tens of thousands of years.

Because of its slow recycling, phosphorus is the chemical element that limits phytoplankton production in the ocean in the long term. However, in some ocean basins such as the Mediterranean Sea, phytoplankton production is also limited by the supply of P in the short, annual timescales. This is different from the situation in most of the ocean where phytoplankton production is limited in the short term by the availability of nitrogen.

9.3.4 Limitation of Phytoplankton Production by N, N_2 Fixation, Denitrification, and Anammox

In the Earth's environment, the main chemical form of nitrogen is molecular nitrogen (N_2), which makes up about 78% of the volume of the atmosphere, and accounts for 65% of all N on Earth. The remaining 35% N are mostly stored in marine sediments and sedimentary rocks as fossil organic nitrogen. A very small part of the global N (less than 1%) is in the forms of dissolved NO_3^- in the ocean, organic nitrogen in organisms, and gaseous nitrogen oxides (N_xO) in the atmosphere. The amount of dissolved NO_3^- in the ocean corresponds to less than 0.02% of the atmospheric N_2 pool.

Many researchers working on the reconstruction of the Earth's primitive environment consider that most of the nitrogen on the planet was initially in the form of N_2, and the present pool of NO_3^- would be the result of the oxidation of part of the original N_2. If so, NO_3^- would be a consequence of the photosynthetic production of molecular O_2 (Eq. 9.7), which would have been another major effect of the Earth's biota on the characteristics of the planet.

On current Earth, the main chemical forms of bioassimilable nitrogen by phytoplankton biomass are the dissolved NO_3^- and NH_4^+ ions. Because the presence of NH_4^+ ions inhibits or repress NO_3^- uptake and assimilation, NH_4^+ is taken first by phytoplankton, and NO_3^- uptake only starts when phytoplankton production has brought the concentration of NH_4^+ to a very low level.

In several areas of the ocean, which include large parts of the warm seas, the supply of dissolved NO_3^- and NH_4^+ is very low, permanently or occasionally (Fig. 9.11a). In such waters, some phytoplanktonic cyanobacteria can compensate for the lack of N-nutrients by using the N_2 gas dissolved in the seawater. The process by which marine cyanobacteria fix N from the N_2 gas into organic matter is called *nitrogen fixation*. The N_2-fixing microorganisms include bacteria and archaea, and N_2 fixation occurs both on continents and in oceans (see Sect. 6.1.2). Marine N_2-fixing cyanobacteria are autotrophs, but some N_2-fixing microorganisms are heterotrophs that live in symbiosis with autotrophs to which they supply N. Examples of the latter are N_2-fixng bacteria that live in the root nodules of terrestrial leguminous plants.

Nitrogen fixation involves a number of reactions that occur sequentially, which can be summarized in a simplified way as:

$$N_2 + 8H^+ \rightarrow 2NH_4^+ \tag{9.13}$$

The NH_4^+ ions produced by this reaction are the form of nitrogen used by organisms to synthesize many organic compounds. In the water column, the N_2-fixing cyanobacteria use the NH_4^+ they produce to build their own biomass, and some of this biomass is consumed by heterotrophic grazers. These in turn excrete the N in the form of ammonia (NH_3), which is rapidly oxidized to NO_3^-, and this NO_3^- is used by phytoplankton. The N_2-fixing cyanobacteria also release NH_4^+ in the surrounding water, where it can be used by other phytoplankton.

Assimilating nitrogen from NO_3^- requires less energy than fixing nitrogen from N_2. As a consequence, phytoplankton that use NO_3^- are dominant in surface waters where the flow of NO_3^- and other nutrients does not limit their production of biomass. Conversely, the large areas of the World Ocean where the supply of NO_3^- is very low are dominated by N_2-fixing cyanobacteria.

The origin of biological N_2 fixation is debated in the scientific literature, and some authors think that it goes back more than 3 billion years whereas others think that it is more recent. If N_2 fixation had operated alone during billions of years, it could have potentially transferred all the N_2 of the planet to the NO_3^- pool and the sediments. This is obviously not the case given that 65% of all the N on Earth is in the atmosphere (see above). The explanation for the persistence of the large atmospheric N_2 pool is that, like most biological processes, N_2 fixation is accompanied by two reverse processes that transform the nitrogenous compounds derived from N_2 fixation back to N_2 gas. These processes are called *denitrification* and *anaerobic ammonium oxidation* (anammox), and they are somewhat analogous to O_2 respiration as the reverse process of O_2-producing photosynthesis (Eq. 9.7). However, denitrification and anammox are not the exact opposites of N_2 fixation, and the two processes are carried out by heterotrophic bacteria.

Denitrification operates on the NO_3^- produced by nitrification (Eqs. 9.10–9.12), through a series of steps whose overall result can be summarized as follows:

$$2NO_3^- + 12H^+ \rightarrow N_2 + 6H_2O \tag{9.14}$$

This reaction releases energy, which supports the metabolism of denitrifying bacteria. In the water column, denitrification mostly occurs under condition of low or no O_2, which are primarily found at depth in the oxygen minima (see Information Box 9.2). Similar conditions are also often found in marine sediments on the seafloor. In environments where dissolved O_2 is low or absent, denitrification also provides heterotrophic bacteria with a source of oxydant, and is therefore also called *nitrate respiration* (*NO₃ respiration*). Denitrification is thus another form of respiration, which provides the oxidative chemical potential required for heterotrophic activities in the same way as O_2 respiration. One form of denitrification can occur in the presence of O_2, but the significance of this process in the ocean's water column is not known.

Anammox produces N_2 by a process very different from denitrification, as it combines dissolved NH_4^+ resulting from the bacterial degradation of organic matter (Eq. 9.9) with dissolved nitrite (NO_2^-). The latter can be abundant in the water column at the base of the euphotic (sunlit) layer (*primary nitrite maximum*) and deeper in low-O_2 areas such as the OMs and the OMZs (*secondary nitrite maximum*). The source of the NO_2^- found in the primary maximum is the leakage of NO_2^- from phytoplankton cells and bacterial nitrification (Eq. 9.10), and the source in the secondary maximum is likely microbial respiration of NO_3^- in absence of O_2 called *dissimilatory nitrate reduction* (the end product of this process is NH_4^+, which is different from denitrification where the end product is N_2, Eq. 9.14). The reactions involved in anammox can be summarized as follows:

$$NH_4^+ + NO_2^- \rightarrow N_2 + 2H_2O \tag{9.15}$$

Anammox occurs only at very low O_2 concentrations or in absence of O_2. It is estimated that anammox could be responsible for 30 to 50% of the N_2 released in the ocean.

The previous paragraphs stressed the key role of oxygen minima below the surface layer in the ocean's production of N_2. This is especially true for the OMZs on the eastern sides of ocean basins, where the concentration of O_2 is very low (see Information Box 9.2). In these mid-depth O_2-deficient zones, the nitrogen originally fixed from environmental N_2 by autotrophic cyanobacteria in surface waters is returned as N_2 to the environment by the action of heterotrophic bacteria.

Denitrification and anammox at depth not only lower the concentrations of NO_3^- and NH_4^+, respectively, but these reactions also produce N_2 gas, which is ultimately lost to the atmosphere (Eq. 9.14). This causes a continual loss of nitrogen from the ocean through the decomposition of organic matter and the biochemical alteration of N-nutrients and organic matter. This loss is compensated by N_2 fixation (Eq. 9.13). The long-term balance between the processes of nitrogen loss and fixation contribute to stabilize the overall concentration of the nitrogen contained in both the dissolved N-nutrients and the organic matter.

The biological N-processes described above include ammonification and nitrification (Sect. 9.3.2), and the bacterial fixation of N_2, denitrification, anammox and dissimilatory nitrate reduction. All these processes take place on continents, mostly in soils, and in aquatic environments, both in water and sediments. The combined actions of these processes keep the global N cycle functioning and balanced.

9.3.5   *Limitation of Phytoplankton Production by Fe, and HNLC Regions*

In three regions of the World Ocean, the concentrations of dissolved inorganic nutrients NO_3^- and PO_4^{3-} are abundant in surface water, but the biomass of phytoplankton is low. The locations of these *High Nutrient Low Chlorophyll* (HNLC) regions are shown in Fig. 9.15, on a map where the chlorophyll *a* concentration is a proxy for phytoplankton biomass. The fact that the biomass of phytoplankton in HNLC regions is low despite the high concentrations of bioavailable N and P indicate that some other element limits phytoplankton production there. This limiting element has been identified as iron (Fe).

As explained in Sect. 6.5, in the ancient, O_2-free ocean earlier than 2.4 billion years ago, dissolved iron was very abundant in the forms of elemental iron (Fe) and ferrous iron (Fe^{2+}) ions, but the release of photosynthetic O_2 in the

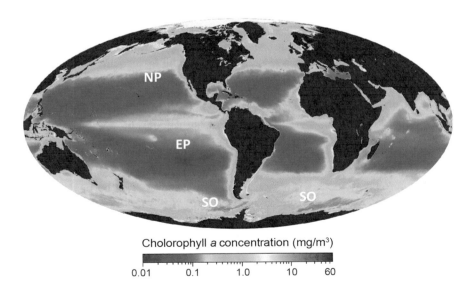

Fig. 9.15 Locations of the three High Nutrient Low Chlorophyll regions in the World Ocean—Northeast Subarctic Pacific (NP), Equatorial Pacific (EP), and Southern Ocean (SO)—represented on a map of the average chlorophyll *a* concentration (NASA ocean colour satellite SeaWIFS). Credits at the end of the chapter

environment favoured the ferric iron form (Fe^{3+}). The combination of dissolved Fe^{2+} and Fe^{3+} ions with water produced ferrous and ferric hydroxides, $Fe(OH)_2$ and $Fe(OH)_3$, respectively. The $Fe(OH)_3$ molecules are insoluble, and the combination of $Fe(OH)_2$ with O_2 produces insoluble Fe_2O_3. In the past, this caused the massive sinking of iron out of the surface waters in the forms of $Fe(OH)_3$ and Fe_2O_3, with the consequence that almost all the dissolved Fe sank to the seafloor. The continuation of these chemical reactions ever since explains the generally low concentrations of dissolved Fe in modern seawater.

In well-oxygenated surface waters, iron is characterized by very low solubility, and thus low bioavailability. However, iron is more soluble when organic Fe-binding *ligands* incorporate it into organic compounds (ligands are organic molecules that attach to metal ions, such as Fe). In various parts on the world, Fe enters the surface ocean from external sources, the most important being desert regions from which the wind transports Fe-containing dust towards the ocean. River input is another important source. In many instances, organic Fe-binding ligands solubilize and stabilize the Fe from these external sources, thus enhancing its bioavailability for phytoplankton. The three HNLC regions cited above are not located in areas downwind from deserts and are far from lands (Fig. 9.15), and thus do not receive Fe from the two major continental sources that exist on the planet, that is, deserts and rivers. Limitation of photosynthesis by the low supply of Fe is the major explanation for the low phytoplankton biomass of HNLC regions.

In surface waters of the ocean, the biological uptake of Fe by phytoplankton and the adsorption of Fe onto organic matter result in the continuous loss of dissolved Fe by the settling of biogenic debris to deeper waters, which contributes to the formation of *Oxygen Minimum Zones*, OMZ (see Information Box 9.2). In OMZs, the concentrations of dissolved Fe may be relatively high because Fe^{2+} can exist under low O_2 conditions, and organic Fe-binding ligands are released in the water by the decomposition of organic matter. Upward diffusion of deep waters or upwelling (see Sect. 3.5.3) brings some of this bioavailable Fe to the surface where it can be used by phytoplankton. In addition, the Fe released from the seafloor by hydrothermal vents contributes significant amounts of dissolved Fe to the surface ocean (see Sect. 3.2.3). The article of Raiswell and Canfield (2012) provides further reading on the cycling of iron in the past and at present.

The N and P nutrients that are not used in the HNLC regions can be transported by ocean currents to areas where iron is sufficient, and be used by phytoplankton there. Hence, the fact that iron limits the magnitude of phytoplankton production in the HNLC regions may not affect the global production of phytoplankton. The article of Underwood (2019) provides further reading on the role of iron in HNLC regions.

9.3.6   The N:P Redfield Ratio: Regulation by Feedback Loops

The N:P ratio of phytoplankton biomass and that of deep-ocean waters are very similar and close to 16:1 (see Sect. 9.3.2, above). The main input of P into the ocean, namely rivers, is under geological control and varies on timescales of

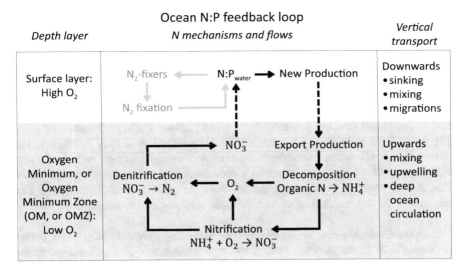

Fig. 9.16 Regulation of both the N:P ratio in seawater and in phytoplankton biomass through negative feedback loops that involve N$_2$ fixation in surface waters (grey) and the N-cycle in the whole water column (black). Solid arrows: biological and chemical flows; dashed arrows: vertical transport of N-containing compounds. Details are given in the text (see Sect. 9.3.6). Credits at the end of the chapter

several tens of thousands of years (see Sect. 9.3.3), whereas the N-nutrients are recycled much more rapidly in the water column (400 and 20 years for NO$_3^-$ and NH$_4^+$, respectively). Hence, the existence of the fixed N:P Redfield ratio implies that environmental variations that could modify this ratio are buffered by mechanisms that involve the most rapidly recycled nutrients, that is, the N-nutrients. The existence of a fixed N:P ratio in both seawater and phytoplankton biomass suggests that this ratio is regulated by negative feedback loops. Although the exact nature of these feedback loops is still under study, it is clear that phytoplankton and bacteria control the feedbacks.

Figure 9.16 illustrates two of the proposed negative feedback loops, which involve N$_2$ fixation by phytoplankton (cyanobacteria) and new phytoplankton production (NPP) in surface waters, and also the biological and chemical oxidation of organic matter at depth. The two feedback loops assume that the production of phytoplankton biomass in surface waters is limited by the availability of nitrogen, and NPP is driven by the supply of NO$_3^-$ (see Sect. 9.3.2, above). The following paragraphs examine the regulation of the N:P ratio in two possible cases, namely increasing and decreasing N relative to P in surface waters, as a consequence of environmental changes.

An *increase in the N:P ratio of dissolved inorganic nutrients in surface waters* has two direct consequences. Firstly, the high availability of dissolved inorganic N in surface waters does not favour the selection of N$_2$-fixing cyanobacteria because their ability to fix nitrogen does not advantage them in a N-rich environment. This reduces N_2 fixation, and thus ultimately the N:P ratio in the water (Eq. 9.13). This is a first negative feedback into the N:P ratio of dissolved nutrients in surface waters. Secondly, the increased dissolved inorganic N causes an increase in NPP and therefore in the downward export of organic matter, which is accompanied by increased microbial decomposition at depth. The decomposition of organic matter in the deep waters consumes O_2 and releases NH_4^+ (Eq. 9.9), whose nitrification produces NO_3^- and further consumes O_2 (Eq. 9.12). The resulting low oxygenated environment favours denitrification, which lowers the concentration of NO_3^- (Eq. 9.14). This reduces the amount of the deep-regenerated NO_3^- brought to the surface by vertical water movements, causing a low supply of new NO_3^- to the surface waters. The latter contributes to a reduction in the surface-water N:P ratio. This second negative feedback, which involves the cycling of nitrogen (Fig. 9.12), contributes to the regulation of the N:P ratio of dissolved nutrients in surface waters.

A *decrease in the N:P ratio of dissolved inorganic nutrients in surface waters* has two direct consequences. Firstly, the low availability of dissolved inorganic N in surface waters selects N_2-fixing cyanobacteria over other groups of phytoplankton because the cyanobacteria can fix their own nitrogen. This increases N_2 fixation, and thus enhances the N:P ratio in the water (Eq. 9.13). This is a first negative feedback that influences the N:P ratio of dissolved nutrients in surface waters. Secondly, there is a decrease in NPP and therefore in the downward export of organic matter, which is accompanied by decreased microbial decomposition at depth. This consumes moderate amounts of O_2 in the deep waters and releases moderate amounts of NH_4^+ (Eq. 9.9), whose nitrification produces some NO_3^- and consumes only moderate amounts of O_2 (Eq. 9.12). In the resulting moderately oxygenated environment, denitrification is not very strong and thus does not affect much the concentration of deep NO_3^- (Eq. 9.12), with the consequence that much of the NO_3^- regenerated at depth is brought to the surface by vertical water movements. The latter contributes to increase the surface-water N:P ratio. This second negative feedback contributes to the regulation of the N:P ratio of dissolved nutrients in surface waters.

Some of the NH_4^+ resulting from the bacterial decomposition of organic matter in deep waters is lost to N_2 by bacterial anammox (Eq. 9.15). This process is not included in Fig. 9.16 for simplicity, but given that anammox is limited by the downward flux of organic matter, its effect on the upward flux of NO_3^- should be similar to that of denitrification (Eq. 9.14).

There are several processes that contribute to the downward export of organic matter produced by phytoplankton in surface waters (dashed arrow new production-to-export production in Fig. 9.16). These include: the sinking of organic particles, which include phytoplankton aggregates and food-web debris; vertical water movements that carry both particulate and dissolved organic matter downwards; and vertical migrations of zooplankton, during which these animals feed at surface, and both excrete NH_4^+ and defecate undigested organic material at depth. The timescales of these processes are rapid, and range between days and months.

The upward supply of new NO_3^- from deep to surface waters is mostly done by three main processes (dashed arrow NO_3^--to-N:P$_{water}$ in Fig. 9.16). The first two are deep vertical mixing and upwelling, which generally take place within one year of the downward export of organic matter. Deep vertical mixing is mostly caused by storms, which occur in different seasons depending on regions, and upwelling is especially active in areas located on the eastern sides of the ocean basins at subtropical and mid-latitudes (see Sect. 3.5.3). The third process is the deep oceanic circulation, which takes place on timescales that range between hundreds of years and a thousand years (see Sect. 3.5.3).

Processes Regulating the N:P Ratio in the Ocean The oceanic N:P ratio is regulated by the combined action of two feedback loops, that is, NO_3^- cycling and N_2 fixation. The timescale of the feedback loop driven by NO_3^- is determined by the cycling of this nutrient on the vertical, which takes place within a thousand years (ocean circulation) or much less in some areas. The timescale of the feedback loop driven by N_2 fixation could be more rapid. However, N_2-fixing cyanobacteria have a high demand for Fe, and given the scarcity of this micronutrient in surface waters, it is possible that the external supply of Fe exerts a strong control on the N_2 fixation feedback in some circumstances. This would depend on how much N_2 fixation is slowed down by Fe-limitation, a point debated among researchers. A strong limitation of N_2 fixation by Fe in surface waters would lengthen the timescale of the corresponding feedback into the N:P ratio.

9.4   Regulation of the Level of Atmospheric Oxygen: The Oxygen Feedback Loops

9.4.1   Free Oxygen in the Atmosphere and the Ocean

A remarkable feature of the Earth System is the fact that the large variations in the volume fraction of atmospheric O_2 that occurred during the last 500 million years remained within the range of 15–35%, with a median value around 20% (Fig. 9.17). As for the environmental CO_2 and the ocean N:P ratio examined above, this narrow range points to the possible action of feedback loops, which may have exerted a long-term regulation on the level of atmospheric O_2. The existence of free O_2 on Earth results from the action of photosynthetic organisms, and this section and the following ones examine how this biological mechanism built up the concentrations of O_2 in the Earth's environment over billions of years, and describes feedback loops that involve the carbon and the sulphur cycles.

The presence of free molecular oxygen (O_2) in the Earth's environment is a very special characteristic of the planet. Indeed, O_2 does not usually exist in the Universe because oxygen is extremely reactive and thus easily combines with other elements

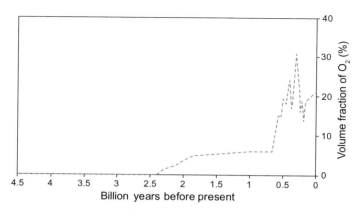

Fig. 9.17 Changes in the volume fraction of oxygen in the atmosphere since 4.5 billion years before present. The O_2 curve is taken from Fig. 8.1b. Credits at the end of the chapter

to form oxides via a chemical reaction called *oxidation*. On Earth, O_2 is constantly resupplied by O_2-producing photosynthesis, of which the O_2 molecule is a by-product (Eq. 9.7 read from left-to-right). A small part of the organic matter resulting from O_2-producing photosynthesis is continually buried on continents and in oceans. The buried organic matter cannot be oxidized because of its lack of contact with oxygen, with the consequence that the rate of burial of organic matter largely determines the long-term changes in the concentration of O_2 in the atmosphere (see Sect. 9.4.5, below). The remaining, non-buried largest part of the organic matter is oxidized, essentially by respiration and forest fires, and the presence of O_2 in the Earth's environment indicates that O_2-producing photosynthesis generally exceeds the oxidation of organic matter and inorganic oxidizable compounds in the Earth's environment at the global scale (Eq. 9.7 read from right-to-left) (see Sect. 9.4.5).

Oxygen is present in the different compartments of the Earth System in very different amounts: 10^{10} Pg in the crust and the mantle, 10 times less (10^9 Pg) in the ocean, still 1,000 times less (10^6 Pg) in the atmosphere, and still more than a 100 times less (less than 10^4 Pg) in the total biomass of organisms. The forms of the oxygen differ among compartments: in the crust and mantle, oxygen exists as oxides of silicon (Si), aluminium (Al), and iron (Fe); in the ocean, it is mostly present as a constitutive part of the water molecules (H_2O), and in a small proportion as dissolved O_2 gas; in the atmosphere, it exists primarily as free O_2 gas, and secondarily as ozone gas, O_3 (see Information Box 1.8); in the biomass of organisms, it is one of the major constituents of organic matter, whose general chemical composition is $C_{106}H_{263}O_{110}N_{16}P_1$ (see Sect. 9.3.1). Water molecules make up the second largest pool of oxygen on Earth, and it is from this pool that photosynthetic organisms extract the hydrogen they require to synthesize organic matter and release the O_2 gas found in the planet's environment. By comparison, the amount of O_2 dissolved in seawater is quite small, that is, about 100 times less than the O_2 gas in the atmosphere (see Sect. 9.4.2). This section and those following focus on the concentration of atmospheric O_2.

Fig. 9.18 Zonal mean dissolved oxygen concentrations in the three ocean basins. In all three oceans, there are O_2 minima at depths of a few hundred metres, and minimum O_2 concentrations have generally lower values in the Northern than in the Southern Hemisphere. Credits at the end of the chapter

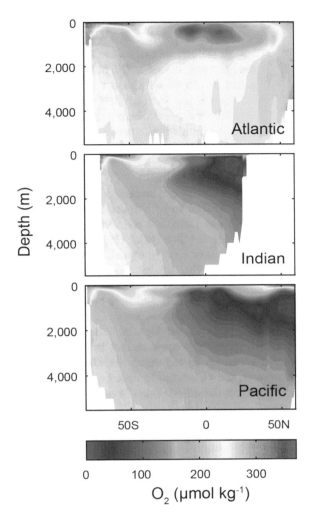

There is a continual exchange of O_2 between the atmosphere and the ocean. As for the other gases, the solubility of O_2 is higher at lower temperature, which explains the generally good oxygenation of deep waters given that these are formed at the surface of cold seas in winter (see Information Box 9.2). In areas or at times of strong photosynthetic activity, the concentration of dissolved O_2 increases in surface waters, and O_2 can be transferred from the surface ocean to the lower atmosphere. Conversely, in areas or at times of strong respiratory activity, the concentration of dissolved O_2 decreases in surface waters, and O_2 can be transferred from the lower atmosphere to the surface ocean. Hence, the surface and deep waters are both generally well oxygenated, and between them most oceans show a minimum in O_2 concentration (Fig. 9.18). The characteristics and causes of the oxygen minima were described above (see Information Box 9.2).

Organisms have complicated relationships with O_2. Indeed, O_2 is a product of biological activity (Eq. 9.7), but it was toxic to most of the early organisms (see Sect. 2.1.5). It follows that when the concentration of O_2 began to increase in the Earth's environment more than 2.5 billion years ago, many of the archaea and bacteria that had developed on a planet without environmental O_2 had to find refuge in habitats without O_2. Such habitats already existed, for example the marine sediments, and others appeared during the course of biological evolution, such as the guts of complex multicellular organisms. Other prokaryotes developed abilities to cope with the increasing levels of environmental O_2, and the adaptations included: O_2 respiration; the advent of eukaryotic cells; and multicellularity in plants, fungi and animals (see Sect. 10.1.6).

Oxygen respiration can be seen as a process in which environmental O_2 oxidizes the molecules taken in by the cells for nourishment, and thus leave alone such key molecules as DNA and RNA which are thus protected from O_2. The *advent of eukaryotic cells* was a major step of biological evolution, which extended considerably the use of oxygen by cells despite the toxicity of this chemical element (see Sect. 2.1.5). The biological innovation was the respiration of oxygen in the cells of eukaryotes by specialized organelles (mitochondria), which consumed the oxygen entering the cells and thus protected to some extent the other organelles and the nucleus from the toxicity of O_2. In prokaryotic cells, the respiration of oxygen takes place in the cytoplasm. The oxygen-respiring prokaryotes and eukaryotes have specialized enzymes that largely destroy the dangerous reactive oxygen species generated by the metabolism of oxygen (see Sect. 2.1.5). In *multicellular eukaryotes*, the cells form tissues, organs and systems (see Information Box 1.1), which protect the cells inside the organisms from direct contact with oxygen and ensure the collection and controlled distribution of oxygen to the cells using it. The eukaryotes evolved in an increasingly O_2-rich environment, and all of them—unicellular (protists) or multicellular (plants and animals)—respire O_2.

9.4.2   The Build-Up of Oxygen in the Atmosphere

Because oxygen is highly reactive, there was no free O_2 in the Earth's environment during the first 2.5 billion years of its history, and there was consequently no O_2 in either the atmosphere or the ocean (Fig. 9.17). The process of O_2-producing photosynthesis (Eq. 9.7 or 9.16 read from left-to-right) was "invented" by the ancestors of the present marine cyanobacteria, probably hundreds of millions of years before O_2 first appeared in the atmosphere. This long delay is explained by the capture of all the photosynthetic O_2 by reactive substances then present in the Earth's environment. Indeed, for hundreds of millions of years, there was massive oxidation of iron ions (Fe^{2+}) to solid iron oxides (Fe_2O_3), of sulphide ions (S^{2-}) to dissolved or solid compounds containing sulphate (SO_4^{2-}), and of methane (CH_4) to CO_2 gas (Eq. 9.6). Progressively, the availability of reactive substances for oxidation decreased, and the biomass of O_2-producing photosynthetic organisms increased, and at a given point the amount of O_2 produced by photosynthesis exceeded the amounts lost to oxidation. From then on, the level of O_2 could increase in the Earth's environment.

Almost all O_2 on Earth comes from O_2-producing photosynthesis. The reaction was summarized above in Eq. 9.7, which is simplified here as follows:

$$CO_2 + H_2O + energy \leftrightharpoons [CH_2O] + O_2 \tag{9.16}$$

where CH_2O represents generic organic matter, and the equation is read from left-to-right. The connections between the cycles of carbon, hydrogen and oxygen in O_2-producing photosynthesis and O_2 respiration are illustrated in Fig. 6.1. The O_2-producing photosynthetic process converts chemically very stable compounds in the Earth's environment—CO_2 and H_2O—into very unstable compounds—[CH_2O] and O_2—using as energy the light from the Sun. The resulting unstable compounds carry the energy captured by photosynthesis. Given that [CH_2O] and O_2 are both chemically unstable, they are very likely to combine together (Eq. 9.16 read from right-to-left), this reaction corresponding to O_2 respiration or combustion (see Information Box 3.4).

The build-up of large amounts of O_2 molecules in the environment since 2.5 billion years ago indicates that a large part of the O_2 released by organisms was not used to oxidize the organic matter formed by various processes, that is, O_2-producing photosynthesis, O_2-less photosynthesis, and chemosynthesis (see Information Box 5.1). In other words, a large part of the organic matter escaped being oxidized by environmental O_2. In addition, some of the O_2 that was not used to oxidize organic matter was not consumed by other oxidable compounds present in the environment. The processes involved in these mechanisms are explained in the following Sects. 9.4.3–9.4.5.

9.4.3   History of Oxygen in the Atmosphere

The history of oxygen in the atmosphere is summarized in Fig. 9.17. However, this figure does not tell the whole story of O_2 on the planet because it does not include the O_2 dissolved in the ocean. In addition, the values in the figure are only semi-quantitative (Information Box 8.1). The history of atmospheric O_2 is one part of the history of the whole atmosphere, whose four phases are described in Chapter 8 (see Sect. 8.1.1 and Table 8.1).

During the *first and second phases of the atmosphere* (that is, earlier than 2.5 billion years before present), there was no free O_2 in the Earth's environment. This does not mean that there was no photosynthetic production of O_2 before 2.5 billion years ago, but it instead means that all the O_2 released by photosynthesis got bound to metals, and this process formed oxides as explained above (see Sect. 9.4.2). This massive oxidation went on during hundreds of millions of years, initially in the ocean, where the O_2 released by marine cyanobacteria oxidized the abundant reactive substances dissolved in seawater and exposed on the seafloor;

and later on continents, when the O_2 that finally escaped from the ocean into the atmosphere oxidized the reactive substances exposed at the surface of emerged lands. The massive formation of iron oxides created the spectacular *banded iron formations*, which are sedimentary rocks rich in silver-to-black magnetite (Fe$_3$O$_4$) and hematite (Fe$_2$O$_3$) that mostly formed before O_2 started to accumulate in the environment 2.4 billion years ago.

The long period before 2.5 billion years ago saw two key biological innovations that had major effects on environmental O_2 as they produced and consumed O_2, respectively (Eq. 9.7). The first innovation was O_2-*producing photosynthesis*, which used solar energy to synthesize organic matter whose C and O atoms came from the abundant CO_2 and H atoms came from the even more abundant liquid water. Simultaneously, this reaction released in the environment the O_2 left over from H_2O, as by-product. The second innovation was O_2 *respiration*, which developed once the concentration of dissolved O_2 grew in the environment. The O_2 respiration released energy from organic matter very efficiently, and re-combined organic H and environmental O_2 into H_2O. The two processes are summarized in Eq. 9.7, read from left-to-right and right-to-left, respectively.

During the *third phase of the atmosphere* (2.5 to 0.5 billion years before present), atmospheric O_2 increased very slowly, and its level reached 5% only around 1 billion years ago. The slow build-up of atmospheric O_2 was due to both the continuing oxidation of reactive environmental substances, whose abundances progressively decreased, and the increasing impact of O_2 respiration. Between 700 and 500 million years ago, atmospheric O_2 rapidly increased up to 15%, for reasons that are not really understood but could be related to the two episodes of snowball Earth that occurred around 710 and 640 million years ago (see Sect. 4.5.4).

The *fourth phase of the atmosphere* (since 500 million years before present, and perhaps earlier) was characterized by the development of the ozone (O_3) layer in the upper atmosphere, which was made possible by the further increase in atmospheric O_2 (Fig. 9.17). The protection provided by the O_3 layer against DNA-damaging solar UV radiation allowed the occupation of emerged lands by organisms and ecosystems, which were derived from aquatic counterparts. This had two major effects on atmospheric O_2. On the one hand, the development of plants on continents increased considerably the biomass of O_2-producing photosynthetic organisms on the planet, and thus the global production of O_2. On the other hand, the presence of organisms on emerged lands increased the amount of organic matter that could be buried, which augmented the amount of photostnthetic O_2 remaining in the atmosphere (see Sect. 9.4.5, below).

The burial of organic matter increased after the occupation of continents because organic matter could then be buried not only in the seafloor as previously, but also in soils, sediments of deltas, and peat that became coal deposits over geological times. There was a massive burial of organic matter during the *Carboniferous* period due to the special conditions that then existed, which caused the atmospheric O_2 level to increase up to 35% by the end of this geological period 300 million years ago (see Sect. 9.4.5).

During the last 500 million years, the atmospheric O_2 concentrations varied between 15 and 35% (Fig. 9.17). The causes of these variations likely included major geological and climatic changes that affected the global photosynthetic production and respiration of O_2, the burial of organic matter, and the oxidation of O_2 by various processes (see Sect. 9.4.5). The median value of these variations was around 20%, which corresponds to 21% O_2 in the current atmosphere. The next section explores negative feedback loops involved in the regulation of the long-term level of atmospheric O_2.

9.4.4 Long-Term Level of Environmental O_2: Regulation by Feedback Loops

The long-term level of atmospheric O_2 has likely been regulated by the two negative feedback loops illustrated in Fig. 9.19, which concern the sulphur and the carbon cycles (left and right parts of Fig. 9.19), respectively. The components of the two feedback loops, detailed below, are homologous: production of O_2, by processes of the sulphur cycle or by photosynthesis; burial of the resulting pyrite minerals or organic carbon; biological or chemical oxidation of these two compounds; and emission of the resulting gases (SO_4^{2-} or CO_2) to the atmosphere and the ocean, where they are oxidized. In Fig. 9.19, the characters and arrows belonging

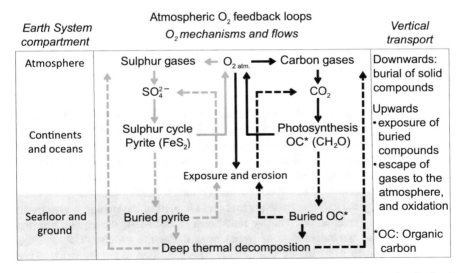

Fig. 9.19 Regulation of the long-term level of oxygen in the atmosphere by negative feedback loops that involve the cycles of sulphur (grey, left part) and carbon (black, right part). Solid arrows: chemical and biological flows of oxygen and sulphur compounds; dashed arrows: vertical transport of gases and solid materials. Details are given in the text (see Sect. 9.4.4). Credits at the end of the chapter

to the cycles of sulphur and carbon (photosynthesis) are in grey and black, respectively, the chemical and biological flows are represented by solid arrows, and the vertical transport mechanisms of gases and solid materials by dashed arrows.

The processes involved in the two feedback loops can be summarized in two chemical equations. The first is the above Eq. 9.16 (repeated here for convenience), which addresses the *carbon cycle*:

$$CO_2 + H_2O \leftrightharpoons CH_2O + O_2 \tag{9.16}$$

where CH_2O represents generic organic carbon. Read from left-to-right, this equation represents the formation of organic carbon by photosynthesis, and the accumulation of O_2 in the environment resulting from the burial of the organic matter (see Sect. 9.4.1). Some of the organic carbon is buried on continents or transported there by plate tectonics, forming fossil sedimentary organic carbon; and some of the organic carbon is buried deep under the ground by subduction into the mantle. Read from right-to-left, the equation represents two processes of alteration of the buried organic carbon: the oxidative erosion on the continents of fossil sedimentary organic carbon (which mostly includes organic carbon dispersed in rocks, and to a much lesser extent petroleum, gaz, coal and peat); and the thermal decomposition of organic carbon in the magma of the mantle.

The resulting carbon gases (such as methane, CH_4) are emitted to the atmosphere and oceans, where they are oxidized into CO_2 (Eq. 9.6).

The second equation addresses the *sulphur cycle*:

$$2Fe_2O_3 + 8SO_4^{2-} + 16H^+ \leftrightharpoons 4FeS_2 + 8H_2O + 15O_2 \tag{9.17}$$

where Fe_2O_3 is iron oxide (a form of *rust*), SO_4^{2-} is the sulphate ion, and FeS_2 is iron sulphide. One of the mineral forms of FeS_2 is called *pyrite*, which is commonly known as *fools gold* because its golden colour sometimes fooled gold-seeking explorers into believing it was the precious metal. Equation 9.17 corresponds to three successive steps of pyrite alteration on continents, regeneration (formation) in the ocean, and alteration at depth. Firstly read from right-to-left, Eq. 9.17 represents the oxidative erosion of pyrite on continents by O_2 dissolved in rainwater. The SO_4^{2-} ions, and thus their oxygen atoms, are carried by running waters to the ocean. Secondly read from left-to-right, Eq. 9.17 represents the regeneration of pyrite (FeS_2) by bacteria from the dissolved SO_4^{2-} in the ocean, where the burial of the regenerated pyrite frees the O_2 molecules from the SO_4^{2-} ions. Some of the buried pyrite is transported to the continents by plate tectonics and some is decomposed at depth by various mechanisms that include melt in the mantle where pyrite is transported by convection. Thirdly read from right-to-left, the equation represents the thermal decomposition of pyrite deep under the ground. The resulting sulphur gases (such as hydrogen sulphide, H_2S) are emitted to the atmosphere and oceans, where they are oxidized into SO_4^{2-}.

The following paragraphs examine the *feedbacks* that regulate the concentration of atmospheric O_2 in two possible cases, that is, an increase in the burial of pyrite

minerals and/or organic carbon in marine sediments and in the ground as a consequence of environmental changes, or a decrease in the burial of these compounds.
The regulation comes from the fact that photosynthesis and the above processes of
the sulphur cycle both release O_2 in the environment (Eqs. 9.16 and 9.17 read from
left-to-right), and the reverse processes both consume O_2 (Eqs. 9.16 and 9.17 read
from right-to-left).

An increase in the burial of the products of organic carbon and pyrite minerals
decreases their oxidation on continents and in the ocean, which favours the accumulation of O_2 in the atmosphere. The buried materials are eventually oxidized
through two different sets of mechanisms. On the one hand, part of the buried
materials are subjected to deep thermal decomposition, and the resulting gases are
emitted in the atmosphere and the ocean. There, the carbon gases resulting from
the decomposition of organic carbon (for example, CH_4) are oxidized to CO_2, and
the sulphur gases resulting from the decomposition of pyrite (for example, H_2S)
are oxidized to SO_4^{2-}. The high level of atmospheric O_2 enhances the oxidation
of these compounds and thus the consumption of O_2, which creates *a first negative feedback* into the level of atmospheric O_2. On the other hand, tectonic uplift
or sea level drop (see a few paragraphs below) eventually bring another part of
the buried materials in contact with the atmosphere, where their erosion releases
the same gases as the above deep decomposition, leading to the same increase in
atmospheric CO_2 and SO_4^{2-}, and thus the same decrease in atmospheric O_2. This
creates *a second negative feedback* into the level of atmospheric O_2. The resulting
CO_2 and SO_4^{2-} are available to fuel again photosynthesis and the processes of the
sulphur cycle, which repeat the feedback loops.

Conversely to the situation described in the previous paragraph, *a decrease in
the burial* of organic carbon and pyrite minerals increases their oxidation on continents and in the ocean, which prevents the accumulation of O_2 in the atmosphere.
The buried materials are eventually oxidized through the same two sets of mechanisms as described in the previous paragraph. The low burial of organic carbon
and pyrite minerals causes a low production of sulphur and carbon gases, and the
low level of atmospheric O_2 slows down the oxidation of these compounds. These
processes promote the accumulation of O_2 in the atmosphere, thus creating *negative feedbacks* into the level of atmospheric O_2.

The following paragraphs examine the timescales of the processes involved
in the regulation of the concentration of atmospheric O_2. Most of these processes
operate in the long term, namely the degradation of organic carbon and pyrite minerals and their burial, and the escape of gases to the atmosphere and the ocean.
However, a difference between the rates of erosion and burial could cause shorter-term changes in atmospheric O_2.

The *degradation* of organic carbon and pyrite minerals (Eqs. 9.16 and 9.17 read
from right-to-left), deep under the ground and at exposed surfaces, involves different types of mechanisms that operate at different timescales. These mechanisms
include: thermal decomposition deep under the ground, which slowly releases

gases later oxidized in the atmosphere; physical and chemical adsorption of O_2 at the surface of exposed materials, which slowly oxidize their compounds; O_2 respiration, by which organisms (mostly bacteria) decompose organic carbon in a controlled manner; and combustion, which destroys organic carbon on land in the same chemical way as respiration but occurs much more rapidly and at higher temperature (see Information Box 3.4). The likelihood of spontaneous combustion is stronger at higher O_2 levels (see Sect. 2.1.5).

In Fig. 9.19, the downward dashed arrows represent the *burial* of both pyrite and organic carbon in the seafloor and in the ground, over long geological times-cales, during which part of the organic carbon is fossilized (see a few paragraphs above). The upward dashed arrows represent the *escape to the atmosphere and the ocean of gases* resulting from thermal decomposition of deeply buried pyrite and organic carbon and from erosion on land of exposed pyrite and organic carbon. The main mechanisms of exposure are sea level drop at times of global cooling (see Sect. 4.5.2) and tectonic uplift, such as mountain building (see Sect. 7.2.3). These two types of processes operate at timescales of hundreds of thousands and millions of years, respectively, and are thought to have contributed to the regu-lation of the variations in the level of atmospheric O_2 during the last 500 million years (Fig. 9.17).

At a shorter timescale, it was observed in gas bubbles trapped in ice cores (see Sect. 9.2.5) that the average level of atmospheric O_2 has decreased by 0.7% over the past 800,000 years, while the average level of atmospheric CO_2 remained quite constant (Fig. 9.8a). One explanation proposed for the observed long-term decrease in atmospheric O_2 lies in the erosion of exposed organic material and pyrite mineral, which would have been higher than their burial in the last 800,000 years and would thus have consumed part of the atmospheric O_2 (Fig. 9.19). During the same period, the tectonic feedback loop illustrated in Fig. 9.6a would have kept constant the long-term average level of atmospheric CO_2. Hence, a relatively short-term *difference between the rates of erosion and burial* could have caused the observed 800,000 year decrease in atmospheric O_2.

9.4.5   Global O_2 Balance, and Major Increases in Atmospheric O_2 Concentration

If all the oxygen produced globally by photosynthesis was consumed by O_2 res-piration, the *global net O_2 production* on Earth would be null. However in real-ity, some fraction of the organic carbon is continually buried in the seafloor and under the ground where it escapes oxidation, so that the global net O_2 production is not null (see Sect. 9.4.2). Presently, between 0.1 and 0.2% of the organic carbon produced annually is buried in sediments (mostly marine). The burial of this very small amount of organic carbon frees a very small amount of O_2, which generates a very small global net O_2 production. However, this tiny value becomes very large over thousands and millions of years because the tiny amount of net O_2 produced annually accrues year after year.

On today's Earth, the net O_2 production is driven by the burial of organic carbon and pyrite (see Sect. 9.4.4), which account for about 60 and 40% of the global net flux of O_2, respectively. The situation was different earlier than 2.4 billion years ago because the ancient ocean was largely devoid of SO_4^{2-} (Eq. 9.17), and the net production of O_2 was then primarily driven by the burial of organic carbon (Eq. 9.16).

The volume fraction of atmospheric O_2 has been varying around a value of 20% since about 300 million years ago, and is considered to be quite constant at 21% today (Fig. 9.17). This long-term steadiness means that the global net O_2 production during the last 300 million years was balanced, overall, by global O_2 losses that took place in the atmosphere and the oceans and on the continents. In the atmosphere, O_2 was consumed by chemical reactions of oxygen with gases released by volcanoes and by geothermal activity. In the oceans, O_2 reacted with gases and minerals released by submarine volcanoes and hydrothermal vents. On the continents, O_2 dissolved in rainwater reacted with minerals during the process of chemical alteration of rocks, creating mineral oxides.

It is generally considered that today's concentration of atmospheric O_2 is steady. However, it actually decreases slightly year after year as the ongoing combustion of fossil organic compounds and other human activities create a small deficit in the Earth's O_2 budget (see Sect. 9.5.1, below). Because of the combustion of fossil organic compounds, the concentration of atmospheric CO_2 has gone from 270 ppmv before the Industrial Revolution to more than 400 ppmv today, which represents a very large increase (about 50%) because this gas makes up only only 0.04% of the atmosphere. In contrast, human activities affected the concentration of O_2 by less than 0.1% due to the high fraction of this gas in the atmosphere (around 21%). The present near constancy of atmospheric O_2 means that there is currently no significant global net gain or net loss of O_2. In other words, the global net O_2 production resulting from the burial of organic carbon and pyrite is entirely lost. It is estimated that the oxidative erosion of minerals on continents is responsible for the loss of about 80 to 90% of the net O_2 production, and the remaining 10–20% is lost by chemical reactions of oxygen with volcanic, geothermal and hydrothermal gases.

During the course of the Earth's history, the concentration of atmospheric O_2 was not always steady, and there were three spectacular increases in atmospheric O_2 in the past. These occurred between 2.4 and 2.2 billion years ago, and around 580 and 300 million years ago (Fig. 9.17). We briefly examine here each of these increases.

The most significant event in the history of Earth's O_2 was the initial rise in atmospheric O_2 between 2.4 and 2.2 billion years ago, which is called the *Great Oxygenation Event* (see Sect. 8.1.2). Despite the adjective "Great", the increase in the concentration of atmospheric O_2 was quite modest, as its volume fraction during the event increased from 0.2 to 2%. Two types of mechanism can explain this initial rise in the O_2 concentration, namely an increase in global net O_2 production or a decrease in global O_2 loss. Some researchers think that the cause was an increase in global net O_2 production, caused by enhanced burial of organic carbon, itself promoted by the growth of continental shelves (which would later develop into continents). Other researchers think that there was instead a decrease in the consumption of O_2 by atmospheric gases.

After the Great Oxygenation Event, the concentration of atmospheric O_2 increased very slowly until a second major increase, during which O_2 went from a few percent to about 15% of the atmospheric gases around 580 million years ago. The causes for this increase are debated among specialists, who invoke either geological or biological mechanisms, which involved or not the enhanced burial of organic carbon. Whatever the causes, this increase in the atmospheric O_2 concentration led to the occupation of continents by complex multicellular organisms (see Sect. 10.1.6).

Atmospheric O_2 was at its highest concentration in the Earth's history about 300 million years ago, when the volume fraction of O_2 reached 35%, that is, almost twice the present value of 21%. This very high O_2 concentration favoured the development of giant insects and amphibians (see Sect. 10.1.9). The high O_2 values of this period were caused by the burial of large amounts of organic carbon on continents. This large burial resulted from the colonization of emerged lands by plants whose lignin—a compound especially abundant in wood and bark—did not easily decompose, which favoured the accumulation of plant biomass underground in the form of coal deposits. Hence the name *Carboniferous* (meaning *coal bearing*) given to this geological period, which spanned the 60 million years from 359 to 299 million years before present; Table 1.3). It has been proposed that the agent responsible for the end of the massive formation of coal was the evolution, about 300 million years ago, of a group of fungi able to break down the lignin. Hence, the agents of the increase in the concentration of atmospheric O_2 and its termination during the Carboniferous period were both biological.

The factors that cause an increase, a decrease or steadiness in the global concentration of atmospheric O_2 are summarized in Fig. 9.20. The *first step* is the net production of O_2, which depends on the rate of burial of organic carbon and pyrite. The rate of burial of organic carbon corresponds to the difference between O_2-producing photosynthesis (by autotrophic organisms) and O_2 respiration (by autotrophic and heterotrophic organisms), and the rate of burial of pyrite corresponds the bacterial regeneration in the ocean of the pyrite chemically oxidized on the continents. The *second step* is the rate of loss of the O_2 produced in the first step, in the atmosphere and the oceans and on the continents. A positive, negative or null difference between net O_2 production and O_2 losses causes the global environmental O_2 to increase, decrease or remain steady, respectively. The book of Canfield (2014) provides further reading on the role of oxygen during the Earth's history.

Biogeochemical processes	Burial	Net O_2 production	O_2 losses	Environmental O_2 concentration
O_2-producing photosynthesis *minus* O_2 respiration by auto- and heterotrophs	→ Organic carbon	+ O_2	Oxidation of gases in the atmosphere, gases and minerals in the ocean, and minerals on the continent	Positive difference → Global O_2 increase
				= Negative difference → Global O_2 decrease
Pyrite oxidation on continents followed by bacterial regeneration in oceans	→ Pyrite	+ O_2		Null difference → Steady global O_2

Fig. 9.20 Balance between the net global production of O_2 resulting from the burial of organic carbon and pyrite (first three columns, light shading) and the global losses of O_2 by oxidative processes (fourth column, intermediate shading). The difference between the two global processes determines whether the global environmental O_2 concentration will increase, decrease or remain steady (last column, heavy shading). Credits at the end of the chapter

> **Processes That Determined the History of Oxygen on Earth** In summary, the history of free oxygen on Earth has been largely determined by three sets of processes. The *first set* included the biological invention of O_2-producing photosynthesis by the ancestors of the present marine cyanobacteria, and the chemical reactions of the resulting O_2 with reactive substances then present in the Earth's environment. The *second set* comprised the geological, chemical and biological processes that drove changes in the burial of organic carbon and pyrite in the seafloor and under the ground, which determined changes in global net O_2 production. The *third set* included the geological and biological processes that caused changes in the losses of O_2 in the atmosphere and ocean and on continents, which brought about increases or decreases in the concentration of atmospheric O_2. The history of O_2 on Earth began with organisms, and these largely determined its course over the next 2.4 billion years and more.

9.5   Interactions Between Feedback Loops and Effects of Forcing Factors

9.5.1   Interactions Between Global Feedback Loops

In this section, we focus on interactions between the negative global feedback loops described above. These interactions involve global pools of key inorganic and organic compounds. They also involve large-scale processes of the Earth System that interact with the pools, and through them with the feedback loops.

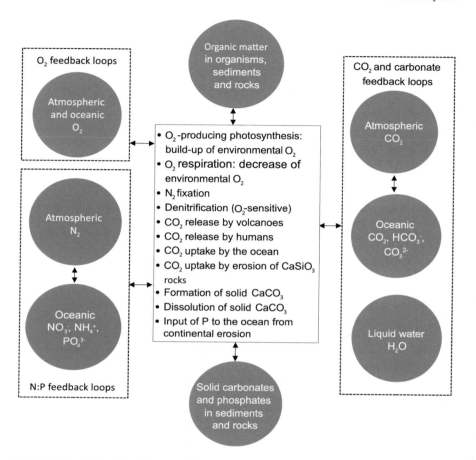

Fig. 9.21 Interactions (double arrows) between three types of components of the Earth System: global pools of key inorganic and organic compounds (eight grey circles); feedback loops that regulate the environmental concentrations of CO_2 and O_2 and the marine N:P ratio (three dashed rectangles); and large-scale processes that interact with the pools, and through them with the feedback loops (central rectangle). Explanations are given in Sect. 9.5.1. Credits at the end of the chapter

Figure 9.21 illustrates in schematic form various interactions (double arrows) between three types of components of the Earth System. The *first group of components* consists of eight global pools of key inorganic and organic compounds (grey circles): liquid water; oceanic dissolved macronutrients (NO_3^-, NH_4^+, and PO_4^{3-}) and inorganic carbon (CO_2, HCO_3^-, and CO_3^{2-}); atmospheric N_2 and CO_2; atmospheric and oceanic O_2; solid carbonates (mainly $CaCO_3$) and phosphates in sediments and rocks; and organic matter in organisms, sediments and rocks. The *second group of components* consists of the feedback loops described in the previous sections (three dotted rectangles), which regulate the following characteristics: concentrations of atmospheric CO_2 and oceanic carbonates; concentration of environmental (atmospheric and oceanic) O_2; and oceanic N:P ratio. The *third group of components* consists of eleven large-scale processes (central rectangle) that interact with the eight pools, and through them with the three feedback loops: O_2-producing photosynthesis and O_2 respiration; N_2 fixation and N_2 loss by denitrification (an O_2-sensitive process); CO_2 release (volcanic and anthropogenic) and uptake (dissolution in ocean, and erosion of $CaSiO_3$); $CaCO_3$ formation and dissolution; and P input to the ocean (from continental erosion). The following paragraphs detail two major interactions between feedback loops.

The *first interaction* described here is between the feedback loops that regulate environmental O_2 and solid oceanic carbonates (see Sects. 9.4.4 and 9.2.3, respectively). This interaction is expected because O_2 is both a product of O_2-producing photosynthesis and an agent of O_2 respiration, and conversely CO_2 is both a product of O_2 respiration and an agent of O_2-producing photosynthesis (Fig. 6.1), and is also involved in the formation and dissolution of solid carbonates (Fig. 7.7).

An example of the global interaction between O_2 and CO_2 is provided by the ongoing combustion by human societies of fossil organic compounds accumulated in sediments and rocks, whose most spectacular effect is the increase in atmospheric CO_2 that causes the ongoing climate change. However, it is often disregarded that combustion also consumes O_2, as evidenced by rewriting Eq. 9.16 from right-to-left:

$$[CH_2O] + O_2 \leftrightharpoons CO_2 + H_2O + \text{energy} \qquad (9.18)$$

In Eq. 9.18, combustion is represented by $[CH_2O]+O_2$, and the equation shows that the process consumes O_2 (left side) and produces CO_2 (right side). Hence it is expected that the ongoing climate change be accompanied by a decrease in environmental O_2. This is indeed the case, and it has been estimated that the concentration of O_2 in the Earth's atmosphere presently decreases by about 0.002% every year as a consequence of the ever-increasing human combustion of fossil organic compounds and other activities. Some models predict that the annual O_2 deficit will increase in the coming years. Most people are not aware of this aspect of climate change because there is much O_2 in the atmosphere, and the cumulated small yearly decreases in this gas do not affect the climate or ecosystems, contrary to the significant enhancement of the greenhouse effect by the cumulated small yearly increases in atmospheric CO_2.

The inverse relationship between environmental O_2 and CO_2 points to an interaction between the feedback loops that regulate O_2 and carbonates. This interaction becomes clear when looking at key equations of the O_2 and carbonate feedback loops (Eqs. 9.16 and 9.4, respectively):

$$\mathbf{CO_2} + H_2O \leftrightharpoons CH_2O + O_2 \qquad (9.16)$$

$$
\begin{aligned}
&CaCO_{3(\text{solid})} + H^+ \rightleftharpoons \\
&Ca^{2+}_{(\text{aqueous})} + x\mathbf{CO_2}_{(\text{aqueous})} + yHCO^2_{3(\text{aqueous})} + zCO^{2-}_{3(\text{aqueous})}
\end{aligned}
\qquad (9.4)
$$

The equations show that CO_2 (in bold characters) is involved in the two processes, which have the following effects in the environment: on the one hand, the O_2 feedback loop creates an inverse relationship between the changes in global O_2 and global CO_2, and on the other hand, the response of the carbonate feedback loop to a global increase (or decrease) in CO_2 is the dissolution (or formation) of solid carbonates on the seafloor, leading to CO_2 uptake by (or release from) the ocean.

In fact, the coupled effects of the two feedback loops predicted by Eqs. 9.16 and 9.4 are currently occurring in the Earth's environment. Indeed, there is a simultaneous global build-up of CO_2 and global decrease in O_2, as noted in the previous paragraph, and $CaCO_3$ sediments are undergoing dissolution in the western North Atlantic Ocean and at various sites in the Southern Atlantic, Indian and Pacific Oceans (see Sect. 9.2.4). These processes are taking place at the relatively short timescale of less than 1,000 years of the carbonate feedback loop (see Sect. 9.2.4).

The tectonic CO_2 feedback loop also involves CO_2, as shown in Eq. 9.3:

$$CaSiO_{3(solid)} + CO_{2(aqueous)} \rightarrow CaCO_{3(solid)} + SiO_{2(solid)} \qquad (9.3)$$

In Eq. 9.3, the chemical alteration of $CaCO_3$ rocks consumes environmental CO_2 released by volcanoes (in bold characters). Hence, it could be expected that this feedback loop will also interact with the environmental O_2 feedback loop (Eq. 9.16). However, the timescale of the tectonic CO_2 feedback loop is very long, on the order of 400,000 years (Sect. 9.4.4), and the different in timescales could decouple the two feedback loops. This is indeed what was observed for the the past 800,000 years, during which the average level of atmospheric CO_2 remained quite constant (Fig. 9.8a), likely reflecting the action of the the tectonic CO_2 feedback loop, while the average level of atmospheric O_2 decreased by 0.7% (Sect. 9.4.4).

The *second interaction* described here is between the feedback loops that regulate environmental O_2 and the oceanic N:P ratio (see Sects. 9.4.4 and 9.3.6, respectively). It is explained above that some of the organic matter produced by phytoplankton in surface waters is exported, directly or through the food web, to deeper waters. There, the decomposition of organic matter by bacteria consumes O_2 (Eqs. 9.9 to 9.12), which creates a local oxygen minimum in the water column (see Information Box 9.2). In some areas and/or at some times, the local consumption of O_2 exceeds its supply by marine currents, and the concentration of O_2 becomes very low. Low O_2 concentration favours such bacterial processes as denitrification and anammox (Eqs. 9.14 and 9.15, respectively, repeated here for convenience):

$$2NO_3^- + 12H^+ \rightarrow N_2 + 6H_2O \qquad (9.14)$$

$$NH_4^+ + NO_2^- \rightarrow N_2 + 2H_2O \qquad (9.15)$$

Photosynthesis in the open ocean is performed by phytoplankton, and these organisms take up nitrogen and phosrorus in the atomic ratio N:P$= 16:1$, called *Redfield ratio* (Fig. 9.10). The recycling of phosphorus occurs very slowly, at geological timescales of several tens of thousands of years (see Sect. 9.3.3). This means that short-term decreases in the ratio of NO_3^- to PO_4^{3-} in seawater can only be corrected by fixation of atmospheric nitrogen by cyanobacteria. Using N_2 fixation, these organisms rapidly produce the ammonium (NH_4^+; Eq. 9.13 is repeated here for convenience) they use to synthesize organic matter (see Sect. 9.3.4):

$$N_2 + 8H^+ \rightarrow 2NH_4^+ \tag{9.13}$$

It was explained in previous sections that most P in the ocean results from the erosion and dissolution of phosphate rocks, whose P is transported to the ocean by rivers, and because of this, the recycling of phosphorus is very slow (see Sect. 9.3.3). Hence the existence of a fixed N:P ratio implies that environmental variations that could modify this ratio are buffered by reactions that involve the N-nutrients (see Sect. 9.3.6).

Interestingly, the Redfeld ratio feedback loops include reactions 9.13–9.15, which are influenced by changes in the concentration of environmental O_2. Indeed, a decrease in environmental O_2 increases the likelihood of low-O_2 conditions in the water column, and thus the likelihood of high denitrification. This results in a decrease in the NO_3^- to PO_4^{3-} (N:P) ratio in seawater. Conversely, an increase in environmental O_2 decreases the likelihood of low-O_2 conditions in the water column, and thus the likelihood of high denitrification. This could lead to an increase in the NO_3^- to PO_4^{3-} (N:P) ratio in seawater. Hence, the stabilization of environmental O_2 by the long-term O_2 feedback loop tends to stabilize the oceanic N:P ratio to 16:1, through its action on the shorter-term oceanic N:P feedback loop.

In the very long term, it is possible that the billion-year increase in environmental O_2 (Fig. 9.17) has influenced the N:P ratio through biological evolution. Indeed, when the O_2 concentration was nil or much lower than at present, 2 or 1 billion years ago, phytoplankton organisms did likely not require as much phosphorus as today. Indeed, phosphorus is largely used by cells today to build such molecules as: phospholipids, which are major constituents of membranes, which themselves offer protection against oxygen; ATP, which is involved in O_2 respiration; and the nucleic acid RNA, which plays a key role in rebuilding proteins damaged or destroyed by oxygen (see Sect. 2.1.5 and Information Box 8.2). If phytoplankton actually required less phosrorus in the past, their N:P ratio might have been higher than the present 16:1. In such a case, the N:P ratio of cells and thus seawater may have progressively decreased from a value higher than 16:1 around 2 billion years ago to the present Redfield value of 16:1.

This section has shown that the Earth System is strongly self-regulated (see Sect. 9.1.4). The agents of the self-regulation are interacting negative feedback loops, which involve global pools of key inorganic and organic compounds and large-scale processes of the Earth System. The next section examines how these large-scale processes respond to forcing factors and the climate, and also to biological evolution.

9.5.2 Effects of Forcing Factors, the Climate and Biological Evolution

In this section, we come back to the large-scale processes of the Earth System considered in the previous section (central rectangle in Fig. 9.21, and examine their responses to forcing factors, the climate and biological evolution. The forcing factors are geological, biological, and anthropogenic.

Figure 9.22 illustrates in schematic form the effects (arrows) of forcing factors and the climate on the eleven large-scale processes of the Earth System invoked in the previous section. These are listed in the central rectangle of Fig. 9.22. The forcing factors illustrated in Fig. 9.22 are of three types. *Firstly*, there are geological factors (black rectangle *Tectonics*), mostly the release of greenhouse gases (CO_2, SO_2) by volcanoes, and the erosion of $CaSiO_3$ (which takes up CO_2) and sedimentary phosphate rocks (which releases phosphorus). *Secondly*, there are results of biological evolution (three grey rectangles), that is, O_2-producing photosynthesis and the associated expansion of photosynthetic organisms; different types of respiration leading to expansion of O_2-respiring species, and retreat of species that did not respire O_2; and production of carbonate structures. *Thirdly*, there are anthropogenic factors (open rectangle), mostly the ongoing release of greenhouse gases, especially CO_2, to the atmosphere.

In addition (rectangles on the right side of Fig. 9.22), the *climate* is determined by the greenhouse effect, which is influenced by large-scale processes of the Earth System and a number of forcing factors. The latter include: the quantity of solar energy that reaches Earth, which is a function of orbital conditions, for example the effect of the Milankovitch cycles on the Quaternary Ice Age; the

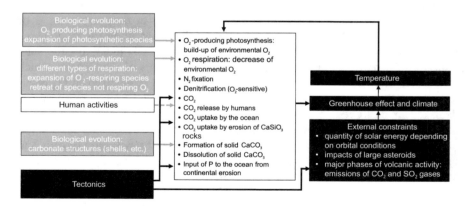

Fig. 9.22 Effects (arrows) of forcing factors (left rectangles) and the climate (right rectangles) on a number of large-scale processes of the Earth System (central rectangle, from Fig. 9.21). The forcing factors are tectonic, evolutionary and anthropogenic (black, grey and white rectangles and corresponding arrows, respectively). Explanations are given in Sect. 9.5.2. Credits at the end of the chapter

occasional impact of large asteroids, for example, the possible role of such an impact in the fifth mass extinction; and major phases of volcanic activity, which can release large amounts of climate-warming CO_2 or climate-cooling SO_2 gases in the atmosphere. The climate sets the temperature of the planet, which affects most of the large-scale processes in the central rectangle given that their rates are generally dependent on temperature.

Some of the factors summarized in Fig. 9.22 are environmental and others are biological. Through their actions on the large-scale processes, all these factors affect the pools of key inorganic and organic compounds in Fig. 9.21, and through them the feedback loops regulating the environmental concentrations of CO_2 and O_2 and the marine N:P ratio. In the next paragraphs, we explain three types of effects on large-scale processes of the Earth System, that is, random and non-random effects of environmental factors, and special effects of biological and anthropogenic factors.

Random effects of environmental factors. Some random astronomical and tectonic events have major effects on the climate, and through it on the large-scale processes of the Earth System. One such event is the impact of a very large asteroid, which causes massive ejection of dust into the atmosphere and widespread wildfires, which may reduce the incoming solar radiation for at least one year, and the resulting large-scale destruction of vegetation can lead to a mass extinction (see Sect. 4.5.3). One example is the asteroid with a diameter of more than 10 km that hit Earth 66 million years ago in the area of the present harbour city of Chicxulub, in Mexico, which may have created a dust cloud that blocked sunlight for up to a year, creating freezing temperatures and inhibiting photosynthesis. The impact of this large asteroid is invoked as the cause of the fifth mass extinction of the Phanerozoic eon, which saw the disappearance of about three-quarters of all species of multicellular organisms, including all dinosaurs except the ancestors of modern birds, and the very diversified and abundant ammonite molluscs in the ocean (see Sect. 4.6.2).

Very large volcanic eruptions, which also occur randomly, can have effects similar to those of hits by large meteoroids (see Sect. 4.5.3). Much more significant are the effects of long phases of massive volcanic eruptions related to the fragmentation of supercontinents, some of which deeply affected the global climate and thus a large number of processes of the Earth System. One example is the fragmentation of supercontinent Rodinia, which was accompanied by very extensive zones of volcanic activity that produced huge amounts of continental lava that solidified into silicate rocks ($CaSiO_3$). The erosion of these rocks over millions of years led to a major reduction of atmospheric CO_2 (Eq. 9.3), and the accompanying greenhouse effect caused a global cooling of the Earth's climate. This initiated the growth of ice sheets and glaciers on land and the ice pack in the ocean, leading to the last snowball Earth episode around 640 million years before present (see Sects. 4.5.4 and 7.4.6).

Non-random effects of environmental factors. The climate is also affected by non-random astronomical cycles that determine the solar radiation received by Earth. The radiation flux at a given point at the top of the atmosphere varies within each day as a result of the rotation of Earth on itself, and seasonally as a result of the motion of the tilted Earth on its orbit around the Sun (Sect. 3.34). In the longer term, the orbital Milankovitch cycles have determined the succession of world-wide glacial and interglacial episodes of the Quaternary period that began 2.6 million years ago (see Sect. 1.3.4). At the Earth's surface, the distribution of the solar radiation flux is more contrasted than at the top of the atmosphere because of geographical differences in both the albedo and the processes involved in the absorption of the solar radiation described in Chapter 3 (see Sect. 3.4.2).

The solar radiation absorbed by Earth fuels the highly complex climate system, which changes under the influence of its own internal dynamics and also because of external forcings. The distribution of the absorbed solar radiation at the surface of Earth is not even, and a major effect of this lack of evenness is the existence of temperature gradients between latitudes in the ocean and the atmosphere, and large seasonal variations in temperature at high latitudes. These latitudinal gradients and the seasonal differences in temperature are among the main causes of the large-scale vertical and horizontal movements in the two fluid envelopes of the Earth System called atmospheric and oceanic circulation (see Sect. 3.5.1). The latter strongly influences both the oceanic carbonate feedback loop and the feedback loops that regulate the oceanic N:P ratio (see Sects. 9.2.4 and 9.3.6).

Effects of biological and anthropogenic factors. There were long-term connections between biological evolution and the large-scale processes of the Earth System. For example, after the evolution of O_2-producing photosynthesis by the ancestors of photosynthetic cyanobacteria, the photosynthetic organisms progressively occupied the surface waters of all aquatic habitats, and their increasingly large biomasses released enough O_2 to start oxygenating the Earth's aquatic, atmospheric and continental environments. Environmental O_2 began to appear between 2.4 and 2.2 billion years before present (Fig. 9.17), that is, hundreds of millions of years after the advent of O_2-producing photosynthesis. Eukaryotic algae, which were initially unicellular and became multicellular later, joined the production of O_2 by aquatic cyanobacteria perhaps 1.2 billion years ago. The development of the ozone (O_3) layer in the upper atmosphere about 600 million years ago allowed the later occupation of emerged lands by organisms, because this layer protected their DNA from damage by harmful solar UV radiation (see Sect. 2.1.5 and Information Box 8.2). This led to the emergence of multicellular plants on continents where their expansion caused a rapid increase in atmospheric O_2, which reached the highest-ever value of 35% about 300 million years ago (Fig. 9.17).

The succession of events described in the previous paragraph indicates that the production of O_2 by aquatic cyanobacteria during hundreds of millions of years, supplemented by that of eukaryotic algae, deeply modified the chemical composition of the atmosphere. The resulting O_3 layer allowed the "greening" of continents by forests, which further increased the environmental O_2 concentration. Over the billions of years that followed the advent of O_2-producing photosynthesis, there was a positive feedback that caused an almost runaway increase in the concentration of environmental O_2 until about 300 million years before present.

Since then, the concentration of atmospheric O_2 went down from 35%, and stabilized at a value close to 20%.

In addition to the above chemical changes to the environment, the oxygenation of the planet created habitats where the newly evolved O_2-respiring organisms could expand and diversify. This was done at the expense of the already-existing organisms that did not respire O_2 or could not tolerate the presence of this gas. The latter organisms continued to develop in O_2-free habitats such as sediments and the guts of some multicellular O_2-respiring organisms (see Sect. 2.1.5). The oxygenation of the planet thus created two broad types of habitats—oxygenated and O_2-free—in which two different types of organisms—respiring O_2 or not—expanded and diversified. These two types of organisms have coexisted on Earth during the last billions of years.

The first O_2-respiring organisms were prokaryotes, after which evolved O_2-respiring eukaryotes, which were initially unicellular and diversified later into multicellular forms. The respiration of O_2 by aquatic and continental organisms is a key process in the feedback loop that presently regulates the concentration of environmental O_2 (Eqs. 9.16 read from right-to-left and 9.18), and O_2 respiration is largely responsible for the fact that the trend to a runaway increase in environmental O_2 described in the previous paragraph gave way to a relatively steady atmospheric O_2 concentration, around the value of 20%, since about 300 millon years (Fig. 9.17).

Similarly since perhaps 3 billion years, some organisms developed, through biological evolution, the ability to form solid carbonates (mostly $CaCO_3$) from substances dissolved in water. This biological process was initiated by microbes, and later involved multicellular organisms (see Sect. 6.4.2). As a result, the formation of solid carbonates on Earth, which was a solely chemical process 4.4 billion years before present, progressively involved organisms and is today about 90% done by unicellular and multicellular organisms. The formation of solid carbonates is a key process in the regulation of long-term variations in the greenhouse effect by the tectonic CO_2 feedback loop (Eq. 9.3).

Human societies also influence some of the large-scale processes of the Earth System. This is especially true of increasing human activities that have released progressively larger amounts CO_2 and other greenhouse gases into the atmosphere since the beginning of the Industrial Age in the late 1700s. The build-up of these gases is causing the ongoing climate change and ocean acidification (see Sects. 11.1.1 and 11.1.3). Contrary to the effects of biological evolution in the previous paragraphs, which have influenced Earth System processes on timescales of tens to hundreds of millions of years, the timescale of the current anthropogenic release of greenhouse gases is very short, that is, at most a few hundred years, although these may have effects lasting thousands of years (see Sects. 11.3.2–11.3.5).

Climate change and ocean acidification are only two among other global anthropogenic changes affecting the Earth System, which also include the loss of biodiversity as well as food and water insecurity (see Sect. 11.1.1). The sequence of events in the last centuries and the last decades that led to the present situation seems to indicates that choices made, consciously or not, by human societies—at the beginning unknowingly, but now with clearer understanding—created positive feedbacks that led to the ongoing runaway climate change, ocean acidification, loss of biodiversity, and food and water insecurity. We are now discovering or predicting—from both daily experience and model forecasts—these runaway effects whose mechanisms have been partly unraveled by researchers in natural and social sciences and philosophers. It is thus time to try turning the positive feedback loops that cause the ongoing runaway effects into negative feedback loops that would bring back stability to the Earth System. The required knowledge, although incomplete, already exists. What is most needed now is the will to take advantage of the existing knowledge to act.

Figure Credits

Fig. 9.1 Original. Figure 9.1 is licensed under CC BY-SA 4.0 by Philippe Bertrand, Louis Legendre and Mohamed Khamla

Fig. 9.2a Original. Figure plotted using the numerical data from page 8 of Johan Mathieu (2017). Figure 9.2a is licensed under CC BY-SA 4.0 by Philippe Bertrand, Louis Legendre and Mohamed Khamla.

Fig. 9.2b Original. Figure 9.2b is licensed under CC BY-SA 4.0 by Philippe Bertrand, Louis Legendre and Mohamed Khamla.

Fig. 9.3 Original. Figure 9.3 is licensed under CC BY-SA 4.0 by Philippe Bertrand, Louis Legendre and Mohamed Khamla.

Fig. 9.4 Original. Figure 9.4 is licensed under CC BY-SA 4.0 by Philippe Bertrand, Louis Legendre and Mohamed Khamla.

Fig. 9.5   https://commons.wikimedia.org/wiki/File:Global_Carbon_Emissions. svg by Autopilot https://en.wikipedia.org/wiki/User:Autopilot?rdfrom=commons: User:Autopilot, used under CC BY-SA 3.0, CC BY-SA 2.5, CC BY-SA 2.0, CC BY-SA 1.0 and GNU FDL FDL. Line styles of the six lines changed, and title of Y-axis modified.

Fig. 9.6a,b Original. Figure 9.6a,b is licensed under CC BY-SA 4.0 by Philippe Bertrand, Louis Legendre and Mohamed Khamla.

Fig. 9.7 Original. Figure 9.7 is licensed under CC BY-SA 4.0 by Philippe Bertrand, Louis Legendre and Mohamed Khamla.

Fig. 9.8a https://www.climate.gov/sites/default/files/paleo_CO2_2018_1500.gif, from https://www.climate.gov/news-features/understanding-climate/climate-change-atmos-pheric-carbon-dioxide by NOAA, in the public domain. Some indications inside the figure removed, numbers on the X-axis changed, and titles of the two axes rewritten

Fig. 9.8b This work, Figure 9.8b is a derivative of https://en.wikipedia.org/wiki/File:Phanerozoic_Carbon_Dioxide.png by Robert A. Rohde https://en.wikipedia.org/wiki/User:Dragons_flight, used under GNU FDL and CC BY-SA 3.0. Figure 9.8b is licensed under GNU FDL and CC BY-SA 3.0 by Mohamed Khamla

Fig. 9.9 This work, Figure 9.9 is a derivative of https://commons.wikimedia.org/wiki/File:Solubility-co2-water.png by The Engineering Toolbox (multiple authors),   https://www.engineeringtoolbox.com/gases-solubility-water-d_1148.html, in the public domain. I, Mohamed Khamla, release this work in the public domain

Fig. 9.10 Adapted from Fig. 1a of Gruber and Deutsch (2014). With permission from Prof. Nicolas Gruber, ETH Zürich, Switzerland

Fig. 9.11a https://commons.wikimedia.org/wiki/File:WOA09_sea-surf_NO3_AYool.png by Plumbago https://en.wikipedia.org/wiki/User:Plumbago, used under CC BY-SA 3.0

Fig. 9.11b https://commons.wikimedia.org/wiki/File:WOA09_sea-surf_PO4_AYool.png by Plumbago https://en.wikipedia.org/wiki/User:Plumbago, used under CC BY-SA 3.0

Fig. 9.12 This work, Figure 9.12, is a derivative of https://commons.wikimedia.org/wiki/File:F_ratio_diagram.gif by Plumbago at English Wikipedia https://en.wikipedia.org/wiki/User:Plumbago, used under GNU FDL and CC BY-SA 3.0. Figure 9.12 is licensed under GNU FDL and CC BY-SA 3.0 by Mohamed Khamla

Fig. 9.13 This work, Figure 9.13, is a derivative of https://commons.wikimedia.org/wiki/File:WOA09_180E_AOU_AYool.png by Plumbago https://en.wikipedia.org/wiki/User:Plumbago, used under CC BY-SA 3.0. Figure 9.13 is licensed under CC BY-SA 3.0 by Mohamed Khamla

Fig. 9.14 Figure of Breitburg et al. (2018). With permission from Prof. Denise Breitburg, Smithsonian Environmental Research Center, USA

Fig. 9.15 Modified after https://earthobservatory.nasa.gov/images/4097/global-chlorophyll by NASA earth observatory, https://eoimages.gsfc.nasa.gov/images/imagerecords/4000/4097/S19972442003273_lrg.jpg, in the public domain

Fig. 9.16 Original. Figure 9.16 is licensed under CC BY-SA 4.0 by Philippe Bertrand, Louis Legendre and Mohamed Khamla

Fig. 9.17 Original. Figure 9.17 is licensed under CC BY-SA 4.0 by Philippe Bertrand, Louis Legendre and Mohamed Khamla

Fig. 9.18 Modified from Fig. 1 of Jaccard et al. (2014). With permissions from Prof. Samuel L. Jaccard a University of Bern, Switzerland, and The Oceanographic Society under its CC BY 4.0

Fig. 9.19 Original. Figure 9.19 is licensed under CC BY-SA 4.0 by Philippe Bertrand, Louis Legendre and Mohamed Khamla

Fig. 9.20 Original. Figure 9.20 is licensed under CC BY-SA 4.0 by Philippe Bertrand, Louis Legendre and Mohamed Khamla

Fig. 9.21 Original. Figure 9.21 is licensed under CC BY-SA 4.0 by Philippe Bertrand, Louis Legendre and Mohamed Khamla

Fig. 9.22 Original. Figure 9.22 is licensed under CC BY-SA 4.0 by Philippe Bertrand, Louis Legendre and Mohamed Khamla

References

Bergman NM, Lenton TM, Watson AJ (2004) COPSE: A new model of biogeochemical cycling over Phanerozoic time. Am J Sci 304:397–437

Berner RA, Kothavala Z (2001) GEOCARB III: A revised model of atmospheric CO_2 over Phanerozoic time. Am J Sci 301:182–204

Breitburg D et al (2018) Declining oxygen in the global ocean and coastal waters. Science 359(6371):eaam7240

Gruber N, Deutsch CA (2014) Redfield's evolving legacy. Nat Geosci 7:853–855

Jaccard SL et al (2014) Ocean (de)oxygenation across the last deglaciation: insights for the future. Oceanography 27(1):26–35. https://doi.org/10.5670/oceanog.2014.05

Lovelock J (1995) The ages of Gaia: a biography of our living Earth. Oxford University Press, Oxford

Mathieu J (2017) La dynamique des populations; Modèle continu, Partie 2: Modèles proies/prédateurs. Available via https://mathemathieu.fr/456, https://www.mathemathieu.fr/

Redfield AC (1934) On the proportions of organic derivatives in sea water and their relation to the composition of plankton. James Johnstone Memorial Volume. Liverpool University Press, Liverpool, pp 176–192

Rothman DH (2001) Atmospheric carbon dioxide levels for the last 500 million years. Proc Natl Acad Sci 99:4167–4171

Royer DL et al (2004) CO_2 as a primary driver of Phanerozoic climate. GSA Today 14:4–10. https://doi.org/10.1130/1052-5173(2004)014<4:CAAPDO>2.0.CO;2

Further Reading

Albertos P, Mareels I (2010) Feedback and control for everyone. Springer, Berlin, Heidelberg

Åström KJ, Murray RJ (2019) Feedback systems. An introduction for scientists and engineers, Second edition (electronic). http://fbsbook.org

Bertrand P (2014) The ocean in the Earth System: Evolution and regulation. In: Monaco A, Prouzet P (eds) Ocean in the Earth System. ISTE, London, and Wiley, Hoboken, NJ, pp 1–53

Canfield DE (2014) Oxygen: a four billion year history. Princeton University Press, Princeton

Ducklow HW, Steinberg DK, Buesseler KO (2001) Upper ocean carbon export and the biological pump. Oceanography 14(4):50–58. https://doi.org/10.5670/oceanog.2001.06

Matsumoto K, Tanioka T, Rickaby R (2020) Linkages Between Dynamic Phytoplankton C:N:P and the Ocean Carbon Cycle Under Climate Change. Oceanography. https://doi.org/10.5670/oceanog.2020.203

Middelburg JJ (2019) Marine carbon biogeochemistry. A primer for Earth System scientists. Springer Briefs in Earth Sciences. Springer, Cham

Pauly D (2019) Gasping fish and panting squids: Oxygen, temperature and the growth of water-breathing animals, 2nd edn. Excellence in Ecology (22), International Ecology Institute, Oldendorf/Luhe, Germany

Raiswell R, Canfield DE (2012) Iron biogeochemical cycle past and present. Geochemical Perspective 1:1–220. https://doi.org/10.7185/geochempersp.1.1

Ramin S (2018) Marine wildlife is starting to suffocate. Knowable Magazine. https://doi.org/10.1146/knowable-041318-080501

Underwood E (2019) The iron ocean. Knowable Magazine. https://doi.org/0.1146/knowable-121919-2

Williams RG, Follows MJ (2011) Ocean dynamics and the carbon cycle: principles and mechanisms. Cambridge University Press, Cambridge

Chapter 10
The Global Earth System: Past and Present

10.1 From Past to Present: The Takeover of Planet Earth by Organisms

10.1.1 The Legend of the Eons

In Chapters 1–8, we found that numerous threads in the fabric of the Earth System tie the organisms and ecosystems to the planet's environment, the Solar System and the whole Universe. These threads were spun by Nature over huge expanses of space and time. In Chapter 9 we also found that negative feedback loops had stabilized key characteristics of the Earth System over its long history. Here, we weave the threads described in the previous chapters into a broad tapestry that depicts the progressive takeover of Earth by its organisms. We call this tapestry *The Legend of the Eons*, a name inspired by Victor Hugo's *La légende des siècles (The Legend of the Ages)*. The tapestry is displayed in Fig. 10.1, where the deep times are to the left and the present time is to the right, and each of the seven horizontal zones (A–G) is identified by its own colour. The different parts of Fig. 10.1 are enlarged in Figs. 10.2–10.7.

In Chapter 1 (see Sect. 1.1.1), we explained that we use *takeover* in this book as a metaphor to stress the roles of organisms in the Earth System, and provided as an example the way Darwin had borrowed in 1859 the word *selection* from animal husbandry and plant breeding to characterize the mechanism of biological evolution, although natural selection is not guided, contrary to animal husbandry and plant breeding where selection is conducted to obtain breeds with some desirable characteristics. Another example is the way astronomers had borrowed the word *attraction* from medicine in the seventeenth century to characterize the interactions among astronomical bodies.

© Springer Nature Switzerland AG 2021
P. Bertrand and L. Legendre, *Earth, Our Living Planet*, The Frontiers Collection,
https://doi.org/10.1007/978-3-030-67773-2_10

Fig. 10.1 *The Legends of the Eons.* The threads described in Chapters 1—9 are woven here into a broad tapestry that depicts the progressive takeover of Earth by its organisms. Past time is to the left, and present time to the right. Vertical arrows: exchanges of materials between layers (for simplicity, these are only illustrated for the most ancient condition of the planet and the most recent). P_{atm}: atmospheric pressure; T: temperature. Zones B to G are further illustrated in Figs. 10.2–10.7. Credits at the end of the chapter

		>4.5 billion	4.5 to 4.4 billion	4.4 to 4.0 billion	4.0 to 3.8 billion	3.8 to 3.5 billion	3.5 to 2.5 billion	2.5 to 2.0 billion	2.0 billion to 540 million	540 to 360 million	360 to 250 million	250 million to present	
A	Years before present	>4.5 billion	4.5 to 4.4 billion	4.4 to 4.0 billion	4.0 to 3.8 billion	3.8 to 3.5 billion	3.5 to 2.5 billion	2.5 to 2.0 billion	2.0 billion to 540 million	540 to 360 million	360 to 250 million	250 million to present	A1
	Geological times	Precambrian Eon: Hadean Era		Precambrian Eon: Archean Era				Precambrian Eon: Proterozoic Era		Phanerozoic Eon			A2
	Phases of the Earth System	Mineral Earth		Cradle Earth						Living Earth			A3
B The Earth System: Components and exchanges		Core (mostly Fe and Ni)						Core (mostly Fe and Ni)					B1
		Mantle						Mantle					B2
			Crust			Oceanic and continental crust						Crust: with land ecosystems	B3
				Chemical formation of carbonate sediments and rocks					Formation of biogenic (carbonate, silica, organic matter) and mineral sediments and rocks				B4
				Ocean (inorganic and organic)		Ocean (early anaerobes)		Ocean (deep waters weakly oxygenated)		Ocean (deep waters progressively well oxygenated)			B5
			Atmosphere phase 1 (H, He, NH₃, CH₄, H₂O)		Atmosphere phase 2			Atmosphere phase 3 (low P_{atm} and T; increasing O₂, minor CO₂)		Atmosphere phase 4 (low pressure, moderate temperature and O₂, O₃ layer, minor CO₂)			B6
C Major external and internal abiotic processes affecting the Earth System	The last giant impact creates the Earth-Moon system												C1
	Increase in Earth's mass; accumulation of heat from gravitational energy		Exchanges of CO₂ between the atmosphere, the ocean and solid carbonates										
			Steady supply of matter (including water and organic molecules) by asteroids and comets										C2
						Production of internal heat by the decay of radioactive heavy chemical elements							C3
						The presence of the ocean allows the very long-term occurrence of plate tectonics, with associated volcanism, formation of the continental crust, mountain uplift, and erosion							C4
						Chemical differentiation of the Earth's core, mantle and crust; retention of the Earth's atmosphere							
D Earth's physical properties; Early acquisition, and effects on the evolution of the planet	Earth structured into stacked layers		Earth's mass (gravity)										D1
	Formation of the ocean		Moon-Earth mass ratio		Stabilization the Earth's axial tilt to a small value								D2
	Onset of the Earth's magnetic field		Earth's axial tilt		Small variations of temperature among seasons								D3
			Earth's rotation		Small variations of temperature within days								D4
			Sun-Earth distance		Physical protection of the Earth's atmosphere, and occurrence of the global liquid-water ocean								D5
			Earth's magnetic field		Protection of the Earth's atmosphere and organisms								D6
E Main steps of biological evolution						Early organisms (anaerobic)	O₂-producing photosynthesis; O₂ respiration	Eukaryotes (mostly aerobic)	Multicellular organisms	Ecosystems on continents		Human activities as a geophysical force	E
F Environmental free oxygen and ecosystems							Increasing O₂, concentration due to O₂-producing photosynthesis and sedimentary storage of organic matter	Increasing O₂ consumption by O₂ respiration and low storage of organic matter		Ozone (O₃) layer and ecosystems on continents		Balanced O₂, release and losses	F1 / F2 / F3
G Self-regulation of the Earth System: feedback loops			Establishment of the tectonic CO₂ feedback loop										G1
			Establishment of the oceanic carbonate feedback loop				Long-term O₂ regulation of atmospheric CO₂, and the greenhouse effect						G2
			Establishment of the oceanic N:P feedback loops				Short-term buffering of the greenhouse effect through the regulation of oceanic carbonates						G3
							Long-term regulation of the N:P Redfield ratio						
							Establishment of the atmospheric oxygen feedback loops					Global oxygen regulation	G4
A1	Years before present	>4.5 billion	4.5 to 4.4 billion	4.4 to 4.0 billion	4.0 to 3.8 billion	3.8 to 2.5 billion		2.5 to 2.0 billion	2.0 billion to 540 million	540 to 360 million	360 to 250 million	250 millions to present	A1

(B6 note, left columns: high to moderate CO₂, P_{atm} and T; R: … ; B: progressive decrease)

In the following eight sections (Sects. 10.1.2–10.1.9), we examine the different zones of Fig. 10.1 in terms of both the co-evolution of the Earth's environment and its organisms and the takeover of the planet by its organisms and ecosystems (Figs. 10.2–10.7). The statements in these eight sections summarize information from previous chapters of the book, reorganized into the conceptual categories and along the historical scheme of Fig. 1.11. Some cross-references to the relevant sections are provided, but not systematically in order to avoid overloading the text with them.

We call our review of the organisms' takeover of the Earth System *The Legend of the Eons*. The work that inspired this name, *The Legend of the Ages*, is a collection of epic poems published between 1859 and 1883 in which Victor Hugo depicts the history and evolution of humanity. The tens of poems of this literary masterpiece begin with a general vision of the history of humanity, and conclude on a dialogue between humans, Earth, the Sun, and the stars. Our book begins with a general concept of the relationships between organisms, Earth, the Sun, and the stars, and concludes in Chapter 11 on the need for humans to become stewards of their unique planet (see Sect. 11.5). In literature, a *legend* is a set of stories about a famous event or person. For example, a central figure in the body of medieval literature and legends called the *Matter of Britain* is King Arthur. In Victor Hugo's *Legend of the Ages*, the central figure is Humankind. In our *Legend of the Eons*, the central figure is Earth. An English edition of Hugo's *La légende des siècles* was published in 2016, and is recommended as further reading.

In the remainder of this chapter, we consider the effects of human activities on the Earth System and their feedbacks into human societies before and during the twentieth century (Sects. 10.2 and 10.3, respectively). Given that the information presented in Sects. 10.1.2–10.1.9 is mostly drawn from previous chapters, some readers may prefer to temporarily skip *The Legend of the Eons* and go immediately to Sects. 10.2 and 10.3, before coming back to Sect. 10.1.

10.1.2   Zone A: Time References

In Fig. 10.1, Zone A provides the time references for the whole figure, from more than 4.5 billion years before present (left) until now (right). It combines the three chronologies of the Earth's history introduced in Chapter 1, which are: the number of years before present, the chronology of geological times, and the three phases of the Earth System.

In the *first chronology* (repeated in Figs. 10.2–10.7), the duration of the different time intervals generally increases from right to left, reflecting the fact that our knowledge of events decreases as we go deeper in the past. Hence the time intervals get generally longer as we go back in time (Fig. 1.1). The *second chronology* consists of the eons and eras detailed in Table 1.3. The boundaries of the different geological time intervals correspond to important environmental and biological events that occurred during the course of Earth's history. These events are depicted in Zones B–G of the Figure.

The *third chronology* reflects the increasing role that organisms and ecosystems played in the Earth System as time progressed (Fig. 1.11). Prior to about 4 billion years ago (Mineral Earth), environmental conditions on the planet gradually stabilized, but these were not yet suitable for organisms. During the following 1.5 billion years (Cradle Earth), organisms progressively occupied the oceans, where they became so numerous that their activity began to affect global environmental conditions. By about 2.5 billion years ago (Living Earth), the activity of photosynthetic organisms had produced so much oxygen that it began to accumulate as free O_2 gas in the ocean and the atmosphere. This major change was the sign that organisms were taking over the Earth's environment.

10.1.3   Zone B: The Earth System at Present and Initially

Zone B of Fig. 10.1 is represented in Fig. 10.2 (where the columns are interchanged with the rows of Fig. 10.1). Figure 10.2 shows the development (from top to bottom) of the main physical and chemical components of the Earth System during the history of the planet. It illustrates key changes that occurred in the six main layers of Earth depicted from the centre of the planet to the mantle and the outer envelopes (crust, sediments and rocks deposited on the crust, liquid water, and gaseous atmosphere), from left (core and mantle) to right (atmosphere) of Fig. 10.2. More information on these envelopes is provided in Table 6.6 and Fig. 6.11. Organisms and ecosystems developed in the Earth's outer envelopes, where they interacted with the remainder of the Earth System. The horizontal arrows indicate exchanges of materials between layers, and although the exchanges are only illustrated for the most ancient condition of the planet and the most recent, the diversity and complexity of these exchanges increased from past to present.

In this Section, we examine successively the present condition of the Earth System, and the state of the planet before and at the time of the formation of the Earth-Moon system 4.5 billion years ago. These two extreme conditions of the Earth System are depicted at the bottom and the top of Fig. 10.2, respectively. The development of the Earth System from its initial condition to the present is described in Sect. 10.1.9. In Fig. 10.2, the number and diversity of horizontal arrows at the bottom and top of the Figure visually stress the magnitude of the changes in the Earth System from relatively simple exchanges of materials between Earth's layers 4.5 billion years ago to a large number of interactions today. These are detailed below.

From 250 million years before present until now. The bottom part of Fig. 10.2 summarizes the current environmental conditions of the planet as follows. The present oceans are brimming with organisms, and the generally well-oxygenated waters from surface to depth are mostly inhabited by O_2-respiring organisms, whereas the sediments below a few centimetres are generally populated by

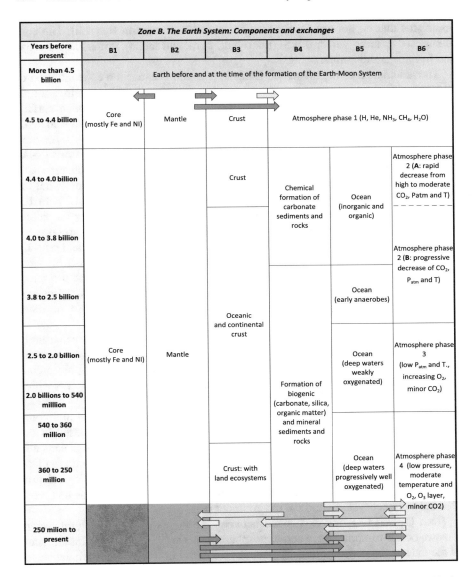

Fig. 10.2 Zone B of *The Legends of the Eons*, where the columns are interchanged with the rows of Fig. 10.1. Past time is at the top, and present time at the bottom. Horizontal arrows: exchanges of materials between layers (for simplicity, these are only illustrated for the most ancient and the most recent conditions of the planet. P_{atm}: atmospheric pressure; T: temperature. The coloured parts of the figure (older than 4.5 billion and younger than 250 million years ago) are examined in Sect. 10.1.2, and the remaining, non-coloured parts of the figure are treated in Sect. 10.1.9 (Fig. 10.7). Credits at the end of the chapter

organisms that do not use O_2 for respiration (see Information Box 5.2). The continents are teeming with wildlife (microbial, flora, and fauna), domesticated crops and livestock, and human beings.

The abundance of organisms on Earth is possible because the concentration of atmospheric carbon dioxide creates a greenhouse effect that keeps temperatures within a range suitable for organisms (see Sect. 2.4.1), that is, a mean surface temperature of about 15 °C and observed extreme values of -93.2 and $+56.7$ °C, respectively. Additional conditions that allow complex organisms to occupy the continents include: a moderate atmospheric pressure; protection from DNA-damaging UV radiation provided by the O_3 layer in the high atmosphere; and a relatively high concentration of atmospheric O_2, but no so high as to be lethal for O_2-respiring organisms or cause spontaneous wildfires (see Sect. 2.1.5). In the oceans, the additional conditions include a range of pH values neither too acidic nor too alkaline. This situation reflects the action of the oxygen and carbon feedback loops that keep the concentrations of atmospheric CO_2 and O_2 and oceanic carbonates at levels suitable for organisms and ecosystems (see Sect. 10.1.8).

The horizontal arrows at the bottom of Fig. 10.2 indicate that there are many different exchanges of materials among the Earth's layers. In the following paragraphs, we distinguish between the exchanges of gases and those of other substances.

Exchanges of gases. For simplicity, we only consider here two gases strongly linked with ecosystems: CO_2 and O_2. The CO_2 molecule consists of two chemical elements, carbon and oxygen atoms, which are both used by autotrophic organisms to synthesize organic matter. The latter also contains hydrogen atoms, which are taken by the autotrophs from H-containing compounds such as H_2O and H_2S (see Information Box 5.1). In O_2-producing photosynthesis, the H atoms come from water molecules (H_2O), and the leftover oxygen atoms are released in the environment as free oxygen (O_2). In return, O_2-respiring organisms in oceans and on continents use environmental O_2 for respiration (oxidation of organic matter), which releases CO_2 and H_2O in the environment. Through these biological processes, CO_2, H_2O and O_2 molecules are continually used by organisms and recycled between ecosystems and the environment (Fig. 6.1).

Both O_2 and CO_2 are exchanged between the atmosphere and natural waters (arrows ocean-to-atmosphere and atmosphere-to-ocean in Fig. 10.2). The direction of the exchanges is determined by both the solubility of each gas, which decreases with increasing temperature, and its partial pressures in the lower atmosphere and the surface ocean. The matter is more complicated for CO_2 because it chemically reacts with H_2O to form HCO_3^- and CO_3^{2-} ions (see Information Box 6.3). Because these ions do not contribute to the partial pressure of the CO_2 gas in water, atmospheric CO_2 easily diffuses from the atmosphere into the ocean, with the consequence that the ocean contains about 50 times more inorganic carbon than the atmosphere (see Sect. 2.1.6). In seawater with a pH between 8.1 and 8.2, the dissolved CO_2 is mostly in the form of HCO_3^- with small proportions of CO_2 and CO_3^{2-} (Fig. 9.7).

Not all the organic matter produced by autotrophs is respired back to CO_2 by heterotrophs. Indeed, a tiny fraction is not respired and progressively accumulates in sediments and soils, where organic carbon is eventually fossilized in rocks and accumulated in fossil fuel deposits (petroleum, natural gas, coal, peat, etc.) (arrow ocean-to-sediment in Fig. 10.2). After millions of years, tectonic processes bring the fossil carbon into contact with free O_2 mostly on continents, causing oxidation of the fossil carbon and release of the resulting CO_2. This slow cycle brings back to the atmosphere and the ocean the CO_2 temporarily stored as fossil carbon (arrows sediment-to-ocean and sediment-to-atmosphere in Fig. 10.2). However, more than 99% of the carbon on Earth is stored in deposits of solid carbonates, a small fraction of which is continually recycled by tectonic processes. As a consequence, the largest natural source of CO_2 to the atmosphere is volcanic activity, which transfers CO_2-rich fluid magma from the mantle to the surface of the seafloor and continents, where the CO_2 gas dissolved in the magma is slowly released into bottom waters and the atmosphere (arrows mantle-to-ocean and mantle-to-atmosphere in Fig. 10.2). The slow and fast carbon cycles are summarized in the previous chapter (see Information Box 9.1).

The massive combustion of organic fossil carbon by humans since the beginning of the Industrial Age in the eighteenth century, and to a lesser extent the calcination of carbonate rocks, releases stored CO_2 much faster than naturally (Fig. 9.5). At present, human societies release as much CO_2 in the atmosphere every year as all volcanoes on Earth normally do during several decades. Human activities strongly accelerate the recycling of fossil carbon, which short-circuits the natural geological cycle.

Exchanges of other substances. When CO_2 dissolves in continental or oceanic waters, these become acidified and thus able to dissolve sedimentary and crustal rocks on continents and on the seafloor (arrows atmosphere-to-sediment and atmosphere-to-crust in Fig. 10.2). The ions resulting from the chemical alteration of rocks are transported to oceans, where they form solid limestone $CaCO_3$ and SiO_2 on the seafloor, which eventually consolidate into limestone rocks and silica minerals (arrow ocean-to-sediment in Fig. 10.2) (see Sect. 7.3.3). The dominant process of $CaCO_3$ formation in the modern ocean is biological calcification, and calcifying organisms produce about 90% the $CaCO_3$ (see Information Box 5.11). The first known forms of biogenic carbonate were microbial stromatolites, which still exist but are largely dominated nowadays by $CaCO_3$ plankton parts, shells, and the spectacular coral reefs (see Sect. 6.4.2). Other organisms use orthosilicic acid dissolved in water ($Si(OH)_4$) to form body parts made of solid silica (SiO_2), which also accumulate in sediments (see Sect. 7.33).

Decreasing pH of seawater may cause part of the solid $CaCO_3$ sediments to dissolve (arrow sediment-to-ocean in Fig. 10.2), which has a negative feedback into the pH of seawater (see Sect. 9.2.4). Modern marine sediments contain two general types of materials: minerals, such as $CaCO_3$, SiO_2, and debris from the erosion of rocks on continents (which are called, in order of decreasing size: *sand, silt* and *clay*); and organic matter (see Sect. 7.33). The latter is derived from debris of aquatic and terrestrial organisms. Organic and mineral particles accumulate not only on the seafloor, but also in soils on continents. Most of the seafloor sediments and resulting rocks are slowly carried by plate tectonics towards subduction zones, where they are subducted into the mantle and become part of the magma (arrow sediment-to-mantle in Fig. 10.2) (see Sect. 7.2.3).

The magma that erupts in continental and submarine volcanoes carries not only CO_2 and other gases (see above), but also the chemical elements that form the silicate rocks (basalt and granite), which make up the bulk of the oceanic and continental crusts (arrow mantle-to-crust in Fig. 10.2). Most of the oceanic volcanoes are located at mid-ocean ridges, and most of the continental volcanoes are found over subduction zones (see Sects. 7.22 and 7.2.3).

The exchanges of gases and other substances recycle chemical elements among the physical reservoirs of the Earth System, namely the mantle, the crustal sediments and rocks, the ocean, and the atmosphere. This way, chemical elements are continually recycled and thus remain available for organisms, even elements whose abundance on the planet is low. This natural recycling is thus a key mechanism for the maintenance of ecosystems on Earth (see Sect. 7.3.1).

The above paragraphs describe the organism-friendly conditions of the Earth System that existed since about 250 million years ago, but it should be understood that the conditions described for this long swath of time are those of the pristine environment before human activities began to deeply transform the planet. One of these major environmental alterations was the extinction of most of the large indigenous fauna that generally accompanied the occupation of new lands by modern humans, *Homo sapiens*, which began 130,000 years ago or later and continued until quite recently. Other major transformations of the planet followed from the advent of agriculture and cities about 10,000 years ago, and the beginning of the Industrial Age in the late 1700s (see Sect. 10.2.2). As a consequence, no environment on present-day Earth is truly pristine.

Before and about 4.5 billion years before present. The top part of Fig. 10.2 shows the conditions of Earth more than 4.5 billion years ago, before the giant collision that formed the Earth-Moon system. In the early Solar System, there were many astronomical bodies of various sizes, which frequently hit each other. Such collisions continue nowadays, but are less frequent now than in the deep past because the smaller bodies got progressively agglomerated into larger ones. Through this process, the large astronomical bodies increasingly incorporated smaller ones into their masses, and as a consequence they encountered less and less material on their orbits (see Sect. 1.5.1). For astronomers, a true planet is an astronomical body that has cleared the neighbourhood of other material around its orbit (Table 1.3).

About 4.5 billion years ago, the astronomical body that would become the present Earth (called proto-Earth or Gaia), had not yet completed the removal of other material around its orbit. At one point, it collided with a large body about the size of planet Mars, known as Theia. This last giant impact created the Earth-Moon system (see Sect. 1.5.1), at which point Earth became a full-fledged planet.

However, even if Earth had cleared its neighbourhood of very large objects, it continued to be hit by smaller bodies, especially during the period of the Late Heavy Bombardment between 4.1 and 3.9 billion years ago. This continued, although at a lower rate, during the remainder of the Earth's long history, as shown by the numerous impact craters at the surface of our planet. The bombardment, although relatively light, continues today as Earth receives between 50,000 and 100,000 tons of solid matter from beyond the atmosphere every year, and could be hit by a very large Near Earth Object someday (see Sects. 1.3.1 and 4.5.3).

It is explained in Chapter 1 (see Sect. 1.5.4) that the last giant impact released a huge amount of energy, which was mostly transformed into heat that caused the initial Earth's temperature to be as high as 2,000 °C. As a consequence, the surface of the planet was then an ocean of hot magma. The exchanges of materials were characterized by planetary differentiation and the formation of an atmosphere (horizontal arrows at the top of Fig. 10.2). Planetary differentiation would progressively lead to the formation of the core, the mantle and the crust (see Sect. 10.1.4). The degassing of volatile compounds from the magma created the first phase of the atmosphere, which mostly consisted hydrogen (H), helium (He), and hydrogen-containing ammonia (NH_3), methane (CH_4) and water vapour (H_2O). This was a truly hellish environment, which explains the name Hadean (meaning *infernal*) given by geologists to this era.

The history of the evolution of planet Earth from the hellish environment that followed the last giant impact 4.5 billion years ago to the current organism-friendly Earth System is full of extraordinary events, which we report here as the *Legend of the Eons*. The following account of this history focuses on the key environmental conditions and events that allowed or led ecosystems to modify the global environment, and thereby help maintain conditions suitable for them. In the next seven sections, we first examine the processes illustrated in Zones C–G of Fig. 10.1 (Sects. 10.1.4–10.18), after which we come back to Zone B and complete the description of the changes that occurred in the Earth System until now (Sect. 10.1.9). The book of Stanley and Luczaj (2014) provides further reading on the history of the Earth System.

10.1.4   *Zone C: Abiotic Processes*

Zone C of Fig. 10.1 is represented in Fig. 10.3 (where the columns are interchanged with the rows of Fig. 10.1). Figure 10.3 addresses major external and internal abiotic processes that have affected the Earth System through its history and continue to currently affect it, from past toward present (from top-to-bottom of the figure). Like a fairy tale, the *Legend of the Eons* begins with "once upon a time".

The Earth-Moon system (Zone C1 in Fig. 10.3). Once upon a time, about 4.6 billion years ago, numerous young planetary bodies of various sizes were born from the accretion of gas, dust particles and ice crystals in the solar nebula.

Zone C. Major external and internal abiotic processes affecting the Earth System			
Years before present	**C1**	**C2**	**C3**	**C4**
More than 4.5 billion	Last giant impact creates the Earth-Moon system	Increase in Earth's mass; accumulation of heat from gravitational energy		
4.5 to 4.4 billion	Earth structured into stacked layers			
	Formation of the ocean	Steady supply of matter (including water and organic molecules) by asteroids and comets		
4.4 to 4.0 billion	Onset of the Earth's magnetic field			
4.0 to 3.8 billion				
3.8 to 2.5 billion				
2.5 to 2.0 billion	Exchanges of CO_2 between the atmosphere, the ocean and solid carbonates		Production of internal heat by the decay of radioactive heavy chemical elements	The presence of the ocean allows the very long-term occurrence of plate tectonics, with associated volcanism, formation of the continental crust, mountain uplift, and erosion
2.0 billion to 540 milllion				
540 to 360 million				
360 to 250 million				
250 million to present				

Fig. 10.3 Zone C of *The Legends of the Eons*, where the columns are interchanged with the rows of Fig. 10.1. Past time is at the top, and present time at the bottom. Credits at the end of the chapter

Because these bodies were all orbiting the Sun, they frequently collided with each other. These impacts caused some of the bodies to merge with others, with the consequence that some gained mass. Through this process, a mother and her daughter, called Gaia and Theia, and Theia's distant cousin Mars progressively grew. Theia and Mars were about the same size, and Gaia (also known as pro-to-Earth) was much bigger than them. One day when Gaia and Theia were rushing together around the Sun, they collided so violently that the core of Theia sunk into the mass of Gaia, and some material originating from the combined two bodies was ejected into orbit. The largest body resulting from this giant impact was planet Earth, and part of the materials ejected in orbit formed the smaller Moon, which was thus the daughter of Gaia and Theia. This created the Earth-Moon system, 4.5 billion years ago (see Sects. 1.3.1 and 1.5.1). Gaia had experienced similar giant impacts in the previous 100 million years, and this one was the last because Earth had then cleared its orbit of other large planetary bodies. Having cleared its orbit, Earth could rightly be called a *planet* (see Information Box 1.3).

The Earth's stacked layers (Zone C1 in Fig. 10.3). The mass of molten magma that made young Earth rapidly differentiated into stacked layers, whose main divisions are the core, the mantle and the crust (Fig. 6.11), and the cooling magma released gases that created the first phase of the atmosphere of the planet (see Sect. 1.5.4). During this planetary differentiation, heavy metals concentrated in the core, and the upper layers (crust and atmosphere) were enriched in lighter chemical elements that included hydrogen, carbon, nitrogen, oxygen, phosphorus and sulphur. When organisms would appear on the planet hundreds of millions of years later, they would use these light elements, whose high abundance in the upper layers of Earth facilitated their emergence and the subsequent success of ecosystems (see Sect. 6.6.2).

The Earth's ocean (Zone C1 in Fig. 10.3). Very old minerals found in Australia indicate that liquid water was present at the surface of Earth as early as 4.4 billion years ago. This initial liquid water would have mainly come from the condensation of water vapour released into the atmosphere by the cooling magma, although most of this vapour had been lost by atmospheric escape of hydrogen to space during the first phase of the atmosphere (see Sect. 5.4.2). After the crust had solidified about 4.4 billion years ago, the release of CO_2 by volcanoes contributed to create the Earth's second phase of the atmosphere, where the concentration of CO_2 was very high. As a consequence of the latter, the early atmospheric pressure was almost 60 times the present value, and the very strong greenhouse effect generated by the high CO_2 concentration generated surface temperatures higher than 200 °C, similar to those that currently exist on planet Venus (Table 5.2). Because atmospheric pressure was very high, surface water could be liquid despite the high temperature (see Sect. 1.5.4), and could thus form a warm ocean.

The Earth's magnetic field (Zone C1 in Fig. 10.3). The differentiation into stacked layers was followed by the establishment of the Earth' magnetic field, perhaps as late as a billion years later. The magnetic field is generated in the deepest parts of the globe, namely the solid inner core and the fluid outer core, which make up a *dynamo* (also called *geodynamo*). The magnetic field in the dynamo is induced by the rotation of an electrically conductive fluid, and currents in the fluid maintain the magnetic field in the long term (see Sect. 8.3.1). On Earth, these conditions are created as follows: the electrically conductive fluid is the iron-rich low-viscosity material of the outer core, as iron conducts electricity; the fluid in the outer core circulates around the inner core because of the Coriolis effect, which is caused by the rotation of the planet (see Sect. 3.5.2); convection in the fluid of the outer core is caused by the heat that escapes from the inner core into the outer core. The dynamo transforms the rotation of Earth and the heat of the inner core into an electric current in the outer core, which in turn induces the magnetic field.

Carbon dioxide (Zone C1 in Fig. 10.3). As soon as the ocean appeared about 4.4 billion years ago, atmospheric CO_2 started to dissolve in the liquid water, where chemical reactions combined the carbon and oxygen atoms with the calcium (or magnesium, iron, or other metals) contained in seafloor silicate rocks to form solid carbonates, mostly calcium carbonate ($CaCO_3$), which accumulated in and on the seafloor. Through this mechanism, the huge concentration of atmospheric CO_2 rapidly decreased, with a parallel decrease in atmospheric pressure and temperature (see Sect. 5.5.4), and relatively low CO_2 concentrations were likely achieved 4.0 billion years ago. After the initial period of massive transfer of CO_2 from the atmosphere to the huge deposits of solid $CaCO_3$ that we presently find on Earth (the initial $CaCO_3$ was replaced many times), exchanges of carbon between the atmosphere and the ocean on the one hand, and the ocean and $CaCO_3$ on the other hand contributed to regulate the concentration of atmospheric CO_2 by the formation or dissolution of solid carbonates (see Sects. 9.2.4 and 10.1.8).

The Earth's mass and initial heat (Zones C2–C4 in Fig. 10.3). Before the last giant impact that created the Earth-Moon system, proto-Earth had been hit by many astronomical bodies whose orbits had intersected its path. The materials supplied by these bodies increased the mass of proto-Earth. In addition, the impacts, including the last giant one, released heat (from gravitational energy) whose accumulation provided the initial heat of Earth (see Sect. 1.5.4). Afterwards, countless impacts of asteroids and comets on Earth continued to progressively increase the mass of the planet.

Asteroids and comets (Zone C2 in Fig. 10.3). The continual bombardment of young Earth by asteroids and comets until 3.9 billion years ago (end of the Late Heavy Bombardment) brought to the planet a steady supply of water and organic molecules from far away in the Solar System.. It is generally thought that these astronomical bodies provided most of the Earth's water, and it is possible that some of the organic molecules they contained contributed to the emergence of organisms on Earth about 3.8 million years ago (see Sect. 1.2.2). Within 100 million years of the formation of the Earth-Moon system, Earth had an ocean, which was a very rapid process considering the hellish conditions that had prevailed at the surface of the planet following the Gaia-Theia collision 4.5 billion years ago.

Decay of radioactive elements (Zone C3 in Fig. 10.3). The Earth's initial heat, described above, accounts for only a small fraction of the present internal heat of the planet. Additional internal heat came from the decay of heavy radioactive elements, which are mainly located in the core and the mantle and account for up to 90% of the Earth's internal (geothermal) heat. The latter is the driving force of plate tectonics, which is a key characteristic of Earth and may be unique in the Solar System (see Sect. 3.2.3). The heavy chemical elements that exist on the planet were formed in the final, gigantic explosions of massive stars located very far away from the Solar System billions of years in the past. This shows that the dynamics of Earth is tightly connected with the remainder of the Universe, with the consequence that the planet is truly a daughter of Stars (see Sect. 1.4.2).

Plate tectonics (Zone C4 in Fig. 10.3). It is thought that the existence of a large volume of surface liquid water is a condition for having plate tectonics on an astronomical body (see Sect. 7.2.5). Hence plate tectonics may have started on Earth as soon as the creation of the ocean, but some researchers think that it began only much later. This was accompanied by volcanism, the formation of continental crust, mountain uplift, and erosion (see Sect. 7.2.1).

10.1.5   Zone D: Earth's Physical Properties

Zone D of Fig. 10.1 is represented in Fig. 10.4 (where the columns are inter-changed with the rows of Fig. 10.1). Figure 10.4 stresses the key roles of six physical properties of the planet in the evolution of the Earth System.

The Earth's mass (gravity): chemical differentiation and retention of the atmosphere (Zone D1 in Fig. 10.4). During the early history of Earth, the increase in gravity resulting from the increasing mass of the planet caused a major rise in the heat content of the planet's interior, which was called *initial heat* above (see Sect. 10.1.4). The action of gravity also caused the heavier chemical elements to sink toward the core and the lighter elements to rise up. This had the consequence that heavy metals concentrated in the Earth's core and the lighter chemical elements moved to the crust and the atmosphere, where they were later available to organisms (see Sect. 10.1.4). Gravity also contributed to retention of the Earth's atmosphere. This is different from what happened on planet Mars, which was likely similar to Earth during its first billion years, after which it lost most of its atmosphere because its mass (one tenth that of Earth) was too small to retain an atmosphere in the long term.

The Moon-Earth mass ratio: small axial tilt (Zone D2 in Fig. 10.4). The Earth-Moon system is characterized by the largest moon:planet mass ratio in all the Solar System (Moon:Earth mass ratio $= 1:81$). This indicates the Moon is an exceptionally large satellite relative to the size of its planet. The presence of the large Moon had the major effect of stabilizing the Earth's axial tilt (angle between the planet's axis of rotation and the plane of its orbit around the Sun) to a small value, which had remained quite constant for perhaps billions of years (see Sect. 1.3.2).

The Earth's small axial tilt: seasonal temperature variations (Zone D3 in Fig. 10.4). The axial tilt of Earth is small, and its present value is $23.4°$ (Table 1.1). Given that the tilt determines the magnitude of the seasonal

Zone D. Earth's physical properties: Early acquisition, and effects on the evolution of the planet						
Years before present	**D1**	**D2**	**D3**	**D4**	**D5**	**D6**
More than 4.5 billion						
4.5 to 4.4 billion	Earth's mass (gravity)	Moon:Earth mass ratio	Earth's axial tilt	Earth's rotation	Sun-Earth distance	Earth's magnetic field
4.4 to 4.0 billion						
4.0 to 3.8 billion						
3.8 to 2.5 billion						
2.5 to 2.0 billion	Chemical differenciation of the Earth's core, mantle and crust; retention of the Earth's atmosphere	Stabilization the Earth's axial tilt to a small value	Small variations of temperature among seasons	Small variations of temperature within days	Physical protection of the Earth's atmosphere, and occurrence of the global liquid-water ocean	Protection of the Earth's atmosphere and organisms
2.0 billion to 540 milllion						
540 to 360 million						
360 to 250 million						
250 million to present						

Fig. 10.4 Zone D of *The Legends of the Eons*, where the columns are interchanged with the rows of Fig. 10.1. Past time is at the top, and present time at the bottom. Credits at the end of the chapter

differences in temperature on Earth, its small value contributed to maintain the temperature at the surface of the planet within a range where most of the water was continually in liquid state (except in polar regions where water may be frozen seasonally or for long periods). This was a key condition for the durable establishment of organisms and the success of ecosystems on Earth. Overall, the range of surface temperatures allowed the build-up of large biomasses and the development of ecosystems over most of the planet (see Sect. 1.3.3).

The Earth's rapid rotation: diurnal temperature variations (Zone D4 in Fig. 10.4). A major characteristic of Earth is its rapid rotation, which causes the shortest day-and-night cycle among the four rocky planets, that is, Mercury, Venus, Earth and Mars (see Sect. 1.3.2). As a result, solar heat is distributed rapidly and quite uniformly around the planet. The diurnal variation in solar radiation is different with latitudes and seasons, but it is such that temperature on most of the planet is suitable for organisms year-round, except at very high latitudes in polar regions (see Sect. 3.3.4).

The Earth's safe distance from the Sun: liquid water and protection of the atmosphere (Zone D5 in Fig. 10.4). The distance of Earth from the Sun is about twice that of Venus and half that of Mars (Table 1.1). On a planetary body close to the Sun such as Mercury, the combination of solar wind (see Information Box 2.6) and photon buffeting prevents the retention of an atmosphere. The same is true of Solar System bodies with a small mass, such as Mars, which cannot retain an atmosphere over the long term because of their low gravitation. Earth could retain a significant part of its atmosphere over its first 4.6 billion years because it was at a safe distance from the Sun and had a relatively large mass (see Sect. 2.3.4).

On current Earth, the moderate atmospheric pressure determines both the freezing and vaporization points of seawater (-2 and $+101$ °C) and the insulation provided by the atmosphere keeps a mild temperature at the surface of the planet (present average of $+15$ °C). Because of these values, water is in liquid phase on most of the surface of the planet (see Sect. 5.6.2). Before the formation of the liquid-water ocean, during the Earth's first phase of the atmosphere, most of the Earth's water was lost to the outer space (see Sect. 10.1.4). The formation of the global ocean, perhaps as early as 4.4 billion years before present, moved water from the gaseous phase where its hydrogen could escape to the outer space, to the liquid phase in the ocean where it was sheltered from escape (see Sect. 8.4.1). Given that liquid water is a key condition for organisms on Earth, the above physical characteristics of Earth were crucial for the durable success of organisms on the planet (see Sect. 5.1.1).

The Earth's magnetic field: protection of the atmosphere and organisms (Zone D6 in Fig. 10.4). There was no magnetic field during the time of the Earth's first phase of the atmosphere, which was lost (see Sect. 8.1.1). It is thought that the Earth's magnetic field began to protect the atmosphere by preventing it from being stripped away by the solar wind during the second phase of the atmosphere. The atmosphere kept the surface water in liquid state, which was essential to organisms, and it also protected the existence of the vast oceans, from which organisms launched their takeover of the planet (see Sect. 8.4.1). Since the development of the third phase of the atmosphere, organisms at the surface of Earth have been largely protected from direct hits by cosmic rays in two ways: first, the magnetic field shielded the atmosphere from most cosmic rays; and second, it protected the atmosphere whose thickness absorbed most of the cosmic radiation that passed the magnetic shield. In addition, organisms on continents and in surface waters largely escaped critical UV damage to their DNA because the protective O_3 layer is continually re-formed by the action of solar UV in the atmosphere, which is protected by the magnetic field (see Sect. 8.4.2).

10.1.6   Zone E: Biological Evolution

Zone E of Fig. 10.1 is represented in Fig. 10.5 (where the columns are interchanged with the rows of Fig. 10.1). Figure 10.5 reviews six crucial steps in biological evolution that played key roles in the takeover of Earth by its organisms.

Early organisms did not use O_2 for respiration, and were thus *anaerobic* (see Information Box 5.2). In this book, we use the conservative value of 3.8 billion years before present for the first occurrence of organisms on Earth, but there are indications that these could date back to almost 4 billion years. Indeed, it is possible that life forms had appeared on Earth a first time before the episode of the Late Heavy Bombardment (4.1 to 3.9 billion years ago), disappeared during it, and appeared a second time afterwards (see Sect. 1.5.4). It was even hypothesised that some of the early life forms had survived the Late Heavy Bombardment, and were thus the ancestors of today's organisms.

There was no free oxygen in the environment of early Earth. Indeed, because of the high chemical reactivity of O_2, all atoms of oxygen were then bound in oxides including molecules of water, H_2O (see Information Box 1.8). The early organisms not only did not need oxygen, but an oxygenated environment would have been lethal for most or all of them. During at least two billion years, all organisms used chemosynthesis to obtain the energy they required for their maintenance, growth and reproduction. Using this biological process, organisms drove chemical reactions of compounds present in the Earth's environment (see Information Box 5.1). The early advent of organisms on Earth was the first step in the takeover of the planet by ecosystems.

Photosynthesis. All organisms on Earth were aquatic until about 0.5 billion years ago. The dominant source of energy at the Earth's surface is solar radiation, and it is thus easy to understand that natural selection favoured organisms that developed the ability to use the energy carried by solar photons. They did it through the process of photosynthesis. The prevalent form of photosynthesis nowadays uses the hydrogen atoms (H) of water molecules (H_2O) to synthesize organic matter, and releases the excess oxygen (O_2) in the environment. However, the initial form of photosynthesis obtained the H atoms from other compounds than water (such as H_2S), and thus did not release O_2 in the environment (see Information Box 5.1).

The evolutionary success of photosynthetic organisms was due to the fact that solar energy is widely available and very abundant at the Earth's surface. A caveat is the fact that light does not penetrate deep in water, with the consequence that autotrophic photosynthetic organisms could only grow in shallow or

Zone E. Biological evolution	
Years before present	E
More than 4.5 billion	
4.5 to 4.4 billion	
4.4 to 4.0 billion	
4.0 to 3.8 billion	
3.8 to 2.5 billion	Early organisms (anaerobic)
	O_2-producing photosynthesis; O_2 respiration
2.5 to 2.0 billion	Eukaryotes (mostly aerobic)
2.0 billion to 540 million	Multicellular aerobic organisms
540 to 360 million	
360 to 250 million	Ecosystems on continents
250 million to present	Human activities as a geophysical force

Zone F. Environmental free oxygen and ecosystems			
Years before present	F1	F2	F3
More than 4.5 billion			
4.5 to 4.4 billion			
4.4 to 4.0 billion			
4.0 to 3.8 billion			
3.8 to 2.5 billion			
2.5 to 2.0 billion	Increasing O_2 concentration due to O_2-producing photosynthesis and sedimentary storage of organic matter	Increasing O_2 consumption by O_2 respiration and low storage of organic matter	
2.0 billion to 540 million			
540 to 360 million			Ozone (O_3) layer and ecosystems on continents
360 to 250 million			
250 million to present	Balanced O2 release and losses		

Fig. 10.5 Zones E and F of *The Legends of the Eons*, where the columns are interchanged with the rows of Fig. 10.1. Past time is at the top, and present time at the bottom. Credits at the end of the chapter

surface waters, typically down to a depth that did not exceed 100 m and was often much smaller depending on water turbidity. This remains true for modern phytoplankton, seaweeds and seagrass. In deeper waters and mostly in the sediment, autotrophic organisms were (and still are) chemosynthetic. The heterotrophs at all depths derive their energy and materials from the photosynthetic or chemosynthetic autotrophs (see Information Box 4.1).

Most bacteria are heterotrophic, but some are autotrophic. Among the latter, all cyanobacteria perform photosynthesis. During the course of biological evolution at some point prior to 2.5 billion years ago, some cyanobacteria developed the O_2-producing form of photosynthesis, which had the advantage of using the infinitely abundant H_2O instead of depending on the availability of such chemical compounds as H_2S (see Information Box 5.1). Cyanobacteria are the only photosynthetic prokaryotes that produce O_2, meaning that all other photosynthetic bacteria and archaea use the form of photosynthesis that does not produce O_2. Another important characteristic of some cyanobacteria is their ability to use the N_2 gas to obtain the nitrogen they need to synthesize their amino acids, proteins and nucleic acids (see Information Box 5.1). The evolution of new forms of autotrophy was not accompanied by the disappearance of the forms that already existed, and the three forms of autotrophy (that is, chemosynthetic autotrophy and the two types of photosynthesis) efficiently co-exist on Earth today. Chemosynthesis and O_2-less photosynthesis are found among the archaea and bacteria, whereas O_2-producing photosynthesis is only performed by cyanobacteria and autotrophic eukaryotes, that is, algae and plants. The advent of photosynthesis was a key step in the takeover of the planet by ecosystems.

Aerobic respiration or *O_2 respiration*. It took hundreds of millions of years before the biologically produced O_2 started to accumulate in the environment, for reasons examined below (see Sect. 10.1.7). After this long period, there was a progressive increase in the concentration of dissolved O_2 in the water surrounding the organisms. As the O_2 concentration built up in natural waters, part of it degassed into the atmosphere, which caused a slow increase in the concentration of atmospheric O_2 over a very long time. The release of O_2 in natural waters caused a major global ecological crisis, because many organisms could not withstand O_2 at any concentration, this gas being a deadly poison for them (see Sect. 2.1.5). The crisis, which occurred between 2.4 and 2.0 billion years ago, is called the *Great Oxygenation Event* (see Sect. 8.1.2).

The presence of O_2 in aquatic environments led to the development of organisms with the ability to use O_2 for respiration. The O_2 respiration has greater energy efficiency than the pathways that do not use O_2 (that is, *anaerobic respiration* and *fermentation*), and this largely explains the subsequent evolutionary success of O_2-respiring organisms (see Information Box 5.2). It is debated among researchers whether O_2 respiration evolved within the context of the increasing environmental O_2, or if pre-existed this increase. The O_2 respiration appeared first in prokaryotes, which passed it on to the eukaryotes when these appeared later (the emergence of eukaryotes is examined a few paragraphs below). Nowadays, some groups of prokaryotes and most eukaryotes use O_2 respiration, but some eukaryotic cells can turn to anaerobic respiration in absence of O_2, and a few species of eukaryotes live in environments without O_2.

On present Earth, the O_2-respiring organisms occupy well-oxygenated aquatic and continental environments, and the organisms that do not use O_2 thrive in environments with low or null O_2 concentration, such as sediments and the guts of many multicellular organisms. There are organisms of the two types among the autotrophs, the heterotrophs and the mixotrophs (the latter are organisms that combine both autotrophy and heterotrophy).

The Proterozoic era, which began 2.5 billion years ago, was marked by the accumulation of O_2 in the Earth's atmosphere. The increase in environmental O_2 was accompanied by a parallel storage of organic matter in sediments, through a process explained in Zone F below (see Sect. 10.1.7). The combination of these two processes deeply modified the atmosphere and the sediments, and made ecosystems major players in the Earth System.

The eukaryotes. Until about 2.0 and even perhaps 1.5 billion years ago, all organisms on Earth were archaea or bacteria (see Information Box 3.1). The organisms in these two domains of life were prokaryotes, meaning that they did not have a nucleus or membrane-bound organelles in their cells. A major innovation of biological evolution was the advent of eukaryotes, that is, organisms with membrane-bound organelles, especially a nucleus and mitochondria (organelles in which cellular respiration and energy conversion take place). The cells of photosynthetic eukaryotes also contained specialized organelles called chloroplasts, in which photosynthetic pigments absorbed the energy of photons and photosynthesis takes place. The eukaryotes form the third domain of life, called *eukarya* (see Information Box 3.1).

Various mechanisms have been proposed to explain the development of membrane-bound organelles during the course of biological evolution. One of them is the invasion of some large prokaryotes by smaller ones able to use O_2 (and became mitochondria) or perform photosynthesis (and became chloroplasts). Another hypothesis is that the prokaryotic host cell surrounded smaller invading cells with its membrane, and thus transported them into its interior. Eukaryotic cells, which were made of different prokaryotic ancestors, acquired the ability to replicate all their organelles when they divided. A third hypothesis is that eukaryotes evolved directly from the common ancestor to the three domains of life (archaea, bacteria, and eukarya), often called *Last Common Universal Ancestor, LUCA*. The three domains of life coexist on present Earth, and although prokaryotes are much more numerous than eukaryotes, it is thought that the two groups may have somewhat similar total biomasses because eukaryotes are generally larger than prokaryotes. Unicellular eukaryotes later evolved into the complex multicellular organisms, which increased the role of ecosystems in the Earth System.

The multicellular organisms. The advent of multicellular organisms, between one billion and 500 million years ago, was a major step in biological evolution. In complex multicellular organisms, specialized cells perform different functions and cooperate and communicate with each other to enable interactions of the organism with its environment. There were many multicellularity attempts during the evolution of eukaryotes but only a few were successful, with the result that multicellular organisms are only found in six groups of eukaryotes: fungi, algae (brown, red, and green), land plants, and animals. All multicellular eukaryotes are O_2-respiring organisms. However, assemblages of cells had existed long before the advent of multicellular organisms, in the forms of biofilms that often included many types of unicellular organisms such as cyanobacteria, heterotrophic bacteria, protozoa, algae and fungi. One example is provided by fossil structures made of carbonates in which microorganisms are locally conspicuous, which may go back more than 3.5 billion years and later evolved into stromatolites and thrombolites (Fig. 6.9) (see Sect. 6.4.2).

Multicellular organisms are able to reproduce their whole organism, which is often very complex, starting from specialised germ cells. The development of this ability was a key evolutionary step. Multicellular organisms can reach much larger sizes than single cells, and this provide them with a key advantage over single cells in some circumstances. On present Earth, the multicellular organisms coexist with the much more numerous unicellular archaea, bacteria and eukarya. The expansion of multicellular organisms in the oceans and later on the continents enhanced the global interactions between ecosystems and the other components of the Earth System.

Occupation of continents. Multicellular organisms started to occupy the emerged lands during the Ordovician period (488–444 million years ago) and this takeover continued during the Silurian and Devonian periods (444–360 million years ago) (see Sect. 1.5.4). This occupation was possible because some of the O_2 that had accumulated in the upper atmosphere as a consequence of the production of O_2 by photosynthetic organisms in surface waters was transformed into ozone (O_3) gas by the action of the solar UV radiation and formed the ozone layer (see Sect. 10.1.7). It is possible that some archaea and/or bacteria began to occupy the continents much earlier because they were not very sensitive to UV radiation, and/or lived in UV-protected niches such as crevices in soil and underground environments.

The occupation of continents considerably increased the global biomass of photosynthetic organisms, which created a positive feedback into the concentration of atmospheric O_2 and O_3, the latter allowing the expansion of multicellular organisms on continents. The increased global production of organic matter by autotrophs in oceans and on continents and the accompanying production of heterotrophs led to the burial of enormous amounts of organic matter in the ground and the sediments. Part of the buried organic matter progressively turned into the present deposits of coal, petroleum and natural gas. Another part was included into sedimentary rocks

Overall, the takeover of continents by multicellular organisms during the last half billion years was possible because of the onset of O_2-producing photosynthesis more than 2.0 billion years before, which had deeply modified the composition of the atmosphere. The increasing photosynthetic biomass on continents enhanced this change by increasing the global production of O_2. The occupation of continents also contributed to accelerate the accumulation of carbon in the large reservoir of fossil organic matter, and the activity of land organisms enhanced continental erosion, which affected the tectonics of mountain building (see Sect. 7.5.2). Ecosystems had become an integral part of the Earth System.

Human activities as a geophysical force. As explained above (see Sect. 10.1.3), the populations of *Homo sapiens* have deeply affected their environment by causing the extinction of most of the large fauna in each land they progressively occupied, beginning 130,000 years ago or later, followed by the advent of agriculture and cities about 10,000 years ago, and the Industrial Age since the late 1700s. As industrial activities developed, there was increasing use of fossil fuels (mostly coal, petroleum and natural gas) to produce heat and synthesize new chemicals, and increasing calcination of calcium carbonate for making cement. The new tools and chemicals supported the development of large-scale agriculture and livestock farming.

The massive combustion of fossil fuels and the calcination of $CaCO_3$ rocks released into the atmosphere as CO_2 carbon that had accumulated underground during hundreds of millions of years. In addition, large amounts of methane (CH_4) were released by livestock and the decay of organic waste. Both CO_2 and CH_4 are powerful greenhouse gases, whose exponential increase in the atmosphere are among the main causes of the ongoing anthropogenic climate change. Furthermore, the dissolution of CO_2 in seawater causes ocean acidification. In this way, human activities became a geophysical force, which is changing the global climate as deeply as a long phase of volcanic eruptions (see Sect. 7.4.6) or the fall of a very large asteroid (see Sect. 4.6.2). The book of Wohlleben (2019) provides further reading, together with many examples, on interconnections in the Earth System.

Anthropogenic climate change and ocean acidification are considered in Chapter 11 (see Sects. 11.1.2 and 11.1.4). In the remainder of this chapter, *anthropogenic climate change* is written *climate change*, for simplicity.

10.1.7   Zone F: Free Oxygen on Earth

Zone F of Fig. 10.1 is represented together with Zone E in Fig. 10.5 (where the columns are interchanged with the rows of Fig. 10.1). As a follow up to the previous section, Fig. 10.5 examines the interactions between environmental free oxygen (O_2) and ecosystems. Indeed, one of the key aspects of the takeover of Earth by its organisms is the interplay between ecosystems and environmental O_2. This gas has strong chemical reactivity, and thus rapidly combines with other substance in the natural environment, which explains that free O_2 has not been observed in the environment of any other Solar System body than Earth. So far, the presence of O_2 in the Earth's environment is a unique phenomenon in the Solar System.

Increasing O_2 concentration (Zone F1 in Fig. 10.5). Everywhere else in the Solar System than Earth, all the O_2 is chemically bound with other substances, and this was also the case on Earth before the advent of O_2-producing photosynthesis. When the latter began to release O_2 in the Earth's environment by splitting water molecules, more than 2.5 billion years ago, all the O_2 produced by photosynthesis combined with iron and other metals dissolved in ocean waters, and this process continued for at least 200 million years, causing the formation of insoluble iron oxides found in sedimentary rocks all over the planet. The same also occurred with the iron and metals on emerged lands after O_2 reached the atmosphere. It is only after all the iron and other metals had been oxidized that the concentration of O_2 began to increase in seawater and the atmosphere, about 2.5 billion years ago (see Sect. 1.5.4).

Oxygen began to accumulate—as dissolved gas in natural waters (where all photosynthetic organisms then lived) and in the overlying atmosphere—when its production exceeded its use (called consumption). The first evidence of global O_2 accumulation in the atmosphere (at a concentration higher than 1 or 2% by volume) goes back more than 2.5 billion years (Fig. 8.1). The photosynthetic production of O_2 is a by-product of the synthesis of organic matter, and the relatively rapid oxidation of this organic matter (through O_2 respiration) generally consumes all or almost all the O_2 produced by photosynthesis (see Information Box 3.4). As long as the consumption of O_2 equals its production, the concentration of O_2 does not change in the environment. This is not what happened in the Earth's environment where free O_2 increased massively from being totally absent until about 2.5 billion years ago to reaching a maximum value of 35% (by volume) about 300 million years ago, a value much higher than the present 21% (Fig. 9.17).

The massive accumulation of O_2 in the ocean and the atmosphere during more than 2 billion years indicates that a large fraction of the O_2 produced by photosynthesis did not react with organic matter or other oxidizable substances that would have consumed the O_2. Indeed, enormous amounts of organic matter were removed from contact with ambient O_2 for a very long time as part of the organic production was progressively buried under the ground and in marine sediments for hundreds of millions of years (see Sect. 9.4.5). A small part of this buried organic carbon chemically evolved to form coal, petroleum and natural gas. The increasing concentration of free O_2 in the Earth's environment since about 2.5 billion years ago is the clear sign of strong global interactions between ecosystems and the aquatic, atmospheric and continental components of the Earth System.

Increasing O_2 consumption (Zone F2 in Fig. 10.5). The increasing environmental O_2 favoured the emergence of O_2-respiring organisms, which used the chemical reactivity of oxygen to meet their energy needs through O_2 respiration (see Sect. 10.1.6). The growth of the global biomass of autotrophic O_2-producing organisms was accompanied by an increase in the global biomass of heterotrophic O_2-respiring organisms, which consumed increasing amounts of O_2 and organic matter. This slowed down both the accumulation of O_2 in the ocean and the atmosphere and the burial of organic matter in sediments. This process gained momentum, and finally reversed the accumulation of O_2 in the atmosphere about 300 million years ago, after which the concentration of atmospheric O_2 decreased from 35% to around 20%. The consumption of O_2 essentially took place in the ocean until the expansion of complex eukaryotes on continents, which began later than 500 million years ago (see Sect. 10.1.6,). Nowadays, the consumption of O_2 is shared between the ocean and on the continents.

Balanced O_2 release and losses (Zones F1-F2 in Fig. 10.5). After the peak in the concentration of O_2 in the atmosphere around 300 million years ago, the environmental O_2 decreased rapidly to a level close to the present one (Fig. 9.17).

The mechanisms of this decrease are not fully understood by researchers, but they imply that massive amounts of environmental O_2 combined with inorganic and/or organic oxidizable substances then present in the environment. Afterwards, the global concentration of atmospheric O_2 progressively stabilized to the present value of about 21%, this stabilization indicating that the production of O_2 (by photosynthesis) became roughly equivalent to the losses through consumption (by O_2 respiration) and oxidation of various substances (Fig. 9.20). The stable O_2 concentration means that organic matter did not globally accumulate in sediments, which does not exclude that accumulation could occur in locally, such local accumulations being compensated by the consumption of organic matter elsewhere including the destruction of sediments by subduction (see Sect. 7.2.3). In reality, today's concentration of atmospheric O_2 decreases slightly year after year as the ongoing combustion of fossil organic compounds and other human activities create a small deficit in the Earth's O_2 budget (see Sect. 9.4.5).

The ozone (O_3) layer: ecosystems on continents (Zone F3 in Fig. 10.5).   In the upper atmosphere (20–40 km above ground), the action of the ultraviolet radiation (UV) from the Sun transforms some O_2 molecules into ozone (O_3). The build-up of O_3 in the upper atmosphere created an initial ozone layer about 500 million years before present, which absorbed most of the solar UV radiation. Because this radiation damages the DNA contained in their cells, the complex multicellular organisms began occupying the continents only after the establishment of the O_3 layer in the upper atmosphere (Sect. 2.1.5). Hence, the occupation of emerged lands depended on the establishment of the O_3 layer 500 million years ago, and the continuing existence of organisms there depended on its maintenance ever since (see Sect. 10.1.6).

Ozone is not chemically stable and decays rapidly, and is thus re-formed continually from atmospheric O_2. The presence of the O_3 layer in the upper atmosphere allows the photosynthetic organisms to produce O_2 on continents and in surface waters, and the action of solar UV on part this O_2 in the upper atmosphere re-creates in turn the O_3 layer. This is another global mechanism by which ecosystems and the global environment are linked together. The establishment of the O_3 layer was a key factor in the occupation of continental surfaces by plants and animals and their subsequent diversification there during the last half billion years (see Sect. 2.1.5).

10.1.8   Zone G: The Earth System's Feedback Loops

Zone G of Fig. 10.1 is represented in Fig. 10.6 (where the columns are interchanged with the rows of Fig. 10.1). Figure 10.6 addresses major self-regulation processes of the Earth System, which are described in Chapter 9 as *negative feedback loops*. Their key effect is to dampen long-term disturbances of the Earth System (see Sect. 9.1.4).

Zone G. Self-regulation of the Earth System: feedback loops				
Years before present	G1	G2	G3	G4
More than 4.5 billion				
4.5 to 4.4 billion				
4.4 to 4.0 biliion	Establishment of the tectonic CO$_2$ feedback loop	Establishment of the oceanic carbonate feedback loop	Establishment of the oceanic N:P feedback loops	
4.0 to 3.8 billion				
3.8 to 2.5 billion				
2.5 to 2.0 billion				
2.0 billion to 540 milllion	Long-term regulation of atmospheric CO$_2$ and the greenhouse effect	Short-term buffering of the greenhouse effect through the regulation of oceanic carbonates	Long-term regulation of the N:P Redfield ratio	Establishment of the atmospheric oxygen feedback loops
540 to 360 million				
360 to 250 million				
250 milion to present				Global oxygen regulation

Fig. 10.6 Zone G of *The Legends of the Eons*, where the columns are interchanged with the rows of Fig. 10.1. Past time is at the top, and present time at the bottom. Credits at the end of the chapter

Tectonic CO_2 feedback loop (Zone G1 in Fig. 10.6). The formation of the liquid ocean around 4.4 billion years before present (see Sect. 10.1.4) led to major drop in the concentration of atmospheric CO_2 from 98% (by volume) initially to less than 1% about 4.0 billion years ago (see Information Box 8.1). The concentration further dropped between 400 and 300 million years ago, after which it rose until about 170 million years ago, and progressively decreased to the present trace value (Fig. 9.8b). The changes in CO_2 concentration caused corresponding changes in the greenhouse effect (see Information Box 2.5) and thus the global temperature, as shown in Fig. 9.8a for the last 800,000 years. While the volume concentration of CO_2 decreased to less than 1%, that of N_2 increased proportionally to more than 99% with no change in the absolute amount of this gas in the atmosphere (Fig. 8.1). Later on, the progressive increase in the amount of atmospheric O_2 (see Sect. 10.1.7) caused the volume concentration of N_2 to progressively decrease to its present value of 78% when that of O_2 reached its present value of 21%.

The existence of a global, negative *tectonic CO_2 feedback loop*, which operated at the timescale of tens of millions of years, generally prevented the global changes in temperature from causing a global runaway warming or cooling of the planet (see Sect. 9.2.3). The mechanism of this feedback loop involves the dissolution of $CaSiO_3$ rocks, which withdraws CO_2 from the atmosphere and whose rate largely depends on global temperature (see Sect. 7.3.3).

Oceanic carbonate feedback loop (Zone G2 in Fig. 10.6). A second mechanism contributes to the regulation of shorter-term variations in the greenhouse effect. The negative *oceanic carbonate feedback loop* involves exchanges of carbon between the atmospheric pool of CO_2 and the much larger pool of $CaCO_3$ on the seafloor, and is mediated by chemical reactions that take place in seawater (see Sect. 9.2.4). Briefly, an excess of atmospheric CO_2 can be compensated by the dissolution of seafloor $CaCO_3$, and conversely, a deficit of atmospheric CO_2 can be compensated by the formation of $CaCO_3$ that sinks to the seafloor. This mechanism operates at the timescale of the deep oceanic circulation, which is of the order of a thousand years. It exerts a powerful, but incomplete negative feedback into the greenhouse effect given that the atmosphere retains a fraction of its excess CO_2 after it equilibrates with the ocean.

Two feedback loops of the oceanic N:P Redfield ratio (Zone G3 in Fig. 10.6). In the ocean, there is a close correspondence between the proportions of nitrogen (N) and phosphorus (P) in phytoplankton biomass and in average seawater, which is known as the Redfield ratio N:P$=16$:1 (see Sect. 9.3.1). The existence of a fixed N:P ratio in both seawater and phytoplankton biomass suggests that this ratio is regulated by negative feedback loops. Two of these feedback loops are summarized here (see Sect. 9.3.6).

A *first feedback loop* is based on *N_2 fixation by phytoplankton (cyanobacteria) in surface waters*. In this feedback loop, the low or high availability of dissolved inorganic N in surface waters selects or not for N_2-fixing cyanobacteria, which exerts a negative feedback into the N:P ratio of dissolved nutrients. The timescale of this feedback loop may be rapid, but N_2-fixing cyanobacteria have a high demand for Fe, and given the scarcity of this micronutrient in surface waters, it is possible that the external supply of Fe slows down the N_2 fixation feedback under certain circumstances.

A *second feedback loop* involves the *cycling of nitrogen*. In this feedback loop, the availability of dissolved inorganic N in surface waters determines the downward export of organic matter, whose biological and chemical oxidation at depth regenerates nitrate (NO_3^-), which is brought back to the surface by vertical water movements. This contributes to regulate the N:P ratio of dissolved nutrients in surface waters. The timescale of this feedback is determined by the vertical circulation of NO_3^-, which takes place within a thousand years (ocean circulation) or less.

Two feedback loops of the atmospheric O_2 concentration (Zone G4 in Fig. 10.6). Following the invention of O_2-producing photosynthesis by cyanobacteria, there was a slow accumulation of atmospheric O_2 from 0.5 to 15% (by volume) between 2.4 and 0.5 billion years ago, after which the O_2 concentration reached a peak of 35% about 300 million years ago, and stabilized at values around 20% (see Sect. 9.4.3 and Fig. 9.17). Two negative feedback loops that contributed to regulate the long-term level of atmospheric O_2 concern the global cycles of sulphur and carbon, respectively (see Sect. 9.4.4).

The two O_2 feedback loops involve the following components: net production of O_2 resulting from the burial of organic carbon and pyrite, biological and chemical oxidation of these two compounds, and emission of the resulting gases to the atmosphere where they combine with O_2. The regulation of the concentration of atmospheric O_2 comes from the fact that burial of organic carbon and pyrite releases O_2 in the environment, whereas the gases resulting from the decomposition of these two compounds consume O_2 (see Sect. 9.4.4). The degradation of pyrite minerals and organic matter, deep under the ground and at exposed surfaces, involves different types of mechanisms that operate at different timescales. Some operate at timescales of millions of years, whereas others contribute to the regulation of variations in the level of atmospheric O_2 at shorter timescales of hundreds of thousands of years.

10.1.9   Back to Zone B: The Earth System (Continued)

We now return to zone B of Fig. 10.1, which is represented in Fig. 10.7 (where the columns are interchanged with the rows of Fig. 10.1). We described above both the present conditions and the most ancient ones represented in Fig. 10.7 (see Sect. 10.1.3). In that description (Fig. 10.2), we contrasted the current state of the Earth System (bottom part of Fig. 10.7) with the situation that had prevailed on the planet before and at the time the giant collision that had created the Earth-Moon system about 4.5 billion years before present (top part of Fig. 10.7). Here we examine, after our above review of zones C–G of *The Legend of the Eons* (see Sects. 10.1.4–10.1.8), the changes that occurred in the components of the Earth System between its early stage and the present conditions. This 4.5 billion-year

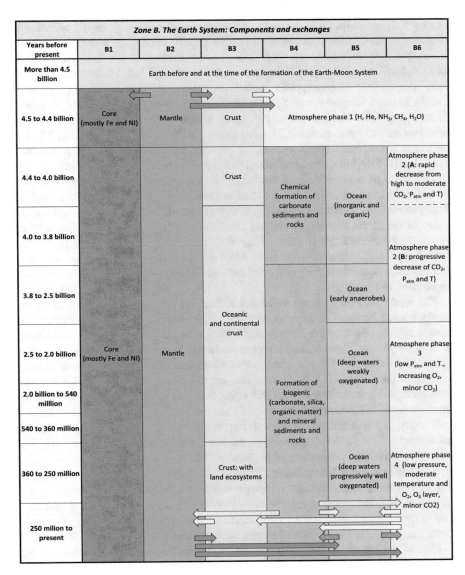

Fig. 10.7 Zone B of *The Legends of the Eons*, where the columns are interchanged with the rows of Fig. 10.1. Past time is at the top, and present time at the bottom. Horizontal arrows: exchanges of materials between layers (for simplicity, these are only illustrated for the most ancient condition of the planet and the most recent). P_{atm}: atmospheric pressure; T: temperature. The top and bottom part of this figure were discussed in Sect. 10.1.2 in relation with Fig. 10.2. Credits at the end of the chapter

journey will highlight the main steps of the takeover of Earth by its organisms, and identify the environmental conditions in which these steps were taken.

From 4.5 to 4.4 billion years before present. Following the last giant impact about 4.5 billion years ago, Earth was a mass of very hot, molten magma (see Sect. 10.1.4). This magma rapidly cooled down, and the gases that escaped from the cooling magma formed the first phase of the atmosphere (arrows mantle-to-atmosphere and crust-to-atmosphere in Fig. 10.7), which is thought to have mostly contained water vapour, methane (CH_4) and ammonia (NH_3) (see Sect. 10.1.3). Most of the initial atmospheric water was lost by gravitational escape of hydrogen to space (see Sect. 10.1.4), but some contributed to the formation of the future ocean.

The mass of molten magma that made up the young Earth rapidly differentiated into the stacked layers of the core, the mantle and the crust (see Sect. 10.1.4). Heavy chemical elements (including iron and nickel) sank by gravity toward the centre of the globe where they formed the core, and the lighter elements moved up by buoyancy toward the surface where they accumulated (see Sect. 10.1.5). The three layers differentiated from each other quite rapidly; say within 100 million years, which is quick at the geological timescale (arrows mantle-to-core and mantle-to-crust in Fig. 10.7).

The crust may have been solid 4.4 billion years before present, and as the planet cooled down, the early crust and mantle released gases that formed the Earth's second phase of the atmosphere, which replaced the lost atmospheric gases of the first phase. Three main factors contributed to protect the atmosphere since these early times: the Earth's gravity, its safe distance from the Sun, and its magnetic field (see Sect. 10.1.5). During the second phase of the atmosphere, gases continued to be transferred from the mantle to the atmosphere across the solid crust through volcanoes (arrow mantle-to-atmosphere in Fig. 10.7). As it is the case today, the volcanic gases were dominated by CO_2, and it is thought that this gas accounted for 98% (by volume) of the Earth atmosphere (see Information Box 8.1). This very high proportion of atmospheric CO_2 is similar to the current value on the sister planet of Earth, Venus. There, CO_2 makes up 96% of the atmosphere, and the elevated concentration of atmospheric CO_2 generates an extreme greenhouse effect that creates a surface temperature of +462 °C (Tables 1.1 and 2.1). The very high concentration of CO_2 at the beginning of the second phase of the Earth's atmosphere likely created similarly high temperatures (see Sect. 5.5.4).

One of the effects of the formation of the Earth-Moon system was to stabilize the Earth's axial tilt to a small value in the long term, which caused the seasonal differences in the surface temperatures of Earth to be generally small. In addition, the rapid rotation of Earth distributed the solar heat quite uniformly among longitudes all around the planet. These combined conditions contributed to maintain the Earth's temperature within a range where water was permanently or temporarily liquid during the annual cycle nearly everywhere on the planet, which was a key condition for the durable establishment of organisms and the success of ecosystems on Earth (see Sect. 5.6.3).

From 4.4 to 4.0 billion years before present. A global ocean appeared on Earth about 4.4 billion years ago, and its initial water came from the condensation of water vapour released from the cooling magma into the atmosphere during its first phase (see Sect. 10.1.4). However, it is thought that most of the ocean's water was brought to Earth by asteroids and/or comets that impacted the planet from 4.4 to 3.9 billion years before present, especially during the Late Heavy Bombardment from 4.1 to 3.9 billion years ago (see Sect. 5.3.2). The latter sources carried not only water but also organic molecules, which may have contributed to the emergence of organisms on Earth (see Sect. 10.1.4). Because of the Earth's safe distance from the Sun, most of the water at the surface of the planet was liquid (see Sect. 10.1.5).

The atmosphere underwent a major transformation during its second phase, in which the ocean played the key role. Indeed, the initially very abundant atmospheric CO_2 progressively dissolved in the liquid water of the ocean, where it formed dissolved bicarbonate ions (HCO_3^-) that combined with dissolved calcium ions (Ca^{2+}) or other metallic ions contained in the seafloor to form solid carbonates (mostly $CaCO_3$). The $CaCO_3$ accumulated in huge deposits in and on the seafloor. This chemical process decreased the very large concentration of CO_2 in the atmosphere, with a parallel reduction in the atmospheric pressure and temperature (see Sect. 10.1.4). This contributed to create environmental conditions favourable to the future emergence of organisms. The water of the ocean contained both inorganic ions—such as Na^+ and Cl^- (sea salt), HCO_3^- and Ca^{2+}—and a number of organic compounds carried from far away in the Solar System by the same meteoroids and comets that brought water to Earth (above).

Another consequence of the presence of the ocean may have been the initiation of plate tectonics. Indeed, it is thought that the existence of a large volume of surface liquid water is a condition for maintaining plate tectonics on a planetary body. On Earth, plate tectonics may have started very early because the mantle contained some of the initial water, which lowered the viscosity of its upper part and thus allowed internal movements of the magma (see Sect. 7.2.1). The low viscosity of the upper mantle was maintained subsequently by the injection of water from the overlying ocean, as soon as it formed or perhaps only much later (see Sect. 10.1.4). The movements of the plates were fuelled by the internal heat of Earth, which resulted from the initial heat acquired by the planet and the decay of radioactive heavy elements contained in the mantle (see Sect. 10.1.4).

Plate tectonics was accompanied by subduction, which carried parts of the crust and their carbonate loads into the mantle. There, the melting crust and carbonates released their chemical constituents into the mantle's magma, including carbon that returned to the atmosphere as CO_2 through volcanoes. However, the net result of the process was the storage of most of the CO_2 that was initially in the Earth's atmosphere into huge $CaCO_3$ rock formations. The initial stock of solid carbonates does not exist any more, but was continually replenished by the formation of new solid carbonates in the ocean (see Sect. 9.2.2). Large pieces of $CaCO_3$ rocks were moved to the surface of continents by tectonic processes. Nowadays, more than 99% of the carbon on Earth is in the form of $CaCO_3$ rocks, and atmospheric CO_2 only represents a minute part of the planet's carbon. However, the tiny amount of CO_2 in the Earth's atmosphere is enough to generate a natural greenhouse effect that keeps the planet suitably warm for the organisms.

From 4.0 to 3.8 billion years before present. The above processes that linked the mantle, the crust, the sediments, the ocean and the atmosphere continued until 3.8 billion years ago. In addition, the period starting 4.0 billion years ago was marked by the enhanced differentiation of the continental crust from the existing oceanic crust.

The density of the continental crust is lower than that of the oceanic crust (Table 6.6), and the lighter continental rocks were formed from the denser oceanic rocks as follows. Pieces of the oceanic crust melted as they subducted in trenches, and some of the resulting partly molten, lighter materials rose by buoyancy towards the surface where they formed continental rocks. The oldest rocks on Earth are about 4 billion years old, and these are relics of the first pieces of continental crust. Since then, oceanic crust has been continually formed at ocean ridges and destroyed in trenches, and continental crust has mostly been formed at the edges of continents and compressed in mountain ranges (see Sect. 7.2.3). This way, plate tectonics led to the progressive differentiation of the oceanic and continental crusts, and is thus responsible for the (slow) growth of continents.

From 3.8 to 2.5 billion years before present. In this book, we use the conservative value of 3.8 billion years before present for the first occurrence of organisms on Earth, but it could have taken place almost 4 billion years ago (see Sect. 10.1.6). However, the significant phenomenon for the eventual takeover of Earth by organisms was the initial development of large biomasses, which occurred later than 3.8 billion years ago.

The relatively mild environmental conditions created during the previous hundreds of millions of years favoured the emergence of the first organisms, which was followed by their evolutionary diversification, the expansion of ecosystems into increasingly numerous aquatic environments, and the build-up of biomasses. These conditions initiated the takeover of the Earth System by organisms, which favoured the long-term existence of ecosystems on the planet despite naturally occurring environmental variations. One example of long-term variations that ecosystems successfully managed is the succession of greenhouses and icehouses that marked the history of the planet (see Sect. 4.5.1).

With the advent of sizeable biomasses of organisms, the transfer of carbon from atmospheric CO_2 gas to solid $CaCO_3$ progressively shifted from the chemical process operating since 4.4 billion years ago to the biological process of calcification (see Sect. 6.4.2).

After the initial period of massive transfer of CO_2 from the atmosphere to huge deposits of solid $CaCO_3$, exchanges of carbon between the atmosphere and the ocean on the one hand and the ocean and seafloor $CaCO_3$ on the other hand contributed to regulate the concentration of atmospheric CO_2 in the relatively short term of thousands of years by the formation or dissolution of solid carbonates (see Sect. 10.1.4). On the longer term of tens of millions of years, the concentration of atmospheric CO_2 was controlled by the tectonic CO_2 feedback loop, which involved the injection of CO_2 in the atmosphere by volcanoes and the chemical alteration of silicate rocks formed by solidification of volcanic lava or magmatic intrusions (see Sects. 9.2.3 and 10.1.8). The progressive accumulation of carbon and other biogenic elements in solid deposits could have deprived organisms of these essential chemical elements. However, the biogenic elements were recycled by the continual subduction of parts of the oceanic crust in trenches and return of their chemical constituents to the Earth's surface and atmosphere by volcanoes, together with the erosion of rocks on continents (see Sect. 7.3.1). These processes continue to be active today. The book of Knoll (2015) provides further reading on the deep history of organisms on Earth from the origin to about half a billion years ago.

From 2.5 to 2.0 billion years before present. The invention of O_2-producing photosynthesis by cyanobacteria was a major event in biological evolution, which took place earlier than 2.5 billion years before present. This did not immediately change the concentrations of O_2 in the ocean and the atmosphere, because to have a global effect the O_2-producing autotrophs had first to spread into most aquatic environments and build up very large biomasses. This occurred in the upper layers of the oceans and lakes, where there was enough light for photosynthesis, which is typically the surface 100 m or less (see Sect. 10.1.6). All the O_2 produced by photosynthesis initially combined with iron and other metals dissolved in waters, a process that continued during at least 200 million years. It is only after all the dissolved iron and other metals had been oxidized that the concentration of O_2 could begin to increase in seawater (see Sect. 10.1.7). When O_2 began to escape from the ocean into the atmosphere, it initially reacted chemically with the continental rocks, which impeded the build-up of its atmospheric concentration. The latter began to take place about 2.5 billion years ago, and this marked the beginning of the third phase of the Earth's atmosphere.

The relatively mild temperatures and other environmental conditions of Earth continued to favour evolutionary diversification, and organisms belonging to a new domain of life, the eukaryotes, evolved from the long-existing prokaryotes. This occurred either earlier than 2.0 billion years ago, or later (see Sect. 10.1.6). The unicellular eukaryotes later diversified into complex multicellular organisms (see Sect. 10.1.6). All organisms then lived in aquatic environments.

The oxygenation of the upper ocean favoured the natural selection of organisms that used O_2 respiration to meet their energy needs, especially that this type of respiration produces more energy per unit sugar than the other types (see Information Box 5.2). The new O_2 respiration was used by both autotrophic and heterotrophic organisms. These could take advantage of the oxygenated surface waters, whereas the organisms that did not use O_2, especially those that could not tolerate any oxygen, occupied the deeper waters and the sediments (see Sect. 10.1.6). Oxygen increased slowly up to 2–3% of the volume of atmospheric gases by 2.0 billion years ago (compared with the present 21%). The increasing accumulation of O_2 in the ocean and the atmosphere was accompanied by an increase in the global consumption of O_2 by the O_2-respiring organisms (see Sect. 10.1.7).

Overall, the release of photosynthetic O_2 in the Earth's environment transformed the characteristics of the planet in a way unparalleled anywhere else in the Solar System, at least as far as we know. Indeed, the presence of free O_2 in the atmosphere of any astronomical body would be, given the high chemical reactivity of oxygen, a clear sign of the presence of a substantial biomass of organisms actively producing this gas, and this phenomenon has not been observed yet anywhere else in the Solar System than Earth. Of course, the absence of free O_2 does not indicate the absence of organisms. For example, O_2 was absent from the Earth's atmosphere until 2.5 billion years ago, but organisms had been present on the planet since at least 3.8 billion years ago. Hence an outside observer could not have detected the presence of organisms on Earth during its first 1.5 billion years or longer, although organisms had already initiated their takeover of the Earth System at the time.

From 2.0 billion to 540 million years before present. The environmental and evolutionary changes that had begun earlier than 2 billion years ago gained momentum during the following 1.5 billion years. Oxygen continued to increase in the ocean and the atmosphere, and the action of solar UV radiation transformed increasing amounts of O_2 into O_3 (ozone) in the high atmosphere, which led to the formation of the O_3 layer in the high atmosphere about 500 million years ago (see Sect. 10.1.7).

In the well-oxygenated upper ocean, the microbial carbonate reefs known as stromatolites reached their greatest abundance and diversity about 1.2 billion years ago (see Sect. 6.4.2). These early structures, which combine bacterial mats and solid $CaCO_3$, are remarkable examples of assemblages of prokaryotes (and also microalgae, which appeared much later), but they did not evolve into multicellular organisms. The peak in stromatolite abundance was followed hundreds of millions of years later by the evolution of multicellular eukaryotic organisms, which appeared between one billion and 500 million years ago (see Sect. 10.1.6). This was a major step in the history of the Earth System because complex multicellular organisms would spread on continents hundreds of millions years later, and thus complete the occupation of the planet by ecosystems. Since their advent, multicellular organisms have shared the aquatic environments with the previously existing single-cells organisms, and they would later share the emerged lands with them (see Sect. 10.1.6).

From 540 to 360 million years before present. The establishment of the UV-protective ozone (O_3) layer) in the high atmosphere allowed complex multicellular organisms to begin the occupation of emerged lands. This marked the beginning of the fourth phase of the atmosphere. Most likely, some UV-resistant unicellular organisms had been present in this environment much earlier (see Sect. 10.1.6). In any case by 360 million years ago, the continents had greened, and were covered by forests all the more luxuriant that large herbivores did not yet exist to keep them in check. Abundant land animals, which were primitive arthropods, had co-evolved with the diversified terrestrial vegetation.

The development of plant roots likely enhanced the erosion of rocks and the deposition of sediments, which contributed to modify the geological environment and create the first soils. Furthermore, the increase in the global production of O_2 by photosynthesis strengthened the O_3 layer. Finally, the growth of the new forests led to the burial of atmospheric carbon in organic matter deposits, which was accompanied by a drop in atmospheric CO_2 (see Sect. 10.1.6). The establishment of multicellular organisms on continents reinforced considerably the role of ecosystems in the Earth System.

Information Box 10.1 The Carboniferous Period Affects the Modern Climate The formation of large coal deposits during the Carboniferous period more than 300 million years ago has a direct bearing on the present global climate crisis. Indeed, coal formed during the Carboniferous provided the energy that fired the Industrial Revolution more than 200 years ago and continues to fuel a large part of the present energy consumption (Fig. 9.5). There was formation of coal deposits in all geological periods since the Carboniferous, the most recent forms of coal being *lignite* and *peat*. Indeed, as time passes and materials are buried deeper, peat is transformed into lignite, and the latter into higher-rank types of coal such as bituminous and anthracite coal.

Here are examples of the role of coal in the development of the Industrial Age: industries used coal to power steam engines and to smelt (extract) iron from minerals; coal provided the fuel for the mass transportation of goods and people on land (trains) and over the seas (steamboats); and rapidly developing cities used coal for heating, cooking and lighting (for lighting, coal was transformed into city gas, which was also used later for heating and cooking). Even in the 2020s, coal is the largest source of energy for the production of electricity in the world and the second-largest global source of primary energy after petroleum, and it accounts for almost half the emissions of CO_2 in the atmosphere.

In the last 200 years, human societies have released into the atmosphere in the form of CO_2 huge amounts of carbon. A large part of this carbon had been sequestered under the ground by the photosynthesis of Carboniferous forests 300 million years ago.

Dissolved oxygen progressively penetrated the deep waters of the ocean, and in the resulting well-oxygenated marine environment, eukaryotes diversified into increasingly large numbers of algae and invertebrates, the latter including the well-known trilobites (arthropods) and also shellfish, fish and amphibians. This diversification was interrupted, between 450 and 440 million years ago, by the first mass extinction during which 60–70% of the marine species of multicellular organisms disappeared (see Sect. 4.6.2). Later on, the fish became so diversified that the period between 416 and 359 million years ago is known as the *Age of Fish*.

From 360 to 250 million years before present. The forests that had greened the planet continued to expand from 360 million years ago onwards. The burial of their wood under the ground created the huge deposits of coal that presently exist on Earth. This unique phenomenon in the history of the planet occurred during the geological period called *Carboniferous* (this word means "coal-bearing"), which began 360 million years ago and lasted 60 million years (Information Box 10.1). As a result of the burial of enormous amounts of organic carbon, the volume fraction of atmospheric O_2 reached the extremely high value of 35%, which is almost twice the present value of 21% (see Sect. 10.1.7).

The very high level of oxygen in the atmosphere may have increased the incidence of wildfires on the planet (see Sect. 2.1.5), and also explains the occurrence of gigantism in insects and amphibians. Indeed, the sizes of these organisms were not constrained any more by their respiratory systems, whose physiological ability to transport and distribute oxygen had previously been limited by the lower O_2 concentration. Examples of this gigantism, which is unrivalled at any other time in the history of Earth, are 2.6 m long millipede arthropods, giant insects with a wingspan of 75 cm, and amphibians over 2 m long. In the shallow waters of the ocean, large reefs harboured a wide variety of organisms, and the well-oxygenated open waters harboured abundant cartilaginous fishes (sharks and rays) and increasingly diversified bony fishes.

Since the initial reduction in the concentration of atmospheric CO_2 more than 4 billion years ago, the resulting decrease in the Earth's greenhouse effect had brought the mean temperature at the surface of Earth down to the present +15 °C (Figs. 4.8 and 4.10b). This was a long-term consequence of the progressive establishment of the tectonic CO_2 and oceanic carbonate feedback loops between 4.4 and 4.0 billion years ago (see Sect. 10.1.8). Simultaneously, the oxygen feedback loops stabilized the volume fraction of atmospheric O_2 at about 21% (Fig. 9.17) (see Sects. 10.1.7 and 10.1.8).

The long-term ranges of surface temperature not only allowed the survival of organisms but also favoured the success of ecosystems. Not all temperatures were equally suitable for ecosystems, but these successfully adapted to most temperature conditions except in cases where water was absent (see Sect. 4.1.3). Of course, some major changes in global temperature—caused by catastrophic events such as the impact of a large meteoroid or a volcanic supereruption—led to mass extinctions. However, mass extinctions have always been followed by outbursts of biological diversification, which produced life forms that were often very different from those that had been erased, and thus launched ecosystems in new directions (see Sect. 4.6.1).

One example of the above catastrophes is the *Carboniferous rainforest collapse,* which marked the end of the millions of years of the exuberant flora and fauna of the Carboniferous period. The causes of this collapse are not understood. The following period, which is called Permian and lasted until 251 million years ago, saw the appearance of the first large land herbivores and carnivores, as well as the ancestors of the dinosaurs and mammals. This period ended with the third and largest mass extinction, during which 90–95% of marine species and 70% of land species of multicellular organisms disappeared (see Sect. 4.6.2).

From 250 million years before present until now. The next era in the Earth's history, called *Mesozoic* (251 to 65 million years before present; Table 1.3), began in the wake of the third mass extinction. In the ocean, algae, corals and other fauna progressively recovered from the destruction of more than 95% of all marine species of multicellular organisms, including the very successful trilobites and 70% of the land species and new, large aquatic reptiles appeared. Forest of conifers occupied the emerged lands, which hosted new groups of insects that included mosquitoes. The most remarkable newcomers on the planet were increasingly larger crocodiles and dinosaurs. This epoch, called *Triassic*, ended with the fourth mass extinction 201 million years ago, during which almost all the large non-dinosaurian fauna disappeared from continents, whereas non-dinosaurian reptiles survived in aquatic environments. These aquatic reptiles included the turtles, crocodiles and snakes (see Sect. 4.6.2).

Dinosaurs filled the space left open by the fourth mass extinction, and dominated the land, waters and even the air for more than 100 million years, during the *Jurassic* and *Cretaceous* epochs (199.6 to 145.5, and to 145.5 to 65.5 million years ago, respectively). Flowering plants became increasingly widespread, and birds were more and more diversified and abundant. Mammals were then represented by small species, which lived in the shadow of dinosaurs. The *Age of Dinosaurs* lasted until 66 million years ago, when the fifth mass extinction destroyed about three-quarters of all species, including all dinosaurs except the ancestors of modern birds. The vacuum created by this mass extinction was filled by mammals, which had been minor players in ecosystems until then. One of these mammals was the distant ancestor of our own species, *Homo sapiens*. The Age of Dinosaurs gave way to the *Age of Mammals* (see Sect. 4.6.2).

The current and most recent era, called *Cenozoic* (since 65 million years before present; Table 1.3), was marked by a rapid diversification of mammals and birds in the absence of the large reptiles that had dominated the scene during the previous 175 million years. The era started with a torrid age, between 65 and 55 million years ago (see Sect. 4.5.5). After this period during which the average surface temperature reached up to 30 °C, the climate progressively cooled to enter the present icehouse 34 million years ago, which led to the Quaternary Ice Age that began 2.6 million years ago (Fig. 4.8 and Table 4.3).

Initially, the Cenozoic food webs were dominated by very large, 1–3 m tall carnivorous birds known as *terror birds*, but the dominance shifted to mammals as these grew increasingly larger on land and in the ocean. The largest mammal in the history of Earth is the extant Blue Whale (*extant* means "still in existence"). However, most of the largest marine and terrestrial mammals are extinct, either from natural causes or as the result of hunting, and the three largest extant species of mammals—the *Blue Whale*, the *African Bush Elephant*, and the *Asian Elephant*—are considered to be endangered or vulnerable (Table 10.1). The most dramatic case of extinction of a very large mammal by humans is that of the *Steller's Sea Cow* (a member of the dugong family), which was hunted to extinction within 27 years of its discovery by Europeans on islands of the Bering Sea in 1741. It is explained below that many other large species were similarly hunted to extinction by people (see Sect. 10.2.1).

The above paragraphs reported the five mass extinctions documented in the fossil record since about half a billion years. No extinctions are documented prior to these five because of the paucity of older fossils (see Sect. 1.5.3). However, mass destructions of organisms likely occurred at frequent intervals earlier than half a billion years ago, but it is not known to what extent such events led to annihilations of entire life forms, that is, extinctions. In fact, mass destruction of organisms likely occurred since the advent of the first life forms on Earth, say on the

Table 10.1 Extant or extinct largest marine and terrestrial mammals, with known period of existence or time of extinction, or present status

Scientific name	Common name	Weight (tons)	Living area	Period (years), or present status
Balaenoptera musculus	Blue Whale	173	World Ocean	Extant, endangered[b]
Livyathan melvillei	Leviathan, Livyathan	50	World Ocean?	10 to 9 million
Paraceratherium transouralicum[a]	Hornless Rhinoceros	15–20	Plains of Asia	35 to 20 million
Mammuthus trogontherii	Steppe Mammoth	9–14.3	Northern Eurasia	600,000–370,000
Basilosaurus (two species)	King Lizard[c], Zeuglodon	10	World Ocean, Tethys Sea	40 to 34 million
Hydrodamalis gigas	Steller's Sea Cow	10	Bering Sea	Hunted to extinction by 1768
Mammuthus primigenius	Woolly Mammoth	6	Northern Asia, Europe and North America	Hunted to extinction 4,000 years ago
Loxodonta africana	African Bush Elephant	4–7	Sub-Saharan Africa	Extant, vulnerable[a]
Elephas maximus	Asian Elephant	2.25–5.5	Indian Subcontinent and Southeast Asia	Extant, endangered[b]
Elasmotherium sibiricum	Giant Rhinoceros, Giant Siberian Unicorn	3.6–4.5	Eastern Europe and Central Asia	2.6 million to 29,000
Megatherium americanum	Giant Ground Sloth	3–4	South America	400,000–8,000 (perhaps hunted to extinction)

[a] Other genus names: Indricotherium, and Baluchitherium
[b] According to the World Widlife Fund, https://www.worldwildlife.org/species/directory?sort=extinction_status&direction=desc
[c] Prehistoric predatory whale, which was originally thought to be a giant reptile

average once every 100 million years or even more frequently. However, it may be more difficult to obliterate groups of simple, unicellular life forms than groups of complex, multicellular organisms.

Mass Extinctions as Times of Rejuvenation of Biodiversity The causes invoked to explain the five documented extinctions include hits by asteroids and large-scale volcanism, which have occurred during the whole history of the planet (see Sect. 4.6.2). Hence mass destruction and perhaps extinctions of organisms and species are natural phenomena on Earth, caused by the action of astronomical and/or geophysical forces. Mass extinctions can be seen as negative events that destroyed a large part of the existing biodiversity every time they occurred, after which it took millions of years for ecosystems to recover. These events can instead be seen as times of rejuvenation of biodiversity, which boosted biological evolution and launched the Earth's ecosystems in new directions. Nature moves on without qualms, within the limits set by the laws of physics and chemistry of the Universe. The 2020s COVID-19 epidemic has reminded modern societies that humans are subjected to rules of Nature, as are all the other components of ecosystems.

Many researchers think that Earth is presently experiencing a sixth mass extinction, caused by the actions of human societies. Within the context of mass extinctions, the combined, global effects of these actions would be similar to the catastrophic effects of the geophysical forces that had caused the previous extinctions (see Sect. 10.1.7). In order to understand the scientific basis of the sixth extinction hypothesis, one should know that species continuously get extinct even when environmental conditions are not catastrophic. It is often stated in the scientific literature that out of one million species, one got extinct every year before the advent of *Homo sapiens*. However, this "background" rate of one (1.0) extinction per million species per year may be overestimated, and some recent studies indicate that typical rates may be closer to 0.1. These values are consistent with the idea that outside the episodes of catastrophic extinctions, a species normally dies out within 10 million years of its appearance or more rapidly (see Information Box 1.6).

The actual value of the background rate of extinction is very important because it provides a benchmark against which to assess the present effects of human activities on biodiversity. Indeed, ecologists conducting research on aquatic and land ecosystems have found that the current rate of species extinction is of the order of at least 100 extinctions per million species per year, which is from 100 to 1,000 times or perhaps more the background extinction rate. It must be noted that other researchers think that the current rate of extinction is much lower than 100, but could nevertheless increase rapidly in the near future. All the above numbers are very approximate, especially that the very notion of *species* is debated among specialists (see Sect. 4.6.1).

The estimated values of the background and current rates of extinction are different for different groups of organisms. Furthermore, the above average rates do not generally consider the microbes, which evolve more rapidly than larger organisms. Despite this variability in estimates, the comparison of the current extinction rates with background values has led many researchers to conclude that a sixth mass extinction was already under way. Different from all previous mass extinctions, the driving factor invoked to explain the sixth mass extinction is the activity of human societies.

In the following Sects. 10.2 and 10.3, we focus on the Anthropocene geological epoch before and during the twentieth century. We examine the following period, that is the Anthropocene during the twenty-first century and beyond, in Chapter 11 within the context of the ongoing global change.

10.2   Anthropocene Geological Epoch: Before the Twentieth Century

10.2.1   Onset of the Anthropocene

In *The Legend of the Eons*, human activities are likened to a geophysical force (Zone E in Figs. 10.1 and 10.5). This is because human activities presently contribute to change the global climate as deeply as a long volcanic eruptive phase or the fall of a large asteroid would do (see Sect. 10.1.6). This major effect of human activities led a number of specialists of Earth Sciences to propose that Earth has now entered a new geological epoch, called the *Anthropocene* (Table 1.3). The proposed Anthropocene would be a new geological epoch, distinct from the *Holocene*, which started at the end of the last glacial episode 11,650 years before present. The Holocene epoch would correspond to the worldwide proliferation, growth and impacts of modern humans, *Homo sapiens*, including all of its written history, technological revolutions, development of major civilizations, and overall transition towards the present urban life.

The beginning of the Anthropocene epoch would correspond to the commencement of significant human impact on the Earth System and its ecosystems including, but not limited to, climate change. The beginning of the Anthropocene is debated among specialists because humans have deeply transformed the Earth's environment since agriculture began thousands of years ago, causing large-scale changes in land characteristics, and their impacts on the environment accelerated with the Industrial Revolution in the mid-1700s, leading to global climate change whose consequences became obvious the 1900s. Hence, the beginning of the Anthropocene could be set thousands, hundreds, or tens of years ago. In this book, we consider that the Anthropocene began about 300 years ago (Table 1.3), and this chapter as well as Chapter 11 examine major human impacts that occurred since the nineteenth century. The article of Ramin (2018) provides further reading on the notion of Anthropocene, and the book of Glikson and Groves (2016) on the conditions that led to the emergence of this new epoch.

It is quite possible that *Homo sapiens* began to deeply influence the Earth's ecosystems as early as 130,000 years ago, given that the occupation of new lands was often accompanied by the extinction of large indigenous fauna (see Sect. 10.1.3). This had not happened with the previous human species including Neanderthals, and the phenomenon has continued until quite recently almost every time people occupied pristine environments. Examples of such extinctions by

Table 10.2 Worldwide human population throughout history. BCE: Before the Common Era. The data for the population up to 7 billion people are from Pariona (2017), and the predicted dates for a population of 8 and 9 billion people are from the United Nations (2019)

Estimated human population worldwide	Year
200,000	130,000 BCE[a]
3 million	10,000 BCE[a]
10 million	6,500 BCE[a]
50 million	2,000 BCE[a]
200 million	0
1 billion	1804
2 billion	1927
3 billion	1959
4 billion	1974
5 billion	1987
6 billion	1999
7 billion	2012
8 billion	2025[a]
9 billion	2040[a]

[a]Approximate years

hunting are cited in Table 10.1. Some studies attribute these extinctions to environmental changes and not human actions, whose effect would then have been secondary. Given that the studies on these and other extinctions (see Sect. 4.6.2) are correlative, that is, they compare the timing of extinctions and external changes (such as the arrival of people and climate change), they do not provide by themselves conclusive information on the causes of the extinctions. Hence, the ongoing debates among specialists on this matter.

Human societies underwent several major transformations, including the transition from hunting-gathering to agriculture, which began around 12,500 years ago (although societies of hunters-gatherers still exist nowadays), and the beginning of the Industrial Age in the late 1700s. The first of these two major steps led to the creation of cities, and the second favoured their major development, and each of these two steps was followed by a major increase in world population (Table 10.2, after 10,000 BCE and since 1804).

10.2.2 Growth of Human Population and Environmental Effects

It is believed that the first humans (genus *Homo*) appeared on Earth around 2.8 million years ago or perhaps earlier. This was much before the divergence between modern *Homo sapiens* and archaic Neanderthal *Homo neanderthalensis*, which may have occurred more than 700,000 years ago. The early humans were hunter-gatherers, and their populations were small. Indeed, it is believed that there existed only 200,000 humans on Earth around 130,000 years before the Common Era (BCE), and these were then largely concentrated in eastern Africa. The various human species migrated to other parts of Africa and into Asia, Australia, Europe and the Americas, and diversified in these new environments, favouring *H. sapiens*, *H. neanderthalensis* and a few other species. By 10,000 BCE, the numbers of humans reached around 3 million people, and *H. sapiens* was then the only human species on Earth. The population estimates cited in this paragraph and the following ones are from Table 10.2.

The human population reached 10 million people by 6,500 BCE, which was after the *Agricultural Revolution*. The latter had begun between 10,000 and 8,000 BCE in different regions, when humans developed ways to control crops and animals, and progressively left the nomadic lifestyle for permanent settlements. The population increased slowly in the following thousands of years, to reach 1 billion in 1804. The increase had not been continuous, as there were strong decreases during periods of pandemics or prolonged social disturbances. For example, the Black Death (plague), which affected Eurasia and peaked in Europe in 1348–1350, reduced the total world population from 450 to 350-375 million people in a few years.

A major change that affected the growth of the population was the *Industrial Revolution*, which occurred between 1760 and 1840 and shifted production methods from being done by hand (of course with tools) to using machines (see Information Box 1.9). From then on, there was sustained, rapid growth of the human population, which numbered in billions of people and was increasingly concentrated in cities.

Different specialists interpret the history of humankind differently. A frequently held view is that human's history represented progress towards more comfort and increasingly varied pursuits. In contrast Harari (2015) argues, in his book on the history of mankind, that as societies went from hunting-gathering to farming, to industrial labour, and to the present connected society, most people had less and less free, happy time (book recommended as further reading). In the remainder of this chapter, we focus not on the functioning of human societies, but instead on their effects on the Earth System and the corresponding positive feedbacks of the latter on societies.

Some human activities initially intended or seen by societies as progress ultimately proved to disturb the Earth System, and these perturbations often had negative feedbacks into human societies. In early times *when the human population was small*, the negative environmental effects of human activities were often local or regional and of short duration, and the feedbacks only affected small groups of people. People then did not necessarily connect the feedbacks to their human causes. For example, the hunter-gatherers who experienced the disappearance of the large wildlife that was an abundant source of food did not necessarily linked the loss of this food source with their hunting pressure. When there were no more large animals to hunt, the small groups of people could either switch locally to other food sources or move to a different area not already occupied by humans.

As the human population grew, the negative environmental effects of some activities became continental and even global, and the resulting positive feedbacks could affect large numbers of people. As a consequence, connections between feedbacks and causes could be suspected. One example of connection between cause and effect was the frequent occurrence of epidemics. As urbanization developed, people realized that in crowded cities, the poor districts that smelled bad were more prone to propagation of diseases such as cholera than the prosperous areas with no bad smells. This led to the *miasma theory* of disease, which ascribed the origin of epidemics to a miasma (a noxious form of bad air) emanating from rotting organic matter. The miasma theory was replaced by the *germ theory* of disease, the validity of which was demonstrated in the 1850s by the French pioneer of microbiology Louis Pasteur. Using experiments, Pasteur demonstrated that several diseases were caused by microorganisms, which were then called *germs* and are now known as *pathogens*. This led to modern hygiene and medicine.

10.3 Anthropocene Geological Epoch: The Twentieth Century

During the twentieth century, increasing knowledge of the processes that govern the Earth System let to connect some human activities with the resulting continental or global environmental changes and the feedbacks of the latter into human societies. There were several instances during the second part of the twentieth century in which actions that had been intended as progress proved to cause perturbations to the environment that outweighed their advantages, and global measures were taken to eliminate or at least reduce the causes of these perturbations. Four examples are described in the following paragraphs.

The first example is the *drastic reduction of phosphorus* (or *phosphates*) *in detergents* in economically developed countries. In the mid-1940s, the soap-based products used for laundering, dishwashing and household cleaning started to be replaced by synthetic detergents that contained large amounts of phosphate compounds. The advantage of the new detergents came from the fact that phosphate compounds bind with dirt and keep it suspended in water, which allows the cleaning agents in the detergent to function very effectively. Because the performance of detergents containing phosphates was excellent, these all but replaced the soap-based products in the 1950s and 1960s. During the same period, the use of synthetic compounds increased rapidly in developed countries, including industry and agriculture. Given that phosphates are not toxic, their wide use did not initially raise special concerns. However, by the mid-twentieth century, freshwaters increasingly suffered from the developments of algal scum, growth of aquatic plants, and periodic fish kills caused by oxygen depletion resulting from the decomposition of the unusually high algal and plant biomasses.

In the 1960s, scientists began to link noxious algal blooms with the increasing supply of anthropogenic nutrients in the catchments of lakes, a phenomenon called *eutrophication*. The analysis of available data suggested that controlling (reducing) phosphorus inputs to lakes could be the key to reducing eutrophication, but some studies came to different conclusions. One spectacular demonstration of the role of phosphorus came from the Canadian Experimental Lake Area, where a whole lake was divided into two equal halves using a large vinyl curtain, after which researchers added an equivalent amount of carbon and nitrogen to both halves of the lake, and phosphorus to only one side. Within a few months, the side of the lake receiving $C + N + P$ was completely covered by a bloom of blue-green algae whereas the other side remained unchanged (Fig. 10.8).

The above and other experiments convinced authorities to curb phosphorus eutrophication. This was achieved by upgrading wastewater treatment plants to remove phosphorus from effluents, and also forcing industries to reduce the amount of phosphate in cleaning products. These measures did not completely resolve the eutrophication problem because there are other sources of nutrient inputs to the environment, including agriculture, but the regulations to control phosphorus emissions in many areas—which include Australia, Canada, Europe, Japan and the USA—have significantly improved the health of freshwaters compared to the 1960s. These regulations banned (at least partly) the use of phosphate in detergents and/or improved sewage treatment to remove phosphorus (at least partly) from wastewaters. Chapters in the textbook of Laws (2017) provide further reading on eutrophication of natural waters, nonpoint source pollution, and sewage treatment.

The above paragraphs show that the creation of phosphate-based detergents was initially considered as a progress, but the fact that phosphorus persists in waste waters created eutrophication, which effect was neither intended nor expected. Once the main cause of this eutrophication was identified by the scientific community, many countries independently took action to reduce the magnitude of the problem.

The second example is the *partial international ban on the insecticide DDT* (dichlorodiphenyltrichloroethane). This organochlorine chemical compound was first synthesized in 1874, and its insecticidal action was discovered by the Swiss chemist Paul Hermann Müller in 1939. Doctor Müller was awarded the Nobel Prize in Physiology or Medicine for this discovery in 1948. During World War II, the Allies used DDT to protect troops and civilians from diseases spread by insects, including malaria and typhus. After the war, DDT was widely used to control pest insects in agriculture in economically developed countries, and to eradicate malaria in many countries worldwide. Malaria was successfully eradicated in economically developed areas, such as North America and Europe, but not in tropical regions due to the continuous life cycle of mosquitoes and poor infrastructures, especially that mosquitoes progressively developed resistance to DDT.

Fig. 10.8 An experimental lake was divided in two parts with a plastic curtain, and phosphorus was added to one half (top): after a few months, the half with added phosphorus turned murky green. Schindler (1974) published the first photo (in black-and-white) of this influential experiment. Credits at the end of the chapter

Ecological research progressively showed that the agricultural use of DDT was a threat to wildlife, and particularly birds because of their dependence on insect food. The bestselling 1962 book *Silent Spring*, by the American marine biologist and conservationist Rachel Carlson (2002), documented the adverse environmental effects of the indiscriminate use of pesticides, of which DDT was a prime example (book recommended as further reading). In addition, medical research indicated that accumulation of DDT in the body had adverse effects on human health because of its interference with reproductive hormones.

The accumulation of scientific information confirming the dangerous effects of DDT on the environment ultimately led to a global ban on the agricultural use of this insecticide under the *Stockholm Convention on Persistent Organic Pollutants*, which is effective since 2004. However, the convention still allows the use of DDT in disease vector control, because its effectiveness in controlling mosquitoes (and other insects) is considered to balance its environmental and health concerns. When the environmental effects of this insecticide outweighed its advantages, the international society took measures to reduce its use.

The third example is the *international action taken to control acid rain*, which designates all forms of precipitation that have unusually low pH and are thus unusually acidic. Acid deposition occurs not only by precipitation, but also via the settling of some aerial particles in the absence of precipitation. Acid rain is caused by the reaction of chemical compounds—mostly sulphur dioxide (SO_2) and nitrogen oxide (N_2O)—with water molecules in the atmosphere, which reaction produces sulphuric acid (H_2SO_4) and nitric acid (and HNO_3).

Both SO_2 and N_2O are naturally released in the atmosphere by volcanic eruptions (see Sect. 7.4.6), and N_2O is also naturally produced in the atmosphere by lightning. The problem examined here is the additional acidity caused by industrial emissions of SO_2 and N_2O. These emissions result from the combustion of fossil fuels, and the generation of electricity using coal is among the greatest contributors to atmospheric pollution causing acid rain. The latter has adverse impacts on forests, where it kills trees, and on soils and fresh waters, where it kills insect and aquatic life forms (Fig. 10.9). In addition, acid rain causes the corrosion of steel structures and the erosion of stone buildings and statues, and has negative impacts on human health.

The detrimental effects of acid rain had been observed since the early days of the Industrial Age. For example, the relationship between acid rain and atmospheric pollution had already been shown in Manchester, England, in 1852, but it is only in the 1960s that scientists began to systematically observe and study the phenomenon. In order to alleviate the local effects of acid rain, factories increased the height of their smoke funnels, but taller smoke funnels caused the pollutants to be carried over hundreds of kilometres in the atmosphere before being converted to acids and deposited. This transformed local acid rain problems into regional and even international issues. Even if it had been shown in 1881 that pollutants were transported over long distance across borders from the United Kingdom to Norway, the general recognition of the cross-border nature of acid rain in North America and Europe had to wait the 1980s.

There are technical solutions to reduce the amount of sulphur emitted by the combustion of coal, and also to reduce the emissions of nitrogen oxides from motor vehicles. These solutions have been implemented by national regulations, and also under the *Convention on Long-Range Transboundary Air Pollution*, which addresses the long-range transport of air pollutants across borders and entered into force in 1983. The convention does not cover all types of atmospheric pollution, but it has nevertheless contributed to a dramatic decline in air pollution emissions in North America, Europe and Russia, particularly for sulphur.

Fig. 10.9 Effect of acid rain on a forest in the Jizera Mountains, Czech Republic, in 2006. Credits at the end of the chapter

The above paragraphs show that the use of fossil fuels was intended to improve human comfort (electricity, cars), and after its acid-rain effects became increasingly alarming, several countries got together to implement technical solutions that dramatically reduced this dangerous effect. The textbook of Laws (2017) has a chapter on the effects of acid rain on aquatic environments.

The fourth example is the *international phasing out of chloroflurocarbons (CFCs)*. Thesecompounds are volatile hydrocarbons that contain carbon, hydrogen, chlorine and fluorine. They were first synthesized in the 1890s, and began to replace in the 1920s the toxic compounds then used as refrigerant in refrigeration systems. A *refrigerant* is a substance or mixture used in a heat pump and refrigeration cycle, where it undergoes phase transitions from a liquid to a gas (absorption of heat) and back again (release of heat). The CFCs were ideal refrigerants because of their low toxicity, low chemical reactivity and low flammability. During the twentieth century, different CFCs were also used as propellants (in aerosol applications), and degreasing solvents. Hence CFCs were widely used by the mid-twentieth century. All these gases were eventually emitted to the atmosphere, either immediately in the case of propellants or solvents, or when refrigeration systems were discarded and dismantled.

In the early 1970s, atmospheric chemists began to suspect that one of the CFCs' most attractive features, that is, their low chemical reactivity, caused them to be very destructive. Indeed, CFC molecules are stable enough to remain in the atmosphere up to 100 years and even more, and can thus reach the middle of the stratosphere (layer located from 10 to 50 km above ground; see Sect. 2.2.1). There, they are broken down by the ultraviolet radiation of the Sun, which releases chlorine atoms (Cl^-). The chlorine act as catalyst in turning ozone (O_3) into oxygen (O_2), thus causing the breakdown of large amounts O_3 in the stratosphere, which destroys the O_3 layer (see Information Box 8.2). Given that this layer protects land organisms from DNA-destructive solar UV radiation (see Sects. 2.1.5 and 10.1.7), its disappearance would be catastrophic.

The frightening hypothesis linking CFCs to the O_3 layer was hotly debated by scientists and CFC-producing industries until the mid-1980s. The hypothesis turned to reality when abnormally low O_3 concentrations were reported over the South Pole, leading to the metaphor of the *ozone hole*. This led to the *Montreal Protocol on Substances that Deplete the Ozone Layer*, which entered into force in 1989 and intends to protect the O_3 layer by phasing out the production of O_3-depleting substances (the protocol originally targeted two substances, and it now covers almost 100). In addition, there is a *Multilateral Fund for the Implementation of the Montreal Protocol* to assist developing countries to comply with the protocol. Although limited amounts of illegal CFCs are still produced in a few areas today, the Montreal Protocol is seen as the most successful international environmental agreement to date. The researchers who had explained the mechanisms that affect the thickness of the ozone layer received the Nobel Prize in Chemistry in 1995. The CFCs are substituted with hydrochlorofluorocarbons (HCFCs, which are partly reactive with O_3, and are thus being phased out) and hydrofluorocarbons (HFCs, which are not reactive with O_3). Unfortunately, these two types of substances are very powerful greenhouse gases, which thus contribute to global warming.

The above paragraphs show that the use of CFCs was initially very successful and thought to be harmless. However, their unexpected O_3-depleting effect threatened the health and even the survival or land organisms, which led within a few years to a highly successful international agreement to phase out the ozone depleting substances.

The four examples explained how, during the second part of the twentieth century, countries acted in concert to reduce or eliminate the causes of four large-scale or global environmental disturbances and their feedbacks into human societies. These examples show that action was taken after researchers had reported environmental disturbances and identified their causes, and once the public had been convinced of the existence and severity of the problem, governments took action. The resolution of some major environmental problems during the twentieth century shows that such actions are possible, and not wild dreams. However, many other developing, large-scale or global environmental problems were not addressed in the twentieth century despite mounting evidence that they were building up. These neglected problems have long-term consequences in the twenty-first century, which are examined in Chapter 11.

Committed citizens have often invoked the motto "Save the Earth" in the last decades, which led many people to engage in environmental action. However, we have seen in previous chapters that the Earth System has overcome innumerable cataclysms since the creation of the Earth-Moon system 4.6 billion years ago, and will likely continue to do so during the billions of years to come. Hence, the main incentive for action should be "Save Ourselves". In the four examples above, governments took action once the scientific evidence and public pressure convinced them that we indeed had to save ourselves and there were ways to do it. The next chapter examines the situation in the early twenty-first century, and what might happen next and in the more distant future.

Figure Credits

Fig. 10.1 Original. Figure 10.1 is licensed under CC BY-SA 4.0 by Philippe Bertrand, Louis Legendre and Mohamed Khamla.

Fig. 10.2 Original. Figure 10.2 is licensed under CC BY-SA 4.0 by Philippe Bertrand, Louis Legendre and Mohamed Khamla.

Fig. 10.3 Original. Figure 10.3 is licensed under CC BY-SA 4.0 by Philippe Bertrand, Louis Legendre and Mohamed Khamla.

Fig. 10.4 Original. Figure 10.4 is licensed under CC BY-SA 4.0 by Philippe Bertrand, Louis Legendre and Mohamed Khamla.

Fig. 10.5 Original. Figure 10.5 is licensed under CC BY-SA 4.0 by Philippe Bertrand, Louis Legendre and Mohamed Khamla.

Fig. 10.6 Original. Figure 10.6 is licensed under CC BY-SA 4.0 by Philippe Bertrand, Louis Legendre and Mohamed Khamla.

Fig. 10.7 Original. Figure 10.7 is licensed under CC BY-SA 4.0 by Philippe Bertrand, Louis Legendre and Mohamed Khamla.

Fig. 10.8 Following the first black-and-white photo published by Prof. David Schindler (1974), the present colour photograph was kindly provided by the International Institute for Sustainable Development (IISD) Experimental Lakes Area (https://www.iisd.org/ela/). With permission from the IISD.

Fig. 10.9 https://commons.wikimedia.org/wiki/File:Acid_rain_woods1.JPG by Lovecz https://commons.wikimedia.org/wiki/User:Lovecz, in the public domain.

References

Darwin C (1859) On the origin of species by means of natural selection, or the preservation of favoured races in the struggle for life. John Murray, London

Pariona A (2017) Worldwide population throughout human history. WorldAtlas. Available via http://www.worldatlas.com/articles/worldwide-population-throughout-human-history.html. Accessed 25 Apr 2017

Schindler DW (1974) Eutrophication and recovery in experimental lakes: implications for lake management. Science 184(4139):897–899
United Nations (2019) World population prospects 2019. Available via https://population.un.org/wpp/

Further Reading

Carlson R (2002) Silent Spring, 40th Anniversary edition. A Mariner Book, Houghton Mifflin, Boston, New York
Glikson AY, Groves C (2016) Climate, fire and human evolution. The deep time dimensions of the anthropocene, Modern Approaches in Solid Earth Sciences, vol 10. Springer, Cham
Harari YN (2015) Sapiens: a brief history of humankind. Vintage, London
Hugo V (2016) La légende des siècles (English Edition) CreateSpace Independent Publishing Platform
Knoll AH (2015) Life on a young planet: the first three billion years of evolution on Earth, Updated edn. Princeton University Press, Princeton
Laws EA (2017) Aquatic pollution. An introductory text, 4th edn. John Wiley & Son, Hoboken
Ramin S (2018) Has humankind driven Earth into a new epoch? Knowable Magazine. https://doi.org/10.1146/knowable-012818-144302
Stanley SM, Luczaj J (2014) Earth system history, 4th edn, W. H. Freeman, New York
Wohlleben P, Billinghurst (Translator) (2019) The secret wisdom of nature: trees, animals, and the extraordinary balance of all living things—stories from science and observation. Greystone Books, Vancouver

Chapter 11
The Global Earth System: Present and Future

11.1 Anthropocene Geological Epoch: The Twenty-First Century

11.1.1 The Twenty-First Century: Global Perturbations of the Earth System

We saw in the previous Chapter 10 that, during the second part of the twentieth century, the human society managed to reduce or eliminate the causes of some of the environmental perturbations that threatened the wellbeing of people. Indeed in at least four major cases, societies took action nationally, regionally, multilaterally or internationally when the scientific case was made, public opinion was mobilized, and solutions were agreed upon (see Sect. 10.3). We now turn to the twenty-first century, and wonder if the same could occur in the coming decades.

In order to explore the present situation and what might happen next, we examine successively the main global perturbations of the Earth System (this section), human activities that affect the Earth System (Sect. 11.1.2), and several feedbacks into the environment and human activities (Sect. 11.1.3). We then explain how adaptation and mitigation approaches provide responses to the ongoing global anthropogenic perturbations of the Earth System (Sects. 11.2), and look at possible states of the Earth System beyond the twenty-first century (Sect. 11.3). We conclude the chapter and the book by considering what planet Earth would be without some of the key biological innovations that shaped the Earth System (Sect. 11.4), and the need for stewardship of the planet (Sect. 11.5).

In the twenty-first century as was the case in the 20th, humans are both the causes of environmental perturbations and the victims of their impacts on societies. In the four examples described in Chapter 10, human societies prevented detrimental impacts of environmental perturbations on people by eliminating or

© Springer Nature Switzerland AG 2021
P. Bertrand and L. Legendre, *Earth, Our Living Planet*, The Frontiers Collection,
https://doi.org/10.1007/978-3-030-67773-2_11

at least controlling their causes. The global problems to which societies are confronted and their possible solutions are investigated nowadays by intergovernmental bodies that bring together independent scientific experts, government representatives, and other stakeholders. Two key bodies that address global environmental problems are the *Intergovernmental Panel on Climate Change* (IPCC) and the *Intergovernmental Science-Policy Platform on Biodiversity and Ecosystem Services* (IPBES), which regularly produce widely publicized reports and assessments, whose conclusions and recommendations are explained in information media all over the world. The documents produced by these bodies can be downloaded free of charge from their websites. Hence, there is no need to dwell here on their analyses and recommendations, which are continually updated by the two bodies. Our purpose in the remainder of this section and in the next two is to examine the ongoing global anthropogenic perturbations of the Earth System within the context of the interactions illustrated in Fig. 10.1, and describe their feedbacks into the environment and human activities.

The Earth System consists of the atmosphere, the liquid water, the ice, the rocks and the organisms, and the physical, chemical and biological processes within and among them (see Sect. 1.1.1). Since the beginning of the Anthropocene epoch (see Sect. 10.2.1), the human component of the biosphere acted more and more independently from natural ecosystems. We consider below the mechanisms by which a range of human activities contribute to global changes in the Earth System and their potentially damaging feedbacks into human societies. We first briefly describe four global perturbations that affect the atmosphere (climate change), the ocean (acidification), the ecosystems (loss of biodiversity), and human societies (water and food insecurity). Table 11.1 summarizes two main characteristics of these global perturbations, that is, the main properties of the Earth System or ecosystems affected by each perturbation, and the general mechanisms of the perturbation. These are explained in the remainder of the section.

1. Climate change refers to changes in the *state of the climate system* (see Information Box 1.4)—that is, the highly complex system consisting of five major components: the atmosphere (air), the hydrosphere (liquid water), the cryosphere

Table 11.1 Four global perturbations of the Earth System: property of the ecosystem or the Earth System affected and general mechanisms. Details are provided in the accompanying text, where the paragraphs are numbered in the same way as the rows in the Table

Global perturbation	Main properties affected	General mechanisms
1. Climate change	State of the climate system	Changes in atmospheric radiation
2. Ocean acidification	Seawater pH and chemical composition	Changes in ocean chemistry
3. Loss of biodiversity	Number of biological species	Changes in land and aquatic habitability
4. Water and food insecurity	Accessibility to freshwater and food	Changes in the environmental conditions of water use and farming

(ice), the lithosphere (rocks), and the biosphere (organisms)—and their interactions (see Sect. 3.5.1). The state of the climate system varies under the influence of its own internal dynamics and because of external forcings. The latter include: modulations of the orbital cycles, with timescales of hundreds of thousands of years (see Sect. 4.5.2); solar cycles, with a decadal timescale (see Sect. 8.3.2); large volcanic eruptions, with a centennial timescale (see Sect. 4.4); and persistent *anthropogenic* changes to the composition of the *atmosphere* and *land use*. Climate variability can thus result from natural causes or human activities that alter directly or indirectly the composition of the global atmosphere. Climate change due to human activities adds to natural climate variability.

Since the beginning of the Industrial Age in the second part of the eighteenth century, the combustion of fossil carbon (coal, petroleum, natural gas, peat) and the calcination of $CaCO_3$ by human activities has released increasing amounts of carbon dioxide (CO_2) in the atmosphere. As a consequence, the concentration of this atmospheric gas has increased from about 290 ppmv (parts per million by volume) in the mid-nineteenth century to more than 415 ppmv in 2019, and the increase is accelerating (Fig. 11.1a). The acceleration is due to the fact that the rate at which CO_2 has been emitted in the last century, and especially the last decades, has grown so fast that the natural carbon sinks on the planet could only partly cope with the excess CO_2.

Carbon sinks are pools that absorb and store the atmosphere's carbon through physical and biological mechanisms, and the main natural sinks for the excess of atmospheric CO_2 are at present humus-storing soils (such as peatlands), actively growing forests in northern mid-latitude areas, and the oceans. Together, these sinks have absorbed and stored about half the carbon from the CO_2 emitted since the beginning of the Industrial Age, the other half being responsible for the ongoing global warming. The continental and oceanic carbon sinks have each taken up about half the stored carbon, and it is expected that the share of the continents will decrease in the future.

One of the results of the increasing atmospheric CO_2 concentration was an intensification of the greenhouse effect, which involves *changes in atmospheric radiation* (see Information Box 2.7). These changes caused the current rise in average temperature, which is 1.5 °C above pre-industrial levels. Indeed, studies conducted with numerical models concluded that the primary cause of global warming is the increasing concentration of atmospheric CO_2. The ongoing global increases in the concentration of atmospheric CO_2 and temperature are larger than any change that occurred during the last 800,000 years. This is also true for the predicted warming by the end of the twenty-first century, in terms of both the maximum temperature of the planet and the rate of temperature increase (Fig. 11.1a, b).

2. Ocean acidification is the *reduction in the pH of seawater*, which is accompanied by changes in other chemical characteristics such as the $CaCO_3$ saturation state (Ω, Eq. 11.5), over decades or longer. It is caused primarily by uptake of CO_2 from the atmosphere, but can also be caused by other chemical additions or subtractions from the ocean. One example is the input of dissolved organic carbon of

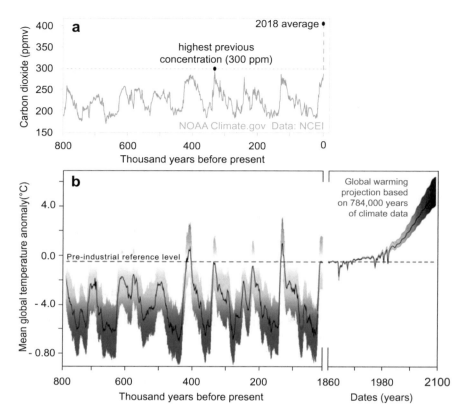

Fig. 11.1 Global changes during the last 800,000 years. **a** Variations in atmospheric CO_2 concentration. **b** Global mean surface temperature expressed as global mean temperature anomaly, that is, the difference with the preindustrial global temperature (reference level); the envelopes around the mean values represent the uncertainty in the data (until 1860), and the range of model predictions (from present until 2100); blue stands for cold, and red for warm; the estimates are from Friedrich et al. (2016). Credits at the end of the chapter

continental origin in some coastal waters, whose remineralization (respiration) can release enough CO_2 to cause acidification. *Anthropogenic ocean acidification* is an additional harmful effect of the anthropogenic increase in atmospheric CO_2 concentration, which is sometimes called the "evil twin" of climate change, described just above.

If all the CO_2 emitted by human activities since the beginning of the Industrial Age had stayed in the atmosphere, the present surface temperature would be much higher than the present 1.5 °C above pre-industrial levels. This is not the case, fortunately, because about half of the anthropogenic CO_2 was taken up by continental and oceanic carbon sinks, as explained above. The atmospheric CO_2 absorbed by the ocean's surface waters (dotted atmosphere-to-ocean arrow in Fig. 9.6b) *changed the chemistry of seawater*, more precisely its carbonate equilibrium (see Information Box 6.3, whose Eqs. 6.3 and 6.4 are repeated here as Eqs. 11.1 and 11.2 for convenience):

$$CO_2 + H_2O \rightleftharpoons HCO_3^- + H^+ \tag{11.1}$$

$$HCO_3^- \rightleftharpoons CO_3^{2-} + H^+ \tag{11.2}$$

Rearranging Eq. 11.1

$$HCO_3^- \rightleftharpoons CO_2 + H_2O - H^+ \tag{11.3}$$

and adding together Eqs. 11.2 and 11.3 gives:

$$CO_2 + CO_3^{2-} + H_2O \rightleftharpoons 2HCO_3^- \tag{11.4}$$

Equation 11.4 indicates that the production of HCO_3^- in seawater consumes both dissolved CO_2 and CO_3^{2-}. Hence, when excess atmospheric CO_2 dissolves in seawater, the carbonate equilibrium is shifted to the right of Eqs. 11.1 and 11.2, and H^+ ions are produced. This excess also causes the relative amounts of HCO_3^- and CO_3^{2-} to increase and decrease, respectively (Eq. 11.4), which is also illustrated in Fig. 9.7. Higher concentration of H^+ means lower pH, and thus acidification (see Information Boxes 6.3 and 6.4). This is what presently happens in the ocean, where pH has decreased from a pre-industrial value of 8.25 to the current value of 8.14, most of this change having taken place in the last decades. Because pH is expressed on a logarithmic scale, the difference of 0.11 pH units represents an increase of almost 30% in the concentration of H^+ ions. The name of this phenomenon, *ocean acidification*, does not mean that marine waters are becoming acidic (acidity is defined as pH < 7.0; see Information Box 6.4) but that their pH is decreasing.

It is expected that ocean acidification will continue and perhaps accelerate in the coming decades (Fig. 11.2). This will have detrimental consequences for the many aquatic organisms and ecosystems sensitive to the effects of multiple stressors, such as combined changes in pH, temperature, and dissolved O_2 concentration. Further reading on ocean acidification is provided by the articles of Feely et al. (2009) and Mackenzie and Andersson (2013).

3. Biodiversity (or biological diversity) refers to the variability among organisms from all sources including, among others, terrestrial, marine and other aquatic ecosystems and the ecological complexes of which they are part; this includes diversity within species, between species, and of ecosystems.

The *number of species* is presently declining faster than at any time in human history. By 2019, for example: natural ecosystems have declined by 47% on average, relative to their earliest estimated states; approximately 25% of species are already threatened with extinction in most animal and plant groups studied; the abundance of naturally-present species has declined by 23% on average in terrestrial communities; and the global biomass of wild mammals has fallen by 82%. During the past 50 years, the *direct drivers* of change in nature with the largest global impact on biodiversity have been (starting with those with most impact): changes in land and sea use; direct exploitation of organisms; climate change; pollution; and invasion of alien species. Overall, these direct drivers led to major

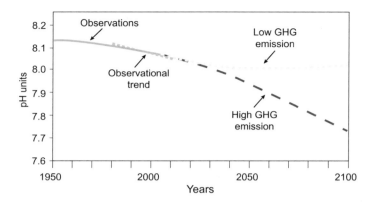

Fig. 11.2 Changes in the global mean surface ocean pH from 1950 to 2100: observations since 1950 (solid curve), observational trend between 1980 and 2012 (dotted straight line), and projected future changes until 2100 from numerical models under scenarios of low and high greenhouse gas (GHG) emissions (dotted and dashed curves, respectively). Credits at the end of the chapter

changes in land and aquatic habitability. The direct drivers result from an array of underlying causes—called the *indirect drivers* of change—which are in turn underpinned by societal values and behaviours that include production and consumption patterns, human population dynamics and trends, trade, technological innovations, and local through global governance.

Cumulative species extinctions since year 1500 provide an example of the accelerating loss of biodiversity in the most recent period (Fig. 11.3). It is known that the background rate of extinction of species of vertebrate ranges between 0.1 and 2 per million species per year. Given this extinction rate, the cumulative % of species driven extinct between 1500 and 2018 should have ranged between 0.05 and 0.1%, whereas it was close to 1.0 for fishes and reptiles, between 1.5 and 2.0 for birds and mammals, and almost 2.5 for amphibians (Fig. 11.3). This is more than ten times the background rate for all groups of vertebrates.

4. Water and food insecurity generally refers to the *accessibility to freshwater and food.* It was explained in previous chapters that water (H_2O) is the major constituent of organisms, and is a key compound in the process by which organic matter is synthesized by O_2-producing photosynthesis (see Sect. 6.1.2, and Fig. 5.1, respectively). Water is also the source of all the free O_2 that exists on Earth, which comes from the splitting of H_2O molecules during the process of O_2-producing photosynthesis (see Information Box 3.4). All organisms including humans require water, and all heterotrophs also require food. In human societies as in the natural environment, the production of food is strongly related to the availability of freshwater, as shown by a few examples in Table 11.2. Hence, water and food are tightly linked, and humans need both (Fig. 11.4).

Table 11.2 indicates that the production of not only food but also industrial goods in modern societies requires lots of water. Most people drink at most 2 or 3 litres of water in a day, but they unknowingly consume thousands of litres of water by eating even simple food and using current industrial products.

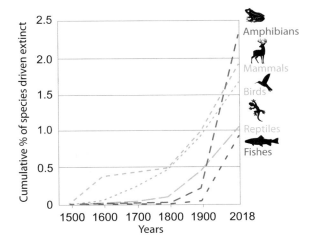

Fig. 11.3 Illustration of the accelerating loss of biodiversity: species extinctions since the year 1500 for different groups of vertebrates (rates for reptiles and fish have not been estimated for all species). Credits at the end of the chapter

Table 11.2 Amount of water required for the production of selected food and industrial products

Food product	Water required (litres)	Industrial product	Water required (litres)
1 kg of beef	15,000	1 ton of steel	210,000
1 kg of lamb	10,000	1 car	52,000-83,000
1 hamburger	2,400	1 smartphone	12,760
1 kg of cereals	1,500	1 pair of leather shoes/ boots	8,000/14,000
1 litre of milk	1,000	1 pair of jeans (cotton)	8,000
1 pizza Margherita	1,260	1 ton of cement	4,700
1 litre of wine	480	1 cotton T-shirt	2,500-3,900
1 cup of coffee	140	1 microchip (2 g)	32
1 egg	135	1 litre of paint	13
1 apple	70	1 sheet of A4 paper (80 g/m²)	10

Water security is the capacity of a population to safeguard sustainable access to adequate quantities of acceptable quality freshwater for sustaining livelihoods, human wellbeing, and socio-economic development. *Water insecurity* occurs when freshwater quality, quantity, availability or accessibility are not achieved.

Food security exists when all people, at all times, have physical, social and economic access to sufficient, safe and nutritious food that meets their dietary needs and food preferences for an active and healthy life. There is *food insecurity* when people lack a secure access to sufficient amounts of safe and nutritious food for normal growth and development and an active and healthy life.

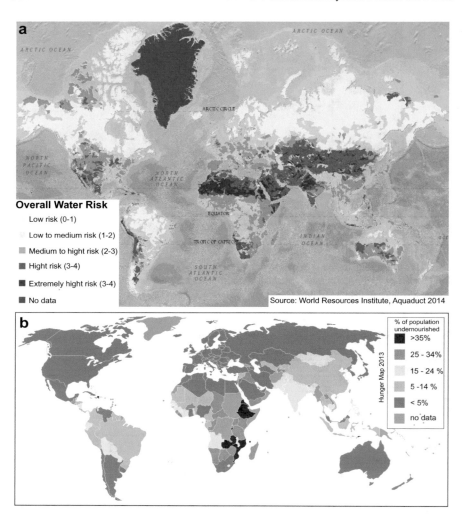

Fig. 11.4 Illustration of the global effects of freshwater and food insecurity: **a** exposure to water-related risk (based on the local availability of adequate water supply and the risk of flooding) in 2014, and **b** percentage of the population suffering from hunger in 2013. Credits at the end of the chapter

Water and food insecurity are not evenly distributed around the world, as illustrated by the two panels of Fig. 11.4. Climate change, urbanization, energy and food requirements, and population growth all put strains on water and food resources. Present *changes in the environmental conditions of water use and farming* increase water and food insecurity.

11.1.2 The Twenty-First Century: Human Activities That Affect the Earth System

Given the large size of human population (Table 10.2) and the efficiency of technologies, human activities increasingly affect several components of the Earth System. Negative effects began to become apparent as early as the second half of the twentieth century, and human societies then modified some of their activities in order to minimize or eliminate certain severe feedbacks into them (see Sect. 10.3). However, many negative effects remain today as well as their associated detrimental feedbacks, because their causes were not corrected. As these effects and feedbacks are better identified, the corresponding causal activities need to be modified in order to achieve *sustainability*. The latter is the dynamic process that could guarantee the persistence of natural and human systems in an equitable manner.

In Table 11.3, we consider successively five types of human activities, which generally contribute to the global wellbeing of humans but have some negative effects on the Earth System in terms of climate change, ocean acidification, loss of biodiversity, and water and food insecurity. The identification of these negative effects is a necessary step towards the improvement of the sustainability of these activities. The five types of activities are listed in Table 11.3 in order of first occurrence during the course of human history, and are examined below in the same order. In addition to the direct effects cited in Table 11.3, there are indirect effects resulting from the fact that some aspects of each global perturbation affect other perturbations. For example, climate change affects both the loss of biodiversity and water and food insecurity through effects of both global warming and global changes in precipitation. For simplicity, such indirect effects of human activities on global perturbations are not included in Table 11.3.

1. Logging natural forests, fishing wild stocks, harvesting wild plants, and hunting wild species has been practiced by hunters-gatherers since humans have appeared on Earth. This sometimes led to destructions of large-sized species as explained above (see Sect. 10.2.1). However, these activities were often conducted in a sustainable way because overexploitation reduced the stocks of the harvested resources, which led to decreased hunting-gathering pressure, thus creating a negative feedback (see Information Box 6.2). This is often not the case anymore because of the rapid increase in human population (Table 10.2) and the ever-increasing consumption of goods and services in the last centuries and especially the last decades. As a consequence, most of these activities have become or are becoming non-sustainable.

Forests more than 20 years old accumulate carbon, so that the rapid logging of forests *destroys a major carbon sink*. In addition, most of the harvested wood is eventually burned, which releases the carbon stored in trees in the form CO_2, which accumulates in the atmosphere. The increase in atmospheric CO_2 resulting from these two effects of forest logging intensifies both the *greenhouse effect* and *ocean acidification*. The harvesting of wild plants or the hunting of wild animals

Table 11.3 Mechanisms by which human activities contribute to the four global perturbations of the Earth System described in Table 11.1, namely climate change, ocean acidification, loss of biodiversity, and water and food insecurity. Details are provided in the accompanying text, where the paragraphs are numbered in the same way as the rows in the Table

Human activities	Climate change	Ocean acidification	Loss of biodiversity	Water and food insecurity
1. Logging, fishing, harvesting, hunting	Destruction of forest carbon sink	Destruction of forest carbon sink	Destruction of species and habitats	Overexploitation of wild stocks
2. Agriculture, silviculture	Destruction of forest carbon sink; emission of $CH_4 + N_2O$	Destruction of forest carbon sink	Destruction and fragmentation of habitats; pollution, and eutrophication	Soil erosion; water scarcity
3. Industrialization, mining, and extraction of industrial aggregates, gas and petroleum	Emission of CO_2, CH_4, N_2O, and SO_2	Emission of CO_2	Land and water pollution, and destruction or perturbation of habitats	Land and water pollution; loss of arable land
4. Urbanization, transportation	Emission of CO_2, CH_4, N_2O, SO_2, and heat	Emission of CO_2	Pollution, destruction and fragmentation of habitats	Land and water pollution; loss of arable land; food lost and waste
5. Growth of information and communication technology	Emission of CO_2 and heat	Emission of CO_2	No direct effect	No direct effect

for such activities as traditional cuisine or medicine, which are now practiced by increasingly numerous populations, can wipe out the target species and thus decrease *biodiversity*. The latter is also threatened by the destruction of forests, which often *obliterates the specific habitats of some plants or animals*. Finally, the *overexploitation of wild stocks* can reduce them to such small populations that they become vulnerable to extreme events (climatic or other, as observed in many commercially fished stocks) or cannot replace themselves, which contributes to increase *food insecurity*.

2. Agriculture, which began between 10,000 and 8000 ago, and more recently *sylviculture* have progressively replaced most of the hunting-gathering activities. In order to cultivate crops and raise domestic animals, humans clear increasing large tracts of forests, an activity that *decreases the efficiency of forests as carbon sinks*, and thus increases both *climate change* and *ocean acidification*. Deforestation was a slow process, which was not continuous given that the same regions were often deforested and reforested several times during the course of history. Presently, agricultural deforestation proceeds at an accelerated pace in some regions of the world, whereas in other regions forests increasingly occupy abandoned marginal farmlands. As a result, most landscapes in the world are deeply marked by humans.

In addition, intensive agriculture produces much *methane (CH_4)* and is the largest producer of anthropogenic *nitrous oxide (N_2O)*, which are both very powerful greenhouse gases, much more powerful than CO_2. The agricultural emissions of CH_4 result from the microbial fermentation of food in the stomachs of ruminant animals such as cows, sheep and goats, and also from rice farming in which CH_4 is emitted by bacteria that live in the waterlogged soil of the paddies. The N_2O gas is emitted from agricultural soils as a consequence of the widespread use of nitrogenous fertilizers, and also from decaying livestock manure. Because of the development of intensive agriculture and sylviculture, the anthropogenic emissions of CH_4 and N_2O contribute significantly to the *global greenhouse effect*.

One of the effects of agriculture is the *destruction and fragmentation of natural habitats*, which is also caused by such human activities as industrialization (for example, the creation of hydroelectric reservoirs), the development of roads and railroads, and urbanization. Fragmentation is the division of formerly continuous habitats into separate fragments. Habitat destruction and fragmentation affect *biodiversity* by reducing the amount of suitable habitats available for wild organisms.

Another negative effect of agriculture and sylviculture is the *pollution of land and water* by herbicides and pesticides. These chemicals are used to improve the productivity of farming, but a side effect is often the mass destruction of wildlife, including insects and birds. Some neonicotinoid pesticides may cause similar threats to biodiversity as DDT did half a century ago (see Sect. 10.3). Another form of water pollution is *eutrophication*, which was described in Chapter 10 for phosphorus (see Sect. 10.3) and is caused by highly abundant nutrients resulting from the excessive use of artificial fertilizers or from large amounts of manure in livestock farming. Another source of nutrients is aquaculture, where the phosphorus and nitrogen contained in effluents can cause the eutrophication of natural waters. These various forms of pollution harm and even kill organisms, so that long-term exposure to pollution can affect *biodiversity*.

In addition, deforestation for agricultural use and farming practices that disrupt soil structure may lead to *soil erosion* (see Sect. 7.3.3), which has become a severe problem worldwide. Erosion reduces the productivity of agricultural soils, and also water quality when the eroded soils mix with surface waters. Finally, agriculture is the largest water user globally and a major source of water pollution, which may cause *water scarcity* for other uses, including human and industrial. Both soil erosion and the heavy use of water by agriculture contribute to *water and food insecurity*.

3. Industrialization, mining, and extraction of aggregates, gas and petroleum have increased considerably since the beginning of the Industrial Age in the late 1700s. This was accompanied by increasing *emissions of CO_2 and other greenhouse gases*. The main anthropogenic sources of CO_2 are the burning of fossil fuels (mostly natural gas, coal and petroleum) for energy production and the calcination of $CaCO_3$ rocks for cement production (Fig. 9.5). The production of energy from fossil fuels also releases CH_4, as do the decomposition of organic matter in landfills and agriculture (above). We have already seen that agriculture is the main source of anthropogenic N_2O, and other sources of this gas are biomass burning and fossil fuel combustion, which also emits SO_2 (see Sect. 10.3). The phasing out of the synthetic O_3-depleting CFC gases under the 1989 *Montreal Protocol* has led to increased use of alternative HCFC and HFC gases, which are very powerful synthetic greenhouse gases supposed to be phased out in the future. The emissions of all these anthropogenic gases increase *climate change*, and the emissions of CO_2 also enhance *ocean acidification*.

During the course of industrialization, human societies have harnessed most wavelengths of the electromagnetic spectrum, including the use of gamma and X rays for industrial and medical purposes, the application of optics to almost all aspects of life, and the use of the longer wavelengths of the spectrum for radio communications, radar and other applications (see Sect. 8.1.2). These applications require much electricity, whose production is largely based on fossil fuels in many countries. The exploitation of the electromagnetic spectrum thus also contributes to augment *climate change* and *ocean acidification*.

The warming caused by the increase in greenhouse gases not only enhances climate change, but also worsens the adverse effects of various environmental changes on *biodiversity*. For example, it is estimated that the fraction of species at risk of climate-related extinction would go from 5 to 16% if warming rose from 2 to 4.3 °C. Coral reefs are especially vulnerable, that is, they are projected to decline to 10–30% of their former cover at 1.5 °C warming and to less than 1% at 2 °C warming.

Many industrial and mining activities cause different forms of *pollution*. We already cited atmospheric pollution by greenhouse gases, but the air can also be polluted by gases that do not contribute to the greenhouse effect such as ozone, a respiratory hazard near ground level, and by fine particles. There are many other types of pollution, for example by light and noise, but we focus here on industrial pollutants that deteriorate the quality of land and water. These include *heat* (for example, from the cooling of power plants), *chemical waste* (for example, hydrocarbons, heavy metals, and toxic organic compounds), and *radioactive compounds* (for example, from nuclear power generation). Mining can cause water pollution problems by the outflow of acidic water from (usually abandoned) metal or coal mines (this is called *acid mine drainage* or *acid rock drainage*), metal contamination, and increased sediment levels in running waters. In addition, the extraction of gas and petroleum has often been accompanied by acute (accidental) and chronic (long-term) release of pollutants. As explained above for agriculture and sylviculture, the pollution from industry and mining can affect *biodiversity*.

The worldwide extraction (mining) of sand and gravel used in industry accounts for about 70–80% of the material mined every year, and represents the highest volume of raw material consumed on Earth after water. These materials are collectively called *aggregates*, and are mostly used for the construction of buildings, roads and other infrastructures, and for land reclamation, and silica sand is the main raw material for the commercial production of glass and electronic chips. The sand from deserts is not suitable for construction because its round grains do not bind well, and most industrial aggregates thus come from freshwater or marine environments. The extraction of large volumes of aggregates destroys or perturbs the natural habitats and thus impacts their *biodiversity*. The extraction of aggregates in coastal areas, combined with sea level rise, accelerates erosion and the loss of land.

Pollution caused by industrial, mining, agricultural (above) and domestic (below) sources limits the use of land for the production of food and affects the overall use of water. In addition, the occupation of large land surfaces by industrial or mining facilities, which are often located on *arable land*, reduces the areas available for farming. This way, the development of industries can increase *food and water insecurity*.

4. Global urbanization (population shift from rural to urban areas) and global transportation are characteristic of modern times. The first urban area to exceed 10,000 inhabitants was Uruk in ancient Mesopotamia (now Iraq), and it perhaps occurred as early as 3500 BCE. Urbanization first developed in Mesopotamia, Iran, India and China, before later reaching the Mediterranean. In 1800 only 3% of the world's population lived in cities, so that global urbanization is a recent phenomenon. Indeed, urbanisation accelerated from the middle of the nineteenth century onwards, together with the rapid growth of the World population (Table 10.2). Nowadays, more than half the human population lives in cities, and mega-metropolises are growing rapidly in Asia, Latin America and Africa.

Concerning transportation, the last two centuries saw the development of modern highways with hard-topped roads, rail transport systems, steamboats and later diesel-powered ships, wide-scale river and canal transport (these waterways had been largely developed before the 1700s), large harbours, and more recently commercial air transportation. Planes were carrying more than one million people at any given time before the COVID-19 epidemic of 2020, at which time the transportation of passengers almost completely stopped. A similar situation had occurred in Europe in 2010 at the time of the eruption of the Eyjafjallajökull volcano.

Urbanization and global transportation emit the *same gases in the atmosphere as industrialization* (above), which thus have the same effects on *climate change* and *ocean acidification*. For example, the production of electricity and heat is responsible for about half the global emissions of CO_2, and transportation for an additional 20%. It is sometimes thought that the shift from fossil fuels to hydrogen or electrical engines in transportation could resolve the present detrimental effects of this activity on the environment. These new modes of transportation would indeed remove noxious exhaust products from cities and roads, but the production of the electricity required to generate hydrogen, refill batteries or power trains emits greenhouse gases, except if this electricity is produced from renewable sources (or nuclear energy). In addition, even if it is possible to produce hydrogen by hydrolysing water, most of the hydrogen is presently produced from natural gas (CH_4), which process releases the greenhouse gases CO or CO_2 to the atmosphere.

In addition, cities increasingly suffer from the phenomenon of *urban heat islands*, whereby heat is absorbed and stored during the day by concrete buildings and black roads, and emitted back into the urban air especially at night. Other sources of heat in cities include vehicles, manufacturing plants, air conditioners and thermal power stations. These combined sources of urban heat increase the effect of heat waves, which are themselves a consequence of *climate change.*

Cities are also major sources of *land and water pollution*. Although an increasing part of domestic solid wastes is recycled, a large part of them is still incinerated or buried in landfills, and the proportion of waste recycled depends on countries. Incineration causes immediate air pollution, and burial can cause long-term water pollution. The disposal of sewage (liquid) can cause water pollution, whose severity depends on the sophistication and efficiency of wastewater treatment plants. The effects of this pollution on *biodiversity* and *water and food security* are similar to those described above for agriculture and industrialization. In addition, the disposal of plastics causes the accumulation of plastic products and microplastics in the environment and ultimately the ocean, whose global effects on wildlife, wildlife habitat and humans are still poorly known but suspected to be adverse. Also, the development of cities and transport infrastructures, such as roads and railroads, contributes to the fragmentation of habitats, with the same negative effects on *biodiversity* as in the case of agriculture above.

Finally, cities often develop on *arable land*, and their rapid growth reduces the surfaces available for farming. This contributes to move the production of food far away from cities where most of the food is consumed, which increases the need for long-haul transportation and the associated pollution. In many countries, transportation systems cannot efficiently cope with moving food from the increasingly distant farming areas to the increasing growing cities, with the result that around 30% the global food production is *lost or wasted* before it reaches the consumers. The combination of these effects with the energy, materials and water required for the production of food increases the global pollution of air land and as well as *water and food insecurity.*

5. Growth of information and communication technology (ICT) is often thought as environment friendly, given that computers, smartphones and other ITC devices do not directly emit noxious products. However, the reality is quite different. Indeed, as for all other manufactured goods, the production of ICT equipment has a number of negative environmental effects, including the release of greenhouse gases for the production of the required electricity. In addition, the ICT products contain high amounts of rare earth metals (scandium, yttrium and the 15 lanthanides; elements number 21, 39 and 57 to 71 in the periodic table of elements; Figs. 1.8 and 6.6), whose mining and processing is very polluting.

The most important effect of ICT on the Earth System is its large use of electricity, for both producing and consuming information and communication. At the production end, data centres—buildings in which the servers are housed—use lots of electricity for both powering the servers and cooling the air to keep the servers from overheating. These facilities release large amounts of *heat* in the environment. At the consumer's end, most of the ITC devices are left to run day long in both offices and homes, and thus continually consume large amounts of electricity and release *heat*. Overall, the ICT sector may be responsible for *around 2% of the global emissions of CO_2*.

The three main ICT categories are: communication networks, personal computers, and data centres. Together, they consume more than 5% of the total electricity worldwide. The upcoming development of ICT is the *Internet of Things*, that is, networks of billions of physical devices, vehicles, home appliances, and other items embedded with electronics, software, sensors, actuators, and connectivity which enables these things to connect and exchange data. It has been estimated that the increasing number of internet-connected devices may drive the electricity consumption of ICT to 20% of the world total in the 2020s. Hence, ICT has significant effects on *climate change* and *ocean acidification*. In addition, the disposal and recycling of increasing amounts of semiconductor-rich devices, replete with a wide variety of heavy metals, rare earth metals and highly toxic synthetic chemicals create environmental problems. This, especially that these devices are often intentionally designed for rapid obsolescence. The article of Williams (2011) provides further reading on the environmental effects of information and communications technologies.

How to Make Human Activities Sustainable? The above review of the contributions of five major types of human activities to climate change, ocean acidification, loss of biodiversity, and water and food insecurity may leave the impression that these activities create almost hopeless environmental problems. Table 11.3 and the accompanying text identify some of the main environmental problems to be resolved. The four examples from the twentieth century examined in Chapter 10 (see Sect. 10.3) showed that the identification of negative environmental effects of human activities can lead to changes in these activities, which eliminate or reduce their detrimental effects and associated feedbacks into human societies. Hence, the intent of the identification of negative effects of human activities on the Earth System in this Sect. 11.1.2 is to favour their transformation into sustainable activities. This could likely not be achieved by technological fixes only, but would require deep changes in the present functioning of human societies.

11.1.3   The Twenty-First Century: Feedbacks into the Environment and Human Activities

The above anthropogenic perturbations of the Earth System—climate change, ocean acidification, loss of biodiversity, and water and food insecurity—affect several environmental characteristics, which in return affect human societies. Table 11.1 and the associated texts summarized two main characteristics of these four global perturbations, that is, the main properties of the Earth System or ecosystems affected, and the general mechanism of the perturbation (see Sect. 11.1.1). Table 11.3 and the associated texts considered mechanisms by which five types of human activities contribute to the four global perturbations (see Sect. 11.1.2). Table 11.4 provides selected examples of environmental characteristics affected by the four global Earth System's perturbations, and some resulting impacts on human activities. The atlas of Govorushko (2016) illustrates the deep impacts of humans on the environment using geographical maps and numerous photos. The books of Shikazono (2012) and Tortell (2020) provide further reading on various effects of humans on the Earth System.

The number of environmental characteristics affected by global perturbations of the Earth System and their related human impacts is very large. These characteristics and impacts are regularly reviewed by intergovernmental and international organizations such as the *Intergovernmental Panel on Climate Change* (IPCC), the *Intergovernmental Science-Policy Platform on Biodiversity and Ecosystem Services* (IPBES), the *United Nation Environment Programme* (UNEP), and the *Food and Agriculture Organization of the United Nations* (FAO). Interested readers could find updated reports on the websites of these organizations, which are free of charge.

The environmental characteristics and human impacts cited in Table 11.4 are only a subset of those known for each of the global perturbations examined in Sects. 11.1.1 and 11.1.2. We selected these examples to provide a general idea of the environmental and human consequences of the ongoing global perturbations of the Earth System. Each item in Table 11.4 is briefly described below, and readers interested in specific aspects could find more information in publications issued by the above organizations.

1. Climate change: examples of environmental effects and their impacts on human activities. The meaning of *climate change* in this book is explained in Chapter 1 (see Information Box 1.4). Briefly, climate change refers to changes in the state of the climate system, which consists of five major components—the atmosphere (air), the hydrosphere (liquid water), the cryosphere (ice), the lithosphere (rocks), and the biosphere (organisms)—and their interactions. A major aspect of the ongoing climate change is *global warming*, which denotes the anthropogenic increase in global temperatures and its projected continuation. Climate change includes both global warming and its observed and predicted effects on the physical environment and the organisms, and how these affect human societies (Fig. 11.5).

Table 11.4 Selected examples of ongoing effects of the four global perturbations of the Earth System in Tables 11.1 and 11.3 on environmental characteristics, and resulting impacts on human activities. Details are provided in the accompanying text, where the paragraphs are numbered in the same way as the items in the Table

Earth System: Global perturbations	Effects of global perturbations on environmental characteristics	Examples of impacted human activities
1. Climate change	A. Extreme weather events B. Sea level rise C. Release of sequestered greenhouse gases (GHG) D. Impacts on biodiversity E. Changes in geographic ranges F. Shifts from sinks to sources of GHG	a. Increased weather-related disasters b. Increased stress on agriculture c. Decreased fisheries d. Increased spread of diseases
2. Ocean acidification	A. Calcification: coral reefs and coastal ecosystems, polar oceans B. Multiple drivers: coral reefs and eastern boundary upwelling systems (EBUS)	a. Damages to ecosystems where calcifying organisms have major roles: coral reefs, coastal zones and polar waters b. Lower fisheries in EBUS
3. Loss of biodiversity	A. Loss of ecosystem services: provisioning, regulation and maintenance, and cultural services B. Global extinction of a large number of species	a. Loss of key ecosystem services, such as pollination b. Threats to public health c. Reduction of future options
4. Water and food insecurity	A. Freshwater overuse: increased salinity of freshwaters and soils, water pollution, loss of floodplains and wetlands, sinking of landforms B. Arable land per person is shrinking: soil erosion, desertification C. Loss of genetic diversity of agricultural crops and livestock breeds D. Hazards from aquaculture	a. Low availability of good-quality freshwater b. Difficult use of the environment c. Lower quality and availability of agricultural land d. Vulnerability of food crops to pests and disease

The following three paragraphs summarize three examples of the *effects of global warming on the physical environment*: increase in the frequency and amplitude of extreme weather events, sea level rise, and the release of sequestered terrestrial and marine greenhouse gases.

1A. Increases in the frequency and amplitude of extreme weather events are expected globally and regionally. These include: more severe and frequent heat waves, and more severe (although perhaps less frequent) cold snaps; stronger

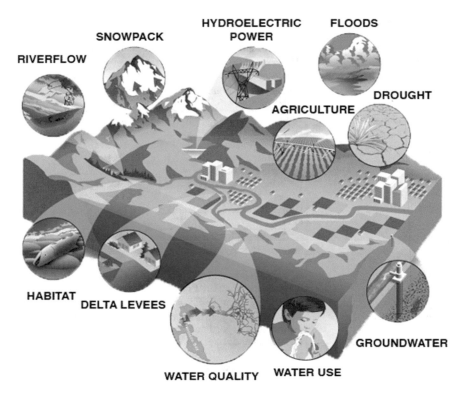

Fig. 11.5 Earth's environments and human activities that are already or are expected to be impacted by global warming. Credits at the end of the chapter

tropical cyclones, fuelled by the warmer surface ocean (see Sect. 4.3.2); heavier precipitations in wet areas, as a result of the higher moisture content of the warmer atmosphere (see Sect. 3.5.2), producing torrential rainstorms in summer and blizzards in winter; extreme droughts in dry areas, as a consequence of increased evapotranspiration (see Sect. 3.5.4), and expansion of drought areas; expansion of areas with flood hazard, as a consequence of heavier precipitations in some regions or seasons; and longer periods of forest fires, as a consequence of more prevalent drought conditions in other regions or seasons.

1B. Sea level rise is examined in some detail later in this chapter within the context of adaptation and mitigation. It is explained there that sea level is rising as a result of the thermal expansion of seawater in the ocean and the melting of ice sheets, glaciers and ice caps on continents (see Sects. 11.2.1 and 11.2.2). The most obvious effect of sea level rise will be increasing floods in coastal areas, but higher sea level will also exacerbate storm surges and coastal erosion. The latter becomes more severe in areas where sea-level rise is combined with aggregate extraction (see Sect. 11.1.3).

1C. The release of sequestered terrestrial and marine greenhouse gases (GHG) is increasing as a consequence of global warming. The three most abundant GHG in the atmosphere (excluding water vapour) are carbon dioxide (CO_2), methane (CH_4) and nitrous oxide (N_2O). Two major natural sources of these gases are the permanently frozen soils, called *permafrost*, and marine sediments (see Sect. 2.1.7). About one fourth of the Northern Hemisphere is covered by permafrost, which is one of the largest natural reservoirs of organic matter on Earth and is thawing at an increasing rate. When permafrost thaws, microbial activity turns organic matter into GHG. Similarly, some of the organic matter buried in marine sediments and permafrost is transformed into CH_4 by archaea in the absence of O_2, forming deposits of frozen CH_4 hydrate. These can be destabilized by increases in seawater temperature, especially on continental slopes, and thus release large amounts of CH_4 gas in the environment. The increase in atmospheric GHG from the two sources identified here will contribute to climate warming.

The following three paragraphs address the fact that global warming will also affect organisms and ecosystems. Three examples of the *effects of global warming on organisms and ecosystems* are: impacts on biodiversity, changes in geographical ranges, and shifts of ecosystems from sink to source of greenhouse gases.

1D. Impacts on biodiversity—including species distribution, population dynamics, community structure and ecosystem function—are consequences of the increasing frequency and intensity of extreme weather events, heavy precipitations, droughts, floods and fires, and the rising sea level. These effects are presently accelerating in marine, terrestrial and freshwater ecosystems. Declines in marine and terrestrial biodiversity are forecasted in boreal, subpolar and polar regions under the combined effects of warming, sea ice retreat and enhanced ocean acidification (see below). Coral reefs are projected to decline to between 10 and 30% of former cover at 1.5 °C warming and to less than 1% of former cover at 2 °C warming.

1E. Changes in the geographical ranges occupied by terrestrial and marine species and ecosystems are already occurring as a consequence of changing environmental conditions, and are expected to continue as the environment continues to change. Changes in geographical range include range contraction, expansion, and shift, which are very often latitudinal or altitudinal. Many populations of terrestrial and marine animals, including fish, have already experienced one of the three types of changes in their ranges, whereas most forest ecosystems cannot move as fast as the pace of the changing physical environment and are thus experiencing range contraction or less suitable conditions. One interesting example is provided by reef-forming corals, whose planktonic larvae are already showing poleward shifts in their patterns of settlement on the seafloor. Even if the settlement of larvae has begun to move poleward, the coral reefs could not escape indefinitely the detrimental effect of the changing marine environment because the solar light available in winter decreases with increasing latitude and would be too low above the mid-latitudes to sustain the photosynthethic activity of the microalgal symbionts of corals (see Information Box 5.3).

1F. Shifts of ecosystem from being sinks to becoming sources of greenhouse gases are expected as a consequence of global warming. Marine and terrestrial ecosystems (including their biomasses and non-living components such as water, sediments and soils) currently sequester together about 60% of the global anthropogenic carbon emissions. It is explained above that permafrost thaw is shifting the northern latitudes from currently being carbon sinks to becoming GHG sources. The same may occur in tropical forests, where the increasing occurrence of droughts and fires may release to the atmosphere the carbon stocks they hold, transforming them from carbon sinks into carbon sources.

The above warming-driven perturbations of the physical environment and the organisms impact *human societies*. Four examples of such impacts are: increased weather-related disasters, increased stress on agriculture, decreased fisheries, and increased spread of diseases. These are summarized in the following four paragraphs.

1a. Increased weather-related disasters already affect human societies in several ways, and will increasingly do so. Here are a few examples. Heat waves kill fragile people, especially when the night-time temperatures do not drop significantly. Changes in precipitation patterns are very damaging to human infrastructures and activities, through increasing droughts and forest fires in dry areas, and extreme rainfall and floods in wet areas. The intensification of tropical cyclones will cause intensified destruction, and their effects in coastal areas will be compounded by the rising sea level.

1b. Increased stress on agriculture will result from high temperatures and frequent droughts, which will cause decline in productivity of crops and livestock in some regions. In other regions, heavy rains and floods will delay planting and harvesting, or ruin crops. Higher temperature will also allow plant and insect pests to spread from lower to higher latitudes, from which they were previously excluded by cold winter temperatures.

1c. Decreasing fisheries as a result of global warming is already apparent in a number of studies. Warming positively impacts some fish populations and negatively impacts others, but overall the losses overweight the gains. The *maximum sustainable yield* of the world fisheries (that is, the amount of fish that could be caught each year without jeopardizing future harvests) has already dropped and is expected to continue to drop as a result of the increasing global warming.

1d. Increased spread of diseases is expected in a warmer world. Indeed, many diseases are spread by vectors such as insects that could invade higher latitudes where winters were previously too cold for them. Other vectors, such as bats and mammals, may proliferate when dry seasons followed by heavy rainfalls would produce an abundance of food. Similarly, warmer waters could favour the spread of cholera and other toxic bacteria through the contamination of drinking water and food (fish and seafood). Another, although perhaps remote possibility, may come from the thawing permafrost, which could release diseases lying dormant in the frozen soils since hundreds or thousands of years.

   2. *Ocean acidification: examples of environmental effects and their impacts on human activities.* Ocean acidification has direct effects on the carbonate chemistry of seawater, which are explained above (see Sect. 11.1.1) where they are summarized in Eqs. 11.1, 11.2 and 11.4. These three equations are repeated here for convenience:

$$CO_2 + H_2O \rightleftharpoons HCO_3^- + H^+ \tag{11.1}$$

$$HCO_3^- \rightleftharpoons CO_3^{2-} + H^+ \tag{11.2}$$

$$CO_2 + CO_3^{2-} + H_2O \rightleftharpoons 2HCO_3^- \tag{11.4}$$

Because the concentration of CO_2 currently increases in the atmosphere, the excess of CO_2 dissolves in seawater, which shifts the carbonate equilibrium to the right of Eqs. 11.1 and 11.2. This causes an increase in H^+ ions, and the higher concentration of H^+ means lower pH, consistent with a higher proportion of HCO_3^- and a lower proportion of CO_3^{2-} (Eq. 11.4 and Fig. 9.7). The present decrease in the pH of seawater is called *ocean acidification* (see Sect. 11.1.1). The above equations and the known and predicted effects of ocean acidification are represented in Fig. 11.6.

   Ocean acidification modifies not only the pH of seawater, but also its *CaCO₃ saturation state* (Ω). The saturation state of seawater for a mineral (here $CaCO_3$) measures the potential for this mineral to form into a solid or dissolve as ions in seawater:

$$\Omega = \left[Ca^{2+}\right]\left[CO_3^{2-}\right]/K \tag{11.5}$$

where the numerator $[Ca^{2+}]\,[CO_3^{2-}]$ is the product of the concentrations of the calcium (Ca^{2+}) and carbonate (CO_3^{2-}) ions in seawater, which form together solid $CaCO_3$, and the denominator K is the theoretical product of $[Ca^{2+}]$ and $[CO_3^{2-}]$ when $CaCO_3$ is at equilibrium (*equilibrium* means here that solid $CaCO_3$ neither forms nor dissolves). When Ω is more than 1, then $[Ca^{2+}]\,[CO_3^{2-}]$ in seawater is larger than theoretical K, meaning that CO_3^{2-} ions are at *supersaturating* concentration in seawater, a condition under which solid $CaCO_3$ can form, or already formed solid $CaCO_3$ does not readily dissolve. Conversely when Ω is less than 1, then $[Ca^{2+}]\,[CO_3^{2-}]$ in seawater is smaller than theoretical K, meaning that CO_3^{2-} ions are at *undersaturating* concentration in seawater, a condition under which solid $CaCO_3$ may dissolve. However, if the production rate of $CaCO_3$ is high enough to offset its dissolution, as a result of strong biological *calcification* (see Information Box 5.11), then the solid mineral can continue to exist even with Ω less than 1. The ongoing ocean acidification lowers $[CO_3^{2-}]$ in seawater and thus Ω (Eq. 11.5).

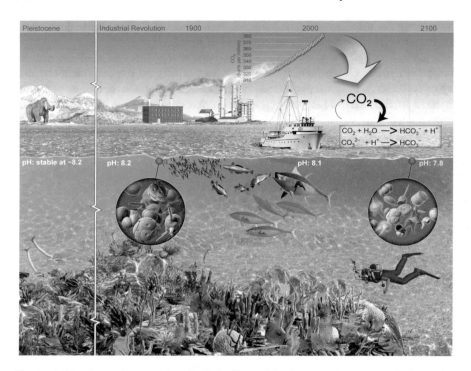

Fig. 11.6 Ongoing and predicted ecological effects of the decrease in seawater pH (increasing acidification) from the past value of about 8.2 (left) to the current value of 8.1 (centre right) and the predicted fall to a value as low as 7.8 (right). The dates are in the top panel of the Figure. The text explains that the decrease in pH will likely affect plankton composition (two inset circles), fish populations and organisms on the bottom especially corals. Credits at the end of the chapter

In the natural environment, solid $CaCO_3$ occurs in two different crystalline forms, *aragonite* and *calcite*, which have the same chemical formula but different physical properties (see Sect. 7.4.2). The saturation state of calcite (Ω_{cal}) is about 1.6 times that of aragonite (Ω_{arg}), so that calcite is much less soluble in water than aragonite. Hence, the calcareous parts of calcifying organisms made of aragonite dissolve more easily than those made of calcite. Aragonite forms the shell of almost all molluscs (including planktonic pteropods), the calcareous skeleton of most reef-building corals, and the tests of some benthic foraminifera. Calcite is the primary constituent of the test of planktonic coccolithophores and foraminifera and most benthic foraminifera, and the hard parts of red algae, some sponges, brachiopods, echinoderms, and a number of other organisms. In some bivalve molluscs, such as oysters, aragonite forms some small areas of the shell otherwise made of calcite.

In the following paragraphs, we examine some of the effects of ocean acidification. We consider effects on *calcification*, and also on marine systems as part of the multiple drivers that affect the organisms in the natural environment.

2A. Effects on calcification. Corals and other calcifying organisms are very sensitive to Ω_{arg}: when Ω_{arg} is greater than 3, they can survive and reproduce; when Ω_{arg} falls below 3, they become stressed; and when Ω_{arg} is less than 1, their shells and other aragonite structures begin to dissolve. In the currently acidifying ocean, there is a decrease in $[CO_3^{2-}]$ and thus in both Ω_{arg} and Ω_{cal}, which may primarily affect the organisms that produce aragonite, including almost all molluscs and most reef-building corals.

In the ocean, the depth in the water column below which the dissolution of solid $CaCO_3$ is more rapid than its formation, which is called the *carbonate compensation depth*, varies among ocean basins and among regions within basins (see Sect. 7.4.2). Because Ω_{arg} is lower than Ω_{calc}, the compensation depth of aragonite is always nearer the surface than that of calcite. The present ocean acidification lowers $[CO_3^{2-}]$ and thus decreases both Ω_{arg} and Ω_{cal}, which raises the compensation depth of the two forms of solid $CaCO_3$ closer to the surface. The decrease in Ω_{arg} and Ω_{cal}, may lead to decreased calcification in marine organisms, especially those with aragonite structures, as the inorganic formation of $CaCO_3$ is directly proportional to Ω. As acidification progresses, it will likely affect all calcifying organisms.

2B. Effects on multiple-driver ecosystems. The organisms in the natural environment are affected by changes that occur in several factors, called *drivers*. A driver is defined as a natural or human-induced factor that directly or indirectly causes a change. Interactions among drivers can be additive, synergistic, or antagonistic. In an *additive* interaction, the combined effect of two or more drivers is equal to the sum of the effects of the individual drivers, whereas in a *synergistic* interaction the effect is greater than this sum. In an *antagonistic* interaction, the effects of the different drivers cancel each other, partly or completely. Hence, the overall effect resulting from the combination of two or more factors cannot be predicted from the results of studies conducted on the effects of individual factors, but requires studies that address all the factors simultaneously.

The drivers of global perturbations of the Earth System, including ocean acidification, affect the marine biota, and knowledge about the interactive effects of such drivers is still limited. Some of the reported effects are positive, but most are negative. Overall, the sensitivity of a number of organisms to warming is exacerbated by acidification and low oxygen concentration (that is, ocean deoxygenation; see Information Box 9.2). Coral reef and coastal ecosystems are especially threatened by multiple drivers, but open-ocean systems are also affected.

In the following paragraphs, we examine the impacts of the above acidification-driven perturbations on *human societies*. These impacts affect the calcifying ecosystems and fisheries.

2a. Impacts of the changing environment on calcifying ecosystems. Several marine food webs may be affected by the decreased calcification of a number of planktonic and benthic organisms, however in ways not easy to forecast. Concerning reef-forming corals, the effect of lower calcification will negatively affect their communities of fish and other organisms (including shellfish), and the people depending on them for food and other activities (including tourism). Reduced calcification will also affect mollusc aquaculture, which accounts for more than 20% (in weight) of the total (*inland and marine*) aquaculture production and more than 60% of the world *marine* aquaculture production. Other effects of ocean acidification are related to the lowering of seawater pH, which can affect physiological processes of organisms, calcifying or not. Concerning the effect of multiple drivers, coral reefs are considered to be the most threatened marine ecosystems by climate-related ocean changes, especially ocean warming and acidification. In the two polar oceans, planktonic and benthic organisms that produce solid $CaCO_3$ structures are increasingly affected by ocean acidification (Fig. 11.6: changing plankton in the two enlargement circles, and changing benthic organisms on the bottom).

2b. Impacts of the changing environment on fisheries. Coastal areas are the site of both dense human populations and most marine aquaculture activities, which are thus particularly threatened by future sea level rise, rising ocean temperature, enhanced coastal erosion, increasing wind, wave height and storm intensity, and ocean acidification. Eastern boundary upwelling systems (often abbreviated EBUS) are amongst the most productive ocean ecosystems (see Sect. 3.5.3), and are especially affected by global warming, ocean acidification and oxygen loss (Fig. 9.14). It is thus expected that the effects of climate change will progressively impact EBUS fisheries (Fig. 11.6: decreasing abundance of fishes with decreasing pH).

3. Loss of biodiversity: examples of environmental effects and their impacts on human activities. Some of the key processes that ensure the functioning of ecosystems and their roles in the Earth System also make essential contributions to human wellbeing. When seen from the viewpoint of humans, these contributions are often called *ecosystem services*. Some ecosystem services depend directly on *biodiversity*, whereas other services include the *non-living* parts of ecosystems that also contribute to human wellbeing, for example water. We consider here only the ecosystem services that depend on biodiversity.

In the following paragraphs, we examine the effects of the loss of biodiversity in terms of ecosystem services and species extinctions. The ecosystem services examined below support either the *provisioning* of material and energy needs or the *regulation and maintenance* of the environment for humans.

3A. Effects on ecosystem services. Provisioning is provided by the terrestrial and aquatic plants that are cultivated and animals that are reared (or fished) for nutrition, materials, or energy. Provisioning is hindered by the fact that human activities have already significantly altered 75% of the land surface, impacted 66% of the ocean area, and caused the loss of over 85% of the wetland areas.

Regulation and maintenance consist of transformations and regulations carried out by ecosystems. Some organisms (both plants and animals) transform wastes or toxic substances by filtration, storage or accumulation, and others (plants) mediate nuisances of human origin by reduction of smell or noise or by providing visual screening. Some ecosystems (mostly plants) regulate the environmental conditions of humans by controlling erosion rates, buffering or attenuating landslides, controlling water flows including floods, and protecting against fire and the effects of storms. Other organisms (animals) have major roles in the regulation of the life cycles of plants, through pollination and seed dispersal. Finally, some organisms (animals) control pests, including invasive species, and diseases. The indicators of the ecosystem regulation services, such as soil organic carbon and pollinator diversity, are declining in most areas.

3B. Effects on species extinctions. Human actions threaten more species with global extinction now than ever before, with around 1 million species already facing extinction, many within decades, unless action is taken to reduce the intensity of drivers of biodiversity loss. As a result, biological communities are becoming ever more similar to each other. In addition, increases in the production of some ecosystem services are causing declines in others. For example, the clearing of forest for agriculture has augmented the supply of food, feed and other materials important for people (such as natural fibres and ornamental flowers), but it has reduced such key services as pollination, climate regulation, water quality regulation, and the maintenance of options for the future. On land, the particularly threatened ecosystems include old-growth forests, island ecosystems, and wetlands.

In the following paragraphs, we examine the impacts of the above perturbations resulting from the loss of biodiversity on *human societies*. These impacts affect the ecosystem services, public health and future options.

3a. Impacts of the changing environment on ecosystem services. Approximately half the cover of live corals in coral reefs has been lost since the 1870s, and losses have accelerated in recent decades. The average abundance of native species in most major terrestrial habitats has fallen by at least 20% and this decline, which has mostly taken place since 1900, may be accelerating. Native biodiversity has often been severely impacted by invasive species, and population sizes of wild vertebrate species have tended to decline over the last 50 years on land, in freshwater and the sea.

Currently, land degradation has reduced productivity in 23% of the global terrestrial area, and 10 to 20% of the annual global crop output is at risk as a result of pollinator loss. Moreover, the decrease in coastal habitats and coral reefs reduces coastal protection, which increases the risk from floods and hurricanes to life and property for people living in coastal zones. Especially threatened are the 100–300 million people living in coastal 100-year flood zones (that is, zones where the probability of a flood in any given year is 1%). Globally, local varieties and breeds of domesticated plants and animals are disappearing. This loss of diversity, including genetic, poses a serious risk to global food security by undermining the resilience of many agricultural systems to threats such as pests, pathogens and climate change.

3b. Impacts of the changing environment on public health. The deterioration of biodiversity and ecosystem functions has implications for public health. The reason is that emerging infectious diseases in wildlife, domestic animals, plants or people can be exacerbated by land clearing and habitat fragmentation.

3c. Impacts of the changing environment on future options. The losses of biodiversity permanently reduce future options. The reason is that wild species that might be domesticated as new crops and/or be used for genetic improvement are disappearing forever.

4. Water and food insecurity: examples of environmental effects and their impacts on human activities. Planet Earth has a limited supply of renewable freshwater in the forms of surface water (lakes and rivers), groundwater (the underground areas where water is naturally collected and stored and where it circulates are known as *aquifers*), and atmospheric water vapour. In addition, seawater is increasingly *desalinated* for agriculture and drinking uses. *Water insecurity* threatens areas in which about 80% of the world's population lives. *Food insecurity* is a measure of the limited or uncertain availability of food, or ability to access it. *Food availability* depends on the supply of food through its production, distribution, and exchanges. Food production is determined by a number of factors that include: land use; soil management; crop selection, breeding, and management; livestock breeding and management; and harvesting. The book of Spellman (2020) provides further reading on the effects of human use, misuse and reuse of fresh and waste water on the overall water supply.

In the following paragraphs, we examine the effects of the increased water and food insecurity on freshwater use, arable land, genetic diversity, and aquaculture.

4A. Effects on freshwater use. The most common threat to water security is *water scarcity*, that is, the geographic and/or temporal mismatch between freshwater demand and availability (Fig. 11.4a). The causes of water scarcity include low rainfall, climate change, high population density, and overuse of the available water. On the one hand, the global demand for water is rising as a consequence of the increasing world population, higher standards of living, changing consumption patterns, and expansion of irrigated agriculture. On the other hand, the water supply cannot meet the demand in many areas or at some times because of the alteration of weather patterns by climate change (which increases the occurrence of droughts or floods), deforestation, increased pollution, and wasteful use of water.

Water scarcity resulting from *consumption* is caused primarily by the extensive use of water in agriculture and livestock farming (80% of the global freshwater use), and industry (Table 11.2). People in developed countries generally use about 10 times more water daily than those in developing countries, and a large part of the water in the latter countries is used and polluted in order to produce goods consumed in developed countries. All the causes of water scarcity are related to human interference with the water cycle (see Sect. 5.6.5).

4B. Effects on arable land. The changing climate increasingly affects crop production, especially through changes in rainfall and temperature. In many countries, agricultural land is used for urbanization, or is lost to desertification, salinization and soil erosion due to unsustainable agricultural practices. The loss of arable land ultimately affects food production. It must be noted that countries do not need crop production to achieve food security, as they can import food as long as their revenues from exports of other products allow it. In fact, most countries import at least part of the food they consume, and some export part of the food they produce.

4C. Effects on genetic diversity. During the last 10,000 years, farmers have selected thousands of varieties and landforms of plants and animals. However, about 30% of *livestock breeds* are close to extinction, and about 75% of the genetic diversity of *agricultural crops* was lost since 1900.

4D. Effects on aquaculture. World fish catches have increased fourfold since 1950, and many fish stocks are fully exploited or overexploited. Aquaculture produces a third of the fish and seafood consumed by people, and the proportion is expected to rise to half the total by 2030.

In the following paragraphs, we examine the impacts of the above perturbations resulting from the increased water and food insecurity on *human societies*. These impacts affect freshwater quality, the use of the environment, the agricultural lands, and the vulnerability of food crops.

4a. Impacts of the changing environment on freshwater quality. Overall, water insecurity has decreased in the last decades. However, the methods used for its prevention often impacted the availability of good quality freshwater and also the use of the environment by the loss of topsoil, reduction in land productivity, sedimentation in reservoirs, and increased risk of flash floods.

4b. Impacts of the changing environment on the use of the environment. The overuse of water related to water scarcity harms the environment in several ways, which include increased salinity of freshwaters and soils, water pollution by fertilizers and farm effluents, and the loss of floodplains and wetlands. Another result of the overuse of water is the gradual sinking of landforms, called *subsidence*. One well-known case of subsiding area is Venice, one of the most beautiful and famous cities in the world, whose historical centre is increasingly flooded by *acqua alta* (high water). The causes of the downward ground displacement in Venice include the pumping of groundwater, which has now stopped, and natural vertical movement of the land. The ongoing sea level rise contributes to increase the level of the *acqua alta*.

4c. Impacts of the changing environment on agricultural land. Because of the combined increase in world population, use of land for urbanization, and loss of agricultural land, the amount of arable land per person is shrinking. *Soil erosion* is responsible for about 40% of land degradation worldwide, and much of it is caused by the time-honoured agricultural practice of tillage. About 20% of irrigated land in the developing world has been damaged to some extent by waterlogging (saturation of the soil by water) or salinity. *Desertification* can be caused by natural processes, such as natural climate cycles, or induced by human activities, such as the current global warming or the overexploitation of soil for farming. As a result, an estimated 250 million people have been directly affected by desertification, and nearly 1 billion are at risk.

4d. Impacts of the changing environment on the vulnerability of food crops.
Food insecurity has decreased in the last decades. However, the increasing
genetic uniformity makes food crops vulnerable to external stress such as pests
and disease. Also, aquaculture generates hazards that include the release of excess
nutrients in the environment leading to eutrophication (see Sect. 11.1.2), the competition or breeding of escaped farmed fish with wild fish, and the spread of disease by farmed fish.

The above examination of the four global anthropogenic perturbations of the
Earth System in Table 11.4 showed that these perturbations affected the environment, the organisms and the ecosystems, and the resulting changes impacted in
return human activities. Table 11.3 shows that the four global perturbations are
themselves caused by human activities. These activities have generally contributed
to the progress of human societies, but they also had the detrimental consequences
summarized in Table 11.3, which mostly emerged during the twentieth century
and whose effects are increasingly stronger in the twenty-first century.

Combining Tables 11.3 and 11.4 shows the existence of positive feedbacks (Fig. 9.1), in which a number of human activities cause global perturbations of the Earth System (Table 11.3) that affect the environment, organisms
and ecosystems, and whose changes impact the same or other human activities
(Table 11.4). Most of these positive feedbacks tend to amplify the undesirable
consequences of human activities, which is the key effect of positive feedbacks
(see Sect. 9.1.4). Breaking this vicious circle, that is, eliminating or reducing
the detrimental positive feedbacks, will require changing the way many human
activities are conducted. The concerted actions taken in the twentieth century to
resolve some global problems that then threatened human societies show that
major changes in the conduct of human activities are possible, although they
may not be easy (see Sect. 10.3). Taking the necessary actions in the twenty-first
century will require major efforts of adaptation and mitigation, as described in
the next sections.

11.2   Adaptation and Mitigation

11.2.1   Adaptation: To Manage the Unavoidable

There are two broad types of responses to the above and other global perturbations of the Earth System. These are *adaptation* and *mitigation*, two words that
mean *adjustment to new conditions* and *reduction of the severity of perturbations*,
respectively. Although the two types of responses are fundamentally different, they
are often confused, and this confusion sometimes prevents people or organizations from taking action. It is explained in the following paragraphs that *adaptation is intended to manage the unavoidable*, and *mitigation is intended to avoid
the unmanageable*. These two types of responses are complementary and both
essential.

Adaptation to environmental changes has always occurred naturally in the Earth System, and it will become an increasingly key activity in human systems in the present context of *climate change. Adaptation to climate change* in *natural systems*—organisms, populations, and communities—is the process by which they adjust to the *current* climate changes and their effects. One example is the ongoing poleward displacement of many populations of plants and animals, both on continents and in the ocean, as the global temperature increases. *Adaptation to climate change* in *human systems* is the process of adjustment to the *current or expected* climate changes and their effects, in order to reduce the resulting harm or exploit potentially beneficial opportunities.

In the present condition of changing climate, human intervention may facilitate the adjustment of natural systems to the *expected* climate change and its effects. One example is *coral gardening*, in which coral seeds are grown in nurseries, where the early growth stages are protected from the high mortality risks that threaten young corals at sea, and fragments of corals are planted on reefs as a way to replenish depleted colonies and thus restore reefs. *Adaptation options* correspond to the array of strategies and measures available and appropriate for addressing adaptation. They include a wide range of actions that can be categorised as structural, institutional, ecological or behavioural (Abram et al. 2019).

We use here *sea level rise* to exemplify adaptation in human systems. We know that sea level is rising as a result of two processes. One is the thermal expansion of seawater, which is the increase in volume of seawater as a result of its increased temperature. The other process is the melt of ice sheets (masses of ice that cover more than 50,000 km^2; in Greenland and Antarctic) and glaciers and ice caps (masses of ice that cover less than 50,000 km^2) on continents. The melt of continental ice masses causes sea level to rise because the water presently frozen as continental ice was extracted from the ocean by evaporation and deposited on continents in the past, which progressively lowered the sea level to its present value. As a consequence, the melt of this continental ice would cause the sea level to rise above its present value. This is not the case of the ongoing melt of the ice pack in the Arctic and Southern Oceans, because the sea ice is made of water that was already present in the ocean, and its melt thus does not contribute to rise the sea level.

In addition, continents in some areas are moving down, which increases the effect of sea level rise in coastal areas. One reason for the vertical movements of continents is the retreat of the huge masses of glaciers that were on the continents in the Northern Hemisphere during the last glacial episode (Fig. 4.5). In response to this retreat thousands of years ago, some continental areas are presently moving up, with corresponding downward movements in other areas (this is called the *post-glacial rebound*). Other causes of downward movement include local phenomena such as the intensive pumping of groundwater, or the natural sinking of delta areas under the mass of accumulating sediments. For these reasons and others, the predicted sea level rise is higher in some regions than others.

One consequence of climate change is an increase in the frequency and intensity of extreme meteorological events such as exceptionally high tides and exceptional storm surges. Because of the increase in the *average* sea level, the intensity of the floods that accompany *extreme* meteorological events will also increase. Other, *long-term* effects of the rising sea level are increased coastal erosion, ecosystem changes (including destruction of coral reefs), and saltwater intrusion. Overall, these effects will be increasingly damaging as the average sea level continues to rise.

Sea level is already rising, and is predicted to continue rising in the coming decades and centuries. Because of the additional, anthropogenic heat already accumulated in the atmosphere and the ocean, some rise in the sea level is unavoidable even if humans stopped emitting greenhouse gases immediately. We know that this will not be the case, and human activities will cause further rise in temperature. In the best-case scenario where the emissions of greenhouse gases would be limited to the level causing only a 1.5 °C increase in temperature, sea level is predicted to rise between 40 cm and 70 on average by 2100. Under less optimistic scenarios, the rise could reach 75 cm or much more. Given that coastal cities and countries with coastlines know that the sea level is rising and will continue to do so for decades and centuries, they should develop and implement long-term *adaptation* plans.

The first reaction to the prospect of sea level rise is often to think of building taller and stronger protective structures. However, this may not always be the best option since the sea level will continue to rise during centuries, meaning that any hard structure would have to be maintained and upraised during centuries. Also, protection is needed not only against the rising sea level, but also against increasingly severe and frequent extreme events, which would require very tall structures. In fact, adaptation includes in addition to possible *protection*, several options of *accommodation* and *retreat* (Fig. 11.7). Protection can be achieved by constructing hard infrastructures, and by improving natural defences. Accommodation makes societies more flexible to sea level rise. Retreat involves moving people and infrastructure to less exposed areas, and preventing further development of infrastructures in areas at risk.

The options to enhance protection against sea level rise include extending the range of sea levels coastal systems that could withstand the predicted conditions without failure. This could be done, for example, by building stronger and higher seawalls, dykes and embankments, and raising the height of docks. It could also be achieved by changing the tolerance of loss or failure, for example by increasing economic reserves or insurance.

Accommodation to sea level rise includes allowing medium-term adjustments, for example progressively changing the activities on the lands that will be flooded (to salt-tolerant crop varieties, or to aquaculture), or avoiding heavy investment or building large infrastructures (such as harbours, airports, new cities) in the areas most at risk. It could also be achieved by reducing economic lifetimes, which would include increasing the depreciation of infrastructures, which could thus be abandoned after a smaller number of years.

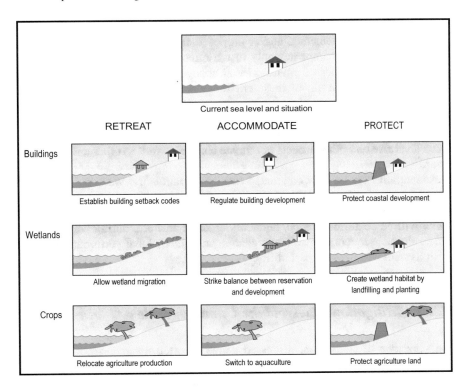

Fig. 11.7 Three main adaptation strategies to sea level rise—retreat, accommodate, or protect—to cope with the effects of coastal erosion on buildings, wetlands and crops. Credits at the end of the chapter

Accommodation and retreat include reducing other stresses than climate change, for example decreasing the use of coastal zones for agriculture or housing. There is also the possibility of removing artificial barriers, including managed retreat and coastal realignment, for example the deliberate removal of existing coastal protection to allow sea water to flood onto agricultural land. Controlled flooding can be accompanied by controlled sedimentation in deltas to raise land levels.

Another type of action is to reverse *maladaptation*, a term that designates actions that increase vulnerability to climate change. Maladaptation is usually an unintended consequence of human activities. For sea level rise, reversing maladaptation includes introducing setbacks for development in vulnerable areas such as coastal floodplains and landwards of eroding cliffs, for example prohibiting the building of houses in areas at risk, which is part of retreat. It also includes stopping land reclamation, which is part of accommodation, and restoring mangroves, sand dunes, and coral reefs, which provide natural protection.

The examples given above are only some of the many adaptation options to sea level rise already implemented and/or proposed around the world. In fact, human societies will have to adapt not only to the effects of sea level rise, but also to those of a large number of environmental changes resulting from global perturbations. Some, but not all, of these environmental changes and their effects are listed in Table 11.4. Successful adaptation would minimize social disturbances and costs, and the key to successful adaptation is to start early and act decisively. The book of Siegel (2020) provides further reading on the adaptation of coastal cities to sea level rise and other aspects of climate change as well as other environmental hazards.

Adaptation may be best conducted locally or nationally. We should be aware that adaptation is not a one-time activity, that is, adaptation will have to continue as long as there will be effects of climate change and other global perturbations, which will be centuries. Hence, humanity enters a new era, during which adaptation to the consequences of climate change and other global perturbations of the Earth System will become part of life.

11.2.2   Mitigation: To Avoid the Unmanageable

The very idea of adaptation, discussed above, is often misunderstood and some people or groups wrongly see adaptation as accepting the inevitability of climate change and giving up efforts to prevent its further development. This is, of course, not the case. As stated above, adaptation is intended to *manage the unavoidable*. However, managing the unavoidable does not mean accepting endless global perturbations of the Earth System. Indeed, if human societies were to continue "business as usual", this would inevitably lead to *unmanageable changes* in the Earth System for which *no adaptation would be possible*. Hence, the development of unmanageable environmental conditions must be prevented at all cost. This requires the active *mitigation* of climate change, which is the *reduction of the severity* of climate change, in order to *avoid the unmanageable*.

It was explained in the previous section that *mitigation* means *reduction of the severity of perturbations*. Hence, *mitigation of anthropogenic climate change* is human intervention to reduce emissions and/or enhance the sinks of greenhouse gases (GHG). *Mitigations options* are technologies or practices that reduce GHG emissions and/or enhances sinks (Abram et al. 2019). Indeed, there are two complementary ways to reduce the severity of anthropogenic climate change.

The first and foremost is to reduce GHG emissions, by changing the manner in which the human activities listed in the first column of Table 11.3 are conducted. Such changes are described below, in the last paragraph of Sect. 11.3.5.

The second is, as a complement to the reduction of GHG emissions, to increase GHG sinks —that is, the processes, activities or mechanisms that *remove* GHG or their chemical precursors from the atmosphere—and *store (or sequester)* them in pools—that is, reservoirs in the Earth System where such elements as carbon and nitrogen would reside for long times. The different methods of removing GHG from the atmosphere through deliberate human activities are known as *negative emission* technologies. This could be done by enhancing the biological and/ or geochemical sinks of CO_2 (for example, planting forests) and using chemical engineering to achieve long-term removal and storage in geological, terrestrial, or ocean reservoirs, or in products with climate-relevant lifetimes. Various methods have been proposed and/or used, and more will likely be proposed to enhance the biological or geochemical sinks, and directly capture GHG from the atmosphere and store them for long times. A complementary approach is to capture carbon from industrial and energy-related sources before it escapes to the atmosphere, which alone does not remove CO_2 from the atmosphere, but can help reduce atmospheric CO_2 if it is combined with bioenergy production. The efficiency and potential risks of each method are under study.

Another group of approaches, called *solar radiation management*, refers to the intentional modification of the Earth's shortwave radiative (heat) budget with the aim to reduce climate change according to a given metric, such as surface temperature or precipitation. The potential feasibility and efficiency of such techniques remains to be demonstrated, and their deployments could entail large risks.

Readers interested in these aspects can consult the most recent reports on the website on the IPCC website, where these technologies are summarized and evaluated. It must be understood that the CO_2 problem can only be resolved by significantly reducing, if not completely stopping, the extraction and use of fossil carbon. However, many specialists think that in order to avoid an excessive concentration of CO_2 in the Earth System by the time this occurs, significant amounts of atmospheric CO_2 should be removed. This could perhaps be done using some of the negative emission approaches cited above, or others to be developed. It should be noted that the time needed to significantly reduce or stop the extraction and use of fossil carbon depends on the level of coordinated political will at the global level, and is therefore not necessarily very long. The chapter of MacMartin and Ricke (2020) provides further reading on this viewpoint.

We use again, to stress the crucial need of mitigation of climate change, the example of *sea level rise*. Indeed, if the release of GHG in the atmosphere continued, this would cause further increase in temperature, which would contribute to enhance both the thermal expansion of seawater and the melting of ice caps (Greenland and Antarctic) and of glaciers on continents. Hence, continuing the release of GHG in the atmosphere would accelerate the ongoing sea level rise.

The thermal expansion of the ocean is very slow because it takes a long time for the whole mass of water in the ocean, which is very large, to absorb heat. Hence, there is a very slow, but inexorable response of the sea level to the increasing temperature. For example, an increase of 1 °C rises the sea level by about 0.1 m within around 10 years, which is the time required to warm up the top well-mixed 500 m of the oceans. The same increase of 1 °C will rise the sea level by about 0.6 m within about 1000 years, which is the time required to warm up the

SEA LEVEL RISE

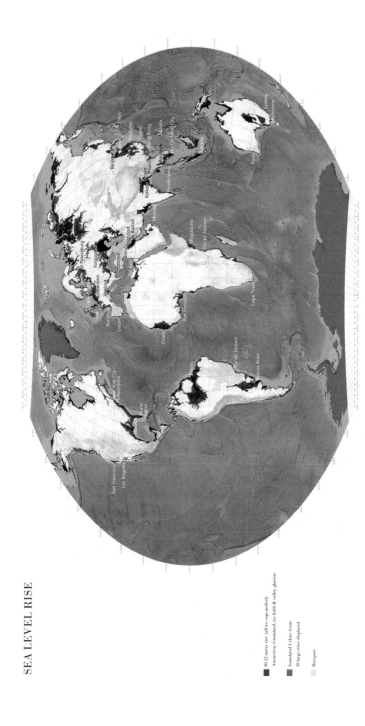

■ 80.32 meter rise (all ice caps melted)
   Antarctica, Greenland, ice fields & valley glaciers

■ Inundated Urban Areas
   50 large cities displaced

   Hotspots

Fig. 11.8 Earth with a 80.3 m rise in sea level. Black areas: lands flooded; red areas: inundated urban areas of 50 large cities; pink areas: present biodiversity hotspots. Readers interested in checking details could consult the map on the website http://atlas-for-the-end-of-the-world.com. Credits at the end of the chapter

3300 m average depth of the oceans. After 500 years, sea level rise from ocean thermal expansion reaches only half of its eventual maximum level. This means that human societies in hundreds of years will experience sea level rise resulting from the present global warming, caused by our release of greenhouse gases in the atmosphere. What will our descendants think of this legacy from our society?

The other cause of sea level rise is the melt of glaciers, ice caps and ice sheets on continents. The complete melt of all glaciers and ice caps would add 0.4 m to the ocean, that of the Greenland Ice Sheet an additional 7.4 m, and that of the Antarctic Ice Sheet the incredibly large value of 58.3 m. Hence the total sea level rise resulting from thermal expansion and the melt of all continental ice could eventually reach about 80 m (Fig. 11.8). In this scenario, vast new coastlines and inland seas would be created and 50 of the world's major cities would become architectural submarine reefs. This could only happen in several centuries, but it would be a terrible legacy of our times for future generations.

If human societies managed to limit or even stop GHG emissions in the coming decades, the twenty-first century should not experience the complete loss of the Greenland Ice Sheet or the West Antarctic Ice Sheet. The latter is separated from the bulk of the Antarctic continent by a mountain range, and is thus more prone to collapse than the much larger East Antarctic Ice Sheet; its collapse would cause a sea level rise of around 5 m. It is fortunate that the complete loss of the Greenland and West Antarctic Ice Sheets will likely not occur in the twenty-first century, because the resulting rapid inflows of large volumes of fresh water in ocean basins would have caused rises in sea level beyond the adaptation abilities of human societies, and would thus have been a major catastrophe.

What is expected for the twenty-first century? There is little doubt that thermal expansion of the ocean will continue, as will the melting of most glaciers and ice caps, and parts of the Greenland and Antarctic Ice Sheets. If the temperature increase reaches 3 °C by year 2100, a large number of cities around the world will be flooded, which could affect hundreds of millions of people. The regional impact of the sea level rise resulting from a 3 °C increase would be highly uneven, with four out of five people affected living in Asia, which does not reflect regional differences in sea level rise but the very large development of coastal cities in that part of the world. The book of Goodell (2017) provides further reading on the likely effects of sea level rise on coastal cities and island states, largely centred on the US but not exclusively, by the end of the twenty-first century.

Of course, the sea level will not rise instantaneously, but once a given temperature will be reached, the corresponding increase in sea level will become irreversible even if warming eventually slowed down. In addition, some temperatures are *tipping points*, that is, thresholds that when exceeded, even by a small value, can lead to very large changes in the climate system. One example is the melt of the Greenland Ice Sheet, which will become inevitable once the increase in temperature reaches a critical temperature between 1 to 4 °C (the value of this tipping point is not precisely known). Given that global temperature has already increased by about 1 °C since the 1880s, the tipping point for the irreversible melt of the

Greenland Ice Sheet may have already been crossed. Beyond the tipping point, the ice sheet may persist for many centuries, but nothing could prevent its long-term melt.

Tipping points also exist for ecosystems, where they correspond to situations in which ecosystems experience irreversible shifts to new states. These are accompanied by significant changes in biodiversity and the services to people it underpins, on regional or global scales. One example is the increasing wildfire activity in western United States caused by climate change. There, regeneration of coniferous forests after wildfires has distinct thresholds for the recruitment (that is, the growth) of seedlings depending on vapour pressure deficit, soil moisture, and maximum surface temperature. The seasonal to annual climate conditions over the past 20 years have crossed these thresholds (that is, tipping points) in some areas, which have thus become unsuitable for postfire regeneration. It is expected that this will result in a shift from conifer forest to non-forest vegetation. Hence in conditions of changing climate where tipping points for forest regeneration have been crossed, stand-replacing wildfires caused an abrupt ecosystem shift to non-forest states.

As already mentioned, the effects of the present increase in atmospheric temperature will affect the sea level during many centuries. At the time scale of 1,000 years, it is estimated that sea level may rise by at least 7 m even if the emissions of GHG were stopped rapidly. It could rise by much more if the emissions continued to increase. This is because of the combination of the following factors: 20% of the emitted CO_2 will remain in the atmosphere for more than 1000 years; the deep ocean will slowly warm up, which will cause thermal expansion; sustained warming above the tipping point of 1 to 4 °C may lead to near-complete loss of the Greenland Ice Sheet; and the Antarctic Ice Sheet will continue to melt. A conservative estimate is that sea level will rise, on average, by approximately 2.3 metres per degree Celsius of temperature increase during the next 2,000 years.

A large rise in sea level resulting from the melt or collapse (that is, slide) into the sea of huge masses of land ice may happen in future centuries, but likely not in the twenty-first century. This is fortunate as human societies could perhaps adapt to sea level rise of one metre of even a few metres (see Sect. 11.2.1, above), but obviously not tens of metres especially if the sea level rise occurred relatively rapidly. This sudden event would be the result of a slow and progressive change in the temperature. It was explained above for adaptation that climate change would cause an increase in the frequency and intensity of the floods that will accompany extreme meteorological events such as exceptionally high tides and exceptional storm surges. Although these strong floods will occur quite suddenly, they will be the result of a slow and progressive change in the average sea level.

One of the main objectives of *mitigation of climate change* is the prevention of catastrophic events resulting from slow and progressive changes in environmental conditions, such as sudden large floods or a rapid, major rise in sea level. Indeed, reaching a global temperature at which coastal regions and islands would frequently experience major floods, or worse at which the melt or collapse of large land-ice masses would cause unmanageable rise in sea level, would be disastrous for human societies.

The above example of sea level rise showed the absolute need of mitigating climate change. In fact, the mitigation of not only climate change but also other global perturbations of the Earth System (Table 11.3) is essential to prevent the numerous detrimental effects of the resulting environmental changes on human societies (Table 11.4). The probability of occurrence of these changes and their effects, some of which could be catastrophic, increases as the perturbations of the Earth System intensify. As explained for sea level rise, mitigation aims at preventing slow and progressive changes in average environmental conditions that could lead to unmanageable effects of these changes on human societies. It follows that we must absolutely avoid reaching and crossing tipping points beyond which climate and ecosystems irreversibly shift to new states. In order to achieve this objective, societies must act together at the international level in order to "save ourselves", as exemplified in the previous chapter (see Sect. 10.3). Adaptation can be conducted locally or nationally, as seen above, but mitigation requires international action.

11.3 Beyond the Twenty-First Century

11.3.1 Predicting the Future?

We explained above that one of the many detrimental effects of the present increase in atmospheric temperature would be a long-term rise in the sea level, which will continue during many centuries after the end of the ongoing anthropogenic global warming (see Sect. 11.2.2). Hence, many effects of the present changes to the Earth' environment by humans will affect the Earth System, including human societies, for a long time. It could thus be instructive to speculate about possible future states of the Earth System. However as pointed out by the American baseball player "Yogi" Berra (1925–2015), who is known for his wit and wisdom: "It's tough to make predictions, especially about the future." Hence, we will not try to make predictions about the Earth's future, but instead report below the results of four exercises in which researchers have used numerical models to explore future aspects of the Earth System. These studies are helpful to better understand the effects of changes presently occurring on the planet.

It was explained in Chapter 4 (see Sect. 4.5.2) that Earth has been experiencing an Ice Age since the beginning of the Quaternary period 2.6 million years ago. The Quaternary Ice Age has been characterized so far by a succession of long glacials and shorter interglacials, the combined duration of each cycle being initially around 40,000 years and presently around 100,000 years. (The words *glacial* and *interglacial* can be used as either names or adjectives, a *glacial* being a *glacial period*.) During each long glacial, the global temperature went down and the ice sheets grew in the Northern Hemisphere, whereas during each shorter interglacial, the global temperature went up and the ice sheets in the Northern Hemisphere retreated. We presently live in an interglacial. Based on their understanding of this global phenomenon, researchers have tried to figure out how long the present interglacial would last, and thus when the Northern Hemisphere ice sheets would start to grow again.

One may think that a way to determine when the present interglacial will end, and thus the next glacial will begin, would be to look at the duration of past interglacials. In the 1970s, researchers believed, based on the geological records then available, that the last two interglacials had lasted around 10,000 years each. Because 10,000 years was about the same duration as the time elapsed between the end of the last glacial episode (which was 11,650 years before present) and 1970, this meant that our current interglacial was about to end and Earth would soon enter a new glacial episode. As a consequence, the question of researchers in the 1970s was: How could humans prevent the upcoming global cooling? As we know, we have now unintentionally overshot the objective of preventing a global cooling, and the question has become: How could humans mitigate the ongoing global warming (see Sect. 11.2.2)?

In fact, we now know that the duration of the last two interglacials was longer than 10,000 years. In addition, climate cycles are driven by variations in the Earth's orbit around the Sun, and the characteristics of these cycles change with time (see Sects. 1.3.2 to 1.3.4). As a consequence of these two factors, the duration of the present interglacial may not necessarily be the same as that of previous interglacials, contrary to what was thought in the 1970s. In fact, researchers found since the 1970s that the amplitude of the variations in solar insolation predicted from *orbital cycles* will be exceptionally small in the coming millennia, and as a consequence global climate changes should also be small. Hence the next glacial episode is not expected to begin for a long time.

Furthermore, studies on the future climate conducted in the 1970s did not include the possibility of a change in the relatively *low concentrations of atmospheric CO_2* that still prevailed at that time, that is, less than 335 ppmv. This low value is unfortunately not the case any more, as the concentrations of atmospheric CO_2 and other greenhouses gases are increasing rapidly (currently above 400 ppmv), and the resulting global temperature of the planet is rising. The present rise in the global temperature of Earth will affect its future climate. In which way will the ongoing temperature increase influence the long-term climate of the Anthropocene period? We briefly examine four possibilities that researchers have investigated for the future climate of Earth, summarized in Table 11.5.

11.3.2   Long Interglacial

The present astronomical conditions of Earth and climatic conditions of the Earth System are very different from those that existed during the previous interglacial, which lasted about 15,000 years between around 130,000 and 115,000 years before present. Indeed, we already cited in the previous paragraph differences in the orbital characteristics and in the concentration of atmospheric CO_2. Concerning the orbital characteristics, the value of the Earth's *eccentricity*

Table 11.5 Four scenarios of long-term effects of human activities on climate discussed in the next sections

Effects of human activities	Effect on the long-term climate	Section
High atmospheric CO_2 followed by return to pre-industrial level	Present interglacial may last 50,000 years	11.3.2
Retention of a significant fraction of anthropogenic CO_2 in atmosphere for 100,000 years	Present interglacial could last 100,000 years, and perhaps be as long as 500,000 years	11.3.3
Burning all the coal supplies: atmospheric CO_2 concentration could reach 2000 ppmv	Earth could shift from the present icehouse to a new greenhouse, with much higher temperature	11.3.4
Crossing the planetary temperature threshold (perhaps 2-3°C above preindustrial level)	Earth may be on a trajectory towards a hothouse, with temperatures 4-5°C higher than present and sea level more than 60 m higher	11.3.5

will be almost zero during the coming 25,000 years, leading to an exceptionally small amplitude of the variations in solar insolation (see Sect. 1.3.2 for an explanation of eccentricity). Concerning the CO_2 *concentration*, the present values are above 400 ppmv, where they climbed from a concentration of 280 ppmv before the Industrial Revolution. The present concentration is much higher than the highest value observed during the last 800,000 years, which did not exceed 300 ppmv, and human activities continue to push the atmospheric CO_2 concentration ever higher (Fig. 11.1a). Because of these differences, the duration of present interglacial will not be the same as that of the previous interglacial.

The combined orbital variations and CO_2 concentrations will deeply influence the glacial-interglacial cycle, but they may not completely control the glacial system. Indeed, stochastic (random) differences in vegetation and the associated *albedo* and in oceanic and atmospheric circulation could also influence the duration of future glacials and interglacials. It is known that the mechanisms of glaciation and deglaciation include fast and slow climate components. The *fast responses* of the climate to orbital forcing are major changes in temperature, precipitation and atmospheric circulation. Over several millennia, these drive—and are also be altered by—*slower components* of the climate system, which are ice sheets, deep-ocean dynamics, ocean biogeochemistry and vegetation.

Given the presently very high concentration of atmospheric CO_2 and unique orbital forcing, models indicate that the present interglacial would last another 50,000 years and perhaps more. The combined atmospheric CO_2 concentration and orbital forcing would delay by tens of thousands of years the positive feedback mechanism of increasing albedo responsible for the runaway growth of Northern Hemisphere ice sheets (see Sect. 9.1.3).

In fact, a long interglacial of 50,000 years may not be so exceptional as a similarly long interglacial may have occurred once in the last 500,000 years, between 424,000 and 374,00 years before present. At the time, the CO_2 concentration was 280 ppmv, which was lower than the present 400 ppmv and much lower than the value that could be reached by the end of the twenty-first century. The present and upcoming higher concentration of CO_2 could thus cause an even longer interglacial than 50,000 years. If after decades of global warming, the atmosphere were to return to the pre-industrial levels (in the range of 280–290 ppmv), then the next glacial may start in about 50,000 years.

11.3.3   Very Long Interglacial

It was mentioned above that the value of Earth's *eccentricity* will be almost zero during the coming 25,000 years (see Sect. 11.3.2), and it is known that other orbital characteristics will also be unusual. Given these unusual orbital parameters, some modelling studies indicate that even the low pre-industrial CO_2 concentration of 280 ppmv would have been high enough to prevent a rapid onset of the next glacial. The models indicate that the value of 280 ppmv of CO_2 in the atmosphere was very critical because, if the pre-industrial CO_2 level had been slightly lower than 280 ppmv, the Earth System would already be on the way towards a new glacial. According to these studies the pre-industrial CO_2 concentration of 280 ppmv was high enough to delay the next glacial by 40,000 years, thus making the present interglacial as long as 50,000 years even without anthropogenic input of CO_2 to the atmosphere (that is, the "usual" duration of 10,000 years plus the delay of 40,000 years).

The situation in the twenty-first century is that human societies are releasing large amounts of CO_2 in the atmosphere, and the atmosphere will retain more than 20% of its excess CO_2 after it will equilibrate with the ocean once the anthropogenic emissions of CO_2 have stopped (see Sect. 9.2.4). Some studies have estimated that 25% of the present anthropogenic emissions will remain in the atmosphere for thousands of years, and about 7% will remain beyond 100,000 years. Owing to the extremely long persistence of anthropogenic CO_2 in the atmosphere, past, present and future anthropogenic CO_2 emissions will have a strong effect on the onset of the next glacial episode, which may not occur within the coming 100,000 years or more. Some studies have even considered that a very large anthropogenic release of CO_2 and CH_4 from fossil carbon and methane hydrate deposits (see Sect. 2.1.7) could prevent the occurrence of glacial episodes during the next 500,000 years. Hence, the human-made interglacial could last 100,000 years, and perhaps be as long as 500,000 years. The occurrence of the next glacial episode, sooner or later, would contribute to compensate the long-term effects of the ongoing anthropogenic global warming.

11.3.4   Beyond the Anthropocene Epoch

If humans used (burned) all the coal deposits, the atmospheric CO_2 concentration could reach 2,000 ppmv. This very high CO_2 level would be similar to the atmospheric concentrations that existed during the 40 million-year period between 305 and 265 million years before present (Fig. 9.8b). This 40 million-year transition period saw the shift from a 100 million-year-long icehouse to the long greenhouse that preceded our present icehouse, which began 34 million years before present (Table 4.3). A study examined the 40 million-year transition period as a possible analogue of the Earth System if the atmospheric CO_2 concentration reached 2,000 ppmv.

During the 40 million-year transition period from 305 to 265 million years ago, the atmospheric CO_2 concentrations went from 280 to 3,500 ppmv, and the sea-surface temperatures experienced an increase of at least 10 °C, which shifted the Earth System from icehouse to greenhouse conditions. By the end of the 40 million years, the atmospheric CO_2 concentration reached a threshold beyond which greenhouse stability precluded the reestablishment of glacial conditions. In other words, the Earth System shifted from the stable icehouse state that existed before 305 million years ago to a new stable greenhouse state after 265 million years ago (see Sect. 4.5.6). Hence, if humans transformed all the coal deposits into CO_2, the condition of the planet could shift from the present icehouse to a new greenhouse, as had naturally occurred over a 40 million-year transition period about 300 million years ago. The next section indicates that such a drastic shift could perhaps occur even without burning all the coal deposits.

11.3.5 Crossing Planetary Thresholds

It was explained above by reference to sea level rise that some temperatures are tipping points, that is, *planetary thresholds* that when exceeded, even by a small value, can lead to very large changes in the climate system (see Sect. 11.2.2). According to the IPCC's special report *Global Warming of 1.5°C* (Hoegh-Guldberg et al. 2018; their Box 3.3), the changes that occurred in the Earth's climate during the last 3.3 million years indicate that the present global warming does not present a serious risk of crossing irreversible thresholds, or amplifying anthropogenic changes under the action positive feedbacks (see Sect. 9.1). Nevertheless, some studies have explored the likelihood that positive feedbacks could push the Earth System towards a planetary threshold that, if crossed, could cause a *runaway warming* that would result in a *hothouse Earth*, also called *greenhouse Earth* (see Sect. 4.5.6). Runaways caused by positive feedbacks are explained in Chapter 9 (see Sect. 9.1.4). The possible hothouse Earth System could have environmental conditions beyond those that prevailed 15–17 million years ago, when the atmospheric CO_2 concentration was up to 300–500 ppmv and temperatures were 4–5 °C warmer than present and the sea level was up to 60 m higher.

The runaway warming could occur if the global temperature increased above a *planetary threshold*, which could be as low as 2 °C above the preindustrial level, although its exact value is not known and could perhaps be 3 °C above the preindustrial level instead of 2 °C. Crossing this temperature threshold would mean the complete melt (and accompanying rise in sea level) of glaciers and ice caps (0.4 m), and the Greenland and West Antarctic Ice Sheets (7.4 and 4.5 m, respectively), which together with partial melt of the huge East Antarctic ice sheet would cause an irreversible sea level rise of more than 12 m. If this temperature threshold were crossed, the planet may find itself on a trajectory towards a hothouse in which temperatures would be 4 to 5 °C higher than present and sea level more than 60 m higher.

In 2015, the global temperature of Earth crossed the threshold of 1 °C above the preindustrial level, and the crucial question is now: Will the temperature of the planet inevitably reach and cross the planetary threshold of 2–3 °C above the pre-industrial level, or will human societies find the collective will to limit the global surface temperature rise to below this threshold? The official international target in the 2015 Paris Agreement is to limit the global surface temperature rise to well below 2 °C above the pre-industrial level, that is, much less than 1 °C above the present level.

An ultimate tipping point for the Earth System would be reached if the greenhouse effect became so extreme that the ocean would boil away and render the planet uninhabitable, thus creating irreversible environmental conditions similar to those prevailing on Venus. Fortunately, anthropogenic activities have virtually no chance of inducing such a runaway greenhouse effect. However, most scientists believe that such a runaway greenhouse effect will occur in the long term, as the Sun gradually gets bigger and hotter as it ages, which should create Venus-like conditions on Earth in one or several billion years.

The ability of ecosystems (see Information Box 4.2), including human societies, to adapt to hothouse conditions will largely depend on the rate of change from the conditions that prevail in the present interglacial icehouse (see Sect. 11.2.1, above). If the magnitude of the climate change remained small or if the change occurred slowly, the shift in ecosystems could track the change in temperature, and the ecosystems may shift gradually, potentially taking up carbon from the atmosphere as fast as it is released. However, if the magnitude of climate change were large or if the change occurred rapidly, extensive disturbances could abruptly remove many existing ecosystems. Examples are destruction of forest ecosystems by combined insect attacks, droughts and wildfires, and of coral reefs by combined increases in seawater temperature, sea level and ocean acidification.

If the Earth System were to experience a shift towards a hothouse, the rate of change of environmental conditions would be rapid. Much more rapid than the change that occurred during the 40 million-year shift from icehouse to hothouse conditions about 300 million years before present, examined in the previous section (see Sect. 11.3.4). During these 40 million years, the ecosystems could adapt to the changing environment though species evolution. In contrast, the possibly upcoming transition from icehouse to hothouse could be as short as a few centuries or millennia. This is much faster than the rate of biological evolution of most plants and animals, whose ecosystems could therefore not adapt to the advent of the new hothouse environment. The inability to adapt to the magnitude and rate of environmental changes would also be true of developed human societies, who could not adapt, for example, to a sea level rise of tens of metres in a few centuries. Alternatively, humans could try to manage the planet in order to avoid crossing a threshold that would lock the Earth System into a trajectory towards a hothouse, and maintain it instead in a state as close as possible to the present interglacial icehouse. This present state has been called *stabilized Earth* in the literature.

Need to Make Deliberate Decisions on Mitigation In order to prevent a shift of the trajectory of the Earth System from the stabilized Earth towards a hothouse, human societies need to make deliberate decisions. These should include fundamental changes in societies, for which there is increasing agreement on the need but not yet on the means, such as: deep cuts in greenhouse gas emissions, to keep the temperature rise lower than 2 °C above preindustrial; and protection and possible enhancement of natural carbon sinks, such as forests, soils, deep-ocean waters, and marine sediments. These could also include: deliberate efforts to remove CO_2 from the atmosphere and store it in geological formations, a concept that is being tested experimentally, but would remain expensive even if the storage was proven to be efficient and permanent; possibly solar radiation management, depending on further research because this technology entails large risks.

11.4   Different Earths

We conclude this chapter with a conceptual test of the central idea of this book that organisms and ecosystems took over the planet during the course of the Earth's billion-year history (Sect. 1.1.1), meaning that organisms, as integral components of the Earth System, determine a large part of its characteristics. In the remainder of this section, we "peel off" successive layers of biological innovations that took place on Earth since more than 3.8 billion years, and try to imagine at each step what Earth would have been without the occurrence the given innovation. In this exercise, the layer peeled off at one step is lost for all subsequent steps, so that losses of biological innovations are cumulative from one version of Earth to the next.

In the previous chapters, we described a large number of biological innovations resulting from biological evolution, and this section could be quite long if we considered all of them here. We instead keep the present exercise reasonably short by limiting the number of biological innovations peeled off to eight. We consider first *Homo sapiens*, as the most recent evolutionary step, and end with the first organisms on Earth, as the earliest step of biological evolution. For convenience, we number the current planet Earth #0, and its successive, increasingly simplified versions Earth #1 to Earth #8, summarized in Table 11.6.

Earth #1. What would planet Earth look like without Homo sapiens? We focus here on our current species, *Homo sapiens*, and do not consider the other human species that once existed on Earth. We only consider *H. sapiens* because the effects of our species on the planet have been unique. Most people generally consider the following effects as being positive: the unravelling of mysteries of Nature, by exploration and scientific research; the improvement of the length and quality of human life, by medicine and social care; the expansion of the human mind, by the arts, literature, science and philosophy; and the gentrification of social interactions, by the combination of culture and politics.

Table 11.6 Conditions and possible consequences of the eight successive versions of Earth described in the text. Earth #0: present Earth

Earth	Condition	Some possible consequences
#1	Present Earth (#0) without *Homo sapiens*	Existence of animals and other species that became extinct on Earth #0
#2	*Earth #1 minus* other multicellular organisms on emerged lands	On emerged lands, no flora or fauna, and small biomass of microorganisms; 6% atmospheric O_2
#3	*Earth #2 minus* all other organisms on emerged lands	Mineral soil on land; low biologically driven fluxes of chemical elements in the Earth System
#4	*Earth #3 minus* multicellular organisms in any environment	In oceans, unicellular prokaryotes and eukaryotes; atmospheric O_2 lower than 6%
#5	*Earth #4 minus* all unicellular eukaryotes	In oceans, bacteria and archaea; atmospheric O_2 much lower than 6%
#6	*Earth #5 minus* O_2 respiring organisms	High concentration of environmental O_2; several environments used only by O_2-tolerant organisms
#7	*Earth #6 minus* O_2-producing photosynthetic organisms	Almost no free environmental O_2; abundant atmospheric methane, causing high temperature
#8	*Earth #7 minus* all other organisms (Earth without organisms)	Atmosphere dominated by N_2, and liquid-water ocean, or very high concentration of atmospheric CO_2, no liquid water, and no plate tectonics

However, other effects of human societies contributed to degrade the environment, and this degradation often generated negative feedbacks into the societies. These effects include, as explained above in this chapter: the destruction of large fauna after groups of *H. sapiens* occupied new lands; the deep transformation of almost all of the planet's landscapes, as a consequence of agriculture and urbanization; the pollution of most environments, which accompanied industrialization and the population explosion (even the space that surrounds Earth is now littered with debris of rockets and satellites); and the combination of climate change, ocean acidification, loss of biodiversity and food and water insecurity, resulting from anthropogenic perturbations of natural processes of the Earth System.

What follows assumes that in the absence of *H. sapiens*, no other human species would have developed equivalent societies. The continents on Earth #1 without *H. sapiens* would likely be largely covered by forests and inhabited by a wide variety of animals that would likely include many large species such as the (extinct) woolly mammoth and giant ground sloth, without forgetting the (extinct) aurochs in Eurasia and North America, passenger pigeon in North America and dodo bird on Mauritius Island. The oceans would be populated by a wide variety of animals that would include many large species such as the (extinct) Steller's sea cow and the Japanese sea lion.

This environment might harbour many groups of hunter-gatherer humans whose individuals would likely live only a few tens of years, whose groups would sometimes experience hunger as they would not know agriculture or husbandry, and would craft and use some tools. They would collectively have concepts for food, work, joy, anger, companionship, love and perhaps art, but contrary to modern *H. sapiens*, these humans would not have concepts for pollution, overpopulation, anthropogenic climate change, ocean acidification, loss of biodiversity, or systemic food and water insecurity as they would not create such situations. The atmosphere and oceans of Earth #1 would likely have similar chemical compositions as those that generally existed on Earth #0 before the Industrial Revolution.

Earth #2. What would planet Earth look like if multicellular organisms did not occupy the emerged lands? Multicellular organisms are all eukaryotes, and they include all the plants and animals, most fungi and many algae (that is, some fungi and most algae are unicellular). The continental surfaces on Earth #2 would look quite naked to a traveller from Earth #0 or Earth #1. No forests, no prairies, no plants whatsoever; no mushrooms or lichens; no fauna on land or in the air. However, the continental surfaces would likely not be devoid of organisms, as they would be probably occupied by a wide variety of microorganisms, such as bacteria, archaea and protozoa, thriving on land, in fresh waters and in wetlands. It must be noted that some researchers think that, on real Earth, microorganisms could occupy emerged lands by moving in together with the multicellular organisms.

As on Earth #0, the microorganisms on continental surfaces would be protected from harmful solar UV radiation by the ozone layer. There would be soils, formed of minerals coming from the erosion of rocks as on current Earth. However, their organic content would likely be lower than that of the current soils because of the absence of the rich source of organic matter provided by the decay of egesta and remains of multicellular organisms. In addition, these soils would be quite compact because of the absence of such burrowing animals as earthworms and insects, and thus less aerated and perhaps less hydrated than the current soils. Overall, the continents would be inhabited by a variety of archaea, bacteria, cyanobacteria and protozoa.

Different groups of microorganisms on land, in soils, in fresh waters and in wetlands would build up organic compounds by chemosynthesis or photosynthesis, and decompose them using O_2 or not; some would produce O_2 and many would respire it; some would cycle nitrogen through N_2 fixation and denitrification; and they would generally perform all biological processes currently known in the Earth System.

The biomasses of continental systems on Earth #2 would be smaller than those on Earths #0 and #1, especially the biomasses of autotrophs (mostly plants). Currently, the total carbon biomass on continents amounts to 470 petagrammes (Pg, one Pg is one billion tonnes) compared to 6 Pg in the oceans, and the carbon biomass of microorganisms in the two environments is 21 and 3 Pg, respectively. The very large difference between continents and oceans is essentially due to the huge biomass of plants on continents compared to oceans, that is, 450 Pg (mostly forests) *versus* 1 Pg (mostly phytoplankton), and the large biomasses of microbial soil decomposers on continents (12 Pg for fungi, and 7 Pg for bacteria and archaea). On Earth #2, there would be no plants and no animals, and the biomass to be decomposed would thus be small, with the consequence that soils would likely contain much fewer microorganisms than on current Earth. Overall, the total carbon biomass on the continents of Earth #2 could be less than 5 Pg, compared to the current 470 Pg.

On Earth #0, the burial of continental biomass under the ground over hundreds of millions of years, which is at the origin of coal deposits, had been a major contributor to the final build-up of atmospheric O_2 (see Sect. 9.4.5). It can thus be expected that the volume fraction of O_2 in the atmosphere of Earth #2 would be similar to the value that existed on Earth before multicellular organisms began to occupy the continents, that is, about 6% instead of the current average of about 20% (Fig. 9.17).

The small *biomass* of autotrophs on the continents of Earth #2 would not necessarily mean that their annual *production* of organic matter would be very low, given that the small biomass of autotrophs in the current oceans (1 Pg of carbon) produce about the same amount of organic carbon as on the continents (48 and 56.4 Pg, respectively). This is because most of the terrestrial plant biomass is non-productive wood, which stores a huge amount of carbon that does not contribute to photosynthesis, meaning that autotrophic production on the continents of Earth #2 could be about the same as that on current Earth #0. As a result, the total amount of carbon fixed annually by autotrophs on the continents and in the oceans of Earth #2 could be similar to the total amount fixed on present Earth. Consequently, the biologically driven fluxes of chemical elements in the Earth Systems of Earths #2, #1 and #0 could also be similar.

Earth #3. What would planet Earth look like if the emerged lands were not occupied by any organism? On Earth #3, the rocky surfaces would be bare and the soils would be entirely mineral, these soils being like the loess sediments on the current continents. To a traveller from Earth #2, this could look somewhat similar to the landscapes on his/her Earth, except that on Earth #2 the rocky surfaces would be coated with microorganisms and the soils would contain some organic matter.

A visual difference between the two Earths that might strike the traveller from Earth #2 could be the colour of the land. Indeed, given that the photosynthetic microorganisms present in the coatings of rocks and soils on Earth #2 would reflect the light not absorbed by their photosynthetic pigments, the land on Earth #2 would look green or blue-green if these microorganisms were similar to cyanobacteria, whereas it would have colours ranging from ochre to red on Earth #3. The latter colours are those of the oxidized forms of the dominant minerals at the surface of continents, which are aluminium silicate ($AlSiO_3$) and iron sulphide (FeS_2, whose mineral form, pyrite, is cited in relation with Eq. 9.17).

A very significant characteristic of Earth #3 would be its much lower total autotrophic production of organic matter than on Earth #2. Indeed, on the latter, there would be as much autotrophic production on continents as in the oceans (see the above paragraphs on Earth #2), whereas there would be no autotrophic production on the continents of Earth #3, and its total autotrophic production would thus be only half that of Earth #2. Given that all biologically driven fluxes of chemical elements in the Earth System are fuelled by autotrophic production, these fluxes would be lower on Earth #3 than on Earth #2. The chemical composition of the atmosphere of two Earths should be similar, with about 6% of O_2 by volume, because there would not be any large-scale burial of organic matter on the continents of either Earth.

Earth #4. What would planet Earth look like if unicellular eukaryotes had not evolved into multicellular organisms? Given that the approach used in this section is to peel off successive layers of selected biological innovations, it is assumed here that all organisms on Earth #4 would be unicellular prokaryotes or eukaryotes and would live in the oceans. The prokaryotic organisms on Earth #4 would belong to bacteria and archaea, and would include groups of organisms able to build organic matter using either chemosynthesis, O_2-less photosynthesis (aerobic anoxygenic phototrophs) or O_2-producing photosynthesis (cyanobacteria), or to decompose organic matter using O_2 or not. The eukaryotic organisms would include groups able to build organic matter using O_2-producing photosynthesis (unicellular algae), or decompose organic matter (unicellular fungi), or feed on organic matter in the forms of other microorganisms and organic debris (protozoa).

On our Earth #0, multicellular eukaryotes began to appear in the oceans not later than 550 million years ago, and possibly very early depending on hypotheses based on different interpretations of fossils. Before the advent of multicellular eukaryotes, the seafloor of ocean shelves was covered by mats of heterotrophic microorganisms, which isolated the underlying O_2-less sediments from the overlying oxygenated waters. The first multicellular eukaryotes were likely sponge-like organisms, which pumped water and filtered out organic particles (mostly unicellular organisms) on which they fed. A net result of this activity was to transfer organic matter from the water into the sediment, and the ensuing burial of carbon contributed to increase the oxygenation of the water column. The advent of mobile, burrowing animals, and their disturbance of sediments contributed to expose buried organic matter, whose oxidation led to decreased concentration of O_2 in the aquatic environments and the atmosphere. The evolution of large zooplankton and other swimming organisms, later on, increased the amount of sinking organic debris, and thus increased the overall burial of organic matter and accompanying increase in environmental O_2.

The above events would not take place on Earth #4 because of the absence of multicellular organisms. A global consequence of the lack of these organisms in the oceans would be a sluggish transfer of carbon to sediments and thus no, or only very small, additional burial of carbon. Hence, the level of atmospheric O_2 en Earth #4 may be lower than the 6% reached about 600 million years ago on Earth #0 (Fig. 9.17). Another consequence would be that multicellular eukaryotes would not occupy the continents at a later stage, which would have the global effects already described for Earth #2.

Earth #5. What would planet Earth look like if biological evolution had not produced eukaryotes? Earth #5 would be similar to Earth #4 except that all its organisms would be prokaryotes. Hence on Earth #5, the continents would be devoid of organisms (as on Earths #3 and #4) and the oceans would only contain bacteria and archaea. These would include groups of organisms able to build organic matter using either chemosynthesis, O_2-less photosynthesis (aerobic anoxygenic phototrophs) or O_2-producing photosynthesis (cyanobacteria), or to decompose organic matter using O_2 or not.

It was mentioned for Earth #4, above, that multicellular eukaryotes began to appear in the oceans about 550 million years ago or earlier. These evolved from unicellular eukaryotes whose origin in the oceans has been traced back to around the first increase in atmospheric oxygen 2.2 billion years ago. The autotrophic eukaryotes divided early in two lineages with different secondary photosynthetic pigments. Members of the first group dominated the oceans until about 100 million years ago, and some of them occupied the continents about 500 million years ago where they evolved into the higher plants. Members of the second group began to dominate the oceans since about 100 million years ago, and include the groups of phytoplankton organisms known as dinoflagellates, coccolithophores and diatoms.

Until the massive development of coccolithophores, $CaCO_3$ in the ocean accumulated on continental shelves. Because the coccolithophores lived in the open ocean where they often produced very large amounts of $CaCO_3$, their advent moved the accumulation of $CaCO_3$ in the sediment to deeper waters. Simultaneously, the strong production of large-sized phytoplankton, especially diatoms, led to the burial of large amounts of organic carbon on the seafloor, which contributed to decrease atmospheric CO_2 and increase atmospheric O_2. Hence Earth #5 without eukaryotes would have buried much less carbon than Earth #0, not only on the continents as explained for Earth #2 above, but also in the oceans, with atmospheric O_2 likely much less than the 6% reached on Earth #0 about 600 million years ago (Fig. 9.17).

Earth #6. What would planet Earth look like if organisms had not invented O_2 respiration? Planet Earth is the only body in the Solar System whose atmosphere contains a significant proportion of O_2. On present Earth #0, several groups of bacteria and archaea do not use O_2 for respiration, but only a few eukaryotes can do without O_2 during their whole life cycle. On Earth #6, the continents would be devoid of organisms (as on Earths #3 to #5), and there would be no eukaryotes (as on Earth #5). All organisms in aquatic environments would be either bacteria or archaea (as on Earth #5), but none of these would use O_2.

Different groups of microorganisms on Earth #6 would be able to build up organic matter using either chemosynthesis, O_2-less photosynthesis (aerobic anoxygenic phototrophs) or O_2-producing photosynthesis (cyanobacteria), or to decompose organic matter without using O_2. The latter characteristic is different from Earth #5 where some groups of microorganisms decomposed organic matter using O_2. We peel off O_2 respiration here, and will peel off O_2-producing photosynthesis separately in Earth #7 for the reason explained below when we examine this next version of Earth.

A net effect of O_2 respiration, or more generally the decomposition of organic matter using O_2, is to decrease the concentration of environmental O_2 and increase that of environmental CO_2. Given that the concentration of environmental O_2 on Earth is much higher than that of environmental CO_2 (for example 21 and 0.04%, respectively, in the current atmosphere; Table 2.1), O_2 respiration has a greater relative effect on the concentration of CO_2 than on that of O_2. Furthermore, dissolved CO_2 combines with water molecules and thus modifies the pH of the aquatic environment, whereas O_2 does not chemically react with water.

When O_2 respiration is very strong, it can severely deplete the environmental O_2. Examples are provided by the eutrophicated waters (see Sects. 10.3 and 11.1.2) and the oceans' O_2 minimum zones (see Information Box 9.2). Significant decrease in O_2 concentration also occurs when O_2 is respired during hundreds of years in the same water masses, as is the case in the ocean's deep waters during their centennial circulation from their sources (located North and South of the Atlantic Ocean) into the Indian and Pacific Oceans (Fig. 3.8). Overall, the initiation on Earth #0 of O_2 respiration hundreds of millions of years after the advent of O_2-producing photosynthesis (see Earth #7) did not prevent O_2 from slowly accumulate in the environment of our planet until about 1.8 billion years ago, but it contributed to keep the concentration of atmospheric O_2 at the low value of 6% during the following 1.2 billion years (Fig. 9.17).

On Earth #6 without O_2 respiration, the concentration of O_2 would likely increase in the environment until feedback mechanisms regulate this increase (see Sect. 9.4.4). The resulting high concentration of environmental O_2 could have two different consequences for organisms on Earth #6. On the one hand if these microorganisms could not tolerate O_2, then only the O_2-free habitats of the planet would be used by organisms, whereas on the other hand if the evolution of O_2-tolerant organisms had accompanied the increase in environmental O_2, then there would be organisms in both the oxygenated and O_2-free habitats of the planet. Organisms that can tolerate environmental O_2 but do not use O_2 for respiration really exist on Earth; they are called *aerotolerant anaerobes*, and are found on current Earth among the bacteria. If we removed the constraint of Earth #3 that there are no organisms on continents, the occupied habitats of Earth #6 would include the continental surfaces because of the presence of an ozone layer as a consequence of the high concentration of atmospheric O_2.

Earth #7. What would planet Earth look like if organisms had not invented O_2-producing photosynthesis? We now consider a planet without O_2-producing photosynthesis. We peeled off O_2 respiration first (Earth #6) and peel off O_2-producing photosynthesis here because the genomic analysis of the evolution of cyanobacteria indicates that these organisms acquired the capability of O_2 respiration following the rise of environmental O_2, that is, hundreds of millions of years after they invented O_2-producing photosynthesis. Assuming that the late acquisition of O_2 respiration was also the case for the other groups of organisms, we decided to peel off O_2 respiration in Earth #6, above, and do the same for O_2-producing photosynthesis next, here.

Peeling off O_2-producing photosynthesis is a very significant step because this biological process creates most of the O_2 found on Earth. Indeed, the breakdown of molecules of water (H_2O) and nitrous oxide (N_2O) by high-energy ultraviolet solar radiation in the atmosphere produces only small amounts of O_2, and almost all the environmental O_2 comes from the splitting of water molecules by photosynthesis. On Earth #7, there would be almost no free O_2. In fact, the oxygen that exists as free O_2 in the environment of Earths #0 to #6 would be combined with hydrogen in molecules of water (H_2O) on Earth #7. All organisms would be aquatic bacteria or archaea, as on Earth #6, and different groups of them would be able to build organic matter using either chemosynthesis or O_2-less photosynthesis, or to decompose organic matter without using O_2, but contrary to Earth #6, none would perform O_2-producing photosynthesis. Some of these microorganisms could be involved in the formation of stromatolite-like structures.

Without environmental O_2 on Earth #7, methane (CH_4) would be abundant in the atmosphere because it would be produced by microorganisms in the oceans, but not continually destroyed in the atmosphere by chemical reactions involving O_2 (Eq. 9.6). Hence without environmental O_2, atmospheric CH_4 would not be oxidized, and would thus be an important atmospheric gas. This would be similar to the Earth's atmosphere before its oxygenation, where CH_4 and CO_2 were both important atmospheric gases (Fig. 8.1b and c). Given that CH_4 and CO_2 are two powerful greenhouse gases, the average surface temperature on Earth #7 could be high, perhaps above 25 °C as it was on Earth #0 about 3 billion years ago (Fig. 7.1). Also without O_2 in the atmosphere, there would be no ozone layer, and the surface of continents would consequently be inhospitable to most organisms (which were peeled off the continents in Earth #3, above). However, all the biogeochemical processes involving carbon, nitrogen, sulphur and other chemical elements including oxygen, except those that require O_2 or produce it, would proceed in the oceans of Earth #7 as they do in the O_2-less marine environments of today's Earth #0.

Earth #8. What would planet Earth look like without organisms? For Earth #8, we peel off all living organisms from the planet, making it similar in this respect to planets Venus and Mars. We assume that Earth #8 would have a magnetic field, contrary to Venus and Mars. The mass of Venus is 82% that of Earth, its atmospheric pressure is 92 times higher than that of Earth, and its surface temperature is +462 °C, whereas the mass of Mars is 11% that of Earth, its atmospheric pressure is almost nil, and its surface temperature ranges between -153 and +20 °C (Tables 1.1 and 2.2). Would Earth #8 be similar to present Venus or present Mars, or be different from these two planets?

On early Earth #0, more than 99% of the initially very abundant atmospheric CO_2 was stored in $CaCO_3$ deposits by about 4 billion years ago. Afterwards, the concentration of methane (CH_4) increased in the atmosphere of the young planet (Fig. 8.1). This build-up was a consequence of the release of CH_4 by geological mechanisms and microorganisms. On modern Earth, the CH_4 generated geologically is released by hydrothermal vent systems and volcanoes, and biogenic CH_4 is a by-product of the decomposition of organic matter by archaea (see Information Box 5.2). This may have been different before the oxygenation of the planet, as it was proposed that CH_4 could then have been produced by microbial ecosystems based on O_2-less photosynthesis.

The build-up of CH_4 in the atmosphere may have been crucial for the global climate because, during the first billion years of the Earth's history, the faint young Sun radiated much less energy than presently, and our young planet could have frozen solid after the initial decrease in atmospheric CO_2. One hypothesis proposed to explain why this did not happen invokes the presence of abundant atmospheric CH_4, which kept the planet warm (see Sect. 2.1.7 and 3.7.1). Methane thus remained a key greenhouse gas until the increasing concentration of atmospheric O_2 oxidized most of it, which may have initiated the first snowball Earth about 2.2 billion years ago (see Sect. 4.5.4). Certain researchers think that the CH_4 generated geologically was enough to keep the planet warm, but it is generally thought that the amount of geological CH_4 was not enough and was supplemented with biogenic CH_4. What about Earth #8?

It can be assumed that the concentration of CH_4 in the atmosphere of Earth #8 would not be high, because of the absence of CH_4-producing organisms. This would have caused Earth #8 to freeze solid during billions of years after the initial decrease in atmospheric CO_2, and perhaps remain frozen until the radiation of the Sun increased enough to melt the iced planet. The fact of being frozen would have preserved the water of Earth #8 from loss to outer space (see Sect. 5.4.2), and the combination of key characteristics of the planet—mass, distance from the warmer Sun, and magnetic field—could have retained its atmosphere and ocean when the Sun warmed up after a few billions of years (see Sects. 2.3.4, 5.6.2, 5.6.3 and 8.4.1). Hence under this hypothesis, Earth #8 could have an atmosphere and a liquid-water ocean, being in this respect somewhat similar to Earth #0 although without organisms and therefore without O_2 or much CH_4 in its environment.

A different, or complementary scenario would follow from the fact that on Earth #0, carbon is continually recycled through the erosion of $CaSiO_3$ rocks by CO_2-laden water followed by the formation of $CaCO_3$, which takes up atmospheric CO_2, and the subduction of seafloor $CaCO_3$ followed by volcanic eruptions, which releases CO_2 to the atmosphere (see Sect. 7.3.3). As a consequence, if the storage of CO_2 by $CaCO_3$ formation were slower than the release of CO_2 by volcanic eruptions, there would be an increase in atmospheric CO_2 and thus in the greenhouse effect. On Earth #8, there is no biological formation of $CaCO_3$, and one may wonder if the sole chemical formation of $CaCO_3$ would be enough to balance its recycling by subduction. One may think that it would be enough given that the chemical formation of $CaCO_3$ exceeded or balanced its destruction on Earth #0 until around 1 billion years ago. Nevertheless, it is interesting to examine the scenario of possible long-term excess in the destruction of $CaCO_3$ and resulting volcanic emissions of CO_2 to the atmosphere over the formation of $CaCO_3$ and accompanying uptake of atmospheric CO_2 on Earth #8.

The latter scenario assumes, for Earth #8, the same initial storage of most of atmospheric CO_2 in $CaCO_3$ deposits as it actually occurred on Earth #0. This step is followed on Earth #8 by the long-term imbalance described in the previous paragraph, which would cause the progressive transfer of carbon from the $CaCO_3$ deposits to atmospheric CO_2. The increased atmospheric CO_2 would cause a very strong greenhouse effect, which would be accompanied by the evaporation of much liquid water, and the resulting very high concentration of atmospheric water vapour would favour the loss of water by escape of hydrogen to space (see Sect. 5.4.2). The rarefaction of liquid water and its ultimate disappearance would stop the uptake of CO_2 by chemical alteration of $CaSiO_3$ rocks while tectonics would continue to feed the volcanic emission of CO_2 from the mantle, which would further increase the concentration of atmospheric CO_2. Finally, the movements of tectonic plates would come to an end because of the lack of liquid water (see Sect. 7.2.5).

Under this scenario, the conditions existing on Earth #8, that is, a very high concentration of atmospheric CO_2 and no plate tectonics, would be similar to those that exist on present Venus. Indeed on planet Venus, CO_2 makes up 96% of the volume of atmosphere (Table 2.1) and as a consequence, the atmospheric pressure at ground level is 92 times that on Earth and the very high greenhouse effect causes the average temperature at the surface of the planet to be around 462°C (Tables 1.1). Likewise, it is possible that the conditions described for Earths #7 to #1 currently exist or have existed in the past on other bodies of the Solar System, or on bodies orbiting other stars in our galaxy or elsewhere.

11.5   Stewardship of Planet Earth

Our journey from Earth #0 to Earth #8 stressed the global significance of eight biological innovations that occurred over the almost 4 billion years of biological evolution. This book describes many more biological innovations, which all contributed to shape the Earth System as we know it. The description of Earth #1 recalled some of the harms that *Homo sapiens* inflicted to the planet's environment and life forms, first unknowingly and now with full knowledge. For some people, Earth #1 may correspond to the ideal concept of a planet without *H. sapiens*. However, as we successively peeled off biological innovations, we realized that planet Earth #0 we know seems better than its counterparts without biological evolution (Earths #1 to #8), and is thus worth preserving as it is. In order to do so, we must become the stewards of our planet instead of trying to become its masters, and the present chapter described ways to achieve this change, especially Sects. 11.2.1 and 11.2.2. Because time is running out, human societies must get their act together without delay.

As we conclude our voyage through space and time, we may have the impression that planet Earth behaves almost like an organism. We saw early in this book that all organisms share some main characteristics, which are hardware, energy flux, software, and ability to replicate or reproduce (see Sect. 1.1.3). Is this also the case of the Earth System? Firstly, organisms are made of interdependent parts that form a unified whole called a system (hardware). This is also the case of the Earth System, whose components are interconnected by negative feedback loops that ensure the stability of its key properties. Secondly, we saw that organisms are open systems, which maintain their high level of internal organization through a continual intake and dissipation of energy (energy flux). Similarly, the functioning of the Earth System is fuelled by fluxes of solar and geothermal energy. Thirdly, organisms have software, namely their genetic information, which is passed down through generations of cells or individuals by replication or reproduction. This is not the case of the Earth System.

Becoming the Stewards of Earth The ability of organisms to replicate or reproduce is a crucial difference with the Earth System, which does not have this ability. However, the Earth System, of which the organisms are part, benefits from their ability to evolve even if it cannot itself reproduce. Indeed, biological evolution has increased the ability of the Earth System, over the course of the history of the planet, to self-regulate through negative feedback mechanisms that involve organisms and maintain conditions favourable to them. However, because the Earth System does not have the ability to replicate, we cannot nurse the dream that it will someday provide us with a new planet, on which we could start afresh. This is an additional reason for humans to become the stewards of their unique planet without delay, taking into account the imperative need to preserve its habitability and biodiversity.

Figure Credits

Fig. 11.1a https://www.climate.gov/sites/default/files/paleo_CO2_2018_1500.gif, from https://www.climate.gov/news-features/understanding-climate/climate-change-atmospheric-carbon-dioxide by NOAA, in the public domain. Some indications inside the figure removed, numbers on the X-axis changed, and titles of the two axes rewritten.

Fig. 11.1b Modified after https://www.hawaii.edu/news/2016/11/09/a-new-study-concludes-warm-climate-is-more-sensitive-to-changes-in-co2/. Courtesy of Drs. Tobias Friedrich and Axel Timmermann, University of Hawaii.

Fig. 11.2 Modified after Fig. 1.5h of Abram et al. (2019). With permission from the Intergovernmental Panel on Climate Change (IPCC).

Fig. 11.3 Modified after Fig. 3B of the Intergovernmental Platform on Biodiversity and Ecosystem Services (Díaz et al. 2019). The IPBES knowledge products are free and open to use.

Fig. 11.4a   https://commons.wikimedia.org/wiki/File:Global_Water_Security.jpg by Sampa, https://nl.wikipedia.org/wiki/Gebruiker:Sampa, used under CC BY-SA 4.0. Some text inside the map removed.

Fig. 11.4b This work, Figure 11.4b, is a derivative of https://commons.wikimedia.org/wiki/File:Percentage_population_undernourished_world_map.PNG by Lobizón at English Wikipedia, use under GNU FDL and CC BY-SA 3.0. Figure 11.4b is licensed under GNU FDL and CC BY-SA 3.0 by Mohamed Khamla.

Fig. 11.5 https://www.pngfuel.com/free-png/efjbt. PNGFuel is an open platform for users to share cutout PNGs; all PNGs in PNGFuel are for Non-Commercial use, no attribution required.

Fig. 11.6 Illustration showing impacts of increasing carbon dioxide on the oceans by David Fierstein © 2007 MBARI. With permission from the Monterey Bay Aquarium Research Institute.

Fig. 11.7 Figure from page 87 of Misdorp (2011), developed in the framework of IPCC by groups of scientists and first published as Fig. 3 (black-and-white) of Bijlsma et al. (1992). With permission from the Coastal & Marine Union (EUCC).

Fig. 11.8 Map "Sea Level Rise" © 2017 Richard J. Weller, Claire Hoch, and Chieh Huang, Atlas for the End of the World http://atlas-for-the-end-of-the-world.com. With permission from Prof. Richard J. Weller, University of Pennsylvania, USA.

References

Abram N et al (eds) (2019) IPCC Special report on the ocean and cryosphere in a changing climate [Can be downloaded free of charge from the IPCC website via https://www.ipcc.ch/]

Díaz S et al (eds) (2019) Summary for policymakers of the global assessment report on biodiversity and ecosystem services of the Intergovernmental Science-Policy Platform on Biodiversity and Ecosystem Services. IPBES secretariat, Bonn, Germany [Can be downloaded free of charge via https://doi.org/10.5281/zenodo.3553579]

Friedrich T, Timmermann A, Tigchelaar M et al (2016) Nonlinear climate sensitivity and its implications for future greenhouse warming. Science Advances 2(11):e1501923

Hoegh-Guldberg O et al (eds) (2018) Global warming of 1.5°C. An IPCC Special Report on the impacts of global warming of 1.5°C above pre-industrial levels and related global greenhouse gas emission pathways, in the context of strengthening the global response to the threat of climate change, sustainable development, and efforts to eradicate poverty [Can be downloaded free of charge from the IPCC website via https://www.ipcc.ch/]

Further Reading

Feely RA, SC Doney, SR Cooley (2009) Ocean acidification: Present conditions and future changes in a high-CO_2 world. *Oceanography* 22(4):36–47. https://doi.org/10.5670/oceanog.2009.95

Goodell J (2017) The water will come. Rising seas, sinking cities, and the remaking of the civilized world. Little, Brown and Company, New York

Govorushko S (2016) Human impact on the environment. Springer, Cham

Mackenzie FT, Andersson AJ (2013) The marine carbon system and ocean acidification during Phanerozoic time. Geochemical Perspective 2:1-227 https://doi.org/10.7185/geochempersp.2.1

MacMartin DG, Ricke KL (2020) Geoengineering. In: Tortell P (ed) Earth 2020: an insider's guide to a rapidly changing planet. OpenBook Publishers, Cambridge, UK, pp 93–100. https://doi.org/10.11647/obp.0193.12

Shikazono N (2012) Introduction to Earth and planetary system science: a new view of Earth, planets, and humans. Springer, Tokyo

Siegel FR (2020) Adaptations of coastal cities to global warming, sea level rise, climate change and endemic hazards. Springer, Cham

Spellman FR (2020) The science of water, concepts and applications, 4th edn. CRC Press, Boca Raton

Tortell P (ed) (2020) Earth 2020. An insider's guide to a rapidly changing planet. OpenBook Publishers, Cambridge, UK. https://doi.org/10.11647/OBP.0193

Williams E (2011) Environmental effects of information and communications technologies. Nature 479: 334–338 [https://files.ifi.uzh.ch/hilty/t/Literature_by_RQs/RQ%20142/2011_Williams_Environmental_effects_of_information_and_communications_technologies.pdf]

Index

© Springer Nature Switzerland AG 2021
P. Bertrand and L. Legendre, *Earth, Our Living Planet*, The Frontiers Collection,
https://doi.org/10.1007/978-3-030-67773-2

Printed in the United States
by Baker & Taylor Publisher Services